Targeted
Molecular Imaging

IMAGING IN MEDICAL DIAGNOSIS AND THERAPY

William R. Hendee, Series Editor

Targeted
Molecular Imaging

Edited by

Michael J. Welch
William C. Eckelman

CRC Press
Taylor & Francis Group
Boca Raton   London   New York

CRC Press is an imprint of the
Taylor & Francis Group, an **informa** business

A TAYLOR & FRANCIS BOOK

Cover Image: From "Targeting Norepinephrine Transporters in Cardiac Sympathetic Nerve Terminals" by David M. Raffel.

CRC Press
Taylor & Francis Group
6000 Broken Sound Parkway NW, Suite 300
Boca Raton, FL 33487-2742

First issued in paperback 2020

© 2012 by Taylor & Francis Group, LLC
CRC Press is an imprint of Taylor & Francis Group, an Informa business

No claim to original U.S. Government works

Version Date: 20110801

ISBN-13: 978-0-367-57677-6 (pbk)
ISBN-13: 978-1-4398-4195-2 (hbk)

Visit the Taylor & Francis Web site at
http://www.taylorandfrancis.com

and the CRC Press Web site at
http://www.crcpress.com

Contents

PART I   *In Vitro* Studies

PART II   Small Animal Studies

PART III   Radiopharmacology and Radiopharmacy

PART IV   Case Studies on Developing Targeted
Nonnuclear Probes

PART V Case Studies on Developing Targeted Radiotracers

Series Preface

The Taylor & Francis Group series of books in medical physics is an effort by the publisher and the series editor to describe and explain scientific and technological developments in medical imaging and radiation therapy. It is the goal of this series to identify and illustrate the improvements that these developments provide and promise for patient care. Several books have been published in the medical physics series, and many more are in progress.

The book *Targeted Molecular Imaging* is an important addition to the medical physics series. It is widely understood that molecular imaging is an emerging frontier of biomedical research with important potential ramifications for medical diagnosis and therapy. Techniques of molecular imaging are revealing new knowledge about normal and abnormal structure and function at the cellular and molecular levels, and are providing promising mechanisms for guiding the delivery of new therapies to aberrant cells and tissues. This book clarifies these techniques and their potential for expanding scientific knowledge and improving the care of patients suffering from a variety of diseases and disabilities. We are very pleased to add *Targeted Molecular Imaging* to the Taylor & Francis Group series in medical physics.

Selecting the editors for this book was relatively easy. We wanted two individuals who are well versed in molecular imaging and have a broad perspective on the discipline and its potential contributions to biomedical science and clinical care. We also wanted editors who are well known and respected as medical researchers, and who have experience in translating research into products and procedures that are useful in patient care. Our first choice as editors was Michael Welch and William Eckelman. We are delighted that they agreed to undertake the challenge of designing and editing this book.

We intend to provide a book that will clarify the principles and techniques of molecular imaging and its applications to biomedical research and, potentially, clinical medicine. We hope that readers will agree that we have achieved our intent.

William Hendee, PhD
Series Editor

Preface

When Bill Hendee, the series editor, discussed with us a book on molecular imaging, we felt it essential to select a format that distinguished this text from the several other publications that have appeared recently on the subject. The majority of contributions to this book discuss the concept, development, preclinical studies, and, in many cases, translation to the clinic of targeted imaging agents. The first four chapters relate to specific *in vitro* approaches needed to develop agents; three chapters discuss animal models available for preclinical development while the fourth relates to the translation of nuclear probes to human use applications. The remaining chapters consist of case studies on the development and translation of targeted agents. The majority of these chapters relate to nuclear probes; one deals with optical approaches, another with ultrasound approaches, and yet another with the application of nanoparticles. If one examines the molecular imaging probes that have been studied in humans, the majority are nuclear probes. There are several reasons for this, one of the major ones being that the amount of mass administered is much less than for the other techniques and so obtaining approval from the U.S. Food and Drug Administration for human studies is less expensive and time consuming than for agents using optical, ultrasound, and magnetic resonance approaches.

The case studies in this publication involve targets for a variety of diseases, from oncology to cardiac disease to neurological disease as well as inflammation and infection. As the number of targets increase in the postgenomic era and proteins that are either mutated or constitutively increased in disease are identified, it is important to validate targeting to those proteins and demonstrate that external imaging will monitor the key process with high sensitivity. These chapters using a common format demonstrate how various investigators approach the comprehensive task of validating a new targeted probe. The value of these case studies should be measured by the degree to which the new investigator can generalize from these studies to their own specific targeted probe. We believe that this will be substantial. It is our hope that this book will provide valuable guidance to scientists planning to translate imaging agents from concept to the clinic and to graduate students, postdoctoral fellows, and junior faculty members who are becoming involved in this rapidly expanding field.

Acknowledgments

We appreciate the efforts of all the contributors to this book; they all have invested a significant amount of time in the preparation of these informative chapters. We also thank Mrs Kari Alca of the Division of Radiological Sciences, Department of Radiology at Washington University St. Louis, for the many hours that she has spent editing and converting these chapters into a uniform format.

Editors

 **Dr. Michael J. Welch** is the codirector of the Division of Radiological Sciences and a professor of radiology at the Washington University School of Medicine's Mallinckrodt Institute of Radiology in St. Louis, Missouri. He is the recipient of several honors, including the Georg Charles de Hevesy Nuclear Medicine Pioneer Award, Paul C. Aebersold Award, American Chemical Society's St. Louis, Midwest and National Awards for Nuclear Chemistry, and the Cassen Award. He has served for many years on the editorial board of the *Journal of Nuclear Medicine*. Dr. Welch received his bachelor's and graduate degrees from Cambridge University, and his doctorate in radiochemistry from the University of London. His primary research involves the development of high-resolution PET imaging techniques, with a focus on the developments of novel PET detectors and systems suitable for imaging of small laboratory animals.

 **Dr. William C. Eckelman** is a professor of radiology at the University of California, San Diego and the president of Molecular Tracer LLC. A pioneer in Tc-99m radiopharmaceutical development, he and colleagues developed the Instant Tc-99m Kit, which became the basis for all subsequent 99mTc radiopharmaceutical kits, and targeted receptor-binding radiotracers for human use for both single-photon emission computed tomography and positron emission tomography, including the first neuroreceptor image in humans. Dr. Eckelman has been the editor-in-chief of *Nuclear Medicine and Biology* since 1985. He received numerous awards including the Paul C. Aebersold Award, the Institute of Clinical PET Distinguished Scientist Award, the American College of Nuclear Physicians Corporate Achievement Award, the Georg deHevesy Nuclear Pioneer Award, the Great Golden Seal of Padua University in recognition of contributions to Radiopharmaceutical Development, and the Society of Radiopharmaceutical Sciences President's Award.

Contributors

Samuel Achilefu
Departments of Radiology and
   Biomedical Engineering
Washington University School of
   Medicine
St. Louis, Missouri

Richard F. Ambinder
Johns Hopkins University School
   of Medicine
Baltimore, Maryland

John W. Babich
Molecular Insight Pharmaceuticals
Cambridge, Massachusetts

Chetan Bettegowda
Johns Hopkins University School
   of Medicine
Baltimore, Maryland

Youngjoo Byun
College of Pharmacy
Korea University
Chungnam, South Korea

Wengen Chen
Diagnostic Radiology and Nuclear Medicine
University of Maryland Medical Center
Baltimore, Maryland

R. Edward Coleman
Department of Radiology
Duke University
Durham, North Carolina

Cathy S. Cutler
Research Reactor Center
University of Missouri
Columbia, Missouri

Brian Davidson
Division of Cardiovascular Medicine
Oregon Health and Science University
Portland, Oregon

Vasken Dilsizian
Diagnostic Radiology and Nuclear
   Medicine
University of Maryland Medical Center
Baltimore, Maryland

William C. Eckelman
Molecular Tracer, LLC
Bethesda, Maryland

Catherine A. Foss
Johns Hopkins University School
   of Medicine
Baltimore, Maryland

Henry Gewirtz
Diagnostic Radiology and Nuclear
   Medicine
University of Maryland Medical Center
Baltimore, Maryland

Raymond E. Gibson
Gibson Imaging Apps, LLC
Holland, Pennsylvania

John R. Grierson
Department of Radiology
University of Washington
Seattle, Washington

Robert J. Gropler
Mallinckrodt Institute of Radiology
Washington University School
   of Medicine
St. Louis, Missouri

Timothy J. Hoffman
Department of Internal Medicine
University of Missouri

and

H.S. Truman Memorial VA Hospital
Columbia, Missouri

John L. Joyal
Molecular Insight Pharmaceuticals
Cambridge, Massachusetts

Michael R. Kilbourn
Department of Radiology
University of Michigan Medical School
Ann Arbor, Michigan

Kenneth A. Krohn
Departments of Radiology and Radiation
    Oncology
University of Washington
Seattle, Washington

Hank F. Kung
Departments of Radiology and
    Pharmacology
University of Pennsylvania
Philadelphia, Pennsylvania

Yali Li
Center for Systems Biology
Harvard Medical School
Massachusetts General Hospital
Charlestown, Massachusetts

Hannah M. Linden
Division of Medical Oncology
University of Washington Medical Center

and

Seattle Cancer Care Alliance
Seattle, Washington

Jonathan R. Lindner
Division of Cardiovascular Medicine and
    Department of Biomedical Engineering
Oregon Health and Science University
Portland, Oregon

Jeanne M. Link
Division of Nuclear Medicine
University of Washington Medical Center
Seattle, Washington

Yang Liu
Departments of Radiology and Biomedical
    Engineering
Washington University School
    of Medicine
St. Louis, Missouri

Robert H. Mach
Department of Radiology
Washington University School of Medicine
St. Louis, Missouri

David A. Mankoff
Division of Nuclear Medicine
University of Washington Medical Center

and

Seattle Cancer Care Alliance
Seattle, Washington

Cheryl L. Marks
Division of Cancer Biology
National Cancer Institute
Bethesda, Maryland

Jason R. McCarthy
Center for Systems Biology
Harvard Medical School
Massachusetts General Hospital
Charlestown, Massachusetts

Mark Muzi
Department of Radiology
University of Washington
Seattle, Washington

Karl Ploessl
Department of Radiology
University of Pennsylvania
Philadelphia, Pennsylvania

Martin G. Pomper
Johns Hopkins University School of
    Medicine
Baltimore, Maryland

Thomas P. Quinn
Department of Biochemistry
University of Missouri

and

H.S. Truman Memorial VA Hospital
Columbia, Missouri

David M. Raffel
Department of Radiology
University of Michigan Medical School
Ann Arbor, Michigan

Jeffrey L. Schwartz
Department of Radiation Oncology
University of Washington
Seattle, Washington

Sally W. Schwarz
Department of Radiology
Washington University School of Medicine
St. Louis, Missouri

George Sgouros
Johns Hopkins University School of
    Medicine
Baltimore, Maryland

Kooresh I. Shoghi
Mallinckrodt Institute of Radiology
Washington University School
    of Medicine
St. Louis, Missouri

Mark Smith
Diagnostic Radiology and Nuclear
    Medicine
University of Maryland Medical Center
Baltimore, Maryland

C. Jeffrey Smith
Department of Radiology
University of Missouri

and

H.S. Truman Memorial VA Hospital
Columbia, Missouri

Arun K. Thukkani
Mallinckrodt Institute of Radiology
Department of Internal Medicine
Washington University School of
    Medicine
St. Louis, Missouri

Wynn A. Volkert
Department of Radiology
University of Missouri

and

H.S. Truman Memorial VA Hospital
Columbia, Missouri

Kenneth T. Wheeler
Wheeler Scientific Consultants, Inc.
Winston-Salem, North Carolina

In Vitro Studies

1. *In Vitro* Approaches to Site-Specific Imaging Agents

Raymond E. Gibson

1.1 Introduction

By now, the information needed to develop site-directed (receptor, transporter, enzyme) radiotracers is well known (Eckelman and Gibson 1993, Eckelman 1981, 2003, Fowler et al. 2003, Patel and Gibson 2008). Although I will focus on developing site-directed radiotracers, the principles apply to developing any site-directed molecular probe (nuclear, optical, and the more speculative site-specific magnetic resonance and ultrasound contrast agents). In general, site-directed radiotracers are based upon receptor antagonists, uptake blockers, and enzyme inhibitors (functionally equivalent to the binding of an antagonist to a soluble receptor).

Receptor agonists can be used (e.g., opiates), but, since many agonists undergo internalization, this will limit the nature of the information that can be obtained (e.g., blockade of receptor binding by unlabeled compounds may be possible, but quantification of receptor *in vivo* may not be). In a similar fashion, while enzyme substrates, for example, [^{11}C]palmitate (Hoffman et al. 1977, Ter-Pogossian et al. 1980), [^{11}C]acetate (Pike et al. 1982), and [^{11}C]glucose (Herrero et al. 2002) can be useful, the metabolism of the radiotracer complicates analysis and certainly these radiotracers cannot be used to quantitate the binding-site concentration. Although these issues are primarily of concern for *in vivo* imaging and thus not treated herein, one should be aware of the caveats before embarking on the *in vitro* methods useful for developing new site-specific probes.

Targeted Molecular Imaging. Edited by Michael J. Welch and William C. Eckelman © 2012 Taylor & Francis Group, LLC. ISBN: 978-1-4398-4195-2

Chapter 1

The main information needed for the design of site-directed probes are the concentration of the site of interest, the affinity of the putative molecular probe for the site of interest, the lipophilicity of the putative probe, whether or not the molecular probe is a substrate for an efflux pump (e.g., the P-glycoprotein pump), and the metabolic fate of the putative probe (stability of the probe and presence or absence of interfering metabolites). For radiotracers including isotopes of short half-life, access to rapid synthetic methods and purifications are also needed. We will discuss in this chapter the various *in vitro* methods that are available to generate the information needed prior to investing time and resources to a new development project.

1.2   Determining the Site Concentration/Affinity of Radiotracer

When a new target is identified as being of interest for the development of a site-specific radiotracer, the most significantly needed information is the concentration of the target. It is axiomatic that if the target-binding site is not present in a tissue, then it is not possible to image it, regardless of how "wonderful" the properties of a radiotracer are (e.g., very high affinity, suitable lipophilicity, etc.). It should also be obvious that very low concentrations of target must require very high affinities, and this relationship has been established over the years, both theoretically (Eckelman et al. 1979a,b) and experimentally (Francis et al. 1982). For example, the dopamine D2 receptor in the caudate nucleus is of sufficiently high concentration that it is well imaged *in vivo* using radiotracers of modest affinity such as raclopride (Dean et al. 1997), while the concentration of receptor present in extrastriatal sites, for example, cortex, is sufficiently low that much higher affinity radiotracers are needed to provide *in vivo* visualization of those receptor populations in humans (Halldin et al. 1995, Olsson et al. 1999). Therefore, not only must one determine the concentration of the target protein in a tissue, but also the affinity of putative ligands that can subsequently be derivatized to produce the imaging probes.

Occasionally, the target is sufficiently mature that there is information concerning receptor concentrations and, perhaps, lead compounds suitable for initial exploration. This is the case for the dopamine D2 receptor ([³H]spiperone: Lyon et al. 1986, Chugani et al. 1988) in brain, muscarinic acetylcholine receptor ([³H]3-quinuclidinyl benzilate) in heart (Gibson et al. 1979) and brain (Pelham and Munsat 1979, Gibson et al. 1984), the β-adrenoceptors ([³H]alprenolol: U'Prichard et al. 1978, [¹²⁵I]Iodohydroxypindolol: Ezrailson et al. 1981) in heart, and the estradiol receptor ([³H]estradiol: Eckelman et al. 1979b) for estradiol receptor-containing tumors However, when the target is new, the characterization of the target becomes an interesting case of boot-strap-lifting. The concentration of target protein cannot be determined without an appropriate tool, usually a tritiated radioligand, and the lack of an *in vitro* radiotracer-binding assay makes it difficult to develop a high-affinity ligand suitable for determining the concentration of the target protein. Initial efforts are focused therefore on finding high-affinity lead compounds that may lead to radioligands.

1.2.1   Physiological Methods

Most initial efforts that lead to radioligands are not oriented toward developing radioligands, but are the result of pharmaceutical exploration of new targets with the intent of finding potent drug candidates, or for the pharmacological characterization of a new receptor. Nonetheless, pharmaceutical companies recognize the utility of radioligands which are often used in moderate-throughput assays to generate compounds with improved properties over those identified in initial screens.

The most common approach to obtaining initial leads for the development of potential high-affinity drugs is to screen compounds for their effect on a physiological response. Historically these have been effects of a response such as muscle contraction, effects on a second messenger for which there is an assay, for example, inhibition of c-AMP production or phosphotidyl inositol turnover, or on subsequent downstream responses, for example, Ca-efflux. The latter is particularly useful since Ca-influx can be monitored using fluorescence technologies that are amenable to high-throughput assays (e.g., FLIPR, Bednar et al. 2004).

Initial leads, most often identified through robotic assays in which entire compound libraries are screened, may exhibit affinities that are in the micromolar (μM) range (i.e., measurable inhibition of response occurs in concentrations of several μM). Using well-established methods in medicinal chemistry (Martin 1981, Evans et al. 1988, Lin et al. 2002, Neamati and Barchi 2002), such modest leads are modified to increase affinity. These increases in affinity are demonstrated using the physiological or biochemical-based screening assays. And herein lies the first caution: apparent affinities determined using nonradiotracer assays do not represent true affinity. One interesting example of this is the development of antagonist of the glutamate receptor subtype, the NR2B receptor (Laube et al. 1997).

The NR2B receptor is a potential therapeutic target for pain (Parsons 2001), and may have positive influence in the treatment of Parkinson's disease (Steece-Collier et al. 2000). Screening and medicinal chemistry refinements, using the fluorescence methodology FLIPR, provided antagonists with affinities <10 nM. Interestingly, despite considerable synthetic effort, an apparent floor to the affinity was found in the nM range. When a radioligand was developed, equilibrium conditions were established and it was found that many of the antagonists in which FLIPR estimates of affinity were near 1 nM exhibited much higher affinity. The kinetics of association of the antagonists is slow, and the fluorescent assay with whole cells did not permit incubation times sufficient to provide accurate estimations of affinity (Kiss et al. 2005).

Estimation of affinity using physiological measures when in error generally errs on the side of underestimating the affinity of compounds. One could reasonably assume that this should not be a problem. After all, if one selects a potential imaging radiotracer on the basis of an assay that underestimates the affinity, then the resulting compound should be better than expected. However, in the case of the NR2B antagonists, many different structural categories varying greatly in physicochemical properties appeared to have the same affinity as determined by the physiological methodology. There was no basis upon which to correctly select reasonable candidates for radiolabeling. Once the radiotracer assay was established, compounds that appeared to have the same affinity sorted themselves out to cover affinities ranging over 100-fold. Thus, the physiological assay did not provide sufficient discrimination to allow rational selection. While other factors may lead to the disagreement between affinities determined via physiological versus radiotracer assays, for example, inability to estimate nonspecific binding in the physiological assay or complexities in the coupling of second effectors as compared to binding at the receptor site, the slow association kinetics and inability to equilibrate binding in the physiological assay are adequate to explain the inability of this assay to provide knowledge of the affinities to allow selection of the best candidate for preparing a site-specific radiotracer.

An interesting aside to the NR2B story is that no *in vivo* imaging radiotracer for this target has been demonstrated. Several potential positron-emitting radiotracers have been reported, but specific *in vivo* binding have not been demonstrated (Haradahira et al. 2002, Årstad et al. 2006, Layton et al. 2006). I would suggest from the very slow kinetics of association demonstrated in the *in vitro* binding of NR2B radiotracers result in *in vivo* radiotracers with binding kinetics too slow to provide specific binding, that is, if significant binding to the receptor-binding sites requires hours of exposure and if either the radiotracer plasma clearance is too rapid or the radiotracer is labeled with carbon-11, site-specific binding *in vivo* will not be observed before clearance or decay obviates the ability to see the specific signal.

1.2.2 Tissue Homogenate Assay

There are excellent publications on radiotracer assays (Braestrup and Squires 1978, Pasternak 1980, Luthin and Wolfe 1984) and thus we will not provide an extensive exposition of these here. The most frequently used *in vitro* target-specific assay is often referred to as the "grind and bind" assay. When the target protein is part of the cell membrane, radioligand binding to the target protein can be assayed using homogenates of the target tissue (e.g., corpus striatum for the dopamine D2 receptor) or cells expressing the receptor (e.g., the NR2B receptor cited earlier). Radioligand is incubated with the membrane homogenates in a suitable physiological medium, in the absence and presence of a competitive unlabeled ligand. Once equilibrated, the radioligand bound to the receptor in the membranes is separated from the unbound usually by filtration over glass fiber filters. The specific signal is enhanced by a washing procedure using several volumes of incubation buffer (with no radioligand or competitive ligand present). The primary limiting factor in this form of *in vitro* radioligand assay is that the dissociation rate of the radioligand must be sufficiently slow so that the washing procedure used to reduce nonspecific binding does not wash away the specific binding. To minimize this, washing buffer usually used is ice cold, the lower temperature reducing the dissociation of specifically bound radioligand.

The most significant information that is obtained from a tissue homogenate assay is the affinity of the radioligand (usually expressed as K_d) and the concentration of binding site (B_{max}). When cell lines are used as the source of receptor, this concentration may not be particularly useful since it may not relate to the concentration expressed in target tissue *in vivo*. Therefore, the receptor concentration needs to be determined in the tissue is of interest. Using tissue from the target of interest, for example, dopamine D2 receptor in the caudate nucleus (of either an appropriate model animal, e.g., rodent, dog, and/or nonhuman primate, or from humans) and the homogenate assay, the effective concentration in the target *in vivo* can be calculated. The B_{max} obtained from the tissue homogenate assay is the number of femtomoles of radioligand that specifically bind to the small amount of tissue that is present in the incubation tube (usually between 1 and 10 mg of tissue, wet weight). The initial B_{max} can therefore be expressed as femtomoles bound per milligram of tissue, or picomoles bound per gram of tissue. If we now assume that the receptor is uniformly distributed throughout the tissue, and that the specific gravity is 1, the result immediately is expressed as nanomolar. While the assumption of uniform distribution of receptor may not be strictly true, for example, membrane receptors do not occupy intracellular space, imaging techniques are resolution-limited, and volume averaging effectively averages the binding protein concentration through the voxel volume. The best example of concordance between *in vitro* measured concentration and *in vivo* determined concentration is the dopamine D2 receptor that has been determined by both methods to exhibit concentrations of 15–20 nM (Wong et al. 1986, Farde et al. 1987).

1.2.3 Soluble Target Assay

For soluble targets, the difficulty is in how one separates the bound radioligand from the free pools. Among the techniques used are the centrifugal assay particularly used in the work on the steroid receptors, and typified by corticosteroid binding (Gardner and Wittliff 1973, Anderson et al. 1980) and equilibrium dialysis (Katzenellenbogen et al. 1982).

Centrifugal assays comprise two varieties: (1) sucrose gradient and (2) activated charcoal. In the former, separation of bound from free radioligand is achieved by ultracentrifugation: Following equilibration of the radioligand with receptor, the incubation mixture is layered onto a density gradient (usually sucrose which is benign to proteins). Separation of radioligand bound to receptor from free radioligand is effected by high *g*-forces which cause the heavy molecular weight species (receptor) to settle through the gradient. Free radiotracer remains on the top of the gradient. The

limitation of the assay is that density-gradient centrifugation usually takes many hours during which time lower affinity radioligands will dissociate.

The second variation on the centrifugal assay involves treating the incubation mixture with dextran-coated activated charcoal (DCC) which absorbs lipophilic compounds, a characteristic of most radioligands. Radioligand bound to receptor is not absorbed onto the DCC (unless too much is used). Centrifugation on a bench-top unit is sufficient to pellet the DCC. Radioligand bound to the receptor is then determined by assaying the supernatant. Similar to the sucrose-gradient assay, low-affinity radioligands are likely to dissociate during incubation with charcoal, thus giving erroneous data. Despite being prone to error with inconsistencies in assay conditions (Thorpe 1987), this assay is the standard one for slowly dissociating radioligands bound to soluble receptors (Johnson et al. 1975, Carola and McGuire 1977).

A viable alternative, particularly for lower affinity radioligands, is the dialysis assay. Dialysis membranes are commercially available which allow small molecular weight species to diffuse through, but retain higher molecular weight species. Dialysis can be accomplished using low-volume cells in which dialysis membrane separates two compartments, one containing the binding site of interest, and the second compartment containing the radioligand solution. An alternate approach is to use dialysis tubing. In this case, the enzyme or soluble receptor is placed inside a tube made by tying each end. The binding-site containing tube is then placed in a large container containing the radioligand in physiological medium. The system is allowed to equilibrate, during which time radioligand migrates across the membrane to bind to the soluble receptor. Unbound radioligand is at the same concentration on either side of the membrane. Bound ligand (representing both specific and nonspecific) is obtained from the difference in radioligand concentration between the binding-site compartment and that not containing the protein. The specifically bound radioligand is determined as in the filtration assay by including saturation concentrations of an unlabeled drug to block the specific binding of the radioligand. Much like the filtration assay, B_{max} and K_d are obtained by plotting the relationship between free radioligand concentration and the concentration of radioligand bound to the specific site in the dialysis tube or chamber, and fitting this binding curve with an appropriate model (e.g., single-site binding isotherm).

Interestingly, equilibrium dialysis is the only technique that provides a measure of unbound radioligand. In other techniques, one either creates conditions in which the initial and final concentrations of radioligand are essentially the same or one must calculate the free radioligand concentration (often with some erroneous assumptions); but, in dialysis, one measures the free radioligand concentration. A second advantage is that there is no separation procedure, that is, no washing, and thus this assay may provide a measure of affinity for lower affinity radioligands which would otherwise be washed away in the tissue homogenate assay or dissociate in the centrifugal assay. For example, the binding of [^{3}H]acetylcholine to acetylcholine receptors from the electric organ of the fish *Torpedo marmorata* with micromolar affinities is easily determined (O'Brien and Gibson 1974).

The primary limitation of this dialysis assay is the time it takes for equilibration. If equilibration times are long, and may be as long as 1–2 days, labile proteins will not survive, denaturing over the time required for equilibration.

1.2.4 Cell Culture Assay

When binding sites are present in cell membranes, the usual method for determining binding parameters is via the tissue homogenate assay. However, when the binding site is internal to the cell, for example, a cytosolic enzyme, maintaining cell integrity offers a method to determine the binding of radioligand to enzyme which would be otherwise quite difficult to accomplish. The centrifugal assay and dialysis assay described earlier may be useful, but obviously have limitations. By contrast, using whole cells provides an assay with similar characteristics to the dialysis assay, but the advantages are that (1) one is assaying the binding site of interest in its natural mileu and (2) times for equilibration of most reasonably lipophilic radioligands are considerably shorter, often being no more than several hours.

An excellent example of this type of binding assay is provided by studies on radiolabeled inhibitors of the enzyme farnesyl transferase (FPTase). This particular enzyme was considered an interesting potential cancer target since the protein which is farnesylated, Ras, has been shown to exhibit mutations in tumors (Bos 1989). Reversible inhibitors for the enzyme were developed which exhibited high affinity, modest lipophilicity, and structures amenable to the synthesis of radiotracers

(Williams et al. 1999). As part of the drug development program, an *in vitro* cell-based assay was developed using a radio-iodinated derivative of a reversible FPTase inhibitor (Lobell et al. 2003). Equilibration of cells with the radioligand and inhibitors at various concentrations provided the relative affinity of the unlabeled inhibitors. The IC$_{50}$ values (concentration of unlabeled inhibitor that blocked the binding of radioligand by 50%) correlated well with the efficacy of these compounds in inhibiting cell proliferation. We conducted saturation assays using the radioligand but the methodology we used did not permit assessing the B_{max} in cell culture, though it clearly can be done. While the apparent affinity of radioligand can thus be determined, the other parameter of interest, B_{max}, has little value since the concentration of enzyme in cells grown in tissue culture may not reflect the concentration enzyme naturally present in targets. Nonetheless, we were able to prepare and characterize radioligands for the enzyme FPTase that provided enzyme-mediated localization in tissues *in vivo*. It should be obvious that, when the concentration of binding site is as high or higher than occurs in target tissues *in vivo*, a radioligand that does not provide a specific signal in a tissue culture assay will have little chance of providing specific binding *in vivo*. However, the relationship between what will provide an *in vitro* signal in cell culture versus localization in a target tissue has not been established.

1.2.5 Autoradiography

Autoradiography is used to examine the binding of radioligands to binding sites in tissues or cells in which an image of the binding is generated by exposing the radioligand–tissue combination on x-ray film or specially design phosphor-imaging plates. Procedures for autoradiography are much like those used in filtration tissue homogenate assays: tissue on slides is incubated with a radioligand of interest to equilibrium, and, subsequently, nonspecific binding is reduced by washing in buffers without radiotracer. In the same manner as other *in vitro* assays, the washing procedures which greatly reduce nonspecific binding may be confounding if the off-rate of the radioligand is too fast and specifically bound radiotracer is lost. After drying, the tissues on slides are juxtaposed to films or phosphor-imaging plates which can be digitally scanned to provide the desired quantifiable images of radioligand binding. Nonspecific binding is determined either by including a high concentration of competitive ligand or by incubation of tissue sections in which the target of interest has been knocked-out (Kitchen et al. 1997, Ramboz et al. 1998, Lopez-Gimenez et al. 2002, Wong et al. 2003).

The advantage of autoradiography is that it can provide information on the binding of a radiotracer to sites that may be too small to be easily dissected for use in tissue homogenate assays. Of course, the obvious caveat is that if the site is too small to be easily isolated, perhaps it will be too small to be imaged, that is, partial volume averaging of very small structures may obviate *in vivo* visualization.

In the development of an *in vivo* imaging radiotracer, a striking feature of autoradiography is that it provides a direct correlate to predict what an image should look like. For example, the *in vitro* autoradiography of the histamine H3 receptor with a new radiotracer provided an exact correlate for the *in vivo* imaging of that receptor population in primates (Figure 1.1; Hamill et al. 2009). We also demonstrated via *in vitro* autoradiographic studies that mGluR5 radiotracer binds to the cerebellum in species other than rat at sufficiently high concentrations (Patel et al. 2007) that one should expect an image of the distribution of these receptors in the cerebellum of primates, both human and nonhuman. The autoradiographic studies prompted us to determine the B_{max} for the mGluR5 in the cerebellum of rat, rhesus monkey, and human brain: 1.4, 11, and 5.1 nM, respectively (Patel et al. 2007). In the rhesus monkey brain, the radiotracer exhibits K_d of 0.1 nM for mGluR5; B_{max}/K_d is 120

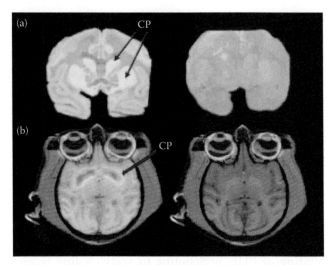

FIGURE 1.1 Comparison of *in vitro* and *in vivo* results of a histamine H3 radiotracer. Left: total binding. Right: nonspecific binding. (a) *In vitro* autoradiography of radioligand binding to rhesus monkey brain sections. Highest specific signal is associated with the basal ganglia—the caudate, putamen, and globus pallidus, with moderate binding in cortex. (b) *In vivo* PET image of the the histamine H3 radiotracer showing highest uptake in the basal ganglia and moderate uptake in cortex, recapitulating the autoradiographic results. (Data from Hamill, T.G. et al. 2009. *Synapse* 63:1122–1132.)

for cerebellum. For comparison, the B_{max}/K_d is 210 for rhesus caudate. Imaging studies in rhesus monkey proved specific binding to the mGluR5 in caudate and cerebellum as predicted by the B_{max}/K_d values (Hamill et al. 2005). By contrast, the B_{max}/K_d for rat cerebellum is 9. The reduction in ratio by more than 10-fold suggests that rat cerebellum mGluR5 would not be imaged *in vivo*.

1.3 Blood–Brain Barrier and Cellular Permeability

The primary characteristics which influence the ability of radiotracers to cross membranes, and thus also to enter the brain, are the radiotracer lipophilicity and the presence of efflux pumps. While there are a number of *in vitro* methods to examine membrane permeability involving the use of cells (Di et al. 2003), easier methods are available based simply on determining the lipophilicity of the new radiotracer and whether or not the radioligand is a substrate of an efflux pump.

1.3.1 Lipophilicity

It is axiomatic the hydrophilic compounds will not cross cell membranes since membranes are composed of bilayer lipids which provide a barrier to the diffusion of hydrophilic compounds. Barring the presence of active transport mechanisms (amino acid transport, glucose transport, or the internalization mechanisms associated with peptide-receptor binding and antibody–antigen binding), radiotracers need to have some lipophilic character which will allow them to enter into and pass through cellular membranes. The primary measure of lipophilicity is log *P* (Waterhouse 2003), the logarithm of the ratio of compound concentration in a lipophilic phase (e.g., 1-butanol or ethylacetate) to that in an aqueous phase (water or buffered physiological medium). The log *P* can be calculated (see below), or determined experimentally, the two most common methods being equilibration between an organic and aqueous phase or by high-performance liquid chromatography (HPLC).

Equilibration between two phases is quite easy to do: compound or radioligand is mixed into the lipophilic phase, then an aqueous phase is added and the bilayer mixed. Once equilibrium is achieved, the concentration of drug or radiotracer is determined in both phases from which the ratio is obtained and its logarithm is calculated. There are a number of cautions to be aware of. A primary concern is that no emulsion is formed during the mixing phase since an emulsion would carry some lipophilic compound into the aqueous phase thus reducing the ratio. Methods that allow equilibration between the two phases to be achieved slowly, for example, shaking gently for 24 h, reduce the likelihood of emulsion formation. Since equilibration may take many hours, the stability of the compound in the two media needs to be known.

When using radiotracer to determine log P, there is another source of error: radiochemical purity. For the most part, radiochemical purity of >98% is considered as quite good. However, if the impurity has different partitioning characteristics than the radiotracer, this can lead to an erroneous estimation of log P. For example, assume a very lipophilic radiotracer (log $P = 3$) which is 99% pure, but the 1% impurity is hydrophilic. Under these circumstances, 99.9% of the authentic radiotracer will be found in the organic phase and 0.1% in the aqueous phase, but essentially all of the hydrophilic impurity will be in the aqueous phase: 1%. The experimentally determined log P will be 2, in error by 10-fold. The solution to this, however, is actually quite simple—replace the aqueous phase and re-equilibrate. This is similar to washing out hydrophilic contaminants in chemical syntheses in which hydrophilic side-products are removed by sequential washes with aqueous phase. If the log P does not change with changes of the aqueous phase, then the value obtained is more likely to represent reality. An obvious caveat for these procedures, whether using unlabeled or labeled compounds, is that they are stable to the conditions of the assay.

Lipophilicity can also be determined from HPLC using a lipophilic column (Veitha et al. 1979). The procedure is, again, quite simple: The compound of interest as well as a number of structurally similar compounds of known but wide-ranging log P is eluted from an HPLC column. The logarithm of the retention time and log P values of the known or standard compounds are plotted, usually giving a linear plot from which the unknown compound's log P can be obtained by interpolation on the standard plot. Extrapolation is risky since breaks in the standard plot can occur. In addition, standard compounds of widely different structural category should be avoided since other

characteristics can influenced elution volume (e.g., van der Waals interactions which may be influenced by compound shape, e.g., linear vs. globular structures).

Calculated log P (Fujita et al. 1964, Leo et al. 1975, Lipinski et al. 1997, Mannhold and van de Waterbeemd 2001) can provide a good first approximation, but this author believes the experimentally determined partition of compound using organic and aqueous phases is the best method.

While the values of log P which are optimum for cell membrane permeability are well known (values between 1 and 3 are recommended with log $P = 2$ often considered to be ideal, Waterhouse 2003), the addition of common sense to the evaluation of results can easily lead to an understanding when *in vivo* results do not make sense. In our early investigations toward identifying a radiotracer for the tetrahydrocannabinoid (CB1) receptor, we were provided estimates of log P for compounds similar to rimanobant (see Figure 1.1 of Lin et al. 2006) ranging between 3.5 and 4.5 using the HPLC methodology. Using a commercially available radio-iodinated analog of rimanobandt, [^{125}I]AM251, we determined CB1-mediated localization in rat brain with a specific-to-nonspecific ratio of approximately 1. These *in vivo* results were not expected based on the very high estimated lipophilicity. We progressed to a structurally novel class of CB1 radiotracers (Burns et al. 2007) in which the HPLC-determined log P values remained significantly above 3. This prompted us to determine the log P via organic/aqueous phase partitioning, and found log $P \sim 2$, more consistent with the *in vivo* behavior of the radiotracers. Again, the determination of log P by organic phase/aqueous phase partitions provides the most reliable result.

1.3.2 Efflux Pumps

The blood–brain barrier, once thought to be comprised of only tight junction epithelial cells that kept hydrophilic drugs out of the brain, is now known to contain an ATP-dependent unidirectional pump that removes a broad range of compounds from the brain (Tatsuta et al. 1992, Leisen et al. 2003). One protein that is responsible, the P-glycoprotein pump (Pgp), is not only a constituent of the brain capillary endothelial cells, but is present in a number of other normal tissues, and more importantly, is often expressed in tumor cells, thus imparting resistance to many antitumor drugs (Ambudkar et al. 1999). For pharmaceutical agents, the effect of the pump can often be overcome just by increasing the plasma concentrations of the drug, provided those plasma concentrations do not lead to adverse effects. Not only has this attitude changed in recent decades, but when it comes to the uptake of radiotracers in tissue such as the brain or Pgp-expressing tumor cells, there is no concomitant method to overcome the pump. That is, one cannot simply increase the concentration of radiotracer to provide the desired uptake. Thus, radiotracers which are targeted to tissues in which the Pgp is a constituent should not be good substrates of the Pgp.

Early *in vitro* assays to determine if a particular compound is a substrate are mostly based on whole cells (Tatsuta et al. 1992, Sharom 1997, Perloff et al. 2003). The primary assay used in pharmaceutical companies has shifted to a monolayer of cell-contained clones, expressing human Pgp. Substrate activity is determined using the basolateral (B) to apical (A) permeability over the apical to basolateral permeability (Polli et al. 2001). When the ratio is 1, there is no preference for transport, that is, passive diffusion. Statistically, a ratio >2 is considered to reflect substrate behavior for the compound under study. The assay is very predictive of *in vivo* behavior with the primary caveat being that there are species differences in the extent to which a compound is or is not a good substrate (Yamazaki et al. 2001). However, since cells are available which express the cloned human Pgp, compounds can be assessed against the pump of primary concern. The utility of such assays rests in the ability to screen a reasonable number of structurally diverse compounds prior to the investment of effort required to synthesize a radioligand. With all other properties of a compound being desirable, for example, high affinity, modest lipophilicity, and so on, a compound that favors A to B transport by 10- or 15-fold over B to A transport should be considered an unlikely prospect as an *in vivo* imaging agent for a target in the CNS or in Pgp-expressing tumors.

There is, as always, a caveat to the issue of Pgp substrates: The substance P receptor PET radiotracer, [^{18}F] SPARQ (Solin et al. 2004), exhibits (A-B)/(B-A) of approximately 5, yet this radiotracer is a superb agent for imaging the substance P receptor in man (Hargreaves 2002 Bottom of Form, Bergström et al. 2004). While the exact reason for this has not been rigorously determined, we hypothesize that the association rate of [^{18}F]SPARQ is sufficiently rapid that binding to receptor occurs more rapidly than the reduction in radiotracer by the Pgp.

If one has a radiotracer, there are good *in vivo* assays for demonstrating that the radiotracer is or is not a Pgp substrate. Among these are Pgp-knockout mice (Tanigawara 2000), and the use of inhibitors of the Pgp (Starling et al. 1997, Kemper et al. 2004). If the uptake of a radiotracer in brain is greater in a Pgp-knockout mouse compared to wild type, or if an inhibitor increases the radiotracer uptake in brain or tumor, the presumptive conclusion is that the radiotracer is a substrate and perhaps a different structural class should be sought for the *in vivo* imaging agent. For evaluation of a limited number of radiotracers, this may be more reasonable than the effort needed to establish an *in vitro* assay for Pgp substrates.

1.4   Metabolism

The issue of metabolism of a new radiotracer generally occurs once *in vivo* studies have begun. In general, there are two issues that arise: (1) clearance of a radiotracer which is too rapid to provide a good signal, and (2) the presence of lipophilic-radiolabeled metabolites may also appear in the tissue of interest, this localization confounding the interpretation of images. The former is an issue when labeling techniques such as fluoromethylation or fluoroethylation are used to synthesize the radiotracer since there is potential for metabolic loss of fluoride. Defluorination not only would reduce the site-specific localization, but can lead to undesirably high bone uptake. The second problem is typified by [^{11}C]WAY100635, a serotonin receptor (5HT1A)-specific radiotracer in which radiolabeled metabolites are sufficiently lipophilic to appear in the brain (Osmun et al. 1996, 1998).

The best use of *in vitro* metabolic studies would be to differentiate amongst potential *in vivo* radiotracers, for example, demonstrating that one of several potential candidates is more metabolically stable or less likely to generate unwanted metabolites. The successful use of the *in vitro* metabolic study of radiotracer development (Lavén et al. 2004, 2006, Giron et al. 2008, Grosse et al. 2009) is based on extensive evidence that the use of isolated microsomal preparations (Garner et al. 1972, Rajaonarison et al. 1991, Paar et al. 1992, Riley and Leeder 1995), isolated hepatocytes (Billings et al. 1977, Maurel 1996, Soars et al. 2002, 2007, Gebhardt et al. 2003, McGinnity et al. 2004), and liver slices (Wormser et al. 1990, Stearns et al. 1992) is quite predictive of the *in vivo* metabolic profile of drugs (Ekins et al. 2000). As such, the application of *in vitro* metabolism to determine the best of multiple possibilities is reasonable, which has been used for the histamine H3 receptor radiotracer (Hamill et al. 2009). The metabolic profile of serotonergic receptor radiotracers was determined via *in vitro* assays for several species (Maa et al. 2006).

Metabolism of drugs differs among species (Quinn et al. 1958, Pearce et al. 1992). Since most drugs, and site-specific radiotracers, are initially tested in small animals, such as rat, then scaled up to larger animals, for example, dogs and primates, it is important to know what the species differences in metabolism are. For radiotracers, it is possible to use these differences in a very positive manner: If a site-specific radiotracer shows specific binding *in vivo* in a species in which the metabolism is rapid, the radiotracer should work equally well, or perhaps better, in species in which the metabolism is considerably slower. For example, the microsomal metabolism of the histamine H3 radiotracer is observed to be fastest in monkey of the three species examined (rat, monkey, and human: Figure 1.2) (Hamill et al. 2009). The agent provides H3 receptor-mediated localization in rhesus monkey. All other factors being equal, this radiotracer should also successfully image the histamine H3 receptor distribution in humans. Unfortunately, at this time,

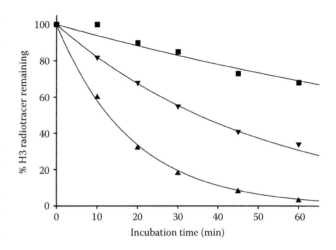

FIGURE 1.2   The *in vitro* metabolic stability of the histamine H3 radiotracer, 2a in rat (▼), monkey (▲), and human (■) liver microsomes. Experimental conditions are described in Hamill et al. (2009). It is interesting to note that the most rapid metabolism occurred in monkey in which the distribution of histamine H3 receptors is well visualized. (Data from Hamill, T.G. et al. 2009. *Synapse* 63:1122–1132.)

the corroborative *in vivo* imaging of human receptor has not been reported.

1.5   Nonwash Autoradiography

Typical autoradiographic methodology involves washing procedures to reduce nonspecific binding. It occurred to us (Patel et al. 2003) that successful imaging radiotracers build up a specific binding signal in the absence of washing procedures and that this behavior *in vivo* should be mimicked *in vitro* in autoradiography. That is, *in vivo* imaging requires that a specific binding signal must be detected in the presence of nonspecific binding—the key feature distinguishing between *in vitro* and *in vivo* studies being the inability to "wash away nonspecific binding" in the *in vivo* imaging study. Thus, a good *in vivo* imaging radiotracer should not need a washing step to observe specific binding when conducting *in vitro* autoradiography. Logically, if the concentration of binding sites and affinity of radiotracer are adequate for *in vivo* imaging, they should also be adequate to provide a specific binding signal *in vitro*.

This simple observation led us to the hypothesis that we could determine what radiotracers should work *in vivo* from an *in vitro* assay in which there are no washing procedures. Clearly, if a radiotracer does not provide a detectable-specific signal *in vitro* in our nonwash assay, it would be hard to understand why you would see such a signal *in vivo*. In order to develop this

as a useful *in vitro* assay, however, we also needed to provide certain limitations that are dictated by *in vivo* radiotracer behavior. When dealing with carbon-11-labeled radiotracer, it is axiomatic that a specific binding signal must develop quickly. If not, the short half-life of the isotope will not lead to useful images. While not limiting in all cases, this suggested that the development of an *in vitro*-specific autoradiographic signal quickly is more likely to be useful *in vivo* than if the signal takes hours to achieve. Using the somewhat arbitrary time of 20 min (obtained from one half-life of carbon-11), we were able to determine what *in vitro* results would be predictive of success in imaging specific binding sites *in vivo*. In essence, when the specific signal is below 20% of the total, the radiotracer is not likely to provide receptor-mediated image *in vivo*.

Since the initial work was conducted using a variety of radioligands for different binding targets, we followed this work with a study looking at radiotracers specifically designed to bind the mGluR5. These radiotracers varied in affinity for the mGluR5 receptor from $K_D = 0.08–11$ nM (Table 1.1). The *in vitro* nonwash autoradiographic assay predicted rather well the results for *in vivo* imaging in rat brain (Patel et al. 2005).

Table 1.1 Relationship between *In Vivo*-Specific Binding of mGluR5 Radiotracers and the Affinity and %-Specific Binding from the *In Vitro* Nonwash Autoradiography Assay

Compound[a]	K_i (nM)[b]	%-Specific Binding[c]	*In Vivo* Signal/Noise[d]
1	2.0	40	2.9
2	0.23	38	2
3	0.08	69	7.2
4	0.2	77	7.1
5	11	15	1

[a] For structures, see Patel et al. (2005).

[b] Apparent affinity determined by competitive blockade of radiotracer binding.

[c] Calculated as $100 \times$ (Total Bound – NS)/Total Bound determined after 20 min of incubation.

[d] S/N = Peak signal in rat corpus striatum/signal in cerebellum at the same time.

Compound 5, which had lowest affinity provided %-specific binding <20%, and did not provide a receptor-mediated signal *in vivo*. A plot of the S/N against %-specific binding in the nonwash autoradiography was reasonably fit with a linear equation ($r^2 = 0.94$). Since the receptor concentration in the target tissue is a constant, we would assume that the apparent affinity should be related to the imaged S/N (i.e., the higher the B_{max}/K_i ratio, the better the S/N). However, a plot of S/N versus K_i was poorly fit by a linear equation ($r^2 = 0.41$). Thus, while a B_{max}/K_i is necessary for developing a good S/N, it is not sufficient.

A potential use of this assay that was not fully examined is that it should provide a measure of nonspecific binding in the tissue of interest. Other interesting modifications would be to include a radiotracer elimination phase, which could be designed to approximate plasma pharmacokinetic curves. This assay would not rely upon a somewhat arbitrary time for the assay, for example, the 20 min we selected.

1.6 Conclusion

In vitro assays for all aspects of radiotracer development that correspond to the parameters needed for *in vivo* receptor- or site-mediated localization are available, though perhaps underused. Clearly, the binding-site concentration and affinity of radiotracer should be determined before extensive effort is invested in radiotracer development. *In vitro* visualization of the binding pattern of a new radiotracer can prevent unwanted surprises as suggested, for example, by the *in vivo* imaging of the cerebellum by mGluR5 radiotracers. The lipophilicity of a new radiotracer is easily determined, though it is equally necessary to demonstrate that the new radiotracer is not a substrate of efflux pumps such as the Pgp. Both of these characteristics can be easily assayed using *in vitro* techniques. Finally, drug metabolism studies in the development of new pharmaceuticals is *de rigueur*, the *in vivo* metabolic fate being predicted from *in vitro* methodologies. The similar use for radiotracer development should also predict quite well the difficulties that may be encountered prior to *in vivo* imaging studies. With the advent of new *in vitro* methodologies such as the nonwash autoradiography assay, prediction of *in vivo* behavior is even more likely. The use of all, or at least most, of these *in vitro* methods can lead to more efficient radiotracer development in which failures are greatly reduced.

References

Ambudkar, S.V., Dey, S., Hrycyna, C.A., Ramachandra, M., Pastan, I., and Gottesman, M.M. 1999. Biochemical, cellular and pharmacological aspects of the multidrug transporter. *Annu. Rev. Pharmacol. Toxicol.* 39:361–398.

Anderson, K.M., Phelan, J., Marogil, M., Hendrickson, C., and Economou, S. 1980. Sodium molybdate increases the amount of progesterone and estrogen receptor detected in certain human breast cancer cytosols. *Steroid* 35:273–280.

Årstad, E., Platzer, S., Berthele, A. et al. 2006. Towards NR2B receptor selective imaging agents for PET—Synthesis and evaluation of N-[11C]-(2-methoxy)benzyl (*E*)-styrene-, 2-naphthyl- and 4-trifluoromethoxyphenylamidine. *Bioorg. Med. Chem.* 14: 6307–6313.

Bednar, B., Cunninghama, M.E., Kiss, L. et al. 2004. Kinetic characterization of novel NR2B antagonists using fluorescence detection of calcium flux. *J. Neurosci. Methods* 137:247–255.

Bergström, M., Hargreaves, R.J., Burns, H. et al. 2004. Human positron emission tomography studies of brain neurokinin 1 receptor occupancy by a precipitant. *Biolog. Psych.* 55: 1007–1012.

Billings, R.E., McMahon, R.E., Ashmore, J., and Wagle, S.R. 1977. The metabolism of drugs in isolated rat hepatocytes. A comparison with *in vivo* drug metabolism and drug metabolism in subcellular liver fractions. *Drug Metab. Disp.* 5:518–526.

Bos, J.L. 1989. Ras oncogenes in human cancer. A review. *Cancer Res.* 49:4682–4689.

Braestrup, C. and Squires, R.F. 1978. Brain specific benzodiazapine receptors. *Br. J. Psych.* 133:249–260.

Burns, H.D., Van Laere, K., Sanabria-Bohorquez, S. et al. 2007. [^{18}F] MK-9470, a positron emission tomography (PET) tracer for *in vivo* human PET brain imaging of the cannabinoid-1 receptor. *Proc. Natl. Acad. Sci. USA* 104:9800–9805.

Carola, R.E. and McGuire, W.L. 1977. An improved assay for nuclear estrogen receptor in experimental and human breast cancer. *Cancer Res.* 37:3333–3337.

Chugani, D.C., Ackermann, R.F., and Phelps, M.E. 1988. *In vivo* [H-3]spiperone binding: Evidence for accumulation in corpus striatum by agonist-mediates receptor internalization. *J. Cereb. Blood Flow Metab.* 8:291–303.

Dean, B., Pavey, G., and Opeskin, K. 1997. [^{3}H]Raclopride binding to brain tissue from subjects with schizophrenia: Methodological aspects. *Neuropharmacology* 36:779–786.

Di, L., Kerns, E.H., Fan, K., McConnell, O.J., and Carter, G.T. 2003. High throughput artificial membrane permeability assay for blood–brain barrier. *Eur. J. Med. Chem.* 38:223–232.

Eckelman, W.C., Gibson, R.E., Rzeszotarski, W.J. et al. 1979a. The design of receptor binding radiotracers. In: *Principles of Radiopharmacology.* ed. L. Colombetti, pp. 251–274. New York: CRC Press.

Eckelman, W.C., Reba, R.C., Gibson, R.E. et al. 1979b. Receptor-binding radiotracers: A class of potential radiopharmaceuticals. *J. Nucl. Med.* 20:350–357.

Eckelman, W.C. 1981. Receptor binding radiotracers. In: *Receptor Binding Radiotracers,* ed. W.C. Eckelman, Vol. 1. pp. 69–91. Boca Raton, FL: CRC Press.

Eckelman, W.C. 2003. Mechanism of target specific uptake. In: *Handbook of Radiopharmaceuticals: Radiochemistry and Applications.* eds. M.J. Welch and C. Redvanly, pp. 487–500. West Sussex, England: John Wiley & Sons, Ltd.

Eckelman, W.C. and Gibson, R.E. 1993. The design of site-directed radiopharmaceuticals for use in drug discovery. In: *Nuclear Imaging and Drug Discovery, Development and Approval.* eds H.D. Burns, R.E. Gibson, R. Dannals, and P. Siegl, pp. 114–134; Boston: Birkhauser.

Ekins, S., Ring, B.J., Grace. J. et al. 2000. Present and future *in vitro* approaches for drug metabolism. *J. Pharmacol. Toxicol. Methods* 44:313–324.

Evans, B.E., Rittle, K.E., Bock, M.G. et al. 1988. Methods for drug discovery: Development of potent, selective, orally effective cholecystokinin antagonists. *J. Med. Chem.* 31:2235–2246.

Ezrailson, E.G., Garber, A.L., Munson, P.J., Swartz, T.L., Birnbaumer, L., and Entman, M.L. 1981. [I-125]iodopindolol: A new beta-adrenergic receptor probe. *J. Cycl. Nucleo. Res.* 7:13–26.

Farde, L., Halldin, C., Stone-Elander, S., and Sedvall, G. 1987. PET analysis of human dopamine receptor subtypes using ^{11}C-SCH 23390 and ^{11}C-raclopride. *Psychopharmacology* 92:278–284.

Fowler, J.S., Ding, Y.S., and Volkow, N.D. 2003. Radiotracers for positron emission tomography. *Sem. Nucl. Med.* 33:14–27.

Francis, B., Eckelman, W.C., Grissom, M.P., Gibson, R.E., and Reba, R.C. 1982. The use of tritium labeled compounds to develop gamma emitting receptor binding radiotracers. *Int. J. Nucl. Med. Biol.* 9:173–179.

Fujita, T., Iwasa, J., and Hansch, C. 1964. A new substituent constant, pi, derived from partition coefficients. *J. Am. Chem. Soc.* 86:5175–5180.

Gardner, D.G. and Wittliff, J.L. 1973. Characterization of a distinct glucocorticoid-binding protein in lactating mammary gland of the rat. *Biochim. Biophys. Acta* 320:617–627.

Garner, R.C., Miller, E.C., and Miller, J.A. 1972. Liver microsomal metabolism of aflatoxin B, to a reactive derivative toxic to *Salmonella typhimurium* TA 1530. *Cancer Res.* 32:2058–2066.

Gebhardt, R., Hengstler, J.G., Müller, D. et al. 2003. New hepatocyte *in vitro* systems for drug metabolism: Metabolic capacity and recommendations for application in basic research and drug development, standard operation procedures. *Drug Metab. Rev.* 35:145–213.

Gibson, R.E., Eckelman, W.C., Vieras, F., and Reba, R.C. 1979. The distribution of the muscarinic acetylcholine receptor antagonists, quinuclidinyl benzilate and quinuclidinyl benzilate methiodide (both tritiated), in rat, guinea pig, and rabbit, *J. Nucl. Med.* 20:865–870.

Gibson, R.E., Weckstein, D.J., Jagoda, E.M. et al. 1984. The characteristics of I-125 4-IQNB and H-3 QNB *in vivo* and *in vitro*. *J. Nucl. Med.* 25:214–222.

Giron, M.C., Portolan, S., Bin, A., Mazzi, U., and Cutler, C.S. 2008. Cytochrome P450 and radiopharmaceutical metabolism. *Q. J. Nucl. Med. Mol. Imaging* 52:254–266.

Grosse, M.E., Wiese, C., Schepmann, D. et al. 2009. Synthesis of spirocyclic sigma1 receptor ligands as potential PET radiotracers, structure-affinity relationships and *in vitro* metabolic stability. *Bioorg. Med. Chem.* 17:3630–3641.

Halldin, C., Farde, L., Hogberg, T. et al. 1995. Carbon-11-FLB 457: A radioligand for extrastriatal D2 dopamine receptors. *J. Nucl. Med.* 36:1275–1281.

Hamill, T.G., Krause, S., Ryan, C. et al. 2005. The synthesis, characterization and first successful monkey imaging studies of metabotropic glutamate receptor subtype 5 (mGluR5) PET radiotracers. *Synapse* 56:205–216.

Hamill, T.G., Sat, N., Jitsuoka, M. et al. 2009. Inverse agonist histamine H3 receptor PET tracers labelled with carbon-11 or fluorine-18. *Synapse* 63:1122–1132.

Haradahira, T., Maeda, J., Okauchi, T. et al. 2002. Synthesis, *in vitro* and *in vivo* pharmacology of a C-11 labeled analog of CP-101,606, (+/-)threo-1-(4-hydroxyphenyl)-2-[4-hydroxy-4-(p-[^{11}C]methoxyphenyl)piperidino]-1-propanol, as a PET tracer for NR2B subunit-containing NMDA receptors. *Nucl. Med. Biol.* 29:517–525.

Hargreaves, R. 2002. Imaging substance P Receptors (NK$_1$) in the living human brain using positron emission tomography. *J. Clin. Psych.* 63 (Suppl 11):18–24.

Hoffman, E.J., Phelps, M.E., Weiss, E.S. et al. 1977. Transaxial tomographic imaging of canine myocardium with "C-palmitic acid. *J. Nucl. Med.* 18:57–61.

Herrero, P., Weinheimer, C.J., Dence, C. et al. 2002. Quantification of myocardial glucose utilization by pet and 1-carbon-11-glucose. *J. Nucl. Cardiol.* 9:5–14.

Johnson, R.B. Jr, Nakamura, R.M., and Libby, R.M. 1975. Simplified Scatchard-plot assay for estrogen receptor in human breast tumor. *Clin. Chem.* 21:1725–1730.

Katzenellenbogen, J.A., McElvany, K.D., Senderoff, S.G. et al. 1982. 16-alpha-[^{77}Br]Bromo-11ß-methoxyestradiol-17ß: A gamma-emitting estrogen imaging agent with high uptake and retention by target organs. *J. Nucl. Med.* 23:411–419.

Kemper, E.M., Verheij, M., Boogard, W., Beijnen, J.H., and van Tellingen, O. 2004. Improved penetration of docetaxel into the brain by co-administration of inhibitor of P-glycoprotein. *Eur. J. Cancer* 40:129–1274.

Kiss, L., Cheng, G., Bednar, B. et al. 2005. *In vitro* characterization of novel NR2B selective NMDA receptor antagonists. *Neurochem. Int.* 46:453–464.

Chapter 1

Kitchen, I., Slowe, S.J., Matthes, H.W.D., and Kieffer, B. 1997. Quantitative autoradiographic mapping of μ-, δ- and κ-opioid receptors in knockout mice lacking the m-opioid receptor gene. *Brain Res.* 778:73–88.

Laube, B., Hirai, H., Sturgess, M., Betz, H., and Kuhse, J. 1997. Molecular determinants of agonist discrimination by NMDA receptor subunits: Analysis of the glutamate binding site on the NR2B subunit. *Neuron* 18:493–503.

Lavén, M., Itsenko, O., Markides, K., and Långström, B. 2006. Determination of metabolic stability of positron emission tomography tracers by LC-MS/MS: An example in WAY-100635 and two analogues. *J. Pharm. Biomed. Anal.* 40:943–945.

Lavén, M., Markides, K., and Långström, B. 2004. Analysis of microsomal metabolic stability using high-flow-rate extraction coupled to capillary liquid chromatography-mass spectrometry. *J. Chromatogr. B Analyt. Technol. Biomed Life Sci.* 806:119–126.

Layton, M.E., Kelly, M.J. III, and Rodzinak, K.J. 2006. Recent advances in the development of NR2B subtype-selective NMDA receptor antagonists. *Current Trends Med. Chem.* 6:697–709.

Leisen, C., Dressler, C., Koggel, A., and Spahn-Langguth, H. 2003. Lipophilicities of baclofen ester prodrugs correlate with affinities to the ATP-dependent efflux pump P-glycoprotein: Relevance for their permeation across the blood–brain barrier? *Pharmacol. Res.* 20:772–778.

Leo, A., Jow, P.Y.C., Silipo, C., and Hansch, C. 1975. Calculation of hydrophobic constant (log *P*) from pi. and f constants. *J. Med. Chem.* 18:865–868.

Lin, J.-H., Perryman, A.L., Schames, J.R., and McCammon, J.A. 2002. Computational drug design accommodating receptor flexibility: The relaxed complex scheme. *J. Am. Chem. Soc.* 124:5632–5633.

Lin, L.S., Lanza, T.J. Jr, Jewell, J.P. et al. 2006. Discovery of N-[(1S,2S)-3-(4-Chlorophenyl)-2-(3-cyanophenyl)-1-methylpropyl]-2-methyl-2-{[5-(trifluoromethyl)pyridin-2-yl]oxy}propanamide (MK-0364), a novel, acyclic cannabinoid-1 receptor inverse agonist for the treatment of obesity. *J. Med. Chem.* 49:7584–7587.

Lipinski, C.A., Lombardo, F., Dominy, B.W., and Feeney, P.J. 1997. Experimental and computational approaches to estimate solubility and permeability in drug discovery and development settings. *Adv. Drug Delivery Rev.* 23:3–25.

Lobell, R.B., Davide, J.P., Kohl, N.E., Burns, H.D., Eng, W-S., and Gibson, R.E. 2003. A cell-based radioligand binding assay for farnesyl: Protein transferase inhibitors. *J. Biomol. Screening* 8:430–438.

Lopez-Gimenez, J.F., Tecott, L.H., Palacios, J.M., Mengod, G., and Vilaro, M.B. 2002. Serotonin 5-HT2C receptor knockout mice: Autoradiographic analysis of multiple serotonin receptors. *J. Neurosci. Res.* 67:69–85.

Luthin, G.R. and Wolfe, B.B. 1984. [H-3] Pirenzepine and [H-3] quinuclidinyl benzilate binding to brain muscarinic cholinergic receptors. *Mol. Pharmacol.* 26:164–169.

Lyon, R.A., Titeler, M., Frost, J.J. et al. 1986. H-3 N-methylspiperone labels D2 dopamine receptors in basal ganglia and S2 serotonin receptors in cerebral cortex. *J. Neurosci.* 6:2941–2949.

Maa, Y., Langa, L., Kiesewetter, D., Jagoda, E., and Eckelman, W.C. 2006. Species differences in metabolites of PET ligands: Serotonergic 5-HT1A receptor antagonists 3-*trans*-FCWAY and 3-*cis*-FCWAY. *Nucl. Med. Biol.* 33:1013–1019.

Mannhold, R. and van de Waterbeemd, H. 2001. Substructure and whole molecule approaches for calculating log *P. J. Computer-Aided Mol. Design.* 15:337–354.

Martin, Y.C. 1981. A practitioner's perspective of the role of quantitative structure–activity analysis in medicinal chemistry. *J. Med. Chem.* 24:229–237.

Maurel, P. 1996. The use of adult human hepatocytes in primary culture and other *in vitro* systems to investigate drug metabolism in man. *Adv. Drug Deliv. Rev.* 22:105–132.

McGinnity, D.F., Soars, M.G., Urbanowicz, R.A., and Riley, R.J. 2004. Evaluation of fresh and cryopreserved hepatocytes as *in vitro* drug metabolism tools for the prediction of metabolic clearance. *Drug Metab. Disp.* 32:1247–1253.

Neamati, N. and Barchi, J.J. Jr. 2002. New paradigms in drug design and discovery. *Curr. Top. Med. Chem.* 2:211–227.

O'Brien, R.D. and Gibson, R.E. 1974. Two binding sites in acetylcholine receptor from *Torpedo marmorata* electroplax. *Arch. Biochem. Biophys.* 165:681–690.

Olsson, H., Halldin, C., Swahn, C-G., and Farde, L. 1999. Quantification of [11C]FLB 457 binding to extrastriatal dopamine receptors in the human brain. *J. Cereb. Blood Flow Metab.* 19:1164–1173.

Osmun, S., Lundkuist, C., Pike, V.W. et al. 1996. Characterization of the radioactive metabolites of the 5HT$_{1A}$ receptor radioligand, [0-methy 11C]WAY-100635, in monkey and human plasma by HPLC: Comparison of the behaviour of an identified radioactive metabolite with parent radioligand in monkey using PET. *Nucl. Med. Biol.* 23:627–534.

Osmun, S., Lundkuist, C., Pike, V.W. et al. 1998. Characterization of the appearance of radioactive metabolites in monkey and human plasma from the 5-HT1A receptor radioligand, [carbonyl-11C]WAY-100635—Explanation of high signal contrast in PET and an aid to biomathematical modelling. *Nucl. Med. Biol.* 25:215–223.

Paar, W.D., Frankus, P., and Dengler, H.J. 1992. The metabolism of tramadol by human liver mierosomes. *Clin. Investig.* 70:708–710.

Parsons, C.G. 2001. NMDA receptor as targets for neuropathic pain. *Eur. J. Pharmacol.* 429:71–78.

Pasternak, G.W. 1980. Multiple opiate receptors: [³H]ethylketocyclazocine receptor binding and ketocyclazocine analgesia. *Proc. Natl. Acad. Sci. USA* 77:3691–3694.

Patel, S., Hamill, T., Hostetler, E., Burns, D., and Gibson, R.E. 2003. An *in vitro* assay for predicting successful imaging radiotracers. *Mol. Imaging Biol.* 5:65–71.

Patel, S., Ndubizu, O., Hamill, T., Chaudhary, A., Burns, H.D., Hargreaves, R., and Gibson, R.E. 2005. Screening cascade and development of potential PET radiotracers for mGluR5: *In vitro* and *in vivo* characterization. *Mol. Imaging Biol.* 7:314–323.

Patel, S., Hamill, T.G., Connolly, B., Jagoda, E., Li, W., and Gibson, R.E. 2007. Species differences in mGluR5 binding sites in mammalian central nervous system determined using *in vitro* binding with [F-18]F-PEB. *Nucl. Med. Biol.* 34:1009–1017.

Patel, S. and Gibson, R. 2008. *In vivo* site-directed radiotracers: A mini-review. *Nucl. Med. Biol.* 35:805–815.

Pearce, R., Greenway, D., and Parkinson, A. 1992. Species differences and interindividual variation in liver microsomal cytochrome P450 2A enzymes: Effects on coumarin, dicumarol, and testosterone oxidation. *Arch. Biochem. Biophys.* 298:211–225.

Pelham, R.W. and Munsat, T.L. 1979. Identification of direct competition for, and indirect influences on, striatal muscarinic cholinergic receptors: *In vivo* [H-3]quinuclidinyl benzilate binding it rats. *Brain Res.* 171:473–480.

Perloff, M.D., Störmer, E.S., von Moltke, L.L., and Greenblatt, D.J. 2003. Rapid assessment of P-glycoprotein inhibition and induction *in vitro*. *Pharm. Res.* 20:1177–1183.

Pike, V.W., Eakins, M.N., Allan, R.M., and Selwyn, A.P. 1982. Preparation of [1–11C] acetate—an agent for the study of myocardial metabolism by positron emission tomography. *Int. J. Appl. Radiat. Isot.* 33:505–512.

Polli, J.W., Wring, S.A., Humphreys, J.E., Huang, K., Morgan, J.B., Webster, L.O., and Serabjit-Singh, C.S. 2001. Rational use of *in vitro* P-glycoprotein assays in drug discovery. *J. Pharmacol. Exptl. Ther.* 299:620–628.

Quinn, G.P., Axelrod, J., and Brodie, B.B. 1958. Species, strain and sex differences in metabolism of hexobarbitone, amidopyrine, antipyrine and aniline. *Biochem. Pharmacol.* 1:152–159.

Rajaonarison, F.J., Lacarelle, B., De Sousa, G., Catalin, J., and Rahmani, R. 1991. *In vitro* glucuronidation of 3′-azido-3′-deoxythymidine by human liver. Role of UDP-Glucuronosyltransferase 2 form. *Drug Metab. Disp.* 19:809–815.

Ramboz, S., Oosting, R., Amara, D.A. et al. 1998. Serotonin receptor 1A knockout: An animal model of anxiety-related disorder. *Proc. Natl. Acad. Sci. USA* 95:14476–14481.

Riley, R.J. and Leeder, J.S. 1995. *In vitro* analysis of metabolic predisposition to drug hypersensitivity reactions. *Clin. Exp. Immunol.* 99:1–6.

Sharom, F.J. 1997. The P-glycoprotein efflux pump: How does it transport drugs? *J. Membrane Biol.* 160:161–175.

Soars, M.G., Burchell, B., and Riley, R.J. 2002. *In vitro* analysis of human drug glucuronidation and prediction of *in vivo* metabolic clearance. *J. Pharmacol. Exp. Ther.* 301:382–390.

Soars, M.G., Grime, K., Sproston, J.L. et. al. 2007. Use of hepatocytes to assess the contribution of hepatic uptake to clearance *in vivo*. *Drug Metab. Disp.* 35:859–865.

Solin, O., Eskola, O., Hamill, T.G. et al. 2004. Synthesis and characterization of a potent, selective, radiolabeled substance-P antagonist for NK1 receptor quantitation: ([^{18}F]SPA-RQ). *Mol. Imaging Biol.* 6:373–384.

Starling, J.J., Shepard, R.L., Cao, J. et al. 1997. Pharmacological characterization of LY335979: A potent cyclopropyldibenzosuberane modulator of P-glycoprotein. *Adv. Enzyme Regul.* 37:335–347.

Stearns, R.A., Miller, R.A, Doss, G.A. et al. 1992. The metabolism of DUP 753, a nonpeptide angiotensin II receptor antagonist, by rat, monkey, and human liver slices. *Drug Metab. Disp.* 20:281–287.

Steece-Collier, K., Chambers, L.K., Jaw-Tsai, S.S., Menniti, F.S., and Greenamyre, J.T. 2000. Antiparkinsonian actions of CP-101,606, an antagonist of NR2B subunit-containing N-methyl-D-aspartate receptors. *Exp. Neurol.* 163:239–243.

Tanigawara, Y. 2000. Role of P-glycoprotein in drug disposition. *Ther. Drug Monit.* 22:137–140.

Tatsuta, T., Naito, M., Oh-hara, T., Sugawara, I., and Tsuruo, T. 1992. Functional involvememt of P-glycoprotein in blood–brain barrier. *J. Biol. Chem.* 267:20363–20391.

Ter-Pogossian, M.M., Klein, M.S., Markham, J. et al. 1980. Regional assessment of myocardial metabolic integrity *in vivo* by positron-emission tomography with "C-labeled" palmitate. *Circulation* 61:24.

Thorpe, S. 1987. Steroid receptors in breast cancer: Sources of inter-laboratory variation in dextran-charcoal assays. *Breast Cancer Res. Treat.* 9:175–189.

U'Prichard, D.C., Byland, D.B., and Snyder, S.H. 1978. (+/-)-[H-3] Epinephrine and (-)-[H-3]dihydroalprenolol binding to beta-1 and beta-2 noradrenergic receptors in brain, heart, and lung membranes. *J. Biol. Chem.* 253:5090–5102.

Veitha, G.D., Austina, N.M., and Morris, R.T. 1979. A rapid method for estimating log *P* for organic chemicals. *Water Res.* 13:43–47.

Waterhouse, R. 2003. Determination of lipophilicity and its use as a predictor of blood–brain barrier penetration of molecular imaging agents. *Mol. Imaging Biol.* 5:367–389.

Williams, T.M., Bergman, J.M., Brashear, K. et al. 1999. N-Arylpiperazinone inhibitors of farnesyltransferase: Discovery and biological activity. *J. Med. Chem.* 42:3779–3784.

Wong, D.F., Wagner H.N. Jr, Tune, L.E. et al. 1986. Positron emission tomography reveals elevated D2 dopamine receptors in drug-naive schizophrenics. *Science* 234:1558–1563.

Wong, J.Y.F., Clifford, J.J., Massalas, J.S., Finkelstein, D.I., Horne, M.K., Waddington, J.L., and Drago, J. 2003. Neurochemical changes in dopamine D1, D3 and D1/D3 receptor knockout mice. *Eur. J. Pharmacol.* 472:39–47.

Wormser, U., Zakine, S.B., Stivelband, E., and Eizen, O. 1990. The liver slice system: A rapid *in vitro* acute toxicity test for primary screening of hepatotoxic agents. *Toxicol. In Vitro* 4:783–789.

Yamazaki, M., Neway, W.E., Ohe, T., Chen, I., Rowe, J.F., Hochman, J.H., Chiba, M., and Lin, J.H. 2001. *In vitro* substrate identification studies for p-glycoprotein-mediated transport: Species difference and predictability of *in vivo* results. *J. Pharmacol. Exp. Ther.* 296:723–735.

Chapter 1

Small Animal Studies

2. Small Animal Tumor Models for Use in Imaging

Cheryl L. Marks

2.1 Overview

For more than a century, rodents, primarily mice (Paigen, 2003a,b) and rats (reviewed in Jacob, 1999), have been key experimental animals for cancer research. Early in the twentieth century, Clarence Cook Little developed the first inbred mouse strain—the origin for today's inbred strain called DBA—to enable his studies on the biology of transplanting malignant and nonmalignant tissues (Little and Tyzzer, 1916; Tyzzer and Little, 1916; Little, 1920, 1924). He employed inbreeding, which is mating of male and female littermates for at least 20 generations, because this reduces genetic heterogeneity and provides an experimental system with higher biological reproducibility. Since then, the numbers of inbred strains and outbred stocks of mice (The Jackson Laboratory, JAX® mice: http://jaxmice.jax.org/findmice/index.html; Federation of International Mouse Resources (FIMRe): http://www.fimre.org/; International Mouse Strain Resource (IMSR): http://www.findmice.org/), other mouse strain collections, databases, and tools for genetics and genetic mapping (Mouse Genome Resources: http://www.ncbi.nlm.nih.gov/projects/genome/guide/mouse/index.html; Wellcome Trust Sanger Institute: http://www.sanger.ac.uk/resources/mouse/) that are available (Grubb et al., 2004; Hill et al., 2006; Churchill, 2007; Peters et al., 2007; Roberts et al., 2007; Szatkiewicz et al., 2008; Cox et al., 2009; Yang et al., 2009; Shao et al., 2010; Zouberakis et al., 2010), the information resources for the care and use of mice (The Jackson Laboratory: http://www.jax.org/index.html; Trans-NIH Mouse Initiatives: http://www.nih.gov/science/models/mouse/resources/index.html), and the range of disease models (NCI Cancer Models Database: http://camod.nci.nih.gov; The Knock-Out Mouse Project (KOMP): http://www.genome.gov/17515708; The Jackson Laboratory; http://jaxmice.jax.org/research/cancer/index.html) have increased dramatically. Researchers also use rats as cancer

Targeted Molecular Imaging. Edited by Michael J. Welch and William C. Eckelman © 2012 Taylor & Francis Group, LLC. ISBN: 978-1-4398-4195-2

models; as there are for mice, there are abundant resources, databases, and strains that enable the use of rats as models (Brown et al., 1998; Jacob, 1999, 2010; Mashimo et al., 2005; Aitman et al., 2008; Twigger et al., 2008; Laulederkind et al., 2011; Rat Genome Database (RGD); http://rgd.mcw.edu/; Rat Resource and Research Center (RRRC): http://www.rrrc.us/; Knock-Out Rat Consortium (KORC): http://www.knockoutrat.org/; NIH Rat Genomics and Genetics: http://www.nih.gov/science/models/rat/resources/index.html; Rat Genome Resources: http://www.ncbi.nlm.nih.gov/projects/genome/guide/rat/; The Wellcome Trust Centre for Human Genetics, Rat Mapping Resources http://www.well.ox.ac.uk/rat_mapping_resources/). Although mice and rats are related species—both are members of the Murinae subfamily of the Muridae family—there are significant differences in their biology, physiology, and size. Because of the differences, rats provide information that is complementary to mouse information, and both are used as comparative experimental systems to inform human cancer biology and translational research (Iannacone and Jacob, 2009; Jacob et al., 2010).

This chapter is not an exhaustive treatment of all rodent cancer models and their applications to cancer investigations. The few examples chosen for in-depth presentation are designed to illustrate some of the exceptional research opportunities afforded by well-designed and characterized animal models, and the appropriate rationale for using them, particularly as translational test-beds. Rodent cancer models bridge basic science discovery to clinical and epidemiological observation; their use *in vivo* enables testing of hypotheses gleaned from *in vitro* experimentation and *in silico* modeling. Observations of outcomes from animal models that are subjected to interventions that parallel clinical standard-of-care, or that mimic human population studies, enable researchers to generate new basic research hypotheses from clinical outcomes and epidemiology. This chapter will also attempt to illustrate that there is not one "best" model for all basic or translational applications. Rather, the most useful translational models result from aggregation and integration of data from experimentation in a variety of systems—*in vitro*, *in vivo*, and *in silico*—using bioinformatics tools that enable cross-comparison with data from clinical and prevention trials. In contrast to a single model system, the aggregated information provides better understanding of clinical and prevention trial outcomes and guidance for improved approaches to therapy and prevention. As critical tools for analysis and assessment of the

various comparative systems, leading edge imaging modalities generate critical *in vitro* and *in vivo* temporal insights into changes in processes. Cancers are complex, dynamic, continually evolving systems whose full understanding requires the means to observe such processes in real time. Otherwise, observations of these complex systems are confined to analyses such as histopathology and "snapshot" molecular assays of the specimens acquired at various stages of the natural histories of tumors and their clinical courses.

2.1.1   Variety of Rodent Tumor Models

Mouse and rat tumor models can be loosely categorized as spontaneous, induced, genetically engineered, or transplantation. Spontaneous models are those that develop tumors because of the innate susceptibility of their genetic backgrounds; rodent modelers employ a variety of breeding schemes to ensure reproducible genetic backgrounds and tumor burden. Induced models result from exposure to environmental agents, such as carcinogens, dietary components, or infectious agents that result in tumors in rat and mouse strains whose background genetics are susceptible or resistant. Genetically engineered models are derived by application of sophisticated genetic techniques to modify the germlines of mice and rats to change expression levels of protein-coding genes, miRNAs, siRNAs, and shRNAs. Transplantation models include those in which human and rodent tumor cells and tumor tissue are injected or surgically placed into rodents; they are often used because of the reliable time course and reproducible biology of tumor development.

2.1.2   Suitable Uses and Limitations of Tumor Model Types

Each model type generates different kinds of information about cancer; each type contributes to the translational toolbox, and has strengths and limitations. Therefore, it is important to employ the appropriate type for specific research questions. For example, spontaneous tumor models are useful to expose the natural variation and interactions among background genetic elements that promote tumor development. They support experiments to examine the stochastic processes that contribute to each stage of progression. However, many spontaneous models develop tumors at a number of sites, unlike the majority of cancer patients. In addition, there is often inconsistency in the numbers and timing of tumors among littermates. Researchers do use various breeding schemes to reduce variability, although this can be costly and doing so may select against informative subtle background genetic contributions.

Another approach is to use a known carcinogen or other environmental perturbant to enhance tumorigenesis in a particular organ site and reduce the time required to tumor development. Use of these approaches generates models from which researchers can generate many hypotheses about initiation, promotion, and progression of human cancers. But there is significant disagreement among researchers as to how representatives of human disease are the tumors that are induced by strong carcinogens, when the levels of carcinogen required may be far above the exposures that people encounter daily, or even over a lifetime.

The limitations of spontaneous and induced tumor models, especially the variable timescale and often incomplete penetrance of tumor development, prompt researchers to use transplantation models. Tumor cell lines, and, more recently, low-passage cells from tumors or tumor explants, are commonly used to derive these models. The sites into which these tumor derivatives are transplanted now include not only subcutaneous locations, but also the appropriate orthotopic sites in which the tumors arose. The resulting transplantation models at present are much improved representations of the tissue architecture of the original tumor, as evidenced by the response of the tumors to clinical agents (Ozawa and James, 2010). These models serve as high-throughput screening tools for molecular leads in drug development and as preclinical tools to test agents and their combinations. However, the information from transplant models of human tumors is somewhat compromised by the fact that the recipient animals are deficient in aspects of immune function (Sughrue et al. 2009). Thus, it is not possible for researchers who use transplant models to explore fully the role of the immune system in tumor progression, in tumor response to therapy, or in development of resistance. Because most transplantation models are derived from a tumor from a single patient, the experimental data may or may not represent what can be expected for a large cohort of patients. Increasingly, the research community is employing panels of orthotopic human tissue explant models for a given disease site to represent patient heterogeneity; this is a significant improvement in xenografting tactics, although these panels of models are expensive to derive and to maintain.

It is also important to note that, although rodent tumor cell lines that researchers transplant into the original tumor organ site (orthotopic transplants) of syngeneic recipients (animals of the same genetic background) most often generate tumor models with properties that reproduce those of the original tumor (Wislez et al., 2005), this is not always the case (Olive et al., 2009). Therefore, to overcome some of the limitations of other model types, the cancer modeling community employs germline genetic engineering to generate tumor models for specific organ sites. The earliest examples of transgenic mouse and rat tumor models, although they were specific for certain organ sites, had the drawback of modeling tumorigenesis that is initiated and sustained by strong, gain-of-function oncogenes in many target cells. This approach fails to mimic the apparently slow, stochastic natural histories of human cancers that are hypothesized to originate in a few target cells. But the early successes in altering rodent genomes (Thomas and Capecchi, 1987) spurred several decades of refinement to reach the present suite of approaches that include the ability to target both gain-of-function and loss-of-function alleles to specific cell types and to do so temporally. Increasingly, genetically engineered mice (GEM) and rats are expanding the range of translational research questions that can be addressed with rodent tumor models. The models are proving to be good representations of the natural history and clinical course of human tumors, although they, too, have limitations. The caveats include the fact that the choice of initiating genetic event limits the range of subsequent stochastic changes that are observed in the final tumor. Consequently, these GEM models generally lack the usual heterogeneity of tumors observed in a human patient population; however, they often represent subtypes of human tumors that arise from specific initiating events in particular cell types.

The cancer research literature contains numerous review articles about the variety of rodent tumor models that are employed to understand the biology and support translational studies; for example, models of breast cancer (Borowsky, 2007; Smits et al., 2007; Szpirer and Szpirer, 2007; Vargo-Gogola and Rosen, 2007; Vernon et al., 2007; Wu et al., 2009; Andrechek and Nevins, 2010; Kim and Baek, 2010), prostate cancer (Pollard and Suckow, 2005; Carver and Pandolfi, 2006; Pienta et al., 2008; Lopez-Barcons 2010; Wang, 2011), lung cancer (Kim et al., 2005; de Seranno and Meuwissen, 2010), or brain cancer (Barth and Kaur, 2009; Sughrue et al., 2009; Hambardzumyan et al., 2011). Researchers who seek rodent tumor models are advised to search the various model databases, such as the Jackson Laboratory (JAX Mice for Cancer Research; http://jaxmice.jax.org/research/cancer/index.html), the Rat Genome Database (http://rgd.mcw.edu/wg/ disease-models2), the NCI Cancer Models Database (http://camod.nci.nih.gov), and the Initiative to Link Animal Models to Human Disease (LAMHDI) (http://www.lamhdi.org/).

2.2   Case Studies: How Researchers Develop and Apply Models of Human Central and Peripheral Nervous System Tumors

2.2.1   Pediatric Nervous System Tumors

Pediatric tumors, although fortunately rare in occurrence, nevertheless are a devastating medical challenge. Although children develop cancers in many of the same sites as adults, the biological characteristics of their tumors often differ considerably from adult tumors, they require different treatment protocols, and the survivors may inherit a lifetime of medical problems because of the treatments used. Pediatric cancers present extraordinary opportunities for use of animal models whose diseases are analogous to their human counterparts. Faithful animal models permit discovering the underlying causes and etiology of pediatric tumors and defining how they are similar to, and differ from, adult tumors. With increasing frequency, new targeted therapies are generating durable responses. However, the therapies used to eliminate the childhood tumors appear in some cases to trigger second malignancies that do not manifest themselves until many decades later. The use of faithful animal models to evolve and test frontline therapeutic strategies that provide durable responses with fewer immediate and long-term side effects in a greater percentage of the afflicted children is a critical translational opportunity. Of equal importance are applications of animal models to evolve new means for early detection of primary malignancies, and for predicting and preventing secondary malignancies.

2.2.1.1   Transgenic Mouse Model of Neuroblastoma

Neuroblastoma is the third most common childhood tumor; it arises in sympathetic peripheral nervous tissue, and is often characterized by amplification of the *MYC* oncogene. The tyrosinase-promoter-targeted overexpression of the human *N-MYC* gene in mice generates a neuroblastoma model with many of the cardinal features of the human tumors (Weiss et al., 1997). In the case of this transgenic animal, the TH-*MYCN* model, the use of a strong oncogene driver is an appropriate surrogate for development of the human malignancy. In fact, analysis of the tumors with genome-wide array CGH (comparative genomic hybridization) shows that many of the chromosomal aberrations in the mouse tumors match those of human tumors (Hackett et al., 2003). The FISH (fluorescence *in situ* hybridization) analysis of the mouse tumors enables narrowing of one rather large region of characteristic genetic copy number changes that is observed in human neuroblastoma, thereby providing a convenient tool for identifying the relevant human genes. Also preserved in the mouse tumors are interactions among three other chromosomal loci that characterize human neuroblastoma (Hackett et al., 2003). Cell lines derived from the TH-*MYCN* tumors—analyzed by CGH, FISH, gene expression, and growth *in vitro* and *in vivo*—are highly similar in neuroblastoma-specific cytogenetic aberrations and biology to clinical features of neuroblastoma (Cheng et al., 2007). These reagents are valuable and reliable tools for studying the etiology and progression of neuroblastoma and for testing potential therapies rapidly. These two studies (Hackett et al., 2003; Cheng et al., 2007) illustrate some of the analyses that animal modelers use to characterize (validate) rodent models as representative of the corresponding human diseases.

The TH-*MYCN* model also provides insight into the possible origins of human neuroblastoma. Programmed cell death is an intricate process in the central and peripheral nervous systems that ensures control over the numbers of neuronal cells during normal development; this process results in the proper physiological function of the nervous system, and depends upon subtle interactions among neurons, neurotrophins, and neurotrophin receptors (reviewed in Huang and Reichardt, 2001). In normal mice at birth, there is neuroblast hyperplasia in paravertebral ganglia that regresses in about 2 weeks (Hansford et al., 2004). However, in the TH-*MYCN* model, although there is increased neuroblast hyperplasia, it fails to regress fully before neuroblastomas appear at about 6 weeks of age. From their analysis of mouse tissues, these authors concluded that tumorigenesis in this model results from inappropriate perinatal expression of MycN in paravertebral ganglia cells, which interferes with the normal process of deletion of neural crest cells. The embryonal neural crest cells persist and undergo amplification of *MYCN* and repression of NGF receptor expression, allowing tumors to initiate and progress. There is evidence from the model that there is a nonsecreted factor that induces spontaneous neuroblast regression; its identification could be an important lead in developing therapies for the most aggressive forms of human neuroblastoma.

The molecular and biological validation of the natural history leads to validation of the clinical course, by testing the animals in a manner that is comparable to the clinical standard-of-care. In one study (Chesler et al., 2008), the focus was on response to standard therapy, and relapse. Neuroblastoma tumors usually present in the clinic without mutations in the *p53* gene and are responsive to cyclophosphamide therapy. However, many patients relapse, and the tumors are found to have acquired mutations in *p53*. In parallel, the TH-*MYCN* mouse model develops aggressive neuroblastoma without mutations in *p53*, and responds to *in vivo* administration of cyclophosphamide through upregulation of *p53* expression and *p53*-mediated apoptosis. These authors also verified that *MYCN*-amplified human neuroblastoma cell lines had the same *p53*-mediated response. To use the TH-*MYCN* mouse to inform mechanistic studies of response/resistance to therapy, Chesler et al. (2008) crossed the TH-*MYCN* model to mice with heterozygous deletion of the *p53* gene (Donehower et al., 1992). To follow the tumor response readily *in vivo*, they crossed the TH-*MYCN* mice with an imaging reporter strain (Momoto and Holland, 2005) in which a luciferase signal emits from proliferating tissue as a surrogate for *MYCN* activity. The *p53*$^{+/-}$; TH-*MYCN* mice had reduced levels of *p53*-mediated response to cyclophosphamide, as did *MYCN*-amplified human tumor cell lines with *p53* suppressed by siRNA; this parallel response of human and mouse indicates that therapy-associated mutations in *p53* are likely to contribute toward resistance to therapy in relapsed neuroblastoma.

Two other studies investigated the effects of angiogenesis inhibition (Chesler et al., 2007) or phosphotidylinositol 3-kinase (PI3K)-targeted agents (Chesler et al., 2006) as potential therapies for neuroblastoma. High levels of *MYCN* are associated with poor outcome, and with a stromal-element-poor phenotype (Chesler et al., 2007). The cumulative evidence about

MYCN-amplified tumors is that the MycN protein is required for tumor angiogenesis, and that neuroblasts signal to the stroma to promote neovascularization. Thus, these authors chose to use the TH-*MYCN* model to test antiangiogenic therapy because, unlike human xenografts, this transgenic model has species-matched interactions among components of the host tissue microenvironment and the evolving tumors. They characterized neuroblastoma progression and angiogenesis in the mouse, and observed that the proliferating tumor cells secrete high levels of vascular endothelial growth factor (VEGF), and stimulate vasculature, which, like human neuroblastoma vasculature, has characteristic cell markers. *In vivo* administration of the clinical agent TNP-470 (*O*-chloroacetylcarbamoylfumagillol) significantly reduces proliferation, enhances apoptosis, and disrupts the vasculature. Chesler et al. (2007) established a phenotypic measure of the neurotoxicity and observed the significant weight loss that is associated with TNP-470 and limits its clinical efficacy. With this evidence of the validity of the TH-*MYCN* model to test antiangiogenic therapy, they administered caplostatin, a water-soluble, copolymer conjugate of TNP-470. The tumors in the model respond robustly to caplostatin without the neurotoxicity or weight loss observed with TNP-470, suggesting that caplostatin may be an effective, well-tolerated therapy for the *MYCN*-amplified subtype of neuroblastoma.

The study of PI3K targeted inhibition in the TH-*MYCN* model (Chesler et al., 2006) was based on several reports that PI3K and its downstream mediators stabilize transcription factors, including the MycN protein (Yeh et al., 2004). Administration *in vivo* of LY294002 (2-morpholin-4-yl-8-phenylchromen-4-one) to the TH-*MYCN* model decreased the mass of established tumors and the quantity of the MycN protein without affecting the levels of *MYCN* mRNA (Chesler et al., 2006). These observations are consistent with the effect of LY294002 applied to human *MYCN*-amplified neuroblastoma cell lines, which also responded with decreased proliferation, increased apoptosis, and lower levels of MycN protein without reducing *MYCN* mRNA. Through experiments designed to provide mechanistic details, these authors observed that blocking phosphorylation and subsequent degradation of MycN dramatically reduced the efficacy of LY294002. From their results, and those of others (Gregory et al., 2003; Yeh et al., 2004), they concluded that a common GSK-3β-dependent mechanism stabilizes and activates a

Chapter 2

variety of transcriptional programs in a number of malignancies, and complements the reports (Sears, 2004; Yaari et al., 2005) that activation of Ras signaling also affects GSK-3β-mediated stabilization of c-myc and MycN proteins. At the time of the Chesler et al.,'s (2006) publication, there were only two PI3K-targeted agents, wortmannin (a furanosteroid metabolite of the fungi *Penicillium funiculosum*) and LY294002. Both of them, although effective, have significant toxicities because they block all of the PI3Ks, a diverse group of proteins with many physiological targets (reviewed in Wyman et al., 2003). The recent development of isoform-specific PI3K inhibitors (Marone et al., 2008) suggests that selective inhibition of PI3K isoforms may be a clinically relevant means to achieve new therapies for *MYCN*-amplified neuroblastoma (Spitzenberg et al., 2010).

This case study of the TH-*MYCN* transgenic mouse model illustrates several principles and further opportunities. The initial development and validation strategies for this type of model represent common approaches and analyses. The techniques of gene expression, CGH, FISH, histopathology, and phenotyping analysis that are used for human tumors are generally applied to many kinds of models. What the published literature shows is that these techniques are often applied in an iterative fashion: The human cell and clinical observations inform the design and validation of the model, and the validated model is a tool for further discoveries that can inform additional human investigations. Thus, a well-characterized transgenic model is a translational tool, suitable for testing clinical agents and their combinations, and understanding the possible mechanisms of response and bases for recurrence.

2.2.1.2 Models of NF1-Related Pediatric Gliomas

Neurofibromatosis type 1 (NF1) (reviewed in Gottfried et al., 2010, and Albers and Gutmann, 2009) is an inherited autosomal dominant disorder; at-risk individuals often develop characteristic cutaneous abnormalities, such as café-au-lait spots, freckling in armpit or groin skin folds, and iris hamartomas. They also develop peripheral nerve sheath tumors, or neurofibromas; these are soft bumps on or under the skin that grow on nerve tissue in the skin and are complex and multicellular benign tumors. NF1 also causes bone deformities, specific learning disabilities, and occasionally short stature. Individuals with NF1 are prone to the development of both benign and malignant tumors. The most common malignant nervous system

tumors in children and adults with NF1 are plexiform neurofibromas, peripheral nerve sheath tumors (MPNSTs) and optic nerve gliomas. MPNSTs are highly malignant, aggressive tumors that are very difficult to treat; the most common CNS tumor in NF1 children is optic glioma.

The genetics and cellular and molecular biology that initiate and sustain tumor formation in NF1 are well studied. Germline mutations in the NF1 tumor suppressor gene confer variable individual risk of malignant tumors. These NF1 mutations generally cause decreased levels of the neurofibromin protein; in turn, this causes increased cell signaling through the *Ras* gene to its downstream effectors. Other key pathways are implicated with NF1 tumorigenesis, including the Ras/mitogen-activated protein kinase (MAPK) and the Akt/mammalian target of rapamycin (mTOR) signaling cascades. Generally, the process of benign tumor formation is initiated by loss of heterozygosity at the NF1 locus in either Schwann cells or their precursors, which occurs in the context of *NF1* gene haplo-insufficiency in other neighboring cells. Complex interactions with other cell types are mediated by abnormal expression of growth factors and their receptors and modification of gene expression. One such example is recruitment and involvement of *NF1*$^{+/-}$ mast cells. What is also reported is that, for malignant transformation to occur, additional mutations are required in genes such as *INK4A/ARF* and *p53*, with the resulting abnormalities in cell signaling. Because mutations of the *NF1* gene with the attendant additional changes in signaling are implicated not only in the etiology of NF1 tumors, but also in some sporadic gliomas, what is learned about treatment of NF1 gliomas may apply to sporadic gliomas as well. The detailed molecular and cellular knowledge associated with NF1 tumorigenesis is now being applied in preclinical models of MPNSTs and gliomas to test for efficacy of such agents as mTOR inhibitors, which are showing promising results.

In childhood, NF1 gliomas are primarily located in the optic pathway, and less frequently in the hypothalamus and brainstem. However, NF1 adults are more likely to develop higher grade gliomas; the treatment for them is similar to nonfamilial adult sporadic gliomas. Although NF1 is a rare (1 in 3000) cancer predisposing condition, studying these NF1-related pediatric gliomas affords a unique opportunity to understand the molecular and cellular underpinnings of all pediatric CNS tumors (Gutmann, 2008).

The World Health Organization classifies NF1-associated optic gliomas as grade I astrocytomas (Louis et al., 2007). Optic gliomas are generally indolent, and so many children with these tumors do not require treatment; however, optic gliomas that interfere with visual function are treated with a genotoxic combination of carboplatin and vincristine chemotherapy, which may be associated with development of secondary malignancies later in life. More recently, researchers have focused on identifying and applying targeted therapies, either as single agents or as combinations (Gottfried et al., 2010).

To develop an appropriate mouse model of NF-1 tumors, Bajenaru et al. (2002) derived a conditional knockout mouse in which the *Nf1* gene was inactivated in astrocytes by embryonic day 14 using cre/loxP technology. The glial fibrillary acidic protein (*GFAP*) *Cre; Nf1^flox/flox^* mice were viable and fertile. The mice were observed for up to 20 months, but they did not develop astrocytomas. However, these investigators found that the mice did have a 1.5-fold increase in astrocyte proliferation *in vivo*. They observed that heterozygosity conferred a growth advantage that was specific to astrocytes and was not observed in oligodendrocytes or microglia. They also showed that astrocytes that were heterozygous for p120-GAP, another p21*ras*-GAP molecule, did not display increased astrocyte proliferation *in vivo*, and that this growth advantage was cell autonomous; the *Nf1-GFAP-CKO* astrocytes proliferate faster *in vitro* than do astrocytes derived from wildtype littermates. From this study, Bajenaru et al. (2002) concluded that *Nf1* inactivation alone was not sufficient for astrocytoma formation in mice, and that the lack of tumors might be due to the need for further genetic alterations, perhaps acquired mutations in oncogenes or loss of expression of other tumor suppressors.

Subsequently, Bajenaru et al. (2003) developed another model, based on the fact that NF1 patients are heterozygous for a germline inactivating *NF1* mutation. They derived *Nf1^+/−^* mice that specifically lacked *Nf1* expression in astrocytes. These GFAP-cre; *Nf1^flox/mut^* lines had areas of gross optic nerve and/or chiasm enlargement that were observed in mice as young as 8 months of age. This phenotype was not observed in either *Nf1^+/−^* or *GFAP-cre; Nf1^flox/flox^* mice. In the GFAP-cre; *Nf1^flox/mut^* lines, they observed unilateral as well as bilateral optic nerve enlargement, with and without chiasmal involvement. Children with NF1 also display such multiplicity of growth pathology patterns. Bajenaru et al. (2003) also sought to determine if

magnetic resonance imaging could detect abnormal optic nerve pathology in this model. At the time of this study, diffusion tensor imaging was used to detect and follow children with NF1; thus these authors used a comparable small animal imaging approach in this model. They observed abnormal tissue between the optic nerves only in the GFAP-cre; *Nf1^flox/mut^* mice, but not in the *Nf1^+/−^* or *GFAP-cre; Nf1^flox/flox^* lines.

On the basis of the similarity to human NF1-associated optic glioma, Hegedus et al. (2008) used this model (now designated Nf1^+/−GFAP^ CKO) in which about 100% of the animals develop prechiasmatic and chiasmatic optic gliomas, for proof-of-principle preclinical studies. Because neurofibromin, the product of the *Nf1* gene, preferentially inhibits K-Ras activity in astrocytes (Wang et al., 2008), they also derived Nf1^+/− KRas*^ mice, which express a constitutively active K-Ras allele in GFAP-positive cells. They closely compared the biology of these two models, and, based on the proliferation rates, locations, and penetrance of the tumors, they chose the Nf1^+/−GFAP^ CKO mice to evaluate the effects of several clinical agents. Previously, Banerjee et al. (2007) had addressed the challenge of imaging small tumors *in vivo*, which is a major limitation of applying mouse models to preclinical therapeutic evaluation. They developed a rapid and reliable method to detect optic glioma in the Nf1^+/−GFAP^ CKO mouse *in vivo* using manganese-enhanced magnetic resonance imaging (MEMRI). They not only correctly identified the presence or absence of tumors in a blinded study, they also accurately measured the tumor size and shape *in vivo*. Hegedus et al. (2008) used MEMRI to detect which littermates had developed optic gliomas, and then randomized the tumor-bearing mice to treatment or control groups. The first agent they tested was temozolomide, an oral alkylating agent that is the conventional single-agent chemotherapy for children with low-grade glioma. Temozolomide alone decreased proliferation and increased apoptosis of the tumor cells *in vivo*, and resulted in tumor shrinkage.

Neurofibromin is a negative regulator of mTOR signaling (Wang et al., 2008). Therefore, Hegedus et al. (2008) also tested rapamycin, an inhibitor of mTOR, and found that rapamycin *in vivo* decreased tumor cell proliferation in a dose-dependent manner, accompanied by decreased tumor volume. However, they did not observe an additive effect of combined rapamycin and temozolomide treatment. They also showed that these treatments, which affect tumor cell proliferation and apoptosis, had no significant effect on progenitor cell proliferation within brain germinal zones. They

Chapter 2

concluded that Nf1$^{+/-\text{GFAP}}$ CKO model will be valuable for identifying and testing novel therapies for pediatric optic glioma.

Because they have high levels of ribosomal S6, NF1-deficient human and mouse cells are assumed to have increased mTOR pathway activation. Based on the prior use of rapamycin in the Nf1$^{+/-\text{GFAP}}$ CKO model (Hegedus et al., 2008), Banerjee et al. (2011) continued experimenting with rapamycin in a preclinical setting, because these studies could have important insights for clinical trials of rapamycin analogs. Durable responses to rapamycin treatment in the Nf1$^{+/-\text{GFAP}}$ CKO optic glioma model required 20 mg/kg/day; only transient tumor growth suppression was observed with 5 mg/kg/day rapamycin, even though that level of agent completely blocked ribosomal S6 activity. Banerjee et al. (2011) used the Nf1-deficient glial cells both *in vitro* and *in vivo*. They observed an exponential relationship between blood and brain rapamycin levels. They also observed that the biomarkers that are currently used as evidence of mTOR pathway inhibition—levels of phospho-S6, phospho-4EBP1, phospho-STAT3, Jagged-1, and Ki67 measures of tumor proliferation—do not accurately reflect either mTOR target inhibition or Nf1-deficient glial cell growth suppression. They found that incomplete suppression of Nf1-deficient glial cell proliferation *in vivo* following 5 mg/kg/day rapamycin treatment is actually due to AKT activation that is mediated by mTOR; combining 5 mg/kg/day rapamycin and PI3K inhibition recapitulates the growth-suppressive effects of 20 mg/kg/day rapamycin. Their results suggest that more accurate biomarkers for rapamycin treatment response are needed, and provide indications of potential therapeutic combinations.

The current chemotherapy for those children whose optic gliomas require treatment is reasonably effective at halting growth of the tumors. However, most of these children are left with substantial vision impairment. Hegedus et al. (2009) chose to use the Nf1$^{+/-\text{GFAP}}$ CKO mice to explore the cellular basis for the reduced visual function resulting from initiation and progression of optic gliomas. They performed visual evoked potential analyses and electroretinogram measurements before and after the appearance of macroscopic tumors. Even before obvious optic gliomas appeared, Nf1$^{+/-\text{GFAP}}$ CKO mice had decreased visual-evoked potential amplitudes and increased optic nerve axon calibers. By the time the mice had attained 3 months of age, they displayed pronounced optic nerve axonopathy and apoptosis of neurons in the retinal ganglion cell layer. Hegedus et al. (2009) used magnetic resonance diffusion tensor imaging on the animals between 6 weeks and 6 months of age and observed progressive increase in radial diffusivity in the optic nerve proximal to the tumor. This observation indicated ongoing deterioration of axons. They concluded from this study that optic glioma formation may result in early axonal disorganization and damage, leading to retinal ganglion cell death. They also suggested that the Nf1$^{+/-\text{GFAP}}$ CKO mice might be a useful model to explore the visual abnormalities in children with NF1, identify the signaling pathways responsible for survival of *Nf1*$^{+/-}$ retinal ganglion cells, and use that information to develop neuron-protecting adjuvant strategies in combination with therapy or novel approaches for intervening early in tumorigenesis to reduce the loss of vision.

Another research opportunity for the Nf1$^{+/-\text{GFAP}}$ CKO mouse model is to endeavor to understand the learning and behavioral abnormalities that are common among NF1 children (Coudé et al., 2006). One of the most difficult medical problems for school-age NF1 children involves learning disabilities and attention system dysfunction, including abnormal visual–motor integration, visual–spatial judgment, and visual–perceptual skills. NF1 children have deficits in receptive and expressive language as well as verbal and visual memory, and attention deficits are more prevalent. Brown et al. (2010) used Nf1$^{+/-\text{GFAP}}$ CKO mice to determine if this model could expose the underlying neurological defects responsible for the learning deficiencies. What they observed was that the Nf1$^{+/-\text{GFAP}}$ CKO mice displayed novel defects in nonselective and selective attention, although they did not have a hyperactivity phenotype. The Nf1$^{+/-\text{GFAP}}$ CKO mice had a reduced rearing response to novel objects and other environmental stimuli in standard laboratory tests of these phenotypes. Because children with NF1 are treated with methylphenidate (the prescription psychoactive medication Ritalin®) for their attention deficits, they tested this same agent in the mice. Methylphenidate reversed the attention system dysfunction in the mice, which indicated that there was a defect in normal levels of brain catecholamines. They observed a link between the attention system abnormality and reduced levels of dopamine in the mouse striatum; administration of methylphenidate or L-dopa (L-3,4-dihydroxyphenylalanine) restored striatal dopamine levels. They found that reduced striatal dopamine levels in the Nf1$^{+/-\text{GFAP}}$ CKO mice correlated with a reduction of tyrosine hydroxylase, the rate-limiting enzyme that synthesizes dopamine in the striatum. However, the reduction in dopamine was not accompanied by dopaminergic cell loss in the substantia nigra. Brown et al. (2010) also observed a cell-autonomous defect in Nf1$^{+/-}$ dopaminergic neuron growth cone areas and neurite extension *in vitro*; this causes decreases in dopaminergic cell projections to the striatum in the intact Nf1$^{+/-GFAP}$ CKO mice. They concluded that this model of optic glioma is useful to understand NF1-associated attention deficits; their discovery of compromised integrity of the dopaminergic pathway that resulted in decreased dopamine levels supplied a possible mechanistic explanation for the observed attention system abnormalities in this mouse model. Their research supports the clinical use of methylphenidate for treating attention system dysfunction in NF1 children, as well as the continued exploration in the Nf1$^{+/-GFAP}$ CKO mice of molecular defects that may contribute to the substantial heterogeneity of the learning deficits in NF1 children.

This case study of models of NF1 pediatric gliomas illustrates design, biological validation, and translational aspects of mouse cancer modeling that are similar to those for the TH-*MYCN* neuroblastoma model. It also underscores that there can be uncommon aspects of cancer etiology and tumor burden that may be approached through innovative uses of the models. The potential exists to apply cognitive, behavioral, or other neurological measures in mice to discover subtle deficits that may be linked to the earliest stages of initiation or progression of occult CNS tumors. If similar deficits are consistently observed in humans prior to brain tumor diagnosis, the studies in mice may pave the way to develop and test new forms of early detection for CNS tumors. Discovering the molecular mechanisms that underlie the deficits could also stimulate development of new imaging methods for early detection based upon discrete and unequivocal changes in CNS biology.

2.2.2 Preclinical Models of Adult Gliomas and Their Translational Use

Brain tumors are devastating malignancies that are difficult to treat and result in significant morbidity and mortality. High-grade gliomas are molecularly heterogeneous; histopathologic criteria generally subdivide them into anaplastic astrocytomas, anaplastic oligodendrogliomas, glioblastoma multiforme, and anaplastic oligoastrocytoma. Although all glioma subtypes progress rapidly, each subtype does have distinct molecular alterations that are still not used effectively to inform the choice of therapies. Gliomas also differ in

their location and biological and clinical behavior (Jeremic et al., 1994).

2.2.2.1 Genetically Engineered Mouse Models of Glioblastoma

To model adult gliomas, Hambardzumyan et al. (2009) used a variation of transgenic technology—termed the RCAS/*t-va* system (Holland and Varmus, 1998)—that permits researchers to test hypotheses about the role in gliomagenesis of specific molecular changes in distinct brain regions. The RCAS/*t-va* system is based on expression in mice of a transgene that encodes the receptor for subgroup A avian leukosis virus. The *t-va* transgene can be controlled by cell-type- or tissue-specific promoters, thereby limiting the sites of tumor development. Delivery of genes is accomplished by injecting the *t-va* mice with subgroup A avian leukosis virus vectors (RCAS) that carry those genes. Adjusting the viral titer can limit the number of cells that are affected so that the resulting model represents the usual human patient who has only a single tumor mass at clinical presentation. Additionally, RCAS vectors carrying different mutant genes can be delivered simultaneously to test hypotheses about the requirement for several genes to drive tumor initiation and progression.

Generally, the genetic alterations cataloged in glioblastoma (Jansen et al., 2010) cluster into two types: Growth factors, their receptors, and their associated signaling pathways, such as platelet-derived growth factor receptors 1 and 2 (Shih and Holland, 2006), epidermal growth factor receptor (Mischel et al., 2003), and PI3 kinase signaling and mTOR (Akhavan et al., 2010); regulators of the cell cycle, such as *Ink4a/ARF* (Schmidt et al., 1994) and *RB* (Ueki et al., 1996). With the release of human glioblastoma data from The Cancer Genome Atlas (TCGA) Research Network (TI Network, 2008), and the results of several other large-scale analyses (Parsons et al., 2008; Yan et al., 2009), it is apparent that there are additional candidate genes that fall into other classes that were previously not known to be implicated in glioblastomagenesis; these can now also be tested as potential "drivers" of these tumors.

To aid in classifying glioblastoma by transduction pathway activation and by mutation in pathway member genes to enable development of targeted therapies, Brennan et al. (2009) performed targeted proteomic analysis of 27 surgical glioma samples to identify patterns of coordinate activation among glioma-relevant signal transduction pathways. They then compared the results with integrated analysis of genomic and expression data of 243 glioblastoma samples from TCGA.

Chapter 2

Brennan et al. (2009) found that, in the pattern of signaling, three subclasses of GBM emerged that appear to be associated with predominance of EGFR activation, PDGFR activation, or loss of the RAS regulator NF1. The EGFR signaling class had prominent Notch pathway activation measured by elevated expression of Notch ligands, cleaved Notch receptor, and downstream target, Hes1. The PDGF class had high levels of PDGFB ligand and phosphorylation of PDGFRb and NFκB. Lower overall MAPK and PI3K activation and relative overexpression of the mesenchymal marker YKL40 was associated with loss of NF1. They found that these three signaling classes appeared to correspond to distinct transcriptome subclasses of primary GBM samples from TCGA for which copy number aberration and mutation of EGFR, PDGFRA, and NF1 were signature events. Proteomic analysis of GBM samples revealed three patterns of expression and activation of proteins in glioma-relevant signaling pathways. These three classes were comprised of roughly equal numbers showing either EGFR activation associated with amplification and mutation of the receptor, PDGF-pathway activation that was primarily ligand-driven, or loss of NF1 expression. The associated signaling activities correlating with these sentinel alterations provided insight into glioma biology and therapeutic strategies.

Although the subventricular zone in the brain is proposed as the region in which gliomas arise, it is important to determine whether some of the heterogeneity in molecular alterations, clinical presentation, and therapy response of gliomas are due to their specific location in the brain. Hambardzumyan et al. (2009) explored this question using nestin/*t-va* and GFAP/*t-va* mice, and an RCAS vector carrying PDGF-B; this approach targeted PDGF-B to either nestin-expressing or GFAP-expressing cells. They delivered the RCAS-PDGFB vector stereotactically into the SVZ, the left and right hemispheres, and the cerebellum. They also performed these experiments in mice that had combinations of deficiencies in *Ink4a*, *Arf*, *p53*, or *PTEN*; these four genes are commonly observed as mutated or silenced in adult gliomas. Their results demonstrated that injecting RCAS-PDGFB into the SVZ induced gliomas in the SVZ and in the frontal lobe, with extension of tumors into the olfactory bulb; this is a common observation in human gliomas (Larjavaara et al., 2007). They found that, when tumors arose in different locations in the cerebral hemispheres and the SVZ, they did not differ in incidence, latency, or mortality. This result contrasted to the result produced by cerebellar overexpression of PDGFB: The

tumors localized to the cerebellum and extended into the brainstem, and they had lower incidence and longer latency compared with tumors that arose from PDGFB injections into the cerebral hemispheres and the SVZ.

Hambardzumyan et al. (2009) observed similar incidence and latency of tumors whether they targeted nestin-positive or GFAP-positive cells by injections in the SVZ. When the same injection protocols were used in mice in which the tumor suppressor genes, *p53*, *Arf*, or *Ink4a-Arf*, were null, the resulting tumors were higher-grade and appeared with shorter latency. Also, 100% of the tumor-suppressor gene null mice developed tumors, compared with their wild-type counterparts, 22% of which developed only low-grade tumors.

In this same study, Hambardzumyan et al. (2009) used a previously generated bioluminescent imaging reporter mouse (Becher et al., 2008). This mouse line, termed Gli-luc, expresses luciferase from a Gli-responsive promoter. Because there is postulated interplay between PDGF and sonic hedgehog (SHH) signaling in glioma development, they reasoned that Gli (Glioma-associated oncogene homolog), a component of the SHH pathway, might be activated in PDGF-induced gliomas; in fact, Becher et al. (2008) used the Gli-luc reporter mouse to image pediatric gliomas; they showed that the Gli-luc mouse gave an accurate readout of Gli activation, and that SHH expression and Gli activity correlated with tumor grade in pediatric, PDGF-induced gliomagenesis. Hambardzumyan et al. (2009) injected RCAS-PDGFB into the SVZ of nestin-tva/*Ink4a*$^{-/-}$*Arf*$^{-/-}$/Gli-luc adult mice. They observed increased bioluminescence at the tumor site in stereotactically injected adult mice. They also imaged the RCAS/tva-generated tumors in adult mice using T2-weighted MRI. They showed that the mouse tumors share many imaging characteristics with T2-weighted images of human gliomas; thus, both bioluminescent imaging and MRI are useful to monitor the emergence of tumors and their *in vivo* response to therapy in preclinical trials.

In related experiments, Bradbury et al. (2008) used nestin-tva/*Ink4a*$^{-/-}$*Arf*$^{-/-}$ mice injected intracranially with RCAS-PDGF vectors and followed the appearance of tumors by ^{18}F-fluorothymidine (FLT) PET. Their goal was to characterize the kinetics of ^{18}F-FLT uptake in this well-validated mouse model of high-grade glioma to develop a practical noninvasive method to assess cellular proliferation *in vivo*. Although several clinical studies reported on compartmental analyses to estimate rate constants to model the kinetics of ^{18}F-FLT uptake on high-grade human gliomas (Jacobs et al., 2005; Muzi et al., 2006; Schiepers et al., 2007), this is the first reported instance in which such compartmental analyses have been used in the preclinical setting. Bradbury et al. (2008) were able to discriminate tracer transport and delivery from tumor proliferation. They estimated compartmental model tracer rate constants using image-derived input and tissue time–activity data (i.e., left ventricular blood clearance and tumor uptake, respectively). Because no previous studies in preclinical brain tumor models had dynamically assessed uptake with evaluation of the relevant rate constants, they also sought to clarify the extent to which tracer transport and metabolism contribute to overall tumor uptake. They correlated static uptake measurements (percentage injected dose per gram [%ID/g]) at 1 h with the metabolic rate constants. Implementing such approaches would address the growing need to elucidate key metabolic and physiological processes in a wide variety of tumor models, not only brain tumors.

2.2.2.2 Surgical Specimen Orthotopic Xenografts of Human Glioblastomas

In addition to genetically engineered or syngeneic transplant, models of human glioblastoma are exceptionally valuable models composed of human surgical specimens that are engrafted into immune-compromised mice or rats. Most often, the location for transplantation is subcutaneous, on the flank of the recipient animal; more recently, xenografting has been orthotopic, which, in the case of gliomas, is in an intracranial location. These models are replacing the use of established glioblastoma cell lines, which even when transplanted into the brain, grow to form solid masses instead of recapitulating the characteristic invasive growth pattern of spontaneously arising glioblastoma. Examples of some of the earlier CNS model systems are reviewed in Goldbrunner et al. (2000).

Because many human cell lines of glioblastoma, even administered intracranially, do not grow invasively, the models are considered to have limited relevance to clinical experience. These lines also lose commonly observed mutations, such as *EGFR* amplification (Pandita et al., 2004), with repeated passage. This seriously limits the ability of the translational research community to study the biology of tumors with this common aberration, or determine its role in response or resistance to clinical agents that target EGFR.

A number of laboratories have developed other methods to establish orthotopic models of glioblastoma. Some involve direct transplantation of patient

surgical material into the brains of nude mice, and this tactic is more successful at maintaining the invasion property of the original tumors. Continuous serial passaging of glioblastoma tumors in subcutaneous locations can preserve the genetic aberrations that were observed in the original patient specimens (Pandita et al., 2004).

Giannini et al. (2005) used a panel of four serially transplanted xenograft glioblastoma lines that they established by directly injecting patient tumor tissue subcutaneously in the flanks of nude mice. They established short-term cultures of cells from xenografts that they excised from the subcutaneous locations, harvested the cells, and then injected exactly the same number of cells (106) intracranially into cohorts of nude mice. The resulting brain tumors had high mitotic rates and exhibited significant invasion. The growth rates of the intracranial tumors in mice that had injections from the same explant culture were internally consistent. In the case of two *EGFR*-amplified flank tumors, the high levels of amplification and overexpression of *EGFR* were retained in the intracranial tumors; this contrasts with the same cells grown in culture, in which *EGFR* amplification is lost. A third intracranial tumor retained patient tumor amplification and high-level expression of *PDGFRA*. Although the intracranial tumors failed to display necrosis or endothelial cell proliferation in all instances, the heterotopic-to-orthotopic transfer and propagation of glioblastoma multiforme does preserve the receptor tyrosine kinase (RTK) gene amplification of patient tumors. Thus, these authors concluded that this type of anatomically correct, direct patient model that preserves important phenotypes of the original tumor is suitable to support investigations that are designed to determine how amplification of a receptor tyrosine kinase influences the response of that tumor to therapies that are directed to that RTK.

In another study of the mutational spectrum of patient specimens and the xenograft models derived from them, Hodgson et al. (2009) examined the suitability of such models to recapitulate the molecular heterogeneity of human glioblastoma multiforme tumors. They cataloged DNA copy number and mRNA expression in 21 glioblastomas maintained as subcutaneous xenografts. They then compared the molecular signatures of the members of this xenograft panel with those that were cataloged in clinical glioblastoma specimens from the Cancer Genome Atlas (TCGA). They found that the predominant copy number changes in the TCGA and xenografts were a gain

Chapter 2

of chromosome 7 and loss of chromosome 10. This is known to be a signature associated with a poor prognosis. In the xenografts, they observed genomic amplification and overexpression of known glioblastoma-associated oncogenes, such as *EGFR*, *MDM2*, *CDK6*, and *MYCN*; they also cataloged novel genes, such as *NUP107*, *SLC35E3*, *MMP1*, *MMP13*, and *DDX1*, at frequencies that resemble those detected in TCGA glioblastomas.

Hodgson et al. (2009) examined the stability of the transcriptional signatures of the xenografts over multiple passages in subcutaneous sites. This consistent signature was characterized by overexpressed M-phase, DNA replication, and chromosome organization genes. They also assessed gene expression in TCGA-derived glioblastomas and documented overexpression of genes such as *AURKB*, *BIRC5*, *CCNB1*, *CCNB2*, *CDC2*, *CDK2*, and *FOXM* that are associated with the G2/M progression.

However, these authors also observed some differences between the glioblastoma xenograft molecular signatures and those of the surgical specimens of glioblastomas from TCGA. For example, the xenografts appeared to lack the characteristic proneural glioblastoma molecular signature. And genes of the proliferation expression signature were significantly overrepresented. The Hodgson et al. (2009) noted that the latter should not be surprising because there may be a selection bias that is conferred when a patient specimen is engrafted: Either the original tumor was more proliferative, or glioblastomas that establish as xenografts must adopt higher proliferation capability. The authors suggested that the possibility that the selection process that is conferred by xenografting somehow alters the ability of the glioblastoma xenograft panels to represent the variability of the response of patients to a given therapy merits further investigation.

Hodgson et al. (2009) noted that there are two reports (Sarkaria et al., 2006, 2007) in which this issue of whether patient heterogeneity is preserved in xenografted glioblastomas was addressed. In the first report, Sarkaria et al. (2006) used intracranial xenografts from 13 different patients to examine the effects of radiation; seven of the patients had amplified *EGFR*. The tumor-bearing mice were randomized to 12-Gy radiation in six fractions delivered over 12 days or to sham treatment. In six cases, four of which had amplified *EGFR*, radiation significantly extended the survival of the mice. In contrast, for the seven remaining xenografted animals, radiation did not extend survival; three of those mice had tumors with amplified *EGFR*. These authors also compared radiation-induced survival with *p53* and *PTEN* mutation status, and found that there was no preferential association with radiation sensitivity or resistance. They did observe that increased radio-resistance was associated with under-phosphorylation of Akt on Ser[473]. In one xenograft with amplified *EGFR*, a line that did not respond to radiation, they coadministered radiation with an *EGFR* inhibitor to determine if inhibition of *EGFR* kinase activity influenced radiation response; concurrent use of both treatments gave an additive survival benefit relative to either agent alone. Sarkaria et al. (2006) concluded from their results that amplified *EGFR* does not by itself predict response of glioblastomas. This study suggests that *EGFR* inhibition and radiation regimens used in combination might achieve additive antitumor effect against a subset of glioblastoma.

In the other report, Sarkaria et al. (2007) addressed the clinical problem of how to interpret the published laboratory correlative studies of *EGFR* kinase inhibitor clinical trials. There are conflicting reports about tumor molecular characteristics that associate individual glioblastoma responsiveness to *EGFR* kinase inhibition. For example, in one study (Haas-Kogan et al., 2005), *EGFR* amplification and tumor phospho-Akt levels were cited as key markers of response. And, in another study (Mellinghoff et al., 2005) found the presence of *EGFRvIII*, a constitutively active mutant of *EGFR*, combined with retained *PTEN* expression, as a positive indicator of responsiveness to *EGFR* kinase inhibition.

Sarkaria et al. (2007) used a panel of serially propagated glioblastoma xenografts from 11 different patients to evaluate how the EGFR inhibitor, erlotinib (Tarceva®), affected survival of the tumor-bearing mice. They found that two of the xenografts were sensitive to erlotinib; each one had amplified *EGFR* and each expressed wild-type *PTEN*. One of the two tumors expressed the truncated *EGFRvIII* mutant, while the other had full-length *EGFR*. However, they found that the full-length *EGFR* was, in fact, a sequence variant, with arginine instead of leucine at amino acid 62. Further analysis showed that this was the only *EGFR* variant, except for the vIII mutant, among the 11 xenografts. Their examination of 12 more xenografts revealed one additional *EGFR* variant, expressing threonine, not alanine, at amino acid 289. Although this glioblastoma had amplified *EGFR*, it was not sensitive to erlotinib; however, it lacked *PTEN* expression. The authors summarized their results as a demonstration that there were two erlotinib-sensitive glioblastoma xenografts, each having wild-type *PTEN* expression in combination with amplified and aberrant *EGFR*. Their correlative molecular analyses of erlotinib-responsive glioblastoma xenografts support tumor expression of wild-type *PTEN* combined with amplification of aberrant *EGFR* as key indicators of erlotinib sensitivity.

The stable genetic characteristics of such human tumor explants make them suitable to test potential therapy for glioblastomas. However, there is one challenge that arises if orthotopic transplantation models of human brain tumors are used to evaluate response to therapy: How to monitor tumor response. Transplants in subcutaneous locations can be monitored easily and directly, through use of calipers; orthotopic xenografts require some form of noninvasive imaging. Although bioluminescence cannot be used as a clinical imaging method, it is a cost-effective, sensitive approach for imaging of growth and treatment response of xenografts in immune-compromised hosts. Ozawa and James (2010) demonstrated procedures for establishing orthotopic brain tumors, and for monitoring tumor growth and response to treatment when testing experimental therapies. They showed that such brain tumor xenografts have a reasonable microenvironment that permits CNS cancer models to be tested for therapeutic response. In addition, these models provide information about access of therapy to normal brain structures and the brain tumor, information that is vital for deciding whether an experimental agent should be advanced to clinical trial evaluation in patients. Ozawa and James showed that bioluminescence imaging is a very practical approach for testing when the primary objective is to assess the extent to which the tumor responds to therapy. Combining bioluminescence imaging results with the survival analysis of the tumor-bearing mice is a reliable means to evaluate the efficacy of experimental therapeutics. Finally, the intracranial brain tumor xenografts can be harvested from euthanized animal subjects, and the morphologic and molecular effects of therapy can be assessed.

Prasad et al. (2011) used a panel of serially passaged glioblastoma xenografts to tackle the challenge of informing the design of clinical trials that use targeted agents in combination with conventional therapy. In this study, they evaluated XL765 (Exelexis Pharmaceutical), a novel PI3K/mTOR dual inhibitor, both *in vitro* and *in vivo*. For the studies *in vitro*, they harvested five *in vivo* serially passaged glioblastoma xenografts and used the disaggregated tumors in cell culture. They evaluated the changes in downstream targets by immunoblot, and measured cytotoxicity by a colorimetric ATP-based assay. For the *in vivo* experiments, they used one of the five xenografted glioblastomas, which had *EGFR vIII* amplification and wildtype *PTEN*. The flank-grown tumor was harvested and disaggregated, and the cells, altered to express luciferase, were used to establish a panel of intracranial xenografts whose response to therapy could be followed by bioluminescence imaging. *In vitro*, the use of XL765 resulted in concentration-dependent decreases in cell viability; doses that were cytotoxic specifically inhibited PI3K signaling. They found that the combination of XL765 with temozolomide gave additive toxicity in four of five xenografts. *In vivo*, administration of XL765 gave a greater than 12-fold reduction in median tumor bioluminescence compared with controls; there was also an improvement in median survival of the mice. Administration of temozolomide alone gave a 30-fold decrease in median bioluminescence; however, the combination XL765 and temozolomide yielded a 140-fold reduction in median bioluminescence, along with an improving trend in median survival when compared with temozolomide alone. Prasad et al. (2011) concluded that XL765 is active as a single agent in genetically diverse glioblastoma xenografts, and that it may be beneficial in combination with conventional therapeutics.

This case study of xenografted models of adult glioblastoma serves to illustrate new opportunities for similar approaches that may be used to develop models for other tumor sites. It also provides a cautionary case in point about the heterogeneity of human malignancies, and about the care that must be taken during establishment and maintenance of these types of models. This approach to establish models is increasingly used; to ensure the validity of this type of model as representative of the original patient material requires careful, in-depth analyses at every stage of development as well as during maintenance of the model. This is costly and time-consuming, but the net result of careful, reproducible protocols for maintenance and use is the availability of very valuable, translational models.

Chapter 2

2.3   Research Opportunities for Imaging Analysis of Preclinical Models

As the case studies in this chapter illustrate, there is abundant rationale to use *in vivo* imaging to tackle the many research problems afforded by an increasing number of well-validated small animal models of human tumors. There are also numerous difficulties to overcome if these important tools are to become sufficiently fast, cost-effective, and highly reliable.

2.3.1   Basic Cancer Biology Research

No issue in basic research is more pressing for its clinical importance than the ability to use *in vivo* models to discover how tumors progress from neoplasia *in situ* to invasive, metastatic tumors. Although there is an impression in the research community that there are few, if any, useful models of metastasis, this is not the case. Necropsies of small animal models often reveal metastatic deposits at many of the same sites to which human tumors metastasize. But the ability to verify that metastasis to distant sites occurs in any model requires that the imaging approach be sufficiently sensitive to locate a few cells. Imaging methods are also required not just to find tumor cells in the circulation and to observe their journey to distant sites, but also to disclose the steps of the process and the temporal sequence.

Other pressing basic research questions involve the roles that the various cells of the immune system have in all stages of cancer natural history. Tantalizing, but often conflicting data about how the immune system contributes to the natural history of tumors abound, yet the application of imaging sciences to understand even normal, nondisease-related immune functioning is not sufficiently advanced to address these questions *in vivo*. However, a concerted effort to develop the tools and reagents required for this field could transform the research community's understanding of many diseases.

There is another significant challenge for imaging that is posed by the derivation of so many new models: How to enable their use for a variety of research questions without the requirement to engineer each one for a specific imaging modality. The proliferation of novel approaches to *in vivo* imaging heightens this challenge. Also important is the need to be able to relate what is visualized by one method to another, perhaps less costly, method so that imaging approaches that are unique to animal use can inform the disease parameters that are visualized by clinical systems.

2.3.2   Translational Opportunities

In preclinical small animal models, the key biological properties, such as the functional parameters of tumor metabolism, must be established. This requires developing and implementing the dynamic imaging and kinetic analysis methods that deliver accurate and specific estimates of those aspects of tumors that are important for testing clinical agents. Very few preclinical studies use dynamic imaging to establish when after injection the maximum uptake of tracer occurs.

Another area for which imaging is appropriate, and which could be significantly improved, is discovering and measuring various biological parameters as surrogate markers of tumor responses and recurrence. These imaging markers are critical if mouse models that are clinically relevant are employed to have their intended benefit for informing and optimizing patient treatment protocols. In addition to histological or serum markers acquired by invasive procedures, imaging biomarkers could enable early changes in protocols, as necessary, to inform different choices of conventional therapies, or a switch to a new single or combination agent strategy. As the case studies in this chapter suggest, appropriate application of imaging is especially crucial in the detection, diagnosis, and treatment of brain tumors; patients with inoperable or recurrent tumors at other sites and distant site metastases will also greatly benefit from the use of suitable animal models to design and test innovative ways to image the biology of tumors.

References

Aitman, T.J., Critser, J.K., Cuppen, E. et al. 2008. Progress and prospects in rat genetics: A community view. *Nature Genetics* 40:516–22.

Akhavan, D., Cloughesy, T.F., and Mischel P.S. 2010. mTOR signaling in glioblastoma: Lessons learned from bench to bedside. *Neuro Oncology* 12:882–9.

Albers, A.C. and Gutmann, D.H. 2009. Gliomas in patients with neurofibromatosis type 1. *Expert Reviews in Neurotherapeutics* 9:535–9.

Andrechek, E.R. and Nevins, J.R. 2010. Mouse models of cancers: Opportunities to address the heterogeneity of human cancer and evaluate therapeutic strategies. *Journal of Molecular Medicine* 88:1095–1100.

Bajenaru, M.L., Hernandez, M.R., Perry, A. et al. 2003. Optic nerve glioma in mice requires astrocyte *Nf1* gene inactivation and *Nf1* brain heterozygosity. *Cancer Research* 63:8573–7.

Bajenaru, M.L., Zhu, Y., Hedrick, N.M., Donahoe, J., Parada, L.F., and Gutmann, D.H. 2002. Astrocyte-specific inactivation of the neurofibromatosis 1 gene (NF1) is insufficient for astrocytoma formation. *Molecular and Cellular Biology* 22:5100–13.

Banerjee, S., Gianino, S.M., Gao, F., Christians, U., and Gutmann, D.H. 2011. Interpreting mammalian target of rapamycin and cell growth inhibition in a genetically engineered mouse model of Nf1-deficient astrocytes. *Molecular Cancer Therapy* 10:279–91.

Banerjee, S., Hegedus, B., Gutmann, D.H., and Garbow, J.R. 2007. Detection and measurement of neurofibromatosis-1 optic glioma *in vivo*. *NeuroImage* 35:1434–7.

Barth, R.F. and Kaur, B. 2009. Rat brain tumor models in experimental neuro-oncology: The C6, 9L, T9, RG2, F98, BT4C, RT-2, and CNS-1 gliomas. *Journal of Neurooncology* 94:299–312.

Becher, O.J., Hambardzumyan, D., Fomchenko, E.I. et al. 2008. Gli activity correlates with tumor grade in platelet-derived growth factor-induced gliomas. *Cancer Research* 68:2241–9.

Borowsky, A. 2007. Special considerations in mouse models of breast cancer. *Breast Disease* 28:29–38.

Bradbury, M.S., Hambardzumyan, D., Zanzonico, P.B. et al. 2008. Dynamic small-animal PET imaging of tumor proliferation with 3′deoxy-3′-18F-fluorothymidine in a genetically engineered mouse model of high-grade gliomas. *Journal of Nuclear Medicine* 49:422–9.

Brennan, C., Momota, H., Hambardzumyan, D. et al. 2009. Glioblastoma subclasses can be defined by activity among signal transduction pathways and associated genomic alterations. *PLoS One* 4: e7752.

Brown, J.A., Emnett, R.J., White, C.R. et al. 2010. Reduced striatal dopamine underlies the attention system dysfunction in neurofibromatosis-1 mutant mice. *Human Molecular Genetics* 15:4515–28.

Brown, D.M., Matise, T.C., Koike, G. et al. 1998. An integrated genetic linkage map of the rat. *Mammalian Genome* 9:521–30.

Carver, B.S. and Pandolfi, P.P. 2006. Mouse modeling in oncologic preclinical and translational research. *Clinical Cancer Research* 15:5305–11.

Cheng, A.J., Cheng, N.C., Ford, J. et al. 2007. Cell lines from *MYCN* transgenic murine tumours reflect the molecular and biological characteristics of human neuroblastoma. *European Journal of Cancer* 43:1467–75.

Chesler, L., Goldenberg, D.D., Collins, R. et al. 2008. Chemotherapy-induced apoptosis in a transgenic model of neuroblastoma proceeds through *p53* induction. *Neoplasia* 10:1268–74.

Chesler, L., Goldenberg, D.D., Seales, I.T. et al. 2007. Malignant progression and blockade of angiogenesis in a murine transgenic model of neuroblastoma. *Cancer Research* 67:9435–42.

Chesler, L., Schlieve, C., Goldenberg, D.D. et al. 2006. Inhibition of phosphatidyinositol 3-phosphate destabilizes Mycn protein and blocks malignant progression in neuroblastoma. *Cancer Research* 66:8139–46.

Churchill, G.A. 2007. Recombinant inbred strain panels: A tool for systems genetics. *Physiological Genomics* 31:174–5.

Coudé, F.X., Mignot, C., Lyonnet, S., and Munnich, A. 2006. Academic impairment is the most frequent complication of neurofibromatosis type-1 (NF1) in children. *Behavior Genetics* 36:660–4.

Cox, A., Ackert-Bicknell, C.L., Dumone, G.A. et al. 2009. A new standard genetic map for the laboratory mouse. *Genetics* 182:1335–44.

de Seranno, S. and Meuwissen, R. 2010. Progress and applications of mouse models for human lung cancer. *European Respiratory Journal* 35:426–43.

Donehower, L.A., Harvey, M., Slagle, B.L. et al. 1992. Mice deficient for p53 are developmentally normal but susceptible to spontaneous tumours. *Nature* 356:215–21.

Giannini, C., Sarkaria, J.N., Saito, A. et al. 2005. Patient tumor *EGFR* and *PDGFRA* gene amplifications retained in an invasive intracranial xenograft model of glioblastoma mutiforme. *Neuro Oncology* 7:164–76.

Gregory, M.A., Qi, Y., and Hann, S.R. 2003. Phosphorylation by glycogen synthase kinase-3 controls c-myc proteolysis and subnuclear localization. *Journal of Biological Chemistry* 278:51606–12.

Grubb, S.C., Churchill, G.A., and Bogue, M. 2004. A collaborative database of inbred mouse strain characteristics. *Bioinformatics* 20:2857–9.

Goldbrunner, R.H., Wagner, S., Roosen, K., and Tonn, J.-C. 2000. Models for assessment of angiogenesis in gliomas. *Journal of Neuro Oncology* 50:53–62.

Gottfried, O.N., Viskochil, D.H., and Couldwell, W.T. 2010. Neurofibromatosis type I and tumorigenesis: Molecular mechanisms and therapeutic implications. *Neurosurgical Focus* 28:1–9.

Gutmann, D.H. 2008. Using neurofibromatosis-1 to better understand and treat pediatric low-grade glioma. *Journal of Child Neurology* 23:1186–94.

Haas-Kogan, D.A., Prados, M.D., Tihan, T. et al. 2005. Epidermal growth factor receptor, protein kinase B/Akt, and glioma response to erlotinib. *Journal of the National Cancer Institute* 97:880–7.

Hackett, C.S., Hodgson, J.G., Law, M.E. et al. 2003. Genome-wide array CGH analysis of murine neuroblastoma reveals distinct genomic aberrations which parallel those in human tumors. *Cancer Research* 63:5266–73.

Hambardzumyan, D., Amankulor, N.M., Helmy, K.Y., Becher, O.J., and Holland, E.C. 2009. Modeling adult gliomas using RCAS/t-va technology. *Translational Oncology* 2:89–95.

Hambardzumyan, D., Parada, L.F., Holland, E.C., and Charest, A. 2011. Genetic modeling of gliomas in mice: New tools to tackle old problems. *Glia* 59:1155–68.

Hansford, L.M., Thomas, W.D., Keating, J.M. et al. 2004. Mechanisms of embryonal tumor initiation: Distinct roles for MycN expression and *MYCN* amplification. *Proceedings of the National Academy of Sciences, U.S.A.* 101:12664–9.

Hegedus, B., Banerjee, D., Yeh, T.-H. et al. 2008. Preclinical cancer therapy in a mouse model of neurofibromatosis-1 optic glioma. *Cancer Research* 58:1520–8.

Hegedus, B., Hughes, W.F., Garbow, J.R. et al. 2009. Optic nerve dysfunction in a mouse model of neurofibromatosis-1 optic glioma. *Journal of Neuropathology and Experimental Neurology* 68:542–51.

Hill, A.E., Lander, E.S., and Nadeau, J.H. 2006. Chromosome substitution strains. A new way to study genetically complex traits. *Methods in Molecular Medicine*, Vol. 128: Cardiovascular Disease; Methods and Protocols; Volume 1. Q. Wang © Humana Press Inc., Totowa, NJ, pp. 153–72.

Hodgson, J.G., Yeh, R.F., Ray, Y. et al. 2009. Comparative analysis of gene copy number and mRNA expression in glioblastoma multiforme tumors and xenografts. *Neuro Oncology* 11:477–87.

Holland, E.C. and Varmus, H.E. 1998. Basic fibroblast growth factor induces cell migration and proliferation after glia-specific gene transfer in mice. *Proceedings of the National Academy of Sciences, U.S.A.* 95:1218–23.

Huang, E.J. and Reichardt, L.F. 2001. Neurotrophins: Roles in neuronal development and function. *Annual Reviews of Neuroscience* 24:677–736.

Chapter 2

Iannaccone, P.M. and Jacob, H.J. 2009. Rats! *Disease Models and Mechanisms* 2:206–10.

Jacob, H.J. 1999. Functional genomics and rat models. *Genome Research* 9:1013–6.

Jacob, H.J., Lazar, J., Dwinell, M.R., Moreno, C., and Geurts, A.M. 2010. Gene targeting in the rat: Advances and opportunities. *Trends in Genetics* 26:510–18.

Jacobs, A.H., Thomas, A., Kracht, L.W. et al. 2005. ^{18}F-Fluoro-L-thymidine and ^{11}C-methylmethionine as markers of increased transport and proliferation in brain tumors. *Journal of Nuclear Medicine* 46:1948–58.

Jansen, M., Yip, S., and Louis, D.N. 2010. Molecular pathology in adult gliomas: Diagnostic, prognostic, and predictive markers. *Lancet Neurology* 7:717–26.

Jeremic, B., Grujicic, D., Antunovic, V., Djuric, L., Stojanovic, M., and Shibamoto, Y. 1994. Influence of extent of surgery and tumor location on treatment outcome of patients with glioblastoma multiforme treated with combined modality approach. *Journal of Neuro-Oncology* 21:177–85.

Kim, C.F.B., Jackson, E.L., Kirsch, D.G. et al. 2005. Mouse models of human non-small-cell lung cancer: Raising the bar. *Cold Spring Harbor Symposia on Quantitative Biology* LXX:241–50.

Kim, I.S. and Baek, S.H. 2010. Mouse models for breast cancer. *Biochemical and Biophysical Research Communications* 394:443–7.

Larjavaara, S., Mäntylä, R., Salminen, T. et al. 2007. Incidence of gliomas by anatomic location. *Neuro Oncology* 9:319–25.

Little, C.C. 1920. The heredity of susceptibility to a transplantable sarcoma (J.W.B.) of the Japanese waltzing mouse. *Science* 51:467–8.

Little, C.C. 1924. The genetics of tissue transplantation in mammals. *Journal of Cancer Research* 8:75–95.

Little, C.C. and Tyzzer, E.E. 1916. Further experimental studies on inheritance of susceptibility to a transplantable tumor, carcinoma (J.w.A.) of the Japanese waltzing mouse. *Journal of Medical Research* 33:393–453.

Lopez-Barcons, L.-A. 2010. Serially heterotransplanted human prostate tumours as an experimental model. *Journal of Cellular and Molecular Medicine* 68:1385–95.

Louis, D.N., Ohgaki, H., Wiestler, O.D. et al. 2007. The 2007 WHO classification of tumours of the central nervous system. *Acta Neuropathology* 114:97–109.

Marone, R., Cmiljanovic, V., Giese, B., and Wymann, M.P. 2008. Targeting phosphoinositide 3-kinase: Moving towards therapy. *Biochimica Biophysica Acta* 1784:159–85.

Mashimo, T., Voigt, B., Kuramoto, T., and Serikawa, T. 2005. Rat phenome project: The untapped potential of existing rat strains. *Journal of Applied Physiology* 98:371–9.

Mellinghoff, I.K., Wang, M.Y., Vivanco, I. et al. 2005. Molecular determinants of the response of glioblastomas to EGFR kinase inhibitors. *New England Journal of Medicine* 353:2012–24.

Mischel, P.S., Shai, R., Shi, T. et al. 2003. Identification of molecular subtypes of glioblastoma by gene expression profiling. *Oncogene* 22:2361–73.

Momoto, H. and Holland, E.C. 2005. Bioluminescence technology for imaging cell proliferation. *Current Opinion in Biotechnology* 16:681–6.

Muzi, M., Spence, A.M., O'Sullivan, F.O. et al. 2006. Kinetic analysis of 3′-deoxy-3′-^{18}F-fluorothymidine in patients with gliomas. *Journal of Nuclear Medicine* 47:1612–21.

Network TI. 2008. Comprehensive genomic characterization defines human glioblastoma genes and core pathways. *Nature* 455:1061–68.

Olive, K.P., Jacobetz, M.A., Davidson, C.J. et al. 2009. Inhibition of hedgehog signaling enhances delivery of chemotherapy in a mouse model of pancreatic cancer. *Science* 324:1457–61.

Ozawa, T. and James, CD. 2010. Establishing intracranial brain tumor xenografts with subsequent analysis of tumor growth and response to therapy using bioluminescent imaging. *Journal of Visualized Experiments* 41: http://www.jove.com/index/Details.stp?ID = 1986.

Paigen, K. 2003a. One hundred years of mouse genetics: An intellectual history. I. The classical period (1902–1980). *Genetics* 163:1–7.

Paigen, K. 2003b. One hundred years of mouse genetics: An intellectual history. II. The molecular revolution (1981–2002). *Genetics* 163:1227–35.

Pandita, A., Aldape, K.D., Zadeh, G., Guha, A., and James, C.D. 2004. Contrasting *in vivo* and *in vitro* fates of glioblastoma cell subpopulations with amplified EGFR. *Genes Chromosomes & Cancer* 239:29–36.

Parsons, D.W., Jones, S., Zhang, X. et al. 2008. An integrated genomic analysis of human glioblastoma multiforme. *Science* 321:1807–12.

Peters, L.L., Robledo, R.F., Bult, C.J., Churchill, G.A., Paigen, B.J., and Svenson, K.L. 2007. The mouse as a model for human biology: A resource guide for complex trait analysis. *Nature Reviews Genetics* 8:58–69.

Pienta, K.J., Abate-Shen, C., Agus, D.B. et al. 2008. The current state of preclinical prostate cancer animal models. *Prostate* 68:629–39.

Pollard, M. and Suckow, M.A. 2005. Hormoone-refractory prostate cancer in the Lobund–Wistar rat. *Experimental Biology and Medicine* 230:520–6.

Prasad, G., Sottero, T., Yang, X. et al. 2011. Inhibition of PI3K/mTOR pathways in glioblastoma and implications for combination therapy with temozolomide. *Neuro Oncology* 13:384–92.

Roberts, A., De Villena, F.P.-M., Wang, W., McMillan, L., and Threadgill, D.W. 2007. The polymorphism architecture of mouse genetic resources elucidated using genome-wide resequencing data: Implications for QTL discovery and systems genetics. *Mammalian Genome* 18:473–81.

Sarkaria, J.N., Carlson, B.L., Schroeder, M.A. et al. 2006. Use of an orthotopic xenograft model for assessing the effect of epidermal growth factor receptor amplification on glioblastoma radiation response. *Clinical Cancer Research* 12:2264–71.

Sarkaria, J.N., Yang, L., Grogan, P.T. et al. 2007. Identification of molecular characteristics correlated with glioblastoma sensitivity to EGFR kinase inhibition through use of an intracranial xenograft test panel. *Molecular Cancer Therapeutics* 6:1167–74.

Schiepers, C., Chen, W., Dahlbom, M., Cloughesy, T., Hoh, C.K., and Huang, S.-C. 2007. ^{18}F-Fluorothymidine kinetics of malignant brain tumors. *European Journal of Nuclear Medicine and Molecular Imaging* 34:1003–11.

Schmidt, E.E., Ichimura, K., Reifenberger, G., and Collins, V.P. 1994. CDKN2 (p16/MTS1) gene deletion or CDK4 amplification occurs in the majority of glioblastomas. *Cancer Research* 54:6321–4.

Sears, R.C. 2004. The life cycle of C-myc: From synthesis to degradation. *Cell Cycle* 3:1133–7.

Shao, H., Sinasac, D.S., Burrage, L.C. et al. 2010. Analyzing complex traits with congenic strains. *Mammalian Genome* 21:276–86.

Shih, A.H. and Holland, E.C. 2006. Platelet-derived growth factor (PDGF) and glial tumorigenesis. *Cancer Letters* 232:139–47.

Smits, B.M.G., Cotroneo, M.S., Haag, J.D., and Gould, M.N. 2007. Genetically engineered rat models for breast cancer. *Breast Disease* 28:53–61.

Spitzenberg, V., König, C., Ulm, S. et al. 2010. Targeting PI3K in neuroblastoma. *Journal of Cancer Research and Clinical Oncology* 136:1881–90.

Sughrue, M.E., Yang, I., Kane, A.J. et al. 2009. Immunological considerations of modern animal models of malignant primary brain tumors. *Journal of Translational Medicine* 7:84–93.

Szpirer, C. and Szpirer, J. 2007. Mammary cancer susceptibility: Human genes and rodent models. *Mammalian Genome* 18:817–31.

Szatkiewicz, J.P., Beane, G.L., Ding, Y., Hutchins, L., de Villena, F.P.-M., and Churchill, G.A. 2008. An imputed genotype resource for the laboratory mouse. *Mammalian Genome* 19:199–208.

Thomas, K.R. and Capecchi, M.R. 1987. Site-directed mutagenesis by gene targeting in mouse embryo-derived stem cells. *Cell* 51:503–12.

Twigger, S.N., Pruitt, K.D., Fernández-Suárez, X.M. et al. 2008. What everybody should know about the rat genome and its online resources. *Nature Genetics* 40:523–7.

Tyzzer, E.E. and Little, C.C. 1916. Studies on the inheritance of susceptibility to a transplantable sarcoma (J.w.B.) of the Japanese waltzing mouse. *Journal of Cancer Research* 1:387–9.

Ueki, K., Ono, Y., Henson, J.W., Efird, J.T., von Deimling, A., and Louis, D.N. 1996. CDKN2/p16 or RB alterations occur in the majority of glioblastomas and are inversely correlated. *Cancer Research* 56:150–3.

Vargo-Gogola, T. and Rosen, J.M. 2007. Modelling breast cancer: One size does not fit all. *Nature Reviews Cancer* 7:659–72.

Vernon, A.E., Bakewell, S.J., and Chodosh, L.A. 2007. Deciphering the molecular basis of breast cancer metastasis with mouse models. *Reviews of Endocrinology and Metabolic Disorders* 8:199–213.

Wang, F. 2011. Modeling human prostate cancer in genetically engineered mice. *Progress in Molecular Biology and Translational Science* 100:1–49.

Wang, X., Fonseca, B.D., Tang, H. et al. 2008. Re-evaluating the roles of proposed modulators of mammalian target of rapamycin complex 1 (mTORC1) signaling. *Journal of Biological Chemistry* 283:30482–92.

Weiss, W.A., Aldape, K., Mohaptra, G., Feuerstein, B.G., and Bishop, J.M. 1997. Targeted expression of *MYCN* causes neuroblastoma in transgenic mice. *The EMBO Journal* 16:2895–95.

Wislez, M., Spencer, M.L., Izzo, J.G. et al. 2005. Inhibition of mammalian target of rapamycin reverses alveolar epithelial neoplasia induced by oncogenic *K-ras*. *Cancer Research* 65:3226–35.

Wu, M., Jung, A.B., Fleet, C. et al. 2009. Dissecting genetic requirements of human breast tumorigenesis in a tissue transgenic model of human breast cancer in mice. *Proceedings of the National Academy of Sciences* 106:7022–7.

Wyman, M.P., Zvelebil, M., and Laffargue, M. 2003. Phosphoinositide 3-kinase signaling—which way to target? *Trends in Pharmacological Sciences* 24:366–76.

Yaari, S., Jacob-Hirsch, J., Amariglio, N., Haklai, R., Rechavi, G., and Kloog, Y. 2005. Disruption of cooperation between Ras and MycN in human neuroblastoma cells promotes growth arrest. *Clinical Cancer Research* 11:4321–30.

Yan, H., Parsons, D.W., Jin, G. et al. 2009. IDH1 and IDH2 mutations in gliomas. *New England Journal of Medicine* 360:765–73.

Yang, H., Ding, Y., Hutchins, L.N. et al. 2009. A customized and versatile high-density genotyping array for the mouse. *Nature Methods* 6:663–6.

Yeh, E., Cunningham, M., Arnold, H. et al. 2004. A signalling pathway controlling c-Myc degradation that impacts oncogenic transformation of human cells. *Nature Cell Biology* 6:308–18.

Zouberakis, M., Chandras, C., Swertz, M. et al. 2010. Mouse resource browser—A database of mouse databases. *Database* 2010: Article ID baq010, doi:10.1093/database/baq010.

Chapter 2

3. Animal Models in Cardiology

Wengen Chen, Henry Gewirtz, Mark Smith, and Vasken Dilsizian

An inherent advantage of nuclear imaging techniques, such as single-photon emission CT (SPECT) or positron emission tomography (PET), is in the synergy that exists between the radionuclide tracers—that by their nature reflect physiological processes at the cellular level—and the underlying pathophysiological states being investigated. Clinical PET and SPECT scanners designed for human imaging are appropriate for cardiac imaging research with larger animals, such as

Targeted Molecular Imaging. Edited by Michael J. Welch and William C. Eckelman © 2012 Taylor & Francis Group, LLC. ISBN: 978-1-4398-4195-2

Chapter 3

Table 3.1   Targeting Myocardial Pathology

Cardiac Pathology	Targeting Radiotracer
Assessment of Myocardial Ischemia	
Assessment of reduction in myocardial perfusion	Thallium 201, sestamibi or tetrofosmin, rubidium 82, nitrogen 13-ammonia
Reduction of myocardial consumption of staple metabolites	BMIPP, IPPA
Myocardial switch to alternative source of metabolism	FDG
Assessment of myocardial hypoxia	Nitroimidazole derivatives
Assessment of ischemic memory	BMIPP
Assessment of Myocardial Damage or Injury	
Detection of loss of sarcolemmal integrity	Antimyosin Fab
Detection of subcellular organelle dysfunction	Glucaric acid, pyrophosphate
Detection of apoptosis	Annexin A5, synaptotagmin
Assessment of Myocardial Inflammation	
Detection of leukocyte infiltration	MCP-1, immunoglobulin G, pentreotide, gallium 68, leukocytes
Assessment of inflammation-related diffuse myocyte necrosis	Antimyosin Fab
Expression of accessory molecules on myocytes	ICAM, HLA-II antibodies
Assessment of Interstitial Alterations	
Assessment of neurohumoral upregulation	ACE inhibitors, AT_1R blockers
Assessment of collagen deposition	RGD peptides
Assessment of collagenolytic activity	MMP inhibitors
Assessment of Myocardial Denervation	
Assessment of receptor alterations	CGP-12177, CGP-12388 (β_1-receptor), ICI-H-89.406 (α_1-receptor)
Assessment of NE reuptake efficiency	MIBG, HED

Note: FDG, fluorodeoxyglucose; ICAM, intsercellular adhesion molecule; HLA, human leukocyte antigen; BMIPP, beta-methyl-iodophenyl-pentadecanoic acid; MIBG, metalodo-benzylguanidine; MCP, monocyte chemoattractant protein; MMP, matrix metalloproteinase; NE, norepinephrine; HED, hydroxyephedrine.

Table 3.2   Molecular Targets of Left Ventricular Remodeling

Myocardial Functional and Structural Target	Imaging Agent
Substrate use	^{18}F-2-Fluoro-2-glucose
Glucose	1-^{11}C-Glucose
Free fatty acid	^{11}C-Palmitate
Oxidative metabolism	^{11}C-Acetate
Ischemic memory	β-Methyl-*p*-[^{123}I]-iodophenyl-pentadecanoic acid
	Selectin ligand-coated lipid microbubbles
Apoptosis	^{99m}Tc-annexin-V
	Ferromagnetic-labeled annexin-V
	Annexin-V-coated gadolinium liposomes
Interstitial Fibrosis	
Angiotensin-converting enzyme	^{18}F-Captopril; ^{18}F-Flurobenzoyl-lisinopril
Angiotensin II type 1 receptor	^{11}C-MK-996; ^{11}C-L-159884; ^{11}C-KR31173; fluoresceinated angiotensin peptide analog; ^{99m}Tc-Iosartan
Matrix metalloproteinases	^{111}Indium-radiolabeled ligand ^{99m}Tc-radiolabeled ligand
Autonomic Innervation	
Sympathetic	
Presynaptic	^{123}I-Metaiodobenzylguanidine
	^{11}C-Meta-hydroxyephedrine
	^{18}F-Fluorodopamine
	^{11}C-Epinephrine
	^{11}C-Phenylephrine
Postsynaptic	^{11}C-CGP 12177
	^{11}C-CGP 12388
	^{18}F-Fluorocarazolol
	^{11}C-ICI-OMe
Parasympathetic	
Acetylcholine transporter	^{18}F-Fluoroethoxybenzovesamicol
Nicotonic receptor	2-Deoxy-2[^{18}F] fluoro-D-glucose A85380
Muscarinic receptors	^{11}C-Methylquinuclidinyl benzilate

canine or swine. Small-animal PET and SPECT imaging systems are now widely available with sufficient resolution and sensitivity for cardiovascular investigations with small animal models, such as mice or rats. Small system sizes make such instrumentation attractive for basic science laboratories and core imaging facilities. In the first section of this chapter, the advantages and limitations of SPECT and PET imaging technologies are presented for studying cardiovascular disorders in animals. The second section examines basic concepts and mechanisms for studying atherosclerosis in animal models. The third section reviews the role of large animal models, particularly domestic swine, to gain insight into the pathophysiology of ischemic myocardial injury and left ventricular hypertrophy (LVH) that provide the basis for identifying targets for molecular imaging (Table 3.1). The fourth and final section addresses new molecular probes for studying interstitial fibrosis and left ventricular remodeling (Table 3.2).

3.1 SPECT and PET Imaging Technologies

3.1.1 Clinical SPECT and PET Cameras for Large-Animal Imaging

Clinical SPECT and PET systems are appropriate for imaging animals whose sizes and weights range between those of small children and adults, with commensurately sized hearts. Canine and porcine models, for example, have been vital in the development and investigation of single gamma and positron-emitting tracers for myocardial blood flow, viability, and other aspects of cardiac metabolism and physiology. The capabilities of these systems for clinical imaging would apply to animal imaging as well.

3.1.1.1 Gamma Camera and SPECT

SPECT scanners and gamma cameras are typically fabricated with large NaI(Tl) crystals, one to three detector heads, and are equipped with parallel hole collimators (Zeng et al. 2004). For SPECT, the detector heads rotate around the patient or animal to acquire two-dimensional projection data, which are used to reconstruct three-dimensional images with typical resolution of 10–15 mm. Gamma camera and SPECT imaging have been important in the development and investigation of the pharmacodynamics of flow agents such as Tl-201 chloride (Steingart et al. 1982, 1986) and Tc-99m-labeled tracers such as Tc-99m sestamibi (Okada et al. 1988, Verani et al. 1988, Li et al. 1990, Glover and Okada 1993) and Tc-99m teboroxime (Stewart et al. 1990).

SPECT cardiac image evaluation is usually visual or semiquantitative, with the region of peak uptake considered normal. To achieve more accurate quantitation, animal studies have been used to investigate the effectiveness of attenuation correction in SPECT. The use of transmission scanning source has been tested in a canine model (Li et al. 1995) and CT-based attenuation correction was investigated in a porcine model (Da Silva et al. 2001).

Novel clinical SPECT scanners that can acquire projection data without detector heads that rotate around the patient provide an opportunity for fast, dynamic imaging of large animals and would facilitate kinetic modeling for single-photon-emitting tracers. Two such SPECT cameras incorporate solid-state cadmium zinc telluride (CZT) detector components. One device has nine CZT detectors with high-sensitivity parallel hole collimators whose holes are matched to the CZT crystal elements. The detectors swivel to increase sensitivity in the region of the heart (Sharir et al. 2008, Gambhir et al. 2009). Another device has 19 pinhole cameras focused on the cardiac region (Herzog et al. 2010). Acquisitions with both imaging systems can be performed approximately five times faster than with conventional gamma cameras. It is likely that new detector technologies will continue to advance cardiac imaging capabilities for single-photon tracers that can be applied to animal as well as human imaging.

3.1.1.2 PET

Clinical PET scanners are built with a ring of detector modules that surrounds the patient or animal. Coincidence circuitry is used to identify individual scintillation crystal elements that are struck simultaneously by 511 keV photons resulting from annihilation of a positron following its emission (Lewellen and Karp 2004). Three-dimensional images of PET tracer biodistribution are reconstructed with typical resolution of 5–8 mm. Attenuation, scatter, and randoms correction permit images of absolute activity concentrations to be formed. Since PET scanners are static, fast dynamic frame rates can be obtained for kinetic modeling of tracer uptake and washout.

PET imaging has been vital to the investigation in canine animal models of PET flow tracers such as O-15 water (Bergmann et al. 1984), N-13 ammonia (Kuhle et al. 1992), and Rb-82 (Coxson et al. 1995) as

Chapter 3

well as F-18 fluorodeoxyglucose (Ratib et al. 1982) for cardiac viability. The study of new imaging agents also commonly uses canine models, for example, F-18 fluorometaraminol (Langer et al. 2000) and F-18 fluorobenzyl triphenyl phosphonium (Madar et al. 2006). PET-CT has been used in a dog model to show that respiratory-averaged CT-based attenuation correction can accurately correct ungated PET cardiac scans (Cook et al. 2007).

Time of flight (TOF) and new scintillation crystal technology offers the potential for improved PET imaging for clinical and animal applications. TOF uses the difference in arrival times between the detected annihilation photons to localize the emission site along the line of response (LOR). This concept was used on prototype PET systems in the 1980s (Ter-Pogossian et al. 1982) and is now available on commercial systems. TOF imaging yields improved contrast to noise trade-offs compared with non-TOF systems (Surti et al. 2007, Karp et al. 2008). This advantage can be used to reduce imaging time, to lower administered activity, or to achieve better-quality images. Cerium-doped lanthanum bromide (LaBr$_3$) is a scintillator with a shorter decay time and higher light output than lutetium oxyorthosilicate (LSO) or lutetium yttrium oxyorthosilicate (LYSO), although with reduced stopping power. A TOF system built with LaBr$_3$ scintillators shows excellent performance, with timing resolution of 375 ps compared with commercial LSO/LYSO systems with 550–600 ps timing resolution (Daube-Witherspoon et al. 2010). The higher light output also enables narrower energy windows that reduce scatter.

There has been a recent investigation into clinical dual-modality PET/MRI imaging systems (Pichler et al. 2010), which would have applications to large-animal cardiovascular imaging. One approach is to develop a PET insert that will fit the inside of an MRI magnet. Technical challenges here are to use nonmagnetic detector components such as avalanche photodiodes (APDs) or silicon photomultipliers (SiPMs) that are minimally influenced by the magnetic field, to ensure that MRI operation is not affected by PET detector components and to achieve robust attenuation correction for PET. A second approach is to place the PET and MRI scanners close to each other and use a common bed, although this precludes simultaneous imaging. Dual-modality PET/MRI will provide an important research tool to study cardiac applications of molecular imaging such as atherosclerotic plaques, angiogenesis, and stem cell therapy in large animal models (Nekolla et al. 2009).

3.1.2 Micro-PET and Micro-SPECT Cameras for Small-Animal Imaging

Though large animal models are valuable in the study of cardiovascular physiology, they are expensive to raise, have a longer life span, and require more care and handling than small animal models such as mice or rats. The advantages of using small animals for molecular imaging have spurred the development of dedicated small-animal PET and SPECT imaging systems.

3.1.2.1 Micro-PET
Small-animal PET scanners are an invaluable tool for biomedical scientists in studying cardiovascular physiology and in the development of new tracers and imaging markers (Cherry and Gambhir 2001, de Kemp et al. 2010). Small-animal PET has the advantage of being able to use tracers for which positron-emitting radionuclides such as O-15, N-13, and C-11 can be substituted for their stable counterparts in biological molecules. These PET systems seek to replicate the capabilities of clinical PET scanners, but on a smaller scale with detector elements and system designs suited for imaging mice and rats at 1–2 mm resolution.

PET systems for dedicated small-animal imaging have included both rotating planar detectors (Jeavons et al. 1999, Weber et al. 1999) and ring designs (Chatziioannou et al. 1999, Seidel et al. 2003, Surti et al. 2005). A design with opposed rotating detectors operating in coincidence mode is easier for researchers to build from a hardware development point of view, while a fixed ring design, though more complex, is preferable for obtaining fast sequential images for kinetic modeling.

The high density avalanche chamber (HIDAC) system developed using gas avalanche detectors offers superb resolution of up to 0.7 mm full width at half maximum (FWHM) with sensitivity of 0.9% (Jeavons et al. 1999), though this technology is technically demanding and has not been used by other groups. The more common system architecture is the use of pixellated scintillation crystals with signals from conventional photomultiplier tubes or position-sensitive photomultiplier tubes and the use of a lookup map to identify crystal elements hit by the 511 keV photons resulting from positron annihilation (Acton 2006, del Guerra and Belcari 2007). Another approach to crystal identification is to directly couple APDs to individual crystal elements in which a scintillation event occurs (Lecomte et al. 1996). Scintillation crystals with high stopping power and fast timing have been used, including bismuth germanate oxide (BGO), gadolinium oxyorthosilicate (GSO), LSO, LYSO, and yttrium aluminum perovskite (YAP). Fabrication of pixellated scintillation crystal arrays with dimensions <0.5 mm on a side is difficult and crystal elements usually range from 1 to 4 mm on a side and 10 to 20 mm long. Typical performance characteristics for commercially available scintillation-based systems are a resolution of 1–2.5 mm FWHM at the center of the field of view, an axial field of view of 5–13 cm, and a sensitivity of 4–10% at the center of the field of view. Capabilities are generally similar to those of human scanners, including scatter, attenuation, and randoms correction, ECG and respiratory gating, list-mode acquisition, and iterative as well as analytic image reconstruction.

Resolution improvements in small-animal PET scanners will come from routine incorporation of detectors with depth-of-interaction (DOI) capability (Yamada et al. 2008), which is the subject of development efforts by several groups. This can be accomplished by the use of layered scintillators that may be offset from each other or from different materials, double-sided scintillator readout, or by side readout of scintillation crystals. The system by Seidel et al. (2003) incorporates phoswich scintillator modules of lutetium gadolinium oxyorthosilicate (LGSO)/GSO layers. The resolution of small-animal PET may be ultimately limited by two physical factors: positron range, since a positron is emitted with some kinetic energy and travels a small distance before undergoing annihilation, and noncolinearity, the property that the annihilation photons are not emitted exactly 180° apart. The depth of interaction and these effects can be partially mitigated by modeling them in image reconstruction.

As is the case for human imaging, fusion of PET with anatomic images is valuable in helping to localize uptake and characterize disease. Small-animal PET/CT systems have been developed that allow acquisition of anatomic and physiological images. Such systems generally translate the small animal on a common bed between the PET and CT subsystems and are commercially available. There is currently much research on the development of small-animal PET systems that are MRI compatible (Catana et al. 2006, Raylman et al. 2006, Judenhofer et al. 2007). Two design philosophies are to use light guides to bring the light to photomultiplier tubes or compact position-sensitive photomultiplier tubes that are outside the scanner and removed from the peak magnetic field, or to construct a scanner using components such as APDs or SiPMs that are insensitive to magnetic fields.

Some representative applications of small-animal PET for preclinical cardiac investigations include gated PET with F-18 fluorodeoxyglucose to evaluate left ventricular viability and function (Figure 3.1) (Croteau et al. 2003, de Kemp et al. 2010), assessment of myocardial blood flow with O-15 water and C-11 acetate (Herrero et al. 2006) and imaging cardiac sympathetic innervation with C-11 metahydroxyephedrine (Law et al. 2010). Hybrid PET/MRI has been used to image cardiac PET viability and anatomic structure (Büscher et al. 2010).

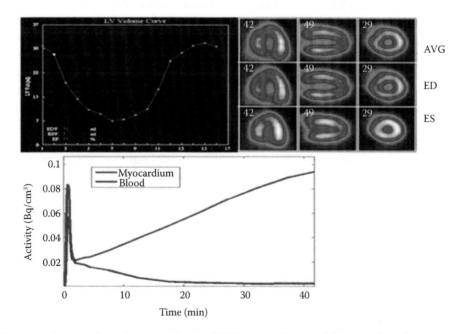

FIGURE 3.1   Gated images of a mouse heart from a small-animal PET system. Dynamic left ventricular volumes can be obtained to calculate ejection fraction. Myocardial and blood pool activity can be extracted for kinetic modeling. (Reprinted by permission of the Society of Nuclear Medicine from de Kemp RA et al. Small-animal molecular imaging methods. *J Nucl Med.* 2010; 51(Suppl 1): 18S–32S. Figure 1.)

Nontargeted 1D targeted 3D targeted

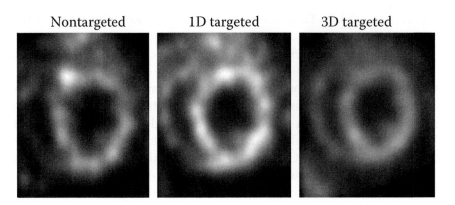

FIGURE 3.2 Reconstructed end-diastolic short-axis slices from gated myocardial perfusion scans for the U-SPECT-II small-animal SPECT system (Branderhorst et al. 2011). Video images of the mouse prior to scanning permits improved targeting of the heart by moving it to the region of the scanner where more of the pinholes are focused and sensitivity is greater. There is improved visualization of the papillary muscles and right ventricle with better targeting. (With kind permission from Springer Science+Business Media: *Eur J Nucl Mol Imaging*, Targeted multi-pinhole SPECT, 38, 2011, 552–561, Branderhorst W. et al.)

3.1.2.2 Micro-SPECT

Small-animal SPECT is an important tool in preclinical molecular imaging (Meikle et al. 2005, Franc et al. 2008, de Kemp et al. 2010). Micro-SPECT imaging of small animals was first achieved with pinhole collimation on clinical SPECT scanners (Jaszczak et al. 1994, Strand et al. 1994, Weber et al. 1994, Ishizu et al. 1995, Yukihiro et al. 1996). High resolution is achieved by obtaining a magnified image of the small animal at each projection so that the resolution of the reconstructed image is dominated by the resolution of the pinhole aperture, not that of the gamma camera (Anger et al. 1958). Images with 1 mm resolution can be easily obtained. Though high-resolution SPECT imaging of small animals is feasible, its sensitivity is about two orders of magnitude less than that of PET due to the use of mechanical collimation if only one pinhole is used for each of a few detector heads; improved sensitivity can only be achieved at the expense of resolution by using wider pinhole apertures.

There has been a recent renaissance in the development of small-animal SPECT systems, galvanized by the use of compact gamma cameras and multipinhole collimators. Compact detectors (McElroy et al. 2002, Weisenberger et al. 2003, Kastis et al. 2004, Furenlid et al. 2004) increase system sensitivity by allowing many cameras to view an animal at once. Detectors built with pixellated scintillation crystals or semiconductor detectors using CZT (Kim et al. 2006) with crystal elements 1–3 mm wide enable high intrinsic spatial resolution. Multipinhole collimation permits multiple projections of a small animal to be obtained simultaneously on one or several detectors, increasing sensitivity by about the number of pinholes (Schramm et al. 2003, Beekman et al. 2005, 2007). The effective sensitivity increase is dependent on projection data overlap. As an example of multipinhole collimation, the U-SPECT-II system developed by the Beekman group in Utrecht uses 75 focused pinholes to achieve high sensitivity and resolution in a small target region (Figure 3.2) (Branderhorst et al. 2011). The guiding principle is to use as much of the detector surface area as possible for imaging the animal. Note that static systems with multipinhole collimation sufficient for tomographic image reconstruction permit acquisition of projection data for kinetic modeling. The use of iterative image reconstruction methods in modeling the pinhole response function has been another significant factor improving the imaging quality of small-animal SPECT systems (Schramm et al. 2003).

The importance of combined functional/anatomic imaging has been recognized and dual-modality small-animal SPECT/CT systems have been built and are commercially available (Franc et al. 2008). Dual-modality SPECT/MRI systems are under investigation as well (Breton et al. 2007, Tsui et al. 2010). Commercial systems enable the researchers to use small-animal SPECT without an instrumentation development effort. Sensitivity and resolution are dependent on the number of pinholes and their aperture size. Current systems can routinely achieve resolutions of 1–2 mm with sensitivities of 40–2500 cps/MBq; best reported resolutions are 0.3–0.8 mm. SPECT has advantages of permitting multiple tracers to be imaged simultaneously if different radionuclides are used as labels.

Representative micro-SPECT imaging of myocardial infarctions in rats may be found in Hirai et al. (2000) and Liu et al. (2002), and a general discussion of cardiovascular micro-SPECT applications in Tsui and Kraitchman (2009).

3.2 Atherosclerosis Models

3.2.1 Initiation of Atherosclerosis Plaque

Vascular atherosclerosis is a chronic inflammatory disease with multiple risk factors. Initiation of atherosclerotic plaque involves interactions between atherogenic lipoproteins such as low-density lipoprotein (LDL) and macrophages. LDL particles deposited in the subendothelial space are subjected to oxidative modification to form modified low-density lipoprotein (mLDL) or extensively oxidized LDL (oxLDL), which induces vascular inflammation and expression of adhesion and migration molecules like vascular cell adhesion molecule-1 (VCAM-1) on the surface of endothelial cells. The surface molecules mediate migration of circulation monocytes into the subendothelial space. Under stimulation of inflammatory cytokines, the monocytes differentiate into macrophages, which phagocyte mLDL or oxLDL and become foam cells with intracellular accumulation of lipid droplets. Foam cells are characteristically present in atherosclerosis lesions, occupying much of the plaque (Figure 3.3) (Glass and Witztum 2001). Balance of the pro- and anti-inflammatory cytokines secreted from plaque macrophages and lymphocytes determines the progression of plaque. A vulnerable plaque is highly inflammatory and has a thin fibrous cap. Proteinases like matrix metalloproteinase (MMP) secreted by macrophages can further weaken the fibrous cap and lead to plaque rupture (Figure 3.4). As a consequence, thrombosis occurs, which causes myocardial infarction and sudden death or stroke clinically. On the other hand, a plaque can advance to its late stage (stable plaque) without rupture (or with small rupture). Different from a vulnerable plaque that does not narrow the lumen, a stable plaque is usually large with significant calcifications within its necrotic core, narrowing the vessel, which could cause ischemia attack like angina clinically (Figure 3.5). Current cardiovascular imaging modalities, such as coronary CT angiography, coronary artery catheterization, or even nuclear medicine myocardial perfusion scan emphasize the anatomical detection of coronary artery luminal narrowing (stable plaque) or its downstream flow-limiting functional consequence in the left ventricular myocardium. There is a lack of an imaging modality that can directly visualize vulnerable plaque, which is critical for clinical intervention. Various methods, including intravascular ultrasonography, functional MRI, and nuclear medicine studies such as SPECT and PET, have been employed for this purpose.

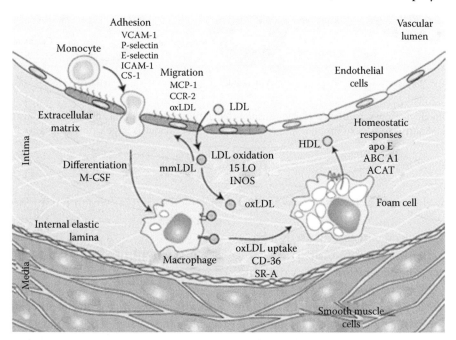

FIGURE 3.3 Initiating events in the development of a fatty streak lesion. LDL is subject to oxidative modifications in the subendothelial space, progressing from minimally modified LDL (mmLDL), to extensively oxidized LDL (oxLDL). Monocytes attach to endothelial cells that have been induced to express cell adhesion molecules by mmLDL and inflammatory cytokines. Adherent monocytes migrate into the subendothelial space and differentiate into macrophages. Uptake of oxLDL via scavenger receptors leads to foam cell formation. (Reproduced from Glass CK, Witztum JL. *Cell*. 2001; 104(4):503–516. With permission.)

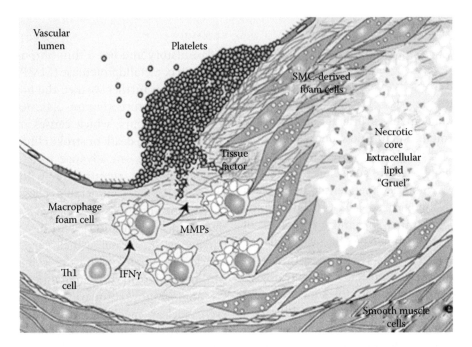

FIGURE 3.4 Plaque rupture and thrombosis. Death of macrophages and smooth muscle cells leads to the formation of a necrotic core and accumulation of extracellular cholesterol. Macrophage secretion of matrix metalloproteinases and neovascularization contribute to weakening of the fibrous plaque. Plaque rupture exposes blood components to tissue factor, initiating coagulation, the recruitment of platelets, and the formation of a thrombus. (Reproduced from Glass CK, Witztum JL. *Cell*. 2001; 104(4):503–516. With permission.)

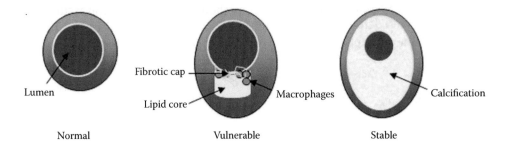

FIGURE 3.5 Development of plaques. A vulnerable plaque usually has a thin cap with significant inflammation and is prone to rupture. It does not cause luminal narrowing due to vascular wall remodeling. A stable plaque is large with significant calcifications narrowing the lumen.

Currently, most of the studies are at the preclinical animal level.

Mouse models have been extensively used in basic research of atherosclerosis because of its small size, easy breeding, short time generation, and availability of inbred strains. There are morphological similarities of atherosclerotic lesions developed in certain mouse models compared with specific stages of human lesions. For the same reasons, mouse models provide the most suitable system for imaging study of atherosclerosis. With current techniques, the small size of mouse is no longer a major challenge for high-quality imaging, particularly with micro-SPECT/CT, which has higher resolution than micro-PET.

3.2.2 Quantification of Atherosclerosis in Mouse

Quantification of atherosclerotic plaque is challenging because of the complex cellular and chemical changes within the lesions, as well as different characteristics of lesions arising from the different parts of the vasculature. To evaluate the extent of atherosclerosis in mouse models, *en face* analysis is well used. The whole aorta and its branches are removed and opened longitudinally.

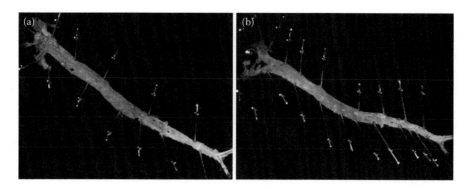

FIGURE 3.6 Example of *en face* analysis of atherosclerosis in an ApoE$^{-/-}$ (a) and control (b) mouse fed chow diet for 4 months. Significant atherosclerotic lesions are seen in the aortic arch and descending aorta. No clear lesions are seen in the control mouse.

The aortas are then pinned on a surface and the area of intima with lesions is measured after visualization by staining with Sudan IV. The extent of lesions is measured as percent of total surface area (Figure 3.6) (Tangirala et al. 1995). For detailed protocol, please refer to National Mouse Metabolic Phenotyping Center: www.mmpc.org.

The extent of atherosclerosis can also be determined in cross-section tissue slides stained in Oil Red O. Under a microscope with a video camera, the edge of the lesion is traced with the aid of an edge-detection feature of the program. The mean area of lipid staining per section is determined (Tangirala et al. 1995). A suitable location for scoring lesion area is in the ascending aorta 400 μm above the coronary ostia. Lesions in the aortic sinus can also be included. The results were reproducible in various strains of mice (Paigen et al. 1987, Tangirala et al. 1995). There has been an attempt to quantify lesions in carotid and coronary arteries, which are most clinically relevant. However, given the size and difficulty of manipulation, no systematic quantitative method has been validated, although there have been descriptions of lesion measurements in randomly selected coronary vessels.

3.2.3 Commonly Used Mouse Models of Atherosclerosis

An ideal mouse model of atherosclerosis should mimic the pathophysiological process of the disease in humans and develop lesions comparable to those found in humans. However, mice have different plasma lipid profile and gene expression compared to human. For example, unlike human, mice have a high level of high-density lipoprotein (HDL), and low LDL and VLDL, which is antiatherosclerotic. Mice express both apoB100 and its truncated form apoB48, while humans only produce apoB100 in the liver. On the other hand, mice do not express cholesterol ester transfer protein (CETP), which plays an important role in mediating cholesterol ester transfer between VLDL and HDL in humans. Hence, wild-type mice are usually highly resistant to developing atherosclerosis. To overcome this problem and make mouse lipid profile more human-like, mice have been fed unphysiological diet (Paigen diet or Western diet), which contains 10–20 times the amount of cholesterol of a human diet and an unnatural constituent, cholate, which induces inflammation. Although certain strains of mice like C57BL/6 show increased plasma cholesterol level and develop foam cell lesion after such a type of diet feeding for 4–5 months, others do not. Thus, application of diet-induced plaque in the wild-type mice is limited for atherosclerosis study.

Genetically manipulated mice with alteration of plasma lipid profiles are then the main models for the study of atherosclerosis. These mouse models generate a range of lesions from simple macrophage foam cell formation to more complex lesions with lipid cores, fibrous cap, and calcifications, which are ideal for early scientific imaging study. Depending on different risk factors and specific experimental interventions, different mouse models may respond differently in developing atherosclerosis. Hence, it is essential to understand the lipid profiles and more importantly the lesion characteristics of different mouse models when designing a specific study. Below are some of the most commonly used mouse models of atherosclerosis.

3.2.3.1 Apolipoprotein E Knock-Out Mouse (ApoE$^{-/-}$)

Apolipoprotein E (apoE) is synthesized in the liver and macrophages and has a profound effect on lipoprotein metabolism. It is associated with chylomicrons, VLDL, and HDL particles. It serves as a ligand for several receptors for the clearance of these lipoproteins, including LDL receptors and LDL-related receptor (Krieger and Herz 1994). In humans, there are three isoforms of apoE, designated apoE2, 3, and 4, with the apoE3 being the most common form. The mutation of apoE is related to increased plasma lipid and lipoprotein levels (de Knijff et al. 1994).

ApoE knock-out (ApoE$^{-/-}$) mouse is the most extensively studied model for atherosclerosis. It shows significant hyperlipidemia most confined to VLDL. ApoE$^{-/-}$ mice develops atherosclerotic lesions spontaneously on chow diet, which is its most obvious phenotype. With chow diet feeding, monocyte adhesion to intact endothelium develops in 2-month-old ApoE$^{-/-}$ mice. Foam cell predominant lesion is seen at 10–30 weeks. Intermediate and fibro-lipid lesions become more prominent from 15 weeks (Nakashima et al. 1994). Feeding ApoE$^{-/-}$ mice with high-fat Western diet would accelerate the progression of these lesions and increase the size by threefold, indicating that atherosclerosis in ApoE$^{-/-}$ mice is diet-responsible (Plump et al. 1992). In addition to foam cell formation, the lesions developed in ApoE$^{-/-}$ mice also show fibroblastic cell and collagen deposition, indicating potential progression of plaque to more complicated human-like lesion. Complex lesions could be seen in many vascular areas of ApoE$^{-/-}$ mice on chow and Western diets. These include the aortic sinus, several regions of aorta-like arch and renal arteries, innominate and even coronary arteries (Rosenfeld et al. 2000). Because atherosclerotic lesion developed in ApoE$^{-/-}$ is rapid, extensive, diet-inducible, and similar to humans, ApoE$^{-/-}$ mouse has been widely used.

The shortcomings of ApoE$^{-/-}$ mice model are that first the major plasma lipoprotein is VLDL rather than LDL, which is different from humans. Second, although the earlier lesions developed in the ApoE$^{-/-}$ mice are similar to those in humans, the advanced lesions are resistant to rupture. Third, in addition to lipoprotein metabolism, apolipoprotein E also shows other antiatherogenic properties, for example, antioxidation, antiproliferation (smooth muscle cell, lymphocytes), anti-inflammation, antiplatelet, immune modulation, cholesterol efflux from macrophages, and others. Thus, studies of any interventions related to the above processes are limited in this model. Fourth, dependent on background strain, ApoE$^{-/-}$ mice show dramatically different degree of plasma cholesterol levels and lesions. For example, ApoE$^{-/-}$ mice in the FVB/NJ strain show higher plasma cholesterol than in the C57BL/6 background. However, lesion in the aortic root is about eight times greater in the C57BL/6 strain compared to the FVB/NJ (Dansky et al. 1999). Thus, selection and use of ApoE$^{-/-}$ mice should be based on specific study design and goal.

3.2.3.2 LDLR Knock-Out Mouse (LDLR$^{-/-}$)

LDL receptor (LDLR) is mainly expressed in the liver. It mediates LDL clearance from the plasma and metabolism in the liver. Mutations of LDLR in humans lead to familiar hypercholesterolemia, mainly from very high LDL. Patients with LDLR mutations have premature cardiovascular disease and succumb to myocardial infarction during the second decade of life. Thus, LDLR knock-out (LDLR$^{-/-}$) mice have been generated for atherosclerosis study. Different from humans, LDLR$^{-/-}$ mice show only a modest hypercholesterolemia when fed a chow diet, and develop atherosclerosis slowly (Ishibashi et al. 1993). On the other hand, when fed Western diet, LDLR$^{-/-}$ mice show highly increased cholesterol level and develop lesions rapidly throughout the aorta, aorta root, and even coronary arteries (Piedrahita et al. 1992). The increased plasma lipoprotein in LDLR$^{-/-}$ mice is mainly confined to LDL, which resembles that of humans. Lesions in LDLR$^{-/-}$ mice fed Western diet develop in a time-dependent manner, starting from the proximal aorta. The lesion predominantly consists of foam cells, with features of advanced lesions like necrotic cores and calcifications after long-term Western diet feeding (Roselaar et al. 1996). Compared to ApoE$^{-/-}$ mice, LDLR$^{-/-}$ mice represent a moderate model of atherosclerosis, mainly because of its modest hyperlipidemia. However, the portion of increased cholesterol is mainly from LDL in LDLR$^{-/-}$ mice, which is similar to humans.

3.2.3.3 Apolipoprotein B Transgenic Mouse (ApoB$^{+/+}$)

Apolipoprotein B (apoB) is the major protein in both VLDL and LDL particles. Increased apoB is associated with increased risk of cardiovascular disease. Thus, transgenic mice with overexpression of human apoB100 (ApoB$^{+/+}$) are generated for atherosclerosis study. Unexpectedly, ApoB$^{+/+}$ mice show plasma lipid concentrations similar to those of normolipidemic humans, without pronounced atherosclerotic lesions developed (Linton et al. 1993). However, when feeding Western diet, the ApoB$^{+/+}$ transgenic mice do develop atherosclerotic lesions that consist mainly of macrophage foam cells (Purcell-Huynh et al. 1995). ApoB$^{+/+}$

transgenic mice are usually crossbred with other mice to generate mixed mouse model for research.

3.2.3.4 Mixed Mouse Models

Mice with different strains or genetic backgrounds can be interbred to produce mixed mouse models that develop lesions more rapidly and larger for specific research intervention. For example, ApoE$^{-/-}$ and LDLR$^{-/-}$ mice can be crossbred to produce ApoE$^{-/-}$ × LDLR$^{-/-}$ double knock-out mice. Similarly, ApoE$^{-/-}$ × ApoB$^{+/+}$ transgenic mice and LDLR$^{-/-}$ × ApoB$^{+/+}$ mice can be established. These mixed mouse models show a wider range of lipid profiles and lesions as well as altered process of the disease. For example, LDLR$^{-/-}$ mice combined with human ApoB100 transgenic mice (LDLR$^{-/-}$ × ApoB$^{+/+}$ mice) (Sanan et al. 1998) show increased LDL and develop lesion on a chow diet, which is different from LDLR$^{-/-}$ or LDLR$^{-/-}$ along mice. Selection and breeding of these mixed models should be dependent on specific study design.

3.2.3.5 Other Mouse Models of Atherosclerosis

In addition to lipoprotein genes as mentioned above, other genes related to lipid metabolism have also been manipulated to generate mouse models of atherosclerosis for research. Among those, the well-studied genes are nuclear transcription factors like peroxisome proliferator-activated receptor (PPAR), liver-X-receptor (LXR), and retinoid-X-receptor (RXR). These transcription factors are involved in lipid metabolism and energy homeostasis. Knock-out and transgenic mouse models of these genes have been extensively used in atherosclerosis and diabetes research. The ATP-binding cassette (ABC) transporter gene family (e.g., ABCA1 and ABCG1) mediates cholesterol and phospholipid efflux from macrophages to form HDL particles. Mice with gene manipulations of the ABC transporter genes are available for studies. These genetically manipulated mice can also be crossbred with lipoprotein gene-manipulated models for specific study aims.

In summary, there are many mouse models of atherosclerosis with different gene manipulations and lesion properties. Depending on specific study design, a model should be chosen appropriately. It should be advised that currently mouse models of plaque rupture and coagulation system that contribute to lesion progression and acute thrombotic events in humans have not been developed (Boisvert et al. 1995). Also, it is still unclear in terms of accuracy of mouse models in representing the disease process in humans.

3.2.4 Examples of Atherosclerosis Imaging in Mouse Models

There is an increasing number of studies in the literature showing successful visualization of mouse atherosclerotic plaques by micro-SPECT/CT and PET/CT with radiotracers targeting different components of plaque. Monocyte recruitment to the vascular wall and its differentiation into macrophage play an important role in foam cell formation and initiation of plaque. Thus, monocytes and macrophages have been the major targets for atherosclerosis imaging. Monocyte trafficking to atherosclerotic lesions has been noninvasively visualized with micro-SPECT/CT in mouse (Kircher et al. 2008). Using 111In-oxine-labeled monocytes, it has been demonstrated that recruitment of monocyte to plaques can be detected *in vivo* within days in ApoE$^{-/-}$ mice (Figure 3.7). Plaque macrophages have also been imaged by PET, MR, and optical imaging with a trimodality nanoparticle (Figure 3.8) (Nahrendorf et al. 2008).

Adhesion molecules like VCAM-1 in the endothelium mediate monocyte adhesion as well as plaque inflammation and are attractive imaging targets. ^{18}F-labeled VCAM-1-specific binding peptide (^{18}F-4V) has been found to accumulate in atherosclerotic plaques (Figure 3.9) (Nahrendorf et al. 2009). When compared to the wild type, uptake in the aortic root is 312% higher in ApoE$^{-/-}$ mice (Figure 3.9a). Treatment with atorvastatin significantly reduces uptake of ^{18}F-4V (Figure 3.9b). Autoradiography and *en face* Oil Red O staining confirm the accumulation of activity in the atherosclerotic plaques (Figure 3.9c).

The production of MMP in inflamed plaques has been proposed to mediate vascular remodeling and breakdown of extracellular matrix, which render the plaque unstable. Noninvasive assessment of MMP activity would allow the identification of lesions that are potentially susceptible to plaque rupture. Mouse plaque MMP can be detected by its inhibitor matrix metalloproteinase inhibitor (MPI) radiolabeled with 99mTc. The radiotracer is predominantly accumulated in the aortic arch and abdominal aorta of ApoE$^{-/-}$ and LDLR$^{-/-}$ mice (Figure 3.10) (Ohshima et al. 2009). No radioactive MPI uptake can be observed in the control mice. Uptake is significantly more intense in the high-cholesterol-fed mice than in the chow diet-fed mice. MMP activity is thought to directly correlate with plaque vulnerability, and noninvasive assessment of MMP activity is expected to allow the identification of unstable plaques.

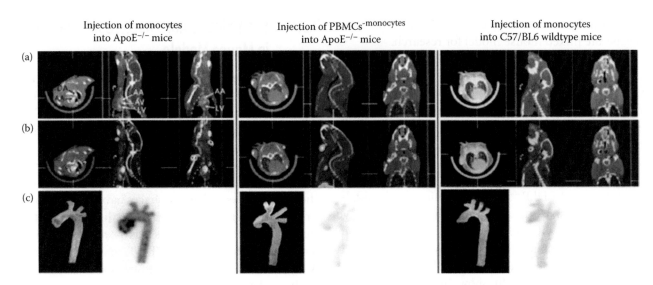

FIGURE 3.7 Micro-SPECT/CT imaging of monocyte recruitment to atherosclerotic plaques in ApoE⁻/⁻ mouse. Monocytes were labeled with 111In-oxine and transferred into ApoE⁻/⁻ mice (left, middle) and C57BL/6 wild-type mice (right). Micro-SPECT/CT imaging was performed 5 days after injection of cells. (a) CT images in axial, sagittal, and coronal views. (b) SPECT/CT overlay images in axial, sagittal, and coronal views. (c) White light images and corresponding autoradiography exposures of aortas excised after *in vivo* imaging. (Reproduced from Kircher MF et al. *Circulation.* 2008; 117(3):388–395. With permission.)

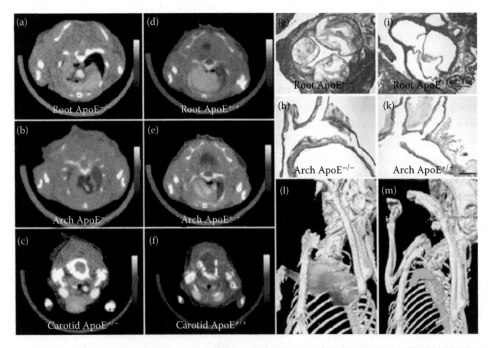

FIGURE 3.8 Multimodality nanoparticle PET/CT imaging of plaque macrophages. 64Cu-TNP (detectable on PET, MRI, and near-infrared fluorescence [NIRF]) facilitates PET-CT imaging of macrophages in atherosclerotic plaque in ApoE⁻/⁻ mice. Fused PET-CT images of the aortic root (a), the arch (b), and carotid artery (c) of aged ApoE⁻/⁻ mice show strong PET signal in these vascular territories with high plaque burden, whereas no activity is observed in the same vasculature of wild-type mice (d–f). (g, h) H&E histology of respective vascular regions, which carry a high plaque burden in ApoE⁻/⁻ but not in wild-type mice (i,k) (magnification 40× for (g) and (i), magnification 20× for (h) and (k), bar equals 0.4 mm). The 3D-maximum intensity reconstruction of the fused dataset (l) demonstrates focal PET signal (red) in the proximal thoracic aorta (blue) of an ApoE⁻/⁻ mouse, but not in wild type (m). (Reproduced from Nahrendorf M et al. *Circulation.* 2008; 117(3):379–387. With permission.)

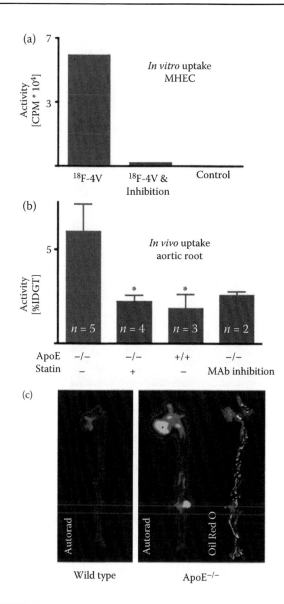

FIGURE 3.9 Visualization of VCAM-1 in atherosclerotic plaque from ApoE$^{-/-}$ mice. (a) Preincubation of the probe with soluble VCAM-1 significantly reduces uptake of ^{18}F-4V into MHEC. CPM: counts per minute. (b) Uptake of ^{18}F-4V in excised aortas by scintillation counting. MAb: preinjection of a monoclonal VCAM-1-targeted antibody, *$P < 0.05$. (c) Exposure of aortas on a phosphorimager corroborates highest uptake of ^{18}F-4V in ApoE$^{-/-}$ mice, with little uptake in wild-type aortas. Oil Red O staining shows peak uptake in plaques. (Reproduced from Nahrendorf M et al. *JACC Cardiovasc Imaging*. 2009; 2(10):1213–1222. With permission.)

Apoptosis is strongly correlated with the vulnerability of plaque. It has been proposed that macrophage apoptosis may promote plaque instability by secreting proinflammatory cytokines. Apoptosis of vascular smooth muscle cells within the fibrous cap may lead to fibrous cap thinning. Annexin A5 imaging of apoptosis has been considered to be a surrogate for

identification of unstable atherosclerotic lesions. *In vivo* micro-SPECT/CT with 99mTc-annexin A5 has allowed noninvasive visualization of apoptotic atherosclerotic lesions. Uptake of the radiotracer has been confirmed in the *ex vivo* imaging of the harvested aorta. The lesions are best visible in the aortic arch of ApoE$^{-/-}$ and LDLR$^{-/-}$ mice. No Annexin A5 uptake can be observed in the control animals (Figure 3.11) (Isobe et al. 2006).

Studies in humans and large animals have shown the potential role of FDG PET in assessing vulnerable atherosclerotic plaque (Chen and Dilsizian 2010). However, FDG micro-PET imaging of plaque in mouse is challenging and results are controversial. One study shows that there is no FDG accumulation of FDG in ApoE$^{-/-}$ mouse aorta, and the uptake seen on imaging is most likely from the interscapular brown fat rather than aorta (Laurberg et al. 2007). However, others have shown that FDG does accumulate in the atherosclerotic plaque (Figure 3.12) (Zhao et al. 2007). The discrepancy could be partially related to the small size of the plaque and different imaging time after FDG injection.

3.3 Ischemic Injury and Left Ventricular Hypertrophy

Large animal models (e.g., primates, swine, and canine) of cardiac disease have been a mainstay of physiological studies since the early part of the twentieth century. These models also have proven very useful in the modern era in a variety of investigations directed at advancing our understanding of molecular processes such as signaling pathways and gene expression, which mediate myocardial and coronary vascular pathophysiological responses to stimuli such as pressure and volume overload as well as myocardial ischemia and infarction. Indeed, insights gained from such studies provide the basis for identifying targets for molecular imaging. The ability to image molecular targets makes possible both acute and chronic noninvasive monitoring of these processes thereby providing the opportunity to greatly advance our understanding of them. Accordingly, this section will focus on the role of large animal models, particularly domestic swine, in molecular imaging of the heart.

3.3.1 Swine Model of Ischemic Heart Disease

The domestic swine is the most commonly employed large animal model of ischemic heart disease (IHD) in use at present. The reasons for this include ready availability, reasonable cost, societal acceptability

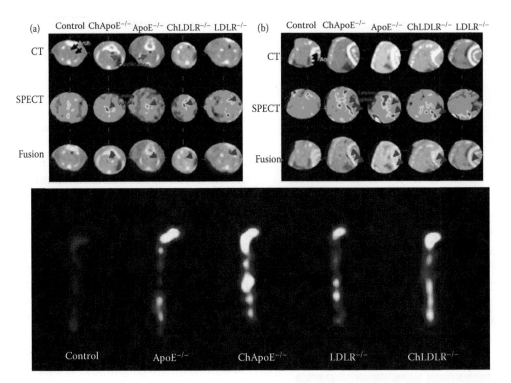

FIGURE 3.10 Visualization of MMP in atherosclerotic plaque from ApoE[−/−] and LDLR[−/−] mice. Top, *in vivo* micro-SPECT/micro-CT transverse (a) and sagittal (b) zoomed and masked images of the five study groups. MPI uptake in atherosclerotic aortic lesions was identified noninvasively in arch and abdominal aorta by micro-SPECT aided by micro-CT; aortic arch calcification was detected by micro-CT. Bottom, MPI uptake in aorta by *ex vivo* imaging in the five study groups. *Ex vivo* imaging reveals background activity in wild-type mice. Uptake in transgenic mice is most prominently seen in aortic arch. (Reproduced from Ohshima S et al. *J Nucl Med.* 2009; 50:612–617. With permission.)

(especially in comparison with the canine model so frequently used in the twentieth century) and most importantly the similarity of the porcine coronary circulation to that of humans (Jonsson et al. 1974). The right coronary artery, for instance, is dominant in the pig and in ~80% of humans. Further, the size of the heart (200–300 g) and LV mass (120–140 g) (Dean et al. 1998) of a 40-kg pig typically employed in laboratory studies is very similar to that of an adult human of average size. Accordingly, for a variety of reasons, the model is ideal for use in physiologically oriented cardiac imaging experiments. Selected important details and more recent developments will be elucidated below.

A very useful model (Gewirtz et al. 1981, Most et al. 1985, Fedele et al. 1989) for acute physiological studies of the coronary circulation was developed in the early 1980s and has a number of features that still make it attractive for translational imaging studies today. A coronary stenosis of known dimension is inserted, using standard cardiac catheterization techniques, in the left anterior descending (LAD, or any other) coronary artery of the pig (Figure 3.13)

(Sun et al. 1983). Accordingly, the animal's chest remains closed, which enhances the stability of the model and greatly facilitates imaging experiments using current modalities such as SPECT, PET, CT, and MR. Further, it is possible to place a catheter retrograde from the aorta into the left atrium for delivery of labeled microspheres to obtain measurements of myocardial blood flow. The stenosis itself has two lumens, one for passage of blood and the other for a catheter (3F), which permits the infusion of drugs and measurements of pressure distal to the stenosis. Metabolic measurements, such as oxygen consumption, lactate production, and venous pH, to name but a few, are also possible by subselective catheterization of the anterior interventricular vein at a level distal to the LAD stenosis (Bier et al. 1991).

The model was further modified to address physiological hypotheses related to myocardial hibernation, stunning, and vascular adaptation distal to a chronic stenosis (Mills et al. 1994). This was done using sterile, open-chest surgical technique to place a constrictor around the LAD of a 40-kg swine. The animal was then kept alive and well for 2–3 months after which it

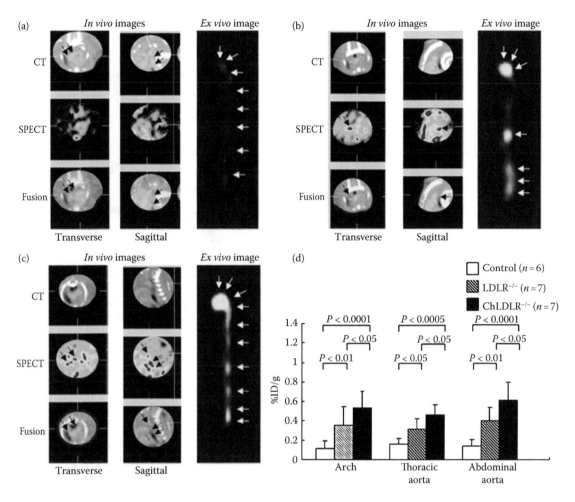

FIGURE 3.11 *In vivo* and *ex vivo* images of macrophage apoptosis in control mice (a) and LDLR$^{-/-}$ mice without cholesterol diet (b) and with cholesterol diet (c). For all images, left panel represents transverse images, middle panel represents sagittal images, and right panel represents *ex vivo* images (a–c). Top panel shows micro-CT, middle panel shows micro-SPECT, and bottom panel shows fusion images. (a) No obvious annexin A5 uptake was seen on either *in vivo* or *ex vivo* images of control animals. (b) Significant uptake was observed in the arch on *in vivo* images and in the arch and abdominal aorta on *ex vivo* image. (c) Distinct uptake was observed in the arch on *in vivo* images and distinct uptake was seen in whole aorta on *ex vivo* image. (d) Quantitative uptake was highest in cholesterol-fed LDLR$^{-/-}$ mice, followed by chow-fed LDLR$^{-/-}$ and control mice in lesions at arch, thoracic, or abdominal level. Ch = cholesterol fed. (Reproduced from Isobe S et al. *J Nucl Med.* 2006; 47:1497–1505. With permission.)

could be brought back to the laboratory for additional catheter-based instrumentation as noted above, followed by detailed physiological and subsequent pathological examination of the excised heart. It should be noted at first that the constrictor was mild and produced little, if any, stenosis. However, as the animal grew, the LAD enlarged and so was progressively narrowed by the band around it (Figure 3.14). The utility of the model not only for physiological studies, but also for PET and other imaging work is readily apparent and has been employed extensively by others (Fallavollita et al. 1997, 2010, Canty and Fallavollita 2000, 2005). Finally, it should be noted that a stent-based, closed-chest catheterization method for the cre-

ation of an intracoronary stenosis has been recently reported for use in studying the coronary collateral circulation of swine (de Groot et al. 2010) (Figure 3.15). It should be possible with current antiplatelet therapy to maintain patency of the stent stenosis and thereby employ the model for molecular imaging studies of hibernation and stunning as well as acute responses to stress-induced ischemia.

The porcine coronary circulation can be made atherosclerotic by combination of selective breeding for familial hypercholesterolemia, dietary manipulation, and balloon injury to the artery (Thim et al. 2010). Such experiments often employ mini-swine in order to overcome the important issue of excess size of the

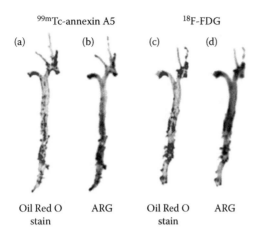

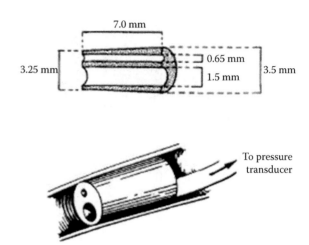

99mTc-annexin A5   18F-FDG

(a)   (b)   (c)   (d)

Oil Red O   ARG   Oil Red O   ARG
stain   stain

FIGURE 3.12 Autoradiograms (ARG) and corresponding Oil Red O staining in aortas of ApoE$^{-/-}$ mice injected with 99mTc-annexin A5 (a, b) or ^{18}F-FDG (c, d). The regions stained red with Oil Red O reveal the presence of atherosclerotic lesions. Corresponding ARG shows tracer accumulation in the lesions matching the Oil Red O staining. (Reproduced from Zhao Y et al. *Eur J Nucl Med Mol Imaging.* 2007; 34(11):1747–55. With permission.)

FIGURE 3.13 A schematic of an artificial (Delrin) stenosis used for insertion by means of cardiac catheterization technique in the coronary artery of domestic swine. (Adapted from Sun Y et al. *Cardiovasc Res.* 1983; 17:499–504. With permission.)

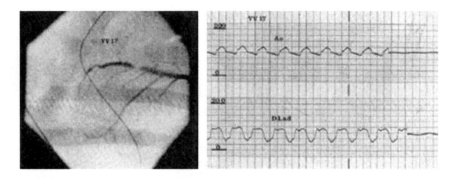

FIGURE 3.14 A coronary angiogram and associated aortic (left) and distal left anterior descending (LAD) coronary pressure tracer (right) from a pig prepared by open-chest surgical technique with a constrictor around the LAD ~2 months prior to the angiogram. (Reproduced from Mills I et al. *Am J Physiol (Heart Circ Physiol 35).* 1994; 266:H447–H457. With permission.)

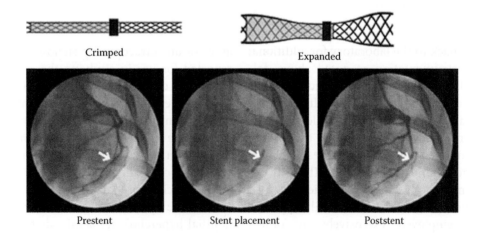

Crimped   Expanded

Prestent   Stent placement   Poststent

FIGURE 3.15 Stenosis model based on stent technology created by de Groot et al. (2010). Stent placement in the left circumflex coronary artery is shown with creation of moderate stenosis (white arrow) poststent. (Reproduced from de Groot D et al. *Am J Physiol Heart Circ Physiol.* 2011; 300(1):H408–H414. With permission.)

animal as it grows during what are often lengthy (months) dietary or drug interventions. The atherosclerotic lesions produced closely mimic those found in human disease, including large necrotic core and thin, inflamed fibrous cap (Virmani et al. 2006). Since balloon injury to the endothelium of the artery is often necessary to create advanced lesions, it is possible to control the location and extent of coronary disease and thereby design more complex and sophisticated experiments. The use of this type of model is very helpful in molecular imaging studies related to pathogenesis of the vulnerable plaque (Johnson et al. 2005) as well as those that test hypotheses regarding response, as noted above, to drug, dietary, or other therapy (Barbeau et al. 1997).

Recently, it has also been possible to genetically engineer swine such that a gene that codes for a protein (alpha 1,3-galactosyltransferase) (Phelps et al. 2003) essential to hyperacute rejection following xenotransplantation of donor heart or other organs is deleted. The cost and effort involved in producing such animals are not trivial, nor are the experiments for which they are intended (e.g., cardiac xenotransplantation to baboons (Kuwaki et al. 2005, Tseng et al. 2005)). Accordingly, the utility of PET is readily apparent for making serial, quantitative measurements of a wide variety of classical physiological parameters of myocardial blood flow, function, and metabolism as well as imaging new molecular probes pertinent to vital cardiac transplantation issues of acute and chronic rejection.

3.3.2 Myocardial Ischemia, Stunning, Hibernation, and Infarction

Since much imaging work in this area currently centers on assessment of myocardial viability, this and closely related issues will be the focus of study. By way of background, the concept that resting myocardial blood flow could be reduced below what would be regarded as normal levels in patients with chronic coronary artery disease but without evidence of acute coronary syndrome was reported as early as the mid-1970s (Cannon et al. 1972, Klocke et al. 1974). Subsequently, a similar observation was made based on the analysis of rest redistribution thallium scans, which demonstrated that such zones were characterized by viable myocardium (Gewirtz et al. 1979). Interest in the concept intensified in the mid-1980s and 1990s, at both the clinical (Rahimtoola 1989, 1994) and experimental (Braunwald and Kloner 1982,

Kloner et al. 1998) level. An animal model with chronic stenosis was reported in the early 1990s and demonstrated evidence of vascular remodeling with reduced protein synthesis and levels of tropoelastin in epicardial vessels distal to the stenosis (1.5 mm diameter) in comparison with that of normal vessels (Mills et al. 1994). While larger vessels tended to atrophy in the stenosis zone, smaller microvessels showed evidence of hypertrophy of the vessel walls. Although PET imaging of the coronary vasculature remains challenging, particularly at the microvascular level, recent developments in nanotechnology (Narula and Dilsizian 2008) coupled with knowledge of specific molecular targets (e.g., tropoelastin) may make this possible in the future.

Considerable work has been done with the above-noted swine model in terms of elucidating not only basic mechanisms active in myocardial ischemia, stunning, and hibernation (Canty and Fallavollita 2005, Heusch 1998, Heusch et al. 2005), but also identifying important molecular targets suitable or already in use for PET imaging (Fallavollita et al. 1999, Canty and Fallavollita 2000, Schulz et al. 2001, Naber et al. 2003, Thijssen et al. 2004, Iyer and Canty 2005, Luisi et al. 2005, Suzuki et al. 2005). Thus, it has been shown that repetitive episodes of myocardial stunning (preserved resting myocardial blood flow (MBF) with impaired contractile function) commonly precede myocardial hibernation (matched reduction of resting blood flow and myocardial contractile function) (Fallavollita et al. 1999). A human study has shown that both states may coexist in different myocardial regions of the same patient (Tawakol et al. 2000). In a porcine model of short-term hibernation, myocardial adaptation both to reduced resting blood flow (Fedele et al. 1988) and superimposed increased myocardial oxygen demand (Berman et al. 1996) occurs and does not necessarily lead to myocardial infarction. Moreover, it has also been demonstrated that chronic hibernation has a number of characteristic pathological findings (e.g., increased glycogen stores, changes in sarcoplasmic reticulum and mitochondrial proteins) (Ausma et al. 1995a, Borgers et al. 1995, Maes et al. 1997, Ausma 1998, Fallavollita et al. 1999, Page et al. 2008). While some human studies indicate such findings may be temporary, and in the absence of revascularization progress to evidence of infarction over time (Flameng et al. 1981, Depre et al. 1997), data in the porcine model suggest they may remain stable at least for a period of a month or more (Fallavollita et al. 2001). Accordingly, the ability to noninvasively and, in theory, quantitatively image these process under controlled conditions in a swine model whose coronary circulation closely mimics that of humans provides an extraordinary means to enhance our understanding of them.

It has been shown, for instance, that repetitive ischemic episodes may activate myocardial apoptosis and thus increase the expression of biomarkers of this process such as annexin-V and caspases (Narayan et al. 2001, Narula and Dilsizian 2008). Further, heat shock proteins, SERCA, phospholamban (Fallavollita et al. 1999), as well as hypoxia-inducing factor (Penna et al. 2008) also appear to be upregulated in the setting of chronic stunning/hibernation or following ischemia/reperfusion (acute stunning). The growth of coronary collaterals is promoted as well (de Groot et al. 2010) and with it the expression of basic fibroblast growth factor (bFGF) and nitric oxide synthase (NOS). The progression to heart failure is accompanied by left ventricular remodeling, which in turn is associated with a plethora of molecular events, including changes in NOS isoform expression (Zhao et al. 1996), induction of fetal gene programs (Taegtmeyer 2002), alterations in mitochondrial function (Huss and Kelly 2005), and increased production of reactive oxidant species (ROS) (Loscalzo et al. 2003, Huss and Kelly 2005) as well as alterations in carbohydrate and fatty acid metabolism (Nikolaidis et al. 2004a,b, Taegtmeyer 2004). All are associated with alterations in levels or activity of various proteins and enzyme systems, some of which have been the focus of molecular imaging studies in large animal models used to test hypotheses related to both pathophysiological mechanisms and potential therapeutic interventions.

3.3.3 From Pathophysiology to Molecular Biology and Molecular Imaging

Chronic hibernating myocardium in a swine model exhibits evidence of impaired presynaptic sympathetic nerve function, which has been imaged serially with the PET tracer ^{11}C-HED and compared to extent of denervation exhibited by myocardial regions with subendocardial necrosis (Fallavollita 2010). Interestingly, regions of hibernation appear to show greater uptake abnormalities (thought to be indicative of presynaptic dysfunction and partial denervation) in comparison with region of necrosis in which sympathetic nerve denervation is more complete. The failure to increase PET ^{11}C-HED uptake following interventions such as pravastatin and percutaneous

transluminal coronary angioplasty (PTCA), which improve regional contractile function in this same swine model, indicates that sympathetic nerve denervation is a persistent feature of hibernation, unresponsive to therapeutic interventions at least in the short term (1 month) (Fallavollita 2010). Molecular changes associated with partial sympathetic denervation in hibernating myocardium include reduced sympathetic nerve density, upregulation both in protein-43, which is associated with sympathetic nerve budding, and in nerve growth factor, which may also promote nerve growth (Fallavollita 2010). Finally, it has been proposed that partial sympathetic nerve denervation may promote ventricular arrhythmia and predispose to sudden cardiac death, an observation made in swine and which has led to a clinical trial involving PET imaging of rest myocardial blood flow (^{13}N-ammonia), ^{18}F-FDG for viability assessment and ^{11}C-HED for sympathetic nerve function (Fallavollita 2010). The trial will test the hypothesis that more extensive ^{11}C-HED defects in regions of viable, hibernating myocardium are more likely to be associated with sudden cardiac death (SCD) than smaller ones.

Insight into the molecular mechanism involved in the well-known observation of [18]FDG accumulation in excess of myocardial blood flow ("mismatch" pattern (Tillisch et al. 1986)) in chronic hibernating myocardium has also been obtained from the porcine chronic stenosis model (McFalls et al. 2004). Animals were sacrificed 3 months after placement of a constrictor around the LAD coronary artery. Tissue samples were analyzed for GLUT-4 translocation, p38 MAP kinase (MAPK), iNOS activity, and glycogen content in the LAD and a remote reference zone. The authors demonstrated increased membrane-bound GLUT-4 as well as increased glycogen content, iNOS activity, and activation of p38 MAPK in the LAD compared with the remote zone. They observed that the degree of p38 MAPK activation correlated reasonably well with the extent of GLUT-4 translocation ($r = 0.81$) as well as glycogen content ($r = 0.70$) and NOS activity ($r = 0.68$) and so concluded that p38 MAPK activation plays an important role in the metabolic adaptations seen in chronic hibernating myocardium (McFalls et al. 2004). The data are particularly noteworthy, although obtained in the fasting state, in that they help explain why glucose uptake is enhanced in hibernating myocardium in the absence of metabolic evidence of ongoing ischemia. The absence of ischemia has been shown in both chronic (Mills et al.

1994) and subacute (Fedele et al. 1988) myocardial hibernation.

Molecular imaging of coronary atherosclerosis, especially identification of the vulnerable plaque, remains difficult for radionuclide tracers, whether SPECT or PET, due to the small size of the lesions involved and potentially high background activity when ^{18}F-FDG is employed (Rudd et al. 2010). Nonetheless, progress is being made with the use of both a swine model for coronary imaging (Johnson et al. 2005) and a murine model for myocardial infarct and aortic atherosclerosis imaging (Nahrendorf et al. 2009). In a proof-of-principle study (Johnson et al. 2005), domestic swine were fed an atherosclerotic diet ×50 days and were subjected to arterial injury (balloon catheter). At the end of that time the animals had coronary angiography and SPECT imaging with ^{99m}Tc-labeled annexin-V after which they were sacrificed and postmortem analysis (immunohistochemistry, well counting, and autoradiography) of the coronary vessels performed. The authors were able to demonstrate focal tracer uptake in 13/22 injured vessels, which exceeded that of control uninjured vessels by roughly twofold (percent injected dose determined by *in vitro* well counting). Injured areas also stained positive for caspase, a marker of apoptosis, which was localized to smooth muscle cells. The authors also noted evidence of focal myocardial ^{99m}Tc-annexin-V uptake in regions corresponding to myocardial infarction as identified by triphenyltetrazolium chloride (TTC) staining. While the confirmation of precise *in vivo* coronary localization could not be obtained due to technical limitations (colocalization with CT was unavailable), the authors nonetheless provided evidence that coronary plaque smooth muscle apoptosis could be imaged *in vivo* with SPECT technology and might provide a marker of plaque vulnerability.

An alternative approach, suitable to PET imaging, for identification of the inflamed and hence potentially vulnerable atherosclerotic plaque has been recently reported (Nahrendorf et al. 2009). The authors developed an ^{18}F-labeled VCAM-1-specific peptide, ^{18}F-4V, and used a murine model to assess its ability to detect inflamed atherosclerotic plaque. VCAM-1 is a cellular adhesion molecule expressed on the surface of such plaques (Liyama et al. 1999). In brief, although the coronary arteries are too small to image in the mouse, the authors were able to demonstrate robust uptake of the tracer in aortic atherosclerotic plaque as well as in areas of myocardial infarction and in cardiac allografts, inflamed as a result of rejection. Tracer uptake was inhibited by 97% with soluble VCAM-1 and was reduced substantially in atherosclerotic mice treated with atorvastatin. Accordingly, the study indicates that the tracer is a potentially useful one for vulnerable plaque detection, especially given very low background myocardial activity. It is noteworthy that inflamed myocytes also accumulate the tracer and thus concomitant CT colocalization methods are likely to be required to distinguish coronary from myocardial uptake. Additional PET/CT studies in a porcine model of coronary atherosclerosis would be especially helpful in further development of the tracer.

Another important arena for PET molecular imaging in conjunction with large animal models, particularly swine, lies at the crossroads of a number of intersecting fields, namely heart failure (HF), IHD, LVH, and stem cell therapy. Myocardial metabolic responses to LVH and the transition to overt HF have been extensively reviewed (Taegtmeyer 2002, 2004, Taegtmeyer and Ballal 2006, Gropler et al. 2010) as have changes in myocardial energetics (Jameel and Zhang 2009). In brief, changes from beta oxidation of free fatty acids to more preferential use of glucose as substrate for ATP production occurs with LVH and is worsened with the onset of heart failure such that even glucose oxidation is impaired (Taegtmeyer et al. 2002, Taegtmeyer 2004, Taegtmeyer and Ballal 2006). Myocardial energetics suffer as a result and the heart is said to "starve in the midst of plenty" (Taegtmeyer and Ballal 2006). The molecular basis for many of these changes is beginning to be unraveled (Taegtmeyer 2002) and with it the opportunity to identify molecular targets for PET imaging to monitor these events.

A closely related area of intense interest concerns efforts to use stems cells to treat heart failure resulting from extensive myocardial infarction. Determining the fate of these cells and the extent to which they actually engraft to the diseased heart (and elsewhere in the body) is a key experimental problem (Rosenzweig et al. 2006, Wu et al. 2008). Recently, PET imaging studies in swine have been reported that address this issue using novel molecular markers (Bengel et al. 2003, Gyongyosi et al. 2008, Rodriguez-Porcel et al. 2008). In brief, the system works as follows. The stem cells are modified by insertion of a reporter gene thereby labeling not only the infused cells but their progeny as well. To permit *in vivo*, noninvasive tracking of the cells, an ^{18}F- or ^{124}I-labeled derivative of either guanine (^{18}F-FHBG) ((Bengel et al. 2003, Gyongyosi et al. 2008, Rodriguez-Porcel et al. 2008) or uracil (^{124}I-FIAU)) has been employed for PET imaging. These studies also use ^{13}N-ammonia both to assess myocardial blood flow and to assist with the localization of the "hot spot" PET marker, which is incorporated selectively in cells expressing the reporter gene (i.e., transplanted stem cells). Additional colocalization imaging is required with either CT (Rodriguez-Porcel et al. 2008) or MRI (Gyongyosi et al. 2008). It should be noted that PET imaging is sensitive to picomolar quantities of the tracer (Rodriguez-Porcel et al. 2008). Further, since tracer-specific activity is known along with a quantitative measurement of its concentration (nCi/mL) in a region

of interest, in theory it is possible to measure the actual number of engrafted cells in a given region and, if serial images are obtained (Gyongyosi et al. 2008), how that number changes with time. Thus, not only is PET imaging capable of localizing and quantifying engrafted stem cells but with serial imaging it can track changes in the population over time. Moreover, using well-established tracers (e.g., ^{13}N-ammonia, ^{11}C-acetate, ^{11}C-palmitate, and ^{18}F-FDG) and kinetic models, it is possible to comprehensively and quantitatively assess the physiological and metabolic effects of such therapy.

3.4 Interstitial Fibrosis and Left Ventricular Remodeling

Congestive heart failure represents late stages of myocardial structural damage and is associated with LV remodeling. At the cellular level, the cardiac myocytes in the remodeled ventricles exhibit decreased expression of α-myosin heavy-chain gene, increased expression of β-myosin heavy chain (Lowes et al. 1997, Nakao et al. 1997), progressive loss of contractile proteins (Schaper et al. 1991) and excitation contraction coupling (Beuckelmann et al. 1992), as well as decreased responsiveness to β-adrenergic stimulation (Bristow et al. 1982). The unfavorable structural changes in cardiac myocytes are compounded by continued muscle cell loss through both necrotic (Mann et al. 1992) and apoptotic pathways (Narula et al. 1996, Olivetti et al. 1997), as well as changes in the composition and volume of the extracellular matrix (Weber 1997) (Figure 3.16). The latter is manifested as interstitial, perivascular, and replacement fibrosis and may contribute to impaired diastolic function, reduced coronary flow reserve, and progressive contractile dysfunction (Figure 3.17) (Weber et al. 1993, 1994, Shirani et al. 2006). Myocardial fibrosis in chronic heart failure is a dynamic process that is determined by a balance between collagen synthesis and its degradation by MMP (Thomas et al. 1998). The activity of MMP is yet regulated by another group of glycoproteins called tissue inhibitors of MMP that bind to and inactivate these enzymes (Li et al. 1998). The dynamic nature of the structural changes in the ventricular remodeling is emphasized by their reversibility in response to appropriate therapy such as coronary revascularization in patients with chronic, antiangiotensin therapy, or β-blockers (Tsutsui et al. 1994, Lowes et al. 2002).

3.4.1 Renin–Angiotensin System and Interstitial Fibrosis

Clinical and experimental studies indicate that the renin–angiotensin system (RAS) and its primary effector peptide, angiotensin II (Ang II), are linked to the pathophysiology of interstitial fibrosis, cardiac remodeling, and heart failure (Cohn et al. 2000). In animal models, increased expression of angiotensin-converting enzyme (ACE) has been associated with cardiac fibrosis (Harada et al. 1999, Dilsizian et al. 2007). In patients with heart failure, inhibition of ACE with ACE inhibitors has proven to have a favorable effect on LV remodeling and patient outcome (SOLVD 1991, 1992). Thus, the development of new radiotracers that target the RAS is an important first step toward image-guided therapy in heart failure patients.

In addition to the traditional systemic or circulating RAS, attention has recently shifted to the tissue RAS, in which Ang II is formed locally within the myocardium, where it can bind to adjacent cells, termed *paracrine signaling*, or to receptors on the same cell that produces an extracellular signal, termed *autocrine signaling* (Figure 3.18) (Paul et al. 2006). In experimental models of systemic hypertension, LVH regression after RAS blockade has correlated with cardiac AII levels but not with systemic blood pressure (Dilsizian et al. 2007). Other evidence for the presence of a tissue RAS comes from the study of transgenic mice with targeted deletion or overexpression of various components of the system (SOLVD 1991). Examples of the latter include transgenic mice overexpressing renin or angiotensinogen that develop LVH independent of systemic hypertension (SOLVD 1992). The tissue RAS also appears to be activated in failing

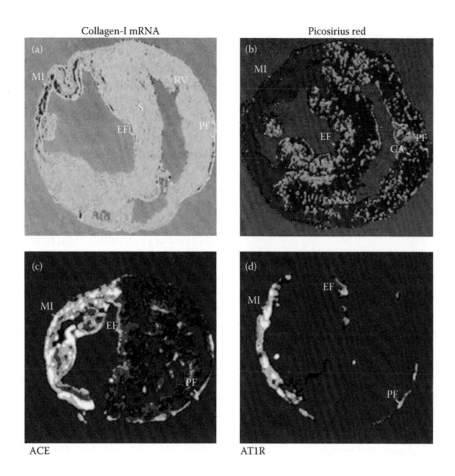

FIGURE 3.16 Tissue repair in infarcted rat heart induced by permanent left coronary artery ligation. Data are presented for week 4 after myocardial infarction. (a) *In situ* hybridization for type I collagen mRNA expression. Increased expression of type I collagen mRNA (yellow and red) is seen at the site of transmural left ventricular MI. Same is true at remote sites that include interventricular septum (S) and right ventricle (RV). This involves endocardial fibrosis (EF) of S, a perivascular fibrosis and microscopic scarring of S and RV, and fibrosis of visceral pericardium (PF). (b) Picosirius red collagen-specific staining demonstrates fibrillar collagen accumulation at site of MI, EF, perivascular fibrosis of intramural vessels (CA), and PF. (c) Autoradiographic detection of ACE-binding density. High-density ACE binding (white, red, and yellow) is seen at site of MI, EF, and PF. (d) Autoradiographic detection of angiotensin II receptor binding. High-density angiotensin II receptor binding is anatomically coincident with high-density ACE binding and sites of type I collagen mRNA expression and fibrous tissue formation. (Adapted from Weber KT. *Circulation.* 1997; 96:4065–4082. With permission.)

human hearts (Zisman et al. 1995, 1998). ACE and ACE messenger RNA are readily detectable in the heart by autoradiography and all the components of the RAS have been found in cardiomyocytes as well as in endothelial cells and vascular smooth muscle cells within the heart (Dostal and Baker 1998, Fleming et al. 2006). In areas of healing myocardial infarction, increased Ang II levels, upregulation of plasma membrane (angiotensin II type 1A [AT1]) receptors, and increased collagen mRNA in myofibroblasts have been described (Weber and Sun 2000, Sun et al. 2004). Noninvasive molecular imaging of neurohumoral upregulation could provide valuable insight into the role of cardiac tissue RAS in pathological LV remodeling as well as monitoring response to heart failure therapy. The two targets within the RAS that have

received the most attention are the ACE and the AT1 receptor.

3.4.2 Animal Models Targeting ACE Binding

To date, there are five radiolabeled ACE inhibitors: an iodotyrosyl derivative of the ACE inhibitor lisinopril (125I-351A), ^{18}F-captopril, ^{18}F-fluorobenzoyl-lisinopril, ^{11}C-zofenoprilat, and 99mTc-lisinopril (Fyhrquist et al. 1984, Lee et al. 2001, Matarrese et al. 2004, Shirani et al. 2005, Dilsizian et al. 2007, Femia et al. 2008). All these radiotracers bind specifically to the active site of ACE, and the binding density correlates closely with literature values for enzyme activity. Autoradiography studies with 125I-351A in rats

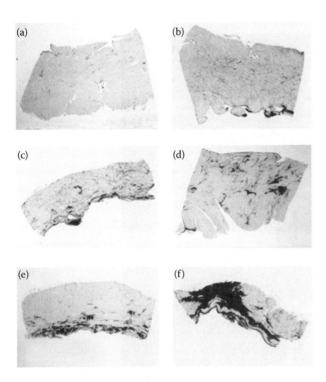

FIGURE 3.17 Significant variations exist in collagen content and distribution of transmural left ventricular sections in ischemic cardiomyopathy. Photomicrographs of left ventricular myocardium stained with Picrosirius red demonstrating normal (a), patchy, scattered areas of replacement fibrosis (b–d), nontransmural infarct (e), and transmural infarct (f) from patients who had stable chronic ischemic heart disease and severe left ventricular dysfunction who underwent orthotopic cardiac transplantation. (Reprinted from *Noninvasive Imaging of Heart Failure, Heart Failure Clinics*, Shirani J et al. Novel imaging strategies for predicting remodeling and evolution of heart failure: Targeting the renin-angiotensin system, pp. 231–247, Copyright (2006), with permission from Elsevier.)

indicated that higher levels of ACE were present in the atria compared with the ventricles and higher levels in the right atrium compared with the left atrium (Johnston et al. 1989). The patterns of ACE labeling on autoradiography revealed that ACE indeed localized in the myocardium and with only low levels in the endocardium. In rats with LVH due to chronic experimental aortic stenosis, autoradiography studies with 125I-351A confirmed the findings of ACE binding in aorta, coronary arteries, atria, and ventricles (Schunkert et al. 1993). Although the distribution of ACE was similar in hypertrophied and normal control rats, quantitative analyses demonstrated that ACE density (counts per minute per cross-sectional area of tissue) was twofold higher within the myocardium of hypertrophied left ventricles compared with controls ($P < 0.005$). Interestingly, the induction of cardiac

ACE was diffusely distributed in the hypertrophied myocardium. Similar experiments were performed with 125I-351A autoradiography in myocardial infarct (MI) rat model using coronary artery ligation) (Kohzuki et al. 1996). At 1 and 8 months after MI, cardiac ACE activity was markedly increased in the infarcted myocardial region and moderately increased in hypertrophied (remodeled) myocardium among rats undergoing coronary artery ligation but not in control hearts.

Recently, lisinopril was successfully labeled with 99mTc with near quantitative great blocking results in rats (Femia et al. 2008). 99mTc was used because of its excellent imaging characteristics and widespread availability (Jurisson and Lydon 1999), as well as the suitability of the previously described technetium-tricarbonyl (Tc[CO]3) core (Alberto 1998) to form robust complexes with recently described single amino acid chelate technology based on the di(pyridylmethyl)amine chelator (Wei et al. 2005). When conjugated to lisinopril by acylation of the e-amine of the lysine residue with a series of di(2-pyridylmethylamino)alkanoic acids where the distance of the chelator from the lisinopril core was investigated by varying the number of methylene spacer groups (*n*) it produced di(2-pyridylmethyl) amine(C_x)lisinopril analogs: $D(C_4)$lisinopril, $D(C_5)$ lisinopril, and $D(C_8)$lisinopril. The inhibitory activity of each rhenium complex was evaluated *in vitro* against purified rabbit lung ACE and was shown to vary directly with the length of the methylene spacer (Femia et al. 2008). Tissue distribution studies of 99mTc-(CO)3D(C_8)lisinopril were performed in male Sprague–Dawley rats after administering 1.85 MBq/kg (50 mCi/kg) bolus injection (~0.37 MBq [10 mCi]/rat) in a constant volume of 0.1 mL via the tail vein. To demonstrate specificity *in vivo*, some rats were injected intravenously with nonradiolabeled lisinopril, 0.6 mg/kg (~272 nmol/rat), 5 min before injection of 99mTc-(CO)3D(C_8)lisinopril. The animals were euthanized by asphyxiation with carbon dioxide at 10 min, 30 min, 1 h, or 2 h after injection. Blood, heart, lungs, liver, spleen, kidneys, large and small intestines (with contents), testes, skeletal muscle, and adipose were excised, weighed wet, transferred to plastic tubes, and counted in an automated gamma-counter. Aliquots of the injected dose were also measured to convert the counts per minute in each tissue sample to percentage injected dose per organ. Tissue radioactivity levels of 99mTc-(CO)3D(C_8)lisinopril expressed as percentage injected dose per gram (%ID/g) were determined by

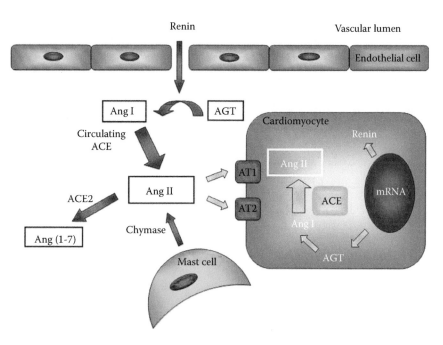

FIGURE 3.18 Schematic representation of the circulating and tissue renin–angiotensin system (RAS) within the heart is shown. Angiotensinogen (AGT) is cleaved by renin to form angiotensin I (Ang I), which is converted by angiotensin-converting enzyme (ACE) to angiotensin II (Ang II). Ang II in turn activates the angiotensin II type 1 (AT1) and type 2 (AT2) receptors. Ang II may also be generated by an alternate pathway, mast-cell-derived chymase. (Adapted from Aras O et al. *Curr Cardiol Rep.* 2007; 9:150–158.)

converting the decay-corrected counts per minute to the percentage dose and dividing by the weight of the tissue or organ sample. Planar anterior imaging was also performed, which consisted of five 1 min consecutive images acquired using a dual-head gamma-camera with a low-energy, all-purpose collimator at 5, 10, 15, 20, 30, and 60 min after injection. Regions of interest (ROIs) were drawn over the lung, liver, intestines, and background (soft tissue) for each animal at each imaging time point. ROIs were quantified and expressed in counts and normalized to the background at that same time point. The most potent compound, 99mTc-(CO)3D(C$_8$)lisinopril, displayed ACE inhibitor activity similar in potency to the starting lisinopril molecule. The tissue distribution studies showed high uptake in organs containing high ACE expression such as the lungs, and pretreatment with lisinopril demonstrated that the uptake was specific.

3.4.3 Explanted Human Heart Model Targeting ACE Binding

In a recent *ex vivo* study of explanted hearts of patients with ischemic cardiomyopathy, the magnitude and distribution of tissue ACE, mast cell chymase, and angiotensin II type 1 plasma membrane receptor (AT1R), in relation to collagen replacement in infarcted and noninfarcted left ventricular myocardial segments, were determined (Dilsizian et al. 2007). ^{18}F-Fluorobenzoyl-lisinopril was synthesized without compromising its affinity for tissue ACE. A 5–10 mm contiguous short-axis slices of explanted hearts from patients with ischemic cardiomyopathy were incubated *in vitro* with ^{18}F-fluorobenzoyl-lisinopril, with and without 10^{-6} M lisinopril. Tissue radioactivity was recorded as a function of position in photostimulating luminescence units. Immunohistochemistry studies were performed with mouse monoclonal antibody against ACE, anti-mast cell chymase, and polyclonal antibody against the human AT1R. The results showed specific binding of ^{18}F-fluorobenzoyl-lisinopril to tissue (myocardial) ACE with the highest activity in regions adjacent to infarcted myocardium (Figure 3.19a–c). Specific ACE binding was about twice as great as the nonspecific binding. Furthermore, the binding of ^{18}F-fluorobenzoyl-lisinopril was nonuniform in infarct, peri-infarct, and remote, noninfarct myocardial segments. ACE binding in peri-infarct segments was about 1.3-fold greater than binding in remote, noninfarct segments. A similar pattern of nonuniform distribution was observed with the AT1 receptor immunoreactivity. There was increased ACE activity and AT1 receptor immunoreactivity in the juxtaposed areas of replacement fibrosis, consistent

Chapter 3

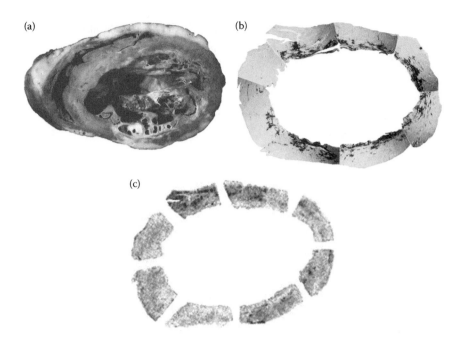

FIGURE 3.19 Presence and distribution of ACE activity in relation to collagen replacement as assessed by picrosirius red stain in human heart tissue removed from cardiac transplant recipient with ischemic cardiomyopathy. Gross pathology of midventricular slice (a), with corresponding contiguous midventricular slices stained with picrosirius red stain (b) and ^{18}F-FBL autoradiographic images (c), is shown. FBL binding to ACE is nonuniform in infarcted, peri-infarcted, and remote, noninfarcted segments. Increased FBL binding can be seen in segments adjacent to collagen replacement. (Reproduced from Dilsizian V et al. *J Nucl Med.* 2007; 48:182–187. With permission.)

with their observed roles in the development of scar and remodeling of the collagen matrix in ischemic cardiomyopathy (Dilsizian et al. 2007).

If reproduced in the heart of patients *in vivo*, the clinical implication of such imaging probes with either PET (for ^{18}F-fluorobenzoyllisinopril) or SPECT (for 99mTc-lisinopril) would be to identify patients with increased myocardial ACE activity, prospectively and in the early stages of heart failure, before the transition to replacement fibrosis and remodeling occurs. Imaging techniques that can identify patients with increased ACE, prospectively, before the transition to replacement fibrosis and remodeling occurs, may result in preserved LV function, thereby improving overall prognosis.

3.4.4 Animal Models Targeting AT1 Receptor Distribution

The role of AT1 receptor in reactive fibrosis and remodeling in noninfarcted myocardium was examined in AT1 receptor knockout and wild-type mice at 1 and 4 weeks after large acute myocardial infarction (Harada et al. 1999). Survival, LV remodeling, cardiac fibrosis, and gene expression were assessed over a 4 week period. At 4 weeks after infarction, control wild-type mice

showed more marked LV remodeling and fibrosis than did AT1 receptor knockout mice. In addition, despite producing similar initial infarct sizes, the cumulative 4 week mortality rate was reduced from 22.7% to 5.9% in AT1 receptor knockout mice, compared with controls. These findings indicate that AT1 receptor plays a pivotal role in the progression of LV remodeling after myocardial infarction. This emerging concept implies a role of local tissue effects of AII on myocytes, LV remodeling, and vascular tone that may exceed those of circulating plasma effects.

The first PET radiotracer for the AT1 receptor was ^{11}C MK-996 (Mathews et al. 1995). This compound, however, was difficult to synthesize and the same group developed its methoxy analog, L-159884 (Hamill et al. 1996). The latter was found to be useful for PET imaging in the canine model (Szabo et al. 2001); however, unpublished human studies indicated rapid metabolism of this compound, thus making it unsuitable for use as a clinical imaging tool (Lee et al. 1999). The same laboratory then examined another class of nonpeptide AT1 receptor-selective antagonists based on the structure of SK-1080 (Lee et al. 1999). An analog of the latter containing an alkyl methoxy group, KR31173, has been developed and tested *ex vivo* after labeling the compound with ^{11}C for PET imaging. It was shown that

[11]C-KR31173 binds selectively to AT1 receptor in various tissues, including the heart, and that the binding is inhibited by other selective AT1 receptor blockers.

Preliminary studies in normal and failing ovine myocardium suggest that upregulation of AT1 receptor occurs predominantly on myofibroblasts while upregulation of ACE is observed in the cardiomyocytes, particularly in the infarct border. Increased release of ACE, and hence AII (a growth factor), from myocytes and myofibroblasts should induce myofibroblastic proliferation in a paracrine and autocrine manner and result in collagen deposition. A myocardial sample obtained from a reliably reproducible ovine heart failure model (Jackson et al. 2002), as compared with another sample from a normal ovine heart, demonstrated the predominance of AT1 receptor upregulation in myofibroblasts in the infarcted zone (Figure 3.20a and b). On the other hand, in the normal sheep myocardium, ACE was confined to the vascular endothelium and AT1 receptor was exclusively localized to vascular smooth muscle cells as expected.

The feasibility of noninvasive imaging of AT1 receptor upregulation in a mouse model of postmyocardial infarction heart failure was recently carried

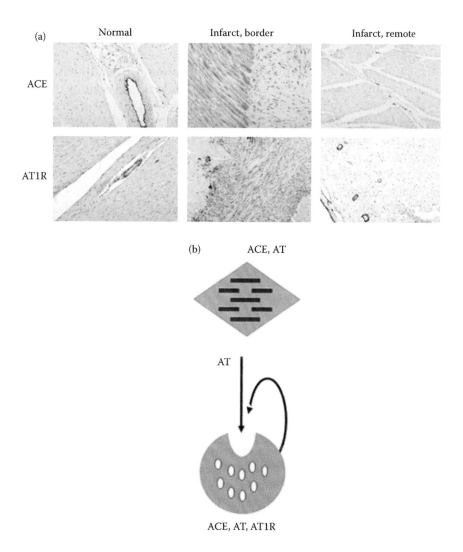

FIGURE 3.20 Distribution of ACE and AT1R in normal and failing ovine myocardium. (a) ACE (*top row*) is localized to the vascular endothelium in the normal heart. The failing myocardium (IB) demonstrates upregulation of ACE in the cardiomyocytes in the border zone and myofibroblasts in the infarct center (IC). On the other hand, AT1R (*bottom row*) is normally observed in the vascular medial layer and is significantly upregulated in myofibroblasts in the infarcted region. AT1R is not observed in the myocytes, at least in the early stages of HF. (b) ACE upregulation in the cardiomyocytes should lead to an increase in AII. Excess release of AII should act as a growth factor for myofibroblasts, which demonstrate an increase in AT1R expression. Myofibroblastic proliferation is accompanied by increased collagen deposition. Targeting of cardiomyocyte ACE and myofibroblast AT1R should offer worthy targets for early imaging of cardiac remodeling (Original magnification ×20). (Reproduced from Shirani J et al. *J Nucl Cardiol.* 2007; 14(1):100–110. With permission.)

Chapter 3

out. Male Swiss Webster mice were anesthetized with pentobarbital and intubated with a 21-G tube for mechanical ventilation (Verjans et al. 2008). The electrocardiogram was monitored continuously, and the body temperature was maintained throughout the procedure at 36.5°C. Ventilation was maintained at a 250 μL tidal volume and a respiratory rate of 20 breaths/min. The heart was exposed through the fourth left intercostal thoracotomy, and the left coronary artery was ligated at its second branching point with a 6-0 Prolene suture. After watching the animal carefully for 30 min, the chest was closed using a 6-0 Monocryl suture. The animals were subjected to optical imaging by real-time fluorescence microscopy of the beating heart or radionuclide imaging at various time points in follow-up after administering a fluoresceinated angiotensin peptide analog or Tc-99m radiolabeled losartan. Before imaging, anesthetized animals were examined by transthoracic echocardiography using a 14 MHz linear probe. B-mode images of LV parasternal long axis and short axis were obtained and digitally analyzed. Tc-99m radiolabeled losartan was used in the mice, 3 weeks after MI and six unmanipulated age-matched control mice, using a dual-head micro-SPECT gamma camera and a micro-CT. Thereafter, each heart was cut into three bread-loaf slices (infarct, peri-infarct, and remote areas). The quantitative radiolabeled losartan uptake was determined with a gamma scintillation counter. After imaging, hearts were harvested for pathological characterization using confocal and two-photon microscopy. The results showed no or little fluoresceinated uptake in control animals or within infarct regions on days 0 and 1 (Verjans et al. 2008). However, distinct uptake occurred in the infarct area at 1–12 weeks after MI; the uptake was maximum at 3 weeks and reduced markedly at 12 weeks after MI. Ultrasonographic examination demonstrated left ventricular remodeling, and pathological characterization revealed localization of the fluoresceinated angiotensin peptide analog with collagen-producing myofibroblasts. Histological, immunohistochemical, and two-photon microscopy confirmed localization of tracer within the myofibroblasts. Nuclear imaging, using a micro-SPECT-CT, showed increased uptake of AT1 receptor ligand in the peri-infarct region as compared with remote, noninfarct regions (Figure 3.21) (Verjans et al. 2008). Several laboratories continue the search for an ideal AT1 receptor-specific radiotracer. To date, no data have been reported for the use of these radiotracers in human.

3.4.5 Transgenic Animal Models for the Study of ACE

Genetically engineered mice or rats carrying gain-of-function or loss-of-function mutations of specific components of RAS offer a unique opportunity to examine potential imaging targets of the cardiac tissue RAS and to address the functional impact of ACE signaling in a defined genetic background. Two lines of ACE-transgenic rats with either a lower or a higher copy number of the ACE transgene have been described to assess gene–dose effects on cardiac architecture (Pokharel et al. 2004, Tian et al. 2004). The cardiac ACE activity was increased 50-fold in L1173 (highest, 5–10 copies of the transgene) and 13-fold in L1172 (2–3 copies of the transgene) rats, compared with controls, within the left ventricle, whereas serum ACE was not increased in these transgenic rats. The results showed that cardiac collagen content increased in parallel with the ACE-gene dose, and that this process could be prevented with ACE inhibition (Pokharel et al. 2004).

More recently, a transgenic rat model with selective overexpression of human ACE in cardiac tissues was generated and its function was assessed under physiological and pathological conditions (Zynda et al. 2009). ACE overexpressing transgenic rats and wild-type Sprague–Dawley control rats were studied and noninvasive images were obtained at 10, 30, 60, and 120 min after 99mTc-(CO)3D(C_8)lisinopril administration using *in vivo* micro-SPECT/CT. A subset of transgenic and control rats also received nonradiolabeled (cold) lisinopril prior to the 99mTc-(CO)3D(C_8)lisinopril administration to determine nonspecific binding. After *in vivo* SPECT/CT imaging, the rat myocardium was explanted, *ex vivo* images were acquired, percent injected dose per gram (%ID/g) gamma-well was counted, followed by an assessment of ELISA-verified ACE activity and mRNA expression. The results showed that 99mTc-(CO)3D(C_8) lisinopril binds specifically to ACE and the activity can be localized in transgenic rat hearts that overexpress human ACE. Whether the localization and intensity of the AT1 receptor and/or ACE signals in the myocardium will be sufficiently high enough to allow external *in vivo* imaging using hybrid SPECT/CT or PET/CT in human subjects is a worthy goal to tackle next.

3.4.6 Imaging Matrix Metalloproteinases

The myocardial extracellular matrix represents a dynamic balance between the synthesis and degradation

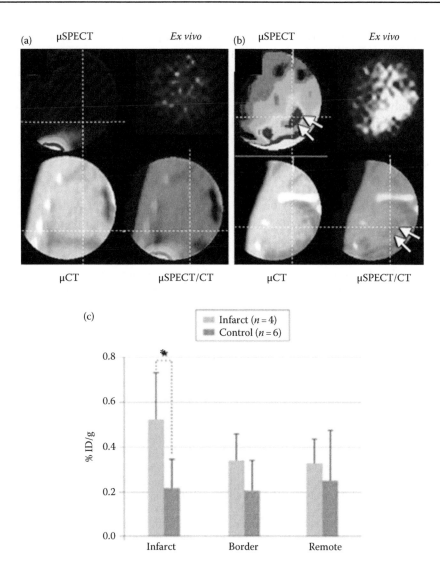

FIGURE 3.21 Noninvasive imaging of AT receptors with radiolabeled Losartan. The micro-SPECT and micro-CT images are shown in a control mouse after technetium Tc-99m losartan administration; no uptake in the heart can be seen (a) in the *in vivo* and *ex vivo* images. There is only some liver uptake on the bottom left of the SPECT image. (b) In the 3-week post-MI animal, significant radiolabeled losartan uptake is observed in the anterolateral wall (arrows). The infarct uptake on the *in vivo* image is confirmed in the *ex vivo* image. The histogram (c) demonstrates significantly (*) higher uptake in the infarcted region (0.524–0.212% ID/g) as compared to control noninfarcted animals (0.215–0.129% ID/g; p _ 0.05). ID, injected dose. (Reproduced from Verjans JWH et al. *J Am Coll Cardiol Img.* 2008; 1:354–362. With permission.)

of collagen in the normal heart. In the setting of LV remodeling, increased myocardial collagen turnover not only allows the repair of damaged tissue, but also permits the muscle fiber slippage and rearrangement that precede chamber enlargement and disfiguration. Myocardial collagen degradation is regulated through the action of MMPs. Both 111indium- and 99mTc-radiolabeled ligands have been synthesized and used to show increased MMP activity in infarcted myocardium by planar and SPECT imaging in a murine model (Su et al. 2005). In this study, MMP activity was also shown in noninfarcted myocardium, thus emphasizing the global nature of LV remodeling.

Recently, a near-infrared fluorescent probe for evaluation of MMP activity was synthesized and evaluated in an experimental model of myocardial infarction (Chen et al. 2005). This "smart" molecular probe was designed to become activated after a peptide sequence was recognized and cleaved by MMP2 and MMP9. After proteolytic cleavage of the peptide sequence by the MMP, the molecule undergoes a configurational change releasing the fluorescent molecules. In a murine model of anterior wall myocardial infarction, the application of this activated near-infrared fluorescent allowed determination of the time course of increased MMP activity following ischemic LV damage. The results were confirmed by gelatinase zymography and quantitative real-time polymerase chain reaction analysis of MMP2 and MMP9 mRNA levels. By confocal microscopy, MMP activity was colocalized with neutrophils within the infarct zone. This exciting method has the potential for further applications to evaluate protease activity in combination with other advanced imaging modalities such as CT or CMR.

3.5 Conclusions

Molecular cardiac imaging and pharmacogenomics are two rapidly growing fields that promise to provide a better understanding of the pathophysiology of myocardial structure in health and disease. Animal models in cardiology have proven to be extremely useful in advancing our understanding of molecular processes such as signaling pathways and gene expression, which mediate vascular as well as myocardial metabolic and neuronal responses to various stimuli. Indeed, insights gained from such studies provide the basis for identifying targets for molecular imaging. Noninvasive interrogation of cellular, molecular, and genetic processes using molecular imaging techniques holds the promise to ultimately allow earlier and more precise disease diagnosis, disease characterization, and assessment of therapeutic response.

References

Acton PD. Animal imaging equipment: Recent advances. *J Nucl Med.* 2006; 47:52N–55N.

Alberto R, Schibli R, Egli A, Schubiger AP, Abram U, Kaden TA. A novel organometallic aqua complex of technetium for the labeling of biomolecules: Synthesis of [99mTc(OH2)3(CO)3]1 from [99mTcO4]2 in aqueous solution and its reaction with a bifunctional ligand. *J Am Chem Soc.* 1998; 120:7987–7988.

Anger HO. Scintillation camera. *Rev Sci Instrum.* 1958; 29:27–33.

Aras O, Messina SA, Shirani J, Eckelman WC, Dilsizian V. The role and regulation of cardiac angiotensin converting enzyme for non-invasive molecular imaging in heart failure. *Curr Cardiol Rep.* 2007; 9:150–158.

Ausma J, Cleutjens J, Thone F, Flameng W, Ramaekers F, Borgers M. Chronic hibernating myocardium: Interstitial changes. *Mol Cell Biochem.* 1995a; 147(1–2):35–42.

Ausma J, Furst D, Thone F, Shivalkar B, Flameng W, Weber K, Ramaekers F, Borgers M. Molecular changes of titin in left ventricular dysfunction as a result of chronic hibernation. *J Mol Cell Cardiol.* 1995b; 27(5):1203–1212.

Ausma J, Thone F, Dispersyn GD, Flameng W, Vanoverschelde JL, Ramaekers FC, Borgers M. Dedifferentiated cardiomyocytes from chronic hibernating myocardium are ischemia-tolerant. *Mol Cell Biochem.* 1998; 186(1–2):159–168.

Barbeau ML, Klemp KF, Guyton JR, Rogers KA. Dietary fish oil. Influence on lesion regression in the porcine model of atherosclerosis. *Arterioscler Thromb Vasc Biol.* 1997; 17(4):688–694.

Beekman FJ, van der Have F, Vastenhouw B, van der Linden AJ, van Rijk PP, Burbach JP, Smidt MP. U-SPECT-I; A novel system for submillimeter-resolution tomography with radiolabelled molecules in mice. *J Nucl Med.* 2005; 46:1194–1200.

Beekman F, van der Have F. The pinhole: Gateway to ultra-high-resolution three-dimensional radionuclide imaging. *Eur J Nucl Med Mol Imaging.* 2007; 34:151–161.

Bengel FM, Anton M, Richter T et al. Noninvasive imaging of transgene expression by use of positron emission tomography in a pig model of myocardial gene transfer. *Circulation.* 2003; 108(17):2127–2133.

Bergmann SK, Fox KA, Rand AL, McElvany KD, Welch MJ, Markham J, Sobel BE. Quantification of regional myocardial blood flow *in vivo* with H$_2$^{15}O. *Circulation.* 1984; 70:724–733.

Berman M, Fischman AJ, Southern J, Carter E, Mirecki F, Strauss HW, Nunn A, Gewirtz H. Myocardial adaptation during and after sustained, demand-induced ischemia. Observations in closed-chest, domestic swine. *Circulation.* 1996; 94(4):755–762.

Beuckelmann DJ, Nabauer M, Erdmann E. Intracellular calcium handling in isolated ventricular myocytes from patients with terminal heart failure. *Circulation.* 1992; 85:1046–1055.

Bier J, Sharaf B, Gewirtz H. Origin of anterior interventricular vein blood in domestic swine. *Am J Physiol. (Heart and Circ Physiol 29).* 1991; 260:H1732–H1736.

Boisvert WA, Spangenberg J, Curtiss LK. Treatment of severe hypercholesterolemia in apolipoprotein E-deficient mice by bone marrow transplantation. *J Clin Invest.* 1995; 96:1118–1124.

Borgers M, Ausma J. Structural aspects of the chronic hibernating myocardium in man. *Basic Res Cardiol.* 1995; 90(1):44–46.

Branderhorst W, Vastenhouw B, van der Have F, Blezer ELA, Bleeker WK, Beekman FJ. Targeted multi-pinhole SPECT. *Eur J Nucl Mol Imaging.* 2011; 38:552–561.

Braunwald E, Kloner RA. The stunned myocardium: Prolonged post ischemic ventricular dysfunction. *Circulation.* 1982; 66:1146–1149.

Breton E, Choquet P, Goetz C, Kintz J, Erbs P, Rooke R, Constantinesco A. Dual SPECT/MR imaging in small animal. *Nucl Instrum Methods Phys Res A.* 2007; 571:446–448.

Bristow MR, Ginsburg R, Minobe W et al. Decreased catecholamine sensitivity and ß-adrenergic-receptor density in failing human hearts. *N Engl J Med.* 1982; 307:205–211.

Büscher K, Judenhofer MS, Kuhlmann MT, Hermann S, Wehrl HF, Schäfers KP, Schäfers M, Pichler BJ, Stegger L. Isochronous assessment of cardiac metabolism and function in mice using hybrid PET/MRI. *J Nucl Med.* 2010; 51:1277–1284.

Cannon PJ, Dell RB, Dwyer EM, Jr. Regional myocardial perfusion rates in patients with coronary artery disease. *J Clin Invest.* 1972; 51:978–988.

Canty JM, Jr., Fallavollita JA. Chronic hibernation and chronic stunning: A continuum. *J Nucl Cardiol.* 2000; 7(5):509–527.

Canty JM, Jr. Nitric oxide and short-term hibernation: Friend or foe? *Circ Res.* 2000; 87(2):85–87.

Canty JM, Jr., Fallavollita JA. Hibernating myocardium. *J Nucl Cardiol.* 2005; 12(1):104–119.

Catana C, Wu Y, Judenhofer MS, Qi J, Pichler BJ, Cherry SR. Simultaneous acquisition of multislice PET and MR images: Initial results with a MR-compatible PET scanner. *J Nucl Med.* 2006; 47:1968–1976.

Chatziioannou AF, Cherry S, Shao Y, Silverman RW, Meadors K, Faquhar TH, Pedarsani M, Phelps ME. Performance evaluation of microPET: A high-resolution lutetium oxyorthosilicate PET scanner for animal imaging. *J Nucl Med.* 1999; 40:1164–1175.

Chen J, Tung C-H, Allport J, Chen S, Weissleder R, Huang PL. Near-infrared fluorescent imaging of matrix metalloproteinase activity after myocardial infarction. *Circulation.* 2005; 111:1800–1805.

Chen W, Dilsizian V. (18)F-fluorodeoxyglucose PET imaging of coronary atherosclerosis and plaque inflammation. *Curr Cardiol Rep.* 2010; 12:179–184.

Cherry SL, Gambhir SS. Use of positron emission tomography in animal research. *ILAR J.* 2001; 42:219–232.

Cohn JN, Ferrari R, Sharpe N. Cardiac remodeling—Concepts and clinical implications: A consensus paper from an international forum on cardiac remodeling. Behalf of an International Forum on Cardiac Remodeling. *J Am Coll Cardiol.* 2000; 35:569–582.

Cook RAH, Carnes G, Lee T-Y, Wells, RG. Respiration-averaged CT for attenuation correction in canine cardiac PET-CT. *J Nucl Med.* 2007; 48:811–818.

Coxson PG, Brennan KM, Huesman RH, Lim S, Budinger TF. Variability and reproducibility of rubidium-82 kinetic parameters in the myocardium of an anesthetized canine. *J Nucl Med.* 1995; 36:287–296.

Croteau E, Bénard F, Cadorette J, Gauthier M-È, Aliaga A, Bentourkia M, Lecomte R. Quantitative gated PET for the assessment of left ventricular function in small animals. *J Nucl Med.* 2003; 44:1655–1661.

Dansky HM, Charlton SA, Sikes JL, Heath SC, Simantov R, Levin LF, Shu P, Moore KJ, Breslow JL, Smith JD. Genetic background determines the extent of atherosclerosis in ApoE-deficient mice. *Arterioscler Thromb Vasc Biol.* 1999; 19:1960–1968.

Da Silva AJ, Tang HR, Wong KH, Wu MC, Dae MW, Hasegawa BH. Absolute quantification of regional myocardial uptake of 99mTc-sestamibi with SPECT: Experimental validation in a porcine model. *J Nucl Med.* 2001; 42:772–779.

Daube-Witherspoon ME, Surti S, Perkins A, Kyba CCM, Wiener R, Werner ME, Kulp R, Karp JS. Imaging performance of a LaBr₃-based PET scanner. *Phys Med Biol.* 2010; 55:45–64.

Dean DA, Amirhamzeh MM, Jia CX, Cabreriza SE, Rabkin DG, Sciacca R, Dickstein ML, Spotnitz HM. Reversal of iatrogenic myocardial edema and related abnormalities of diastolic properties in the pig left ventricle. *J Thorac Cardiovasc Surg.* 1998; 115(5):1209–1214.

de Groot D, Grundmann S, Timmers L, Pasterkamp G, Hoefer IE. Assessment of collateral artery function and growth in a pig model of stepwise coronary occlusion. *Am J Physiol Heart Circ Physiol.* 2010; 300(1):H408–H414.

de Kemp RA, Epstein FH, Catana C, Tsui BMW, Ritman EL. Small-animal molecular imaging methods. *J Nucl Med.* 2010; 51 (suppl 1):18S–32S.

de Knijff P, van den Maagdenberg AM, Frants RR, Havekes LM. Genetic heterogeneity of apolipoprotein E and its influence on plasma lipid and lipoprotein levels. *Hum Mutat.* 1994; 4:178–194.

del Guerra A, Belcari N. State-of-the-art of PET, SPECT and CT for small animal imaging. *Nucl Instrum Methods Phys Res A.* 2007; 583:119–124.

Depre C, Vanoverschelde JL, Gerber B, Borgers M, Melin JA, Dion R. Correlation of functional recovery with myocardial blood flow, glucose uptake, and morphologic features in patients with chronic left ventricular ischemic dysfunction undergoing coronary artery bypass grafting. *J Thorac Cardiovasc Surg.* 1997; 113(2):371–378.

Dilsizian V, Eckelman WC, Loredo ML et al. Evidence for tissue angiotensin-converting-enzyme in explanted hearts of ischemic cardiomyopathy using targeted radiotracer technique. *J Nucl Med.* 2007; 48:182–187.

Dostal DE, Baker KM. The cardiac renin-angiotensin system: Conceptual, or a regulator of cardiac function? *Circ Res.* 1999; 85:643–650.

Fallavollita J, Perry B, Canty JJ. 18-F-2-deoxyglucose deposition and regional flow in pigs with chronically dysfunctional myocardium: Evidence for transmural variations in chronic hibernating myocardium. *Circulation.* 1997; 95:1900–1909.

Fallavollita J, Canty JJ. Differential 18-F-2-deoxyglucose uptake in viable dysfunctional myocardium with normal resting perfusion: Evidence for chronic stunning in pigs. *Circulation.* 1999; 99:2798–2805.

Fallavollita JA, Jacob S, Young RF, Canty JM, Jr. Regional alterations in SR Ca(2+)-ATPase, phospholamban, and HSP-70 expression in chronic hibernating myocardium. *Am J Physiol.* 1999; 277(4 Pt 2):H1418–H1428.

Fallavollita JA, Logue M, Canty JM, Jr. Stability of hibernating myocardium in pigs with a chronic left anterior descending coronary artery stenosis: Absence of progressive fibrosis in the setting of stable reductions in flow, function and coronary flow reserve. *J Am Coll Cardiol.* 2001; 37(7):1989–1995.

Fallavollita JA, Banas MD, Suzuki G, deKemp RA, Sajjad M, Canty JM, Jr. ¹¹C-meta-hydroxyephedrine defects persist despite functional improvement in hibernating myocardium. *J Nucl Cardiol.* 2010; 17(1):85–96.

Fallavollita JA, Canty JM, Jr. Dysinnervated but viable myocardium in ischemic heart disease. *J Nucl Cardiol.* 2010; 17(6):1107–1115.

Fedele FA, Gewirtz H, Capone RJ, Sharaf B, Most AS. Metabolic response to prolonged reduction of myocardial blood flow distal to a severe coronary artery stenosis. *Circulation.* 1988; 78(3):729–735.

Fedele FA, Sharaf B, Most AS, Gewirtz H. Details of stenosis morphology influence its hemodynamic severity and distal flow reserve. *Circulation.* 1989; 80:636–642.

Femia FJ, Maresca KP, Hillier M, Zimmerman CN, Joyal JL, Barrett J, Coleman T, Aras O, Dilsizian V, Eckelman EC, Babich JW. Synthesis and evaluation of a series of 99mTc(CO)3+ lisinopril complexes for *in vivo* imaging angiotensin converting enzyme expression. *J Nucl Med.* 2008; 49:970–977.

Flameng W, Suy R, Schwarz F, Borgers M, Piessens J, Thone F, Van Ermen H, De Geest H. Ultrastructural correlates of left ventricular contraction abnormalities in patients with chronic ischemic heart disease: Determinants of reversible segmental asynergy postrevascularization surgery. *Am Heart J.* 1981; 102(5):846–857.

Fleming I, Kohlstedt K, Busse R. The tissue renin-angiotensin system and intracellular signalling. *Curr Opin Nephrol Hypertens.* 2006; 15:8–13.

Franc BL, Acton PD, Mari C, Hasegawa BH. Small-animal SPECT and SPECT-CT: Important tools for preclinical investigation. *J Nucl Med.* 2008; 49:1651–1663.

Chapter 3

Furenlid LR, Wilson DW, Chen Y-C, Hyunki K, Pietraski PJ, Crawford MJ, Barrett HH. Fast SPECT II: A second-generation high-resolution dynamic SPECT imager. *IEEE Trans Nucl Sci.* 2004; 51:631–635.

Fyhrquist F, Tikkanen I, Gronhagen-Riska C et al. Inhibitor binding assay for angiotensin-converting enzyme. *Clin Chem.* 1984; 30:696–700.

Gambhir SS, Berman DS, Ziffer Z, Nagler M, Sandler M, Patton J, Hutton B, Sharir T, Ben Haim S, Ben Haim S. A novel high-sensitivity rapid-acquisition single-photon cardiac imaging camera. *J Nucl Med.* 2009; 50:635–643.

Gewirtz H, Beller GA, Strauss HW, Dinsmore RE, Zir LM, McKusick KA, Pohost GM. Transient defects of resting thallium scans in patients with coronary artery disease. *Circulation.* 1979; 59:707–713.

Gewirtz H, Most AS. Production of a critical coronary artery stenosis in closed chest laboratory animals: Description of a new non-surgical method based on standard cardiac catheterization techniques. *Am J Cardiol.* 1981; 47:589–596.

Glass CK, Witztum JL. Atherosclerosis, the road ahead. *Cell.* 2001; 104(4):503–516.

Glover DK, Okada RD. Myocardial technetium 99m sestamibi kinetics after reperfusion in a canine model. *Am Heart J.* 1993; 125:657–666.

Gropler RJ, Beanlands RS, Dilsizian V, Lewandowski ED, Villanueva FS, Ziadi MC. Imaging myocardial metabolic remodeling. *J Nucl Med.* 2010; 51 Suppl 1:88S–101S.

Gyongyosi M, Blanco J, Marian T et al. Serial noninvasive *in vivo* positron emission tomographic tracking of percutaneously intramyocardially injected autologous porcine mesenchymal stem cells modified for transgene reporter gene expression. *Circ Cardiovasc Imaging.* 2008; 1(2):94–103.

Hamill TG, Burns HD, Dannals RF et al. Development of [^{11}C] L-159,884: A radiolabelled, nonpeptide angiotensin II antagonist that is useful for angiotensin II, AT1 receptor imaging. *Appl Radiat Isot.* 1996; 47:211–218.

Harada K, Sugaya T, Murakami K et al. Angiotensin II type 1A receptor knockout mice display less left ventricular remodeling and improved survival after myocardial infarction. *Circulation.* 1999; 100:2093–2099.

Herrero P, Kim J, Sharp TL, Engelbach JA, Lewis JS, Gropler RJ, Welch MJ. Assessment of myocardial blood flow using ^{15}O-water and 1-^{11}C-acetate in rats with small animal PET. *J Nucl Med.* 2006; 47:477–485.

Herzog BA, Buechel RR, Katz R et al. Nuclear myocardial perfusion imaging with a cadmium-zinc-telluride detector technique: Optimized protocol for scan time reduction. *J Nucl Med.* 2010; 51:46–51.

Heusch G. Hibernating myocardium. *Phys Rev.* 1998; 78:1055–1085.

Heusch G, Schulz R, Rahimtoola SH. Myocardial hibernation: A delicate balance. *Am J Physiol Heart Circ Physiol.* 2005; 288(3):H984–H999.

Hirai T, Nohara R, Hosokawa R, Tanaka M, Inada H, Fujibayashi Y, Fujita M, Konishi J, Sasayama S. Evaluation of myocardial infarct size in rat heart by pinhole SPECT. *J Nucl Cardiol.* 2000; 7:107–111.

Huss JM, Kelly DP. Mitochondrial energy metabolism in heart failure: A question of balance. *J Clin Invest.* 2005; 115(3):547–555.

Ishibashi S, Brown MS, Goldstein JL, Gerard RD, Hammer RE, Herz J. Hypercholesterolemia in low density lipoprotein receptor knockout mice and its reversal by adenovirus-mediated gene delivery. *J Clin Invest.* 1993; 92:883–893.

Ishizu K, Mukai T, Yonekura Y, Pagani M, Fujita T, Magata Y, Nishizawa S, Tamaki N, Shibasaki H, Konishi J. Ultra-high resolution SPECT system using four pinhole collimators for small animal studies. *J Nucl Med.* 1995; 36:2282–2287.

Isobe S, Tsimikas S, Zhou J et al. Noninvasive imaging of atherosclerotic lesions in apolipoprotein E-deficient and low-density-lipoprotein receptor-deficient mice with annexin A5. *J Nucl Med.* 2006; 47:1497–1505.

Iyer VS, Canty JM, Jr. Regional desensitization of beta-adrenergic receptor signaling in swine with chronic hibernating myocardium. *Circ Res.* 2005; 97(8):789–795.

Jackson BM, Gorman JH, Moainie SL et al. Extension of borderzone myocardium in postinfarction dilated cardiomyopathy. *J Am Coll Cardiol.* 2002; 40:1160–1167.

Jameel MN, Zhang J. Myocardial energetics in left ventricular hypertrophy. *Curr Cardiol Rev.* 2009; 5(3):243–250.

Jaszczak RJ, Li J, Wang H, Zalutsky MR, Coleman RE. Pinhole collimation for ultra-high-resolution, small-field-of-view SPECT. *Phys Med Biol.* 1994; 39:425–437.

Jeavons AP, Chandler RA, Dettmar CAR. A 3D HIDAC-PET camera with sub-millimetre resolution for imaging small animals. *IEEE Trans Nucl Sci.* 1999; 46:468–473.

Johnson LL, Schofield L, Donahay T, Narula N, Narula J. 99mTc-annexin V imaging for *in vivo* detection of atherosclerotic lesions in porcine coronary arteries. *J Nucl Med.* 2005; 46(7):1186–1193.

Johnston CI, Fabris B, Yamada H et al. Comparative studies of tissue inhibition by angiotensin converting enzyme inhibitors. *J Hypertens Suppl.* 1989; 7:S11–S16.

Jonsson L, Johansson G, Lannek N, Lindberg P. Intramural blood supply of the porcine heart. A postmortem angiographic study. *Anat Rec.* 1974; 178:647.

Judenhofer MS, Catana C, Swann BK, Siegel SB, Jung W-I, Nutt RE, Cherry SR, Claussen CD, Pichler BJ. PET/MR images acquired with a compact MR-compatible PET detector in a 7-T magnet. *Radiology.* 2007; 244:807–814.

Jurisson SS, Lydon JD. Potential technetium small molecule radiopharmaceuticals. *Chem Rev.* 1999; 99:2205–2218.

Karp JS, Surti S, Daube-Witherspoon ME, Muehllehner G. Benefit of time-of-flight in PET: Experimental and clinical results. *J Nucl Med.* 2008; 49:462–470.

Kastis GA, Furenlid LR, Wilson DW, Peterson TE, Barber HB, Barrett HH. Compact CT/SPECT small-animal imaging system. *IEEE Trans Nucl Sci.* 2004; 51:63–77.

Kim H, Furenlid LR, Crawford MJ, Wilson DW, Barber HB, Peterson TE, Hunter WC, Liu Z, Woolfenden JM, Barrett HH. SemiSPECT: A small-animal single-photon emission computed tomography (SPECT) imager based on eight cadmium zinc telluride (CZT) detector arrays. *Med Phys.* 2006; 33:465–474.

Kircher MF, Grimm J, Swirski FK, Libby P, Gerszten RE, Allport JR, Weissleder R. Noninvasive *in vivo* imaging of monocyte trafficking to atherosclerotic lesions. *Circulation.* 2008; 117(3):388–395.

Klocke FJ, Bunnell IL, Greene DG, Whittenberg SM, Visco JP. Average coronary blood flow per unit weight of left ventricle in patients with and without coronary artery disease. *Circulation.* 1974; 50:547–559.

Kloner RA, Bolli R, Marban E, Reinlib L, Braunwald E. Medical and cellular implications of stunning, hibernation, and preconditioning: An NHLBI workshop. *Circulation.* 1998; 97(18):1848–1867.

Kohzuki M, Kanazawa M, Yoshida K et al. Cardiac angiotensin converting enzyme and endothelin receptor in rats with chronic myocardial infarction. *Jpn Circ J.* 1996; 60:972–980.

Krieger M, Herz J. Structures and functions of multiligand lipoprotein receptors: Macrophage scavenger receptors and LDL receptor-related protein (LRP). *Annu Rev Biochem.* 1994; 63:601–637.

Kuhle WG, Porenta G, Huang SC, Buxton D, Gambhir SS, Hansen H, Phelps ME, Schelbert HR. Quantification of regional myocardial blood flow using ^{13}N-ammonia and reoriented dynamic positron emission tomographic imaging. *Circulation.* 1992; 86:1004–1017.

Kuwaki K, Tseng YL, Dor FJ et al. Heart transplantation in baboons using alpha1,3-galactosyltransferase gene-knockout pigs as donors: Initial experience. *Nat Med.* 2005; 11(1):29–31.

Langer O, Valette H, Dollé F et al. High specific radioactivity (1R,2S)-4-[(18)F]fluorometaraminol: A PET radiotracer for mapping sympathetic nerves of the heart. *Nucl Med Biol.* 2000; 27:233–238.

Laurberg JM, Olsen AK, Hansen SB, Bottcher M, Morrison M, Ricketts SA, Falk E. Imaging of vulnerable atherosclerotic plaques with FDG-microPET: No FDG accumulation. *Atherosclerosis.* 2007; 192:275–282.

Law MP, Schäfers K, Kopka K, Wagner S, Schober O, Schäfers M. Molecular imaging of cardiac sympathetic innervation by ^{11}C-mHED and PET: From man to mouse? *J Nucl Med.* 2010; 51:1269–1276.

Lecomte R, Cadorette J, Rodrigue S, Lapointe D, Rouleau D, Bentourkia M, Yao R, Msaki P. Initial results from the Sherbrooke avalanche photodiode positron tomography. *IEEE Trans Nucl Sci.* 1996; 43:1952–1957.

Lee SH, Jung YS, Lee BH et al. Characterization of angiotensin II antagonism displayed by SK-1080, a novel nonpeptide AT1-receptor antagonist. *J Cardiovasc Pharmacol.* 1999; 33:367–374.

Lee YHC, Kiesewetter DO, Lang L et al. Synthesis of 4-[^{18}F] fluorobenzoyllisinopril: A radioligand for angiotensin converting enzyme (ACE) imaging with positron emission tomography. *J Labelled Comp Radiopharm.* 2001; 44:S268–S270.

Lewellen T, Karp J. PET systems, In *Emission Tomography: The Fundamentals of PET and SPECT,* eds. Wernick MN, Aarsvold JN. San Diego, California, USA: Elsevier Academic Press 2004.

Li Q-S, Solot G, Frank TL, Wagner Jr HN, Becker LC. Myocardial redistribution of technetium-99m-methoxyisobutyl isonitrile (SESTAMIBI). *J Nucl Med.* 1990; 31:1069–1076.

Li J, Jaszczak RJ, Greer KL, Gilland DR, DeLong DM, Coleman RE. Evaluation of SPECT quantification of radiopharmaceutical distribution in canine myocardium. *J Nucl Med.* 1995; 36:278–286.

Li YY, Feldman AM, Sun Y, McTiernan CF. Differential expression of tissue inhibitors of metalloproteinases in the failing human heart. *Circulation.* 1998; 98:1728–1734.

Linton MF, Farese RV Jr, Chiesa G, Grass DS, Chin P, Hammer RE, Hobbs HH, Young SG. Transgenic mice expressing high plasma concentrations of human apolipoprotein B100 and lipoprotein(a). *J Clin Invest.* 1993; 92:3029–3037.

Liu Z, Kastis GA, Stevenson GD, Barrett HH, Furenlid LR, Kupinski MA, Patton DD, Wilson DW. Quantitative analysis of acute myocardial infarct in hearts with ischemia-reperfusion using a high-resolution stationary SPECT system. *J Nucl Med.* 2002; 43:933–939.

Liyama K, Hajra L, Iiyama M, Li H, DiChiara M, Medoff BD, Cybulsky MI. Patterns of vascular cell adhesion molecule-1 and intercellular adhesion molecule-1 expression in rabbit and mouse atherosclerotic lesions and at sites predisposed to lesion formation. *Circ Res.* 1999; 85(2):199–207.

Loscalzo J. Oxidant stress: A key determinant of atherothrombosis. *Biochem Soc Trans.* 2003; 31(Pt 5):1059–1061.

Lowes BD, Minobe W, Abraham WT et al. Changes in gene expression in the intact human heart: Down-regulation of α-myosin heavy chain in hypertrophied, failing ventricular myocardium. *J Clin Invest.* 1997; 100:2315–2324.

Lowes BD, Gilbert EM, Abraham WT et al. Myocardial gene expression in dilated cardiomyopathy treated with beta-blocking agents. *N Engl J Med.* 2002; 346:1357–1365.

Luisi AJ, Jr., Suzuki G, Dekemp R, Haka MS, Toorongian SA, Canty JM, Jr., Fallavollita JA. Regional ^{11}C-hydroxyephedrine retention in hibernating myocardium: Chronic inhomogeneity of sympathetic innervation in the absence of infarction. *J Nucl Med.* 2005; 46(8):1368–1374.

Madar I, Ravert HT, Du Y, Hilton J, Volokh L, Dannals RF, Frost JJ, Hare JM. Characterization of uptake of the new PET imaging compound ^{18}F-fluorobenzyl triphenyl phosphonium in dog myocardium. *J Nucl Med.* 2006; 47:1359–1366.

Maes A, Borgers M, Fleming W, Nuyts J, van de Werf F, Ausma J, Sergeant P, Mortelmans L. Assessment of myocardial viability in chronic coronary artery disease using technetium-99m sestamibi SPECT: Correlation with histological and positron emission tomographic studies and functional follow-up. *J Am Coll Cardiol.* 1997; 29:62–68.

Mann DL, Kent RL, Parsons B, Cooper GIV. Adrenergic effects on the biology of the adult mammalian cardiocyte. *Circulation.* 1992; 85:790–804.

Matarrese M, Salimbeni A, Turolla EA et al. ^{11}CRadiosynthesis and preliminary human evaluation of the disposition of the ACE inhibitor [^{11}C]zofenoprilat. *Bioorg Med Chem.* 2004; 12:603–611.

Mathews WB, Burns HD, Dannals RF et al. Carbon-11 labeling of the potent nonpeptide angiotensin-II antagonist MK-996. *J Labelled Comp Radiopharm.* 1995; 36:729–737.

McElroy DP, MacDonald LR, Beekman FJ, Wang Y, Patt BE, Iwanczyk JS, Tsui BMW, Hoffman EJ. Performance evaluation of A-SPECT: A high resolution desktop pinhole system for imaging small animals. *IEEE Trans Nucl Sci.* 2002; 49:2139–2147.

McFalls EO, Hou M, Bache RJ, Best A, Marx D, Sikora J, Ward HB. Activation of p38 MAPK and increased glucose transport in chronic hibernating swine myocardium. *Am J Physiol Heart Circ Physiol.* 2004; 287(3):H1328–H1334.

Meikle SR, Kench P, Kassiou M, Banati RB. Small animal SPECT and its place in the matrix of molecular imaging technologies. *Phys Med Biol.* 2005; 50:R45–R61.

Mills I, Fallon JT, Wrenn D, Sasken HF, Gray W, Bier J, Levine D, Berman S, Gilson M, Gewirtz H. Adaptive responses of the coronary circulation and myocardium to chronic reduction in perfusion pressure and flow. *Am J Physiol (Heart Circ Physiol 35).* 1994; 266:H447–H457.

Most AS, Williams DO, Gewirtz H. Elevated coronary vascular resistance in the presence of reduced resting blood flow distal to a severe coronary stenosis. *Cardiovasc Res.* 1985; 19:599–605.

Naber CK, Baumgart D, Heusch G, Siffert W, Oldenburg O, Huesing J, Erbel R. Role of the eNOS Glu298Asp variant on the GNB3825T allele dependent determination of alpha-adrenergic coronary constriction. *Pharmacogenetics.* 2003; 13(5):279–284.

Chapter 3

Nahrendorf M, Zhang H, Hembrador S, Panizzi P, Sosnovik DE, Aikawa E, Libby P, Swirski FK, Weissleder R. Nanoparticle PET-CT imaging of macrophages in inflammatory atherosclerosis. *Circulation*. 2008; 22; 117(3):379–387.

Nahrendorf M, Keliher E, Panizzi P, Zhang H, Hembrador S, Figueiredo JL, Aikawa E, Kelly K, Libby P, Weissleder R. [18]F-4V for PET-CT imaging of VCAM-1 expression in atherosclerosis. *JACC Cardiovasc Imaging*. 2009; 2:1213–1222.

Nakashima Y, Plump AS, Raines EW, Breslow JL, Ross R. ApoE-deficient mice develop lesions of all phases of atherosclerosis throughout the arterial tree. *Arterioscler Thromb*. 1994; 14:133–140.

Nakao K, Minobe W, Roden R et al. Myosin heavy chain gene expression in human heart failure. *J Clin Invest*. 1997; 100:2362–2370.

Narayan P, Mentzer RM, Jr., Lasley RD. Annexin V staining during reperfusion detects cardiomyocytes with unique properties. *Am J Physiol Heart Circ Physiol*. 2001; 281(5):H1931–H1937.

Narula J, Haider N, Virmani R et al. Apoptosis in myocytes in end-stage heart failure. *N Engl J Med*. 1996; 335:1182–1189.

Narula J, Dilsizian V. From better understood pathogenesis to superior molecular imaging, and back. *JACC Cardiovasc Imaging*. 2008; 1(3):406–409.

Nekolla SG, Martinez-Moeller A, Saraste A. PET and MRI in cardiac imaging: From validation studies to integrated applications. *Eur J Nucl Med Mol Imaging*. 2009; 36(Suppl): S121–S130.

Nikolaidis LA, Mankad S, Sokos GG, Miske G, Shah A, Elahi D, Shannon RP. Effects of glucagon-like peptide-1 in patients with acute myocardial infarction and left ventricular dysfunction after successful reperfusion. *Circulation*. 2004a; 109(8):962–965.

Nikolaidis LA, Sturzu A, Stolarski C, Elahi D, Shen YT, Shannon RP. The development of myocardial insulin resistance in conscious dogs with advanced dilated cardiomyopathy. *Cardiovasc Res*. 2004b;61(2):297–306.

Ohshima S, Petrov A, Fujimoto S, Zhou J, Azure M, Edwards DS, Murohara T, Narula N, Tsimikas S, Narula J. Molecular imaging of matrix metalloproteinase expression in atherosclerotic plaques of mice deficient in apolipoprotein e or low-density-lipoprotein receptor. *J Nucl Med*. 2009; 50:612–617.

Okada RD, Glover D, Gaffney T, Williams S. Myocardial kinetics of technetium-99m-hexakis-2-methoxy-2-methylpropyl-isonitrile. *Circulation*. 1988; 77:491–498.

Olivetti G, Abbi R, Quaini F et al. Apoptosis in the failing human heart. *N Engl J Med*. 1997; 336:1131–1141.

Page B, Young R, Iyer V, Suzuki G, Lis M, Korotchkina L, Patel MS, Blumenthal KM, Fallavollita JA, Canty JM, Jr. Persistent regional downregulation in mitochondrial enzymes and upregulation of stress proteins in swine with chronic hibernating myocardium. *Circ Res*. 2008; 102(1):103–112.

Paigen B, Morrow A, Holmes PA, Mitchell D, Williams RA. Quantitative assessment of atherosclerotic lesions in mice. *Atherosclerosis*. 1987; 68(3):231–240.

Paul M, Poyan MA, Kreutz R. Physiology of local reninangiotensin systems. *Physiol Rev*. 2006; 86:747–803.

Penna C, Mancardi D, Raimondo S, Geuna S, Pagliaro P. The paradigm of postconditioning to protect the heart. *J Cell Mol Med*. 2008; 12(2):435–458.

Phelps CJ, Koike C, Vaught TD et al. Production of alpha 1,3-galacto syltransferase-deficient pigs. *Science*. 2003; 299(5605):411–414.

Pichler BJ, Kolb A, Nägele T. Schlemmer H-P. PET/MRI: Paving the way for the next generation of clinical multimodality imaging applications. *J Nucl Med*. 2010; 51:333–336.

Piedrahita JA, Zhang SH, Hagaman JR, Oliver PM, Maeda N. Generation of mice carrying a mutant apolipoprotein E gene inactivated by gene targeting in embryonic stem cells. *Proc Natl Acad Sci USA*. 1992; 89:4471–4475.

Plump AS, Smith JD, Hayek T, Aalto-Setälä K, Walsh A, Verstuyft JG, Rubin EM, Breslow JL. Severe hypercholesterolemia and atherosclerosis in apolipoprotein E-deficient mice created by homologous recombination in ES cells. *Cell*. 1992; 71:343–353.

Pokharel S, van Geel PP, Sharma UC et al. Increased myocardial collagen content in transgenic rats overexpressing cardiac angiotensin-converting enzyme is related to enhanced breakdown of N-acetyl-Ser-Asp-Lys-Pro and increased phosphorylation of Smad2/3. *Circulation*. 2004; 110:3129–3135.

Purcell-Huynh DA, Farese RV Jr, Johnson DF, Flynn LM, Pierotti V, Newland DL, Linton MF, Sanan DA, Young SG. Transgenic mice expressing high levels of human apolipoprotein B develop severe atherosclerotic lesions in response to a high-fat diet. *J Clin Invest*. 1995; 95:2246–2257.

Rahimtoola SH. The hibernating myocardium. *Am Heart J*. 1989; 117:211–221.

Rahimtoola SH. Chronic myocardial hibernation. *Circulation*. 1994; 89(4):1907–1908.

Raylman RR, Majewski S, Lemieux SK, Velan SS, Kross B, Popov V, Smith MF, Weisenberger AG, Zorn C, Marano GD. Simultaneous MRI and PET imaging of a mouse brain. *Phys Med Biol*. 2006; 51:6371–6379.

Ratib O, Phelps ME, Huang S-C, Henze E, Selin CE, Schelbert HR. Positron tomography with deoxyglucose for estimating local myocardial glucose metabolism. *J Nucl Med*. 1982; 23:577–586.

Rodriguez-Porcel M, Brinton TJ, Chen IY et al. Reporter gene imaging following percutaneous delivery in swine moving toward clinical applications. *J Am Coll Cardiol*. 2008; 51(5):595–597.

Roselaar SE, Kakkanathu PX, Daugherty A. Lymphocyte populations in atherosclerotic lesions of apoE -/- and LDL receptor -/- mice. Decreasing density with disease progression. *Arterioscler Thromb Vasc Biol*. 1996; 16:1013–1018.

Rosenfeld ME, Polinsky P, Virmani R, Kauser K, Rubanyi G, Schwartz SM. Advanced atherosclerotic lesions in the innominate artery of the ApoE knockout mouse. *Arterioscler Thromb Vasc Biol*. 2000; 20:2587–2592.

Rosenzweig A. Cardiac cell therapy—Mixed results from mixed cells. *N Engl J Med*. 2006; 355(12):1274–1277.

Rudd JH, Narula J, Strauss HW et al. Imaging atherosclerotic plaque inflammation by fluorodeoxyglucose with positron emission tomography: Ready for prime time? *J Am Coll Cardiol*. 2010; 55(23):2527–2535.

Sanan DA, Newland DL, Tao R, Marcovina S, Wang J, Mooser V, Hammer RE, Hobbs HH. Low density lipoprotein receptor-negative mice expressing human apolipoprotein B-100 develop complex atherosclerotic lesions on a chow diet: No accentuation by apolipoprotein(a). *Proc Natl Acad Sci USA*. 1998; 14; 95:4544–4549.

Schaper J, Froede R, Hein St et al. Impairment of the myocardial ultrastructure and changes of the cytoskeleton in dilated cardiomyopathy. *Circulation*. 1991; 83:504–514.

Schramm NU, Ebel G, Engeland U, Schurrat T, Behe M, Behr TM. High-resolution SPECT using multipinhole collimation. *IEEE Trans Nucl Sci*. 2003; 50:315–320.

Schulz R, Cohen MV, Behrends M, Downey JM, Heusch G. Signal transduction of ischemic preconditioning. *Cardiovasc Res*. 2001; 52(2):181–198.

Schunkert H, Jackson B, Tang SS et al. Distribution and functional significance of cardiac angiotensin converting enzyme in hypertrophied rat hearts. *Circulation.* 1993; 87:1328–1339.

Seidel J, Vaquero JJ, Green MV. Resolution uniformity and sensitivity of the NIH ATLAS small animal PET scanner: Comparison to simulated LSO scanners without depth-of-interaction capability. *IEEE Trans Nucl Sci.* 2003; 50:1347–1350.

Sharir T, Ben-Haim S, Merzon K, Prochorov V, Dickman D, Ben-Haim S, Berman DS. High-speed myocardial perfusion imaging: Initial clinical comparison with conventional dual detector Anger camera imaging. *J Am Coll Cardiol Img.* 2008; 1:156–163.

Shirani J, Loredo ML, Eckelman WC et al. Imaging the renin-angiotensin–aldosterone system in the heart. *Curr Heart Fail Rep.* 2005:2:78–86.

Shirani J, Narula J, Eckelman W, Dilsizian V. Novel imaging strategies for predicting remodeling and evolution of heart failure: Targeting the renin-angiotensin system, In *Noninvasive Imaging of Heart Failure, Heart Failure Clinics*, eds. Dilsizian V, Garcia MJ, Bello D. Philadelphia, Pennsylvania, USA: Elsevier-Saunders Publishing Company, Inc., 2006, pp. 231–247.

Shirani J, Narula J, Eckelman WC, Narula N, Dilsizian V. Early imaging in heart failure: Exploring novel molecular targets. *J Nucl Cardiol.* 2007; 14(1):100–110.

SOLVD Investigators. Effect of enalapril on survival in patients with reduced left ventricular ejection fractions and congestive heart failure. *N Engl J Med.* 1991; 325:293–302.

SOLVD Investigators. Effect of enalapril on mortality and the development of heart failure in asymptomatic patients with reduced left ventricular ejection fractions. *N Engl J Med.* 1992; 327:685–691.

Steingart RM, Bontemps R, Scheuer J, Yipintosi T. Gamma camera quantitation of thallium-201 redistribution at rest in a dog model. *Circulation.* 1982; 65:542–550.

Steingart RM, Cohen MV. Thallium-201 scintigraphic quantitation of regional flow disparity and subsequent redistribution in dogs. *J Nucl Med.* 1986; 27:75–83.

Stewart RE, Schwaiger M, Hutchins GD, Chiao P-C, Gallagher KP, Nguyen N, Petry N, Rogers WL. Myocardial clearance kinetics of technetium-99m-SQ30217: A marker of regional myocardial blood flow. *J Nucl Med.* 1990; 31:1183–1190.

Strand S-E, Ivanovic M, Erlandsson K, Franceschi D, Button T, Sjögren K, Weber DA. Small animal imaging with pinhole single-photon emission computed tomography. *Cancer Suppl.* 1994; 73:981–984.

Su H, Spinale FG, Dobrucki LW et al. Noninvasive targeted imaging of matrix metalloproteinase activation in a murine model of postinfarction remodeling. *Circulation.* 2005; 112:3157–3167.

Sun Y, Most AS, Ohley W, Gewirtz H. Estimation of instantaneous blood flow through a rigid, coronary artery stenosis in anaesthetised domestic swine. *Cardiovasc Res.* 1983; 17:499–504.

Sun Y, Zhang J, Lu L et al. Tissue angiotensin II in the regulation of inflammatory and fibrogenic components of repair in the rat heart. *J Lab Clin Med.* 2004; 143:41–51.

Surti S, Karp JS, Perkins AE, Cardi CA, Daube-Witherspoon ME, Kuhn A, Muehllehner G. Imaging performance of A-PET: A small animal PET camera. *IEEE Trans Med Imaging.* 2005; 24:844–852.

Surti S, Kuhn A, Werner ME, Perkins AE, Kolthammer J, Karp JS. Performance of Philips Gemini TF PET/CT scanner with special consideration for its time-of-flight imaging capabilities. *J Nucl Med.* 2007; 48:471–480.

Suzuki G, Lee TC, Fallavollita JA, Canty JM, Jr. Adenoviral gene transfer of FGF-5 to hibernating myocardium improves function and stimulates myocytes to hypertrophy and reenter the cell cycle. *Circ Res.* 2005; 96(7):767–775.

Szabo Z, Speth RC, Brown PR et al. Use of positron emission tomography to study AT1 receptor regulation in vivo. *J Am Soc Nephrol.* 2001; 12:1350–1358.

Taegtmeyer H. Switching metabolic genes to build a better heart. *Circulation.* 2002; 106(16):2043–2045.

Taegtmeyer H, McNulty P, Young ME. Adaptation and maladaptation of the heart in diabetes: Part I: General concepts. *Circulation.* 2002; 105(14):1727–1733.

Taegtmeyer H. Cardiac metabolism as a target for the treatment of heart failure. *Circulation.* 2004; 110(8):894–896.

Taegtmeyer H, Ballal K. No low-fat diet for the failing heart? *Circulation.* 2006; 114(20):2092–2093.

Tangirala RK, Rubin EM, Palinski W. Quantitation of atherosclerosis in murine models: Correlation between lesions in the aortic origin and in the entire aorta, and differences in the extent of lesions between sexes in LDL receptor-deficient and apolipoprotein E-deficient mice. *J Lipid Res.* 1995; 36(11):2320–2328.

Tawakol A, Skopicki HA, Abraham SA, Alpert NM, Fischman AJ, Picard MH, Gewirtz H. Evidence of reduced resting blood flow in viable myocardial regions with chronic asynergy. *J Am Coll Cardiol.* 2000; 36(7):2146–2153.

Ter-Pogossian MM, Ficke DC, Hood Sr JT, Yamamoto M, Mullani NA. PETT VI: A positron emission tomograph utilizing cesium fluoride scintillation detectors. *J Comput Assist Tomogr.* 1982; 6:125–133.

Thijssen VL, Borgers M, Lenders MH, Ramaekers FC, Suzuki G, Palka B, Fallavollita JA, Thomas SA, Canty JM, Jr. Temporal and spatial variations in structural protein expression during the progression from stunned to hibernating myocardium. *Circulation.* 2004; 110(21):3313–3321.

Thim T, Hagensen MK, Drouet L, Bal Dit Sollier C, Bonneau M, Granada JF, Nielsen LB, Paaske WP, Botker HE, Falk E. Familial hypercholesterolaemic downsized pig with human-like coronary atherosclerosis: A model for preclinical studies. *EuroIntervention.* 2010; 6(2):261–268.

Thomas CV, Coker ML, Zellner JL et al. Increased matrix metalloproteinase activity and selective upregulation in LV myocardium from patients with end-stage heart failure. *Circulation.* 1998; 97:1708–1715.

Tian XL, Pinto YM, Costerousse O et al. Over-expression of angiotensin converting enzyme-1 augments cardiac hypertrophy in transgenic rats. *Hum Mol Genet.* 2004; 13:1441–1450.

Tillisch J, Brunken R, Marshall R, Schwaiger M, Mandelkern M, Phelps M, Schelbert H. Reversibility of cardiac wall-motion abnormalities predicted by positron tomography. *N Engl J Med.* 1986; 314(14):884–888.

Tseng YL, Kuwaki K, Dor FJ et al. Alpha1,3-Galactosyltransferase gene-knockout pig heart transplantation in baboons with survival approaching 6 months. *Transplantation.* 2005; 80(10):1493–1500.

Tsui BMW, Kraitchman DL. Recent advances in small-animal cardiovascular imaging. *J Nucl Med.* 2009; 50:667–670.

Tsui B, Xu J, Chen S, Meier D, Yu J, Patt B, Wagenaar D. The application of a compact MR-compatible SPECT system for small animal SPECT/MR imaging and tracer kinetic studies. *J Nucl Med.* 2010; 51 (Suppl 2):409.

Chapter 3

Tsutsui H, Spinale FG, Nagatsu M et al. Effects of chronic ß-adrenergic blockade on the left ventricular and cardiocyte abnormalities of chronic canine mitral regurgitation. *J Clin Invest.* 1994; 93:2639–2648.

Verani MS, Jeroudi MO, Mahmarian JJ, Boyce TM, Borges-Neto S, Patel B, Bolli R. Quantification of myocardial infarction during coronary occlusion and myocardial salvage after reperfusion using cardiac imaging with technetium-99m hexakis 2-methoxyisobutyl isonitrile. *J Am Coll Cardiol.* 1988; 12:1573–1581.

Verjans JWH, Lovhaug D, Narula N et al. Noninvasive imaging of angiotensin receptors after myocardial infarction. *J Am Coll Cardiol Img.* 2008; 1:354–362.

Virmani R, Burke AP, Farb A, Kolodgie FD. Pathology of the vulnerable plaque. *J Am Coll Cardiol.* 2006; 47 (Suppl 8):C13–C18.

Weber D, Ivanovic M, Franceschi D et al. SPECT: An approach to *in vivo* high resolution SPECT imaging in small laboratory animals. *J Nucl Med.* 1994; 35:342–348.

Weber KT, Brilla CG, Janicki JS. Myocardial fibrosis: Functional significance and regulatory factors. *Cardiovasc Res.* 1993; 27:341–348.

Weber KT, Sun Y, Guarda E. Structural remodeling in hypertensive heart disease and the role of hormones. *Hypertension.* 1994; 23:869–877.

Weber KT. Extracellular matrix remodeling in heart failure: A role for de novo angiotensin II generation. *Circulation.* 1997; 96:4065–4082.

Weber KT, Sun Y. Recruitable ACE and tissue repair in the infarcted heart. *J Renin Angiotensin Aldosterone Syst.* 2000; 1:295–303.

Weber S, Herzog H, Cremer M et al. Evaluation of the TierPET system. *IEEE Trans Nucl Sci.* 1999; 46:1177–1183.

Wei L, Babich JW, Eckelman WC, Zubieta JA. Rhenium tricarbonyl core complexes of thymidine and uridine dervatives. *Inorg Chem.* 2005; 44:2198–2209.

Weisenberger AG, Wojcik R, Bradley EL et al. SPECT-CT system for small animal imaging. *IEEE Trans Nucl Sci.* 2003; 50:74–79.

Wu JC. Molecular imaging: Antidote to cardiac stem cell controversy. *J Am Coll Cardiol.* 2008; 52(20):1661–1664.

Yamada R, Watanabe M, Omura T et al. Development of a small animal PET scanner using DOI detectors. *IEEE Trans Nucl Sci.* 2008; 55:906–911.

Yukihiro M, Inoue T, Iwasaki T, Tomiyoshi K, Erlandsson K, Endo K. Myocardial infarction in rats: High resolution single-photon-emission tomographic imaging with a pinhole collimator. *Eur J Nucl Med.* 1996; 23:896–900.

Zhao G, Shen W, Zhang X, Smith CJ, Hintze TH. Loss of nitric oxide production in the coronary circulation after the development of dilated cardiomyopathy: A specific defect in the neural regulation of coronary blood flow. *Clin Exp Pharmacol Physiol.* 1996; 23(8):715–721.

Zhao Y, Kuge Y, Zhao S, Morita K, Inubushi M, Strauss HW, Blankenberg FG, Tamaki N. Comparison of 99mTc-annexin A5 with ^{18}F-FDG for the detection of atherosclerosis in ApoE-/- mice. *Eur J Nucl Med Mol Imaging.* 2007; 34:1747–1755.

Zeng GL, Galt JR, Wernick MN, Mintzer RA, Aarsvold J. Single-photon emission computed tomography, In *Emission Tomography: The Fundamentals of PET and SPECT*, eds. Wernick MN, Aarsvold JN. San Diego, California, USA: Elsevier Academic Press 2004.

Zisman LS, Abraham WT, Meixell GE et al. Angiotensin II formation in the intact human heart. Predominance of the angiotensin-converting enzyme pathway. *J Clin Invest.* 1995; 96:1490–1498.

Zisman LS, Asano K, Dutcher DL et al. Differential regulation of cardiac angiotensin converting enzyme binding sites and AT1 receptor density in the failing human heart. *Circulation.* 1998; 98:1735–1741.

Zynda TK, Petrov A, Ohshima S et al. Molecular imaging of angiotensin converting enzyme-1 expression in the myocardium of angiotensin converting enzyme-1 overexpressing transgenic rats. *J Am Coll Cardiol.* 2009; 53:A145.

4. Imaging Diabetes in Small Animals
Models, Methods, and Targets

Kooresh I. Shoghi

4.1   Introduction

Diabetes is a metabolic condition in which the body fails to produce or respond to insulin. Type-1 diabetes (T1D) is an autoimmune disease in which the immune system targets insulin-producing β-cells in the pancreas leading to their destruction. Type-2 diabetes (T2D), on the other hand, occurs when cellular resistance to insulin is accompanied by the failure of β-cells to produce sufficient insulin. Current epidemiology suggests that T2D accounts for nearly 95% of all cases (Permutt et al. 2005). In 2005, the total prevalence of diabetes in the United States was approximately 7% representing 20.8 million people; by 2050 it is projected that as many as 1 in 3

Targeted Molecular Imaging. Edited by Michael J. Welch and William C. Eckelman © 2012 Taylor & Francis Group, LLC. ISBN: 978-1-4398-4195-2

Chapter 4

adults in the United States could have diabetes if current trends continue (National Center for Health Statistics, Centers for Disease Control and Prevention). The medical burden of diabetes is equally astounding. In the United States, heart disease and stroke account for nearly 65% of deaths associated with diabetes (National Center for Health Statistics, Centers for Disease Control and Prevention). Diabetes is also associated with retinopathy, renal diseases, peripheral vasculature disease, and neuropathy. With that in mind, animal models are extensively used to study underlying mechanisms in the pathogenesis of diabetes as well as in the development and validation of imaging probes and targets. Needless to say, the topic is vast. This chapter focuses on the available animal models of T1D and T2D with related imaging targets.

4.1.1 Animal Models of T1D

T1D is characterized by a specific destruction of the pancreatic β-cells commonly associated with autoimmune-mediated damages (Dahlquist 1998). β-Cells secrete the hormone insulin which stimulates peripheral uptake of glucose as well as suppression of hepatic glucose production among other mechanisms. In the pathological progression of T1D, β-cells are destroyed over time such that at clinical presentation there is little surviving β-cell mass. Available animal models which attempt to replicate the destruction of β-cell mass are classified as chemically induced or spontaneous models of T1D. The most common spontaneous animal models of T1D are the nonobese diabetic (NOD) mouse and the biobreeding (BB) rats.

4.1.2 Streptozotocin–Induced T1D

Streptozotocin (STZ) is a nitrosurea derivative isolated from *Streptomyces achromogenes* with broad-spectrum antibiotic activity which interferes with cellular mechanisms including glucose transport (Wang and Gleichmann 1998), glucokinase function

(Zahner and Malaisse 1990), and which induces DNA-strand breaks (Bolzan and Bianchi 2002). Thus, the STZ-induced T1D represents an example of chemically induced diabetes. The administration of STZ can be given either as a single dose or multiple doses. A single dose of STZ, which may be toxic, will induce diabetes in rodents. Traditionally, however, multiple doses of STZ are used (e.g., 40 mg/kg over 5 days) inducing immune destruction of pancreatic cells such as in T1D. The multiple-dose STZ model has been used extensively to study underlying mechanisms leading to β-cell death.

4.1.3 Nonobese Diabetic Mouse

The nonobese diabetic (NOD) mouse was established as an inbred mouse strain of spontaneous development of autoimmune T1D by Dr. Susumu Makino (Makino et al. 1980). Insulitis—the penetration of mononuclear cells into pancreatic islets—is not observed until the age of 3 weeks when the frequency reaches 70–90%; by 20 weeks, 100% of mice will have developed insulitis (Makino et al. 1985). The majority of mononuclear cells infiltrating the islets are T-cells (CD4+ and CD8+) but B-cells, dendritic cells, and macrophages have also been observed. Insulitis is followed by a subclinical destruction of β-cells and decreasing insulin concentrations. Despite massive infiltration of mononuclear cells, β-cells remain intact until about 12–15 weeks of age at which point the destruction of β-cells becomes aggressive and frank diabetes presents itself. Interestingly, although almost all NOD mice develop insulitis, only a fraction develops diabetes. In fact, 90% of female mice and only 60% of male mice develop diabetes in some colonies (Atkinson and Leiter 1999) in contrast to human T1D, suggesting that several gender-related factors may modify the process from insulitis to overt diabetes. After the onset of diabetes, NOD mice lose weight and generally die within 1–2 months unless treated with daily injections of insulin, much like human T1D (Makino et al. 1980).

4.2 Animal Models of T2D

T2D is a complex, heterogenous, polygenic disease that affects more than 150 million people worldwide, and a large increase in these numbers is expected within the coming years (Stumvoll et al. 2005). Recent epidemiological and experimental evidence suggests that there is a close link between the pathogenesis of

obesity and T2D (Flier 2001), thus confounding the bleak outlook for T2D. A common feature to both obesity and T2D is insulin resistance (Kahn and Flier 2000). Numerous animal models of obesity and T2D diabetes have been developed to recapitulate the pathogenesis of insulin resistance in T2D. Herein, we will focus on a select few that are actively utilized in imaging diabetes.

4.2.1 Leptin–Based Animal Models

Leptin is a 16 kDa adipose-derived hormone which plays a key role in regulating energy intake and expenditure, including appetite and metabolism (Brennan and Mantzoros 2006). The effects of leptin were discovered at the Jackson Laboratories by observing obese mice that arose at random within a mouse colony (Ingalls et al. 1950). The leptin-resistant animal models, including both the Zucker diabetic fatty (ZDF) rat and the db/db mouse, are well-characterized models of insulin resistance. In one hypothetical scenario proposed by McGarry (2002) (Figure 4.1), leptin deficiency results in decreased oxidative capacity by the muscle, resulting in increasing intramuscular triglyceride (IMTG). Increased IMTG results in reduced glucose uptake and thus insulin resistance. To compensate for the insulin resistance, the pancreas increases insulin secretion to maintain glycemic control. Increased insulin secretion leads to hyperinsulinemia, which further diminishes fatty acid oxidation resulting in increased fatty acyl-CoA, increased IMTG, and inhibition of processes involved in glucose metabolism. In parallel, high levels in insulin also increase VLDL secretion from the liver, which deposit in muscle and adipose tissue thus increasing fat deposition. By late stages, the ability of the adipose tissue to reesterify FFA is overwhelmed resulting in increased FFA into the

circulation, increased IMTG, and β-cell failure. Below, the three most common leptin-based animal models are described.

4.2.1.1 ob/ob Mouse

The mutation in ob/ob mice has been identified in the leptin gene, which encodes for leptin. The mutation occurs on chromosome 7 and is inherited as a monogenic autosomal recessive mutation (Ingalls et al. 1950). Homozygous mutant mice exhibit rapid growth gain accompanied by decreased energy expenditure resulting in mice that are considered obese weighting on average three times as much as control mice. Ob/ob mice also exhibit diabetes-like syndromes, including hyperglycemia, insulin resistance, impaired glucose tolerance, and severe hyperinsulinemia potentially as a direct result of increased body weight. Additionally, obese mice exhibit both hypertrophy and hyperplasia of pancreatic islets. Interestingly, glucose levels in ob/ob mice are maintained at normal levels owing to sustained hyperinsulinemia. At the molecular level, insulin resistance in ob/ob mice is attributed to deficiencies in insulin signaling (Herberg and Coleman 1977).

4.2.1.2 db/db Mouse

The mutation on db/db mice has been traced to the db gene which encodes for leptin receptors in contrast to ob/ob mice in which the mutation is attributed to deficiencies in the hormone leptin. To compensate for the lack of insulin signaling, db/db mice over-secrete

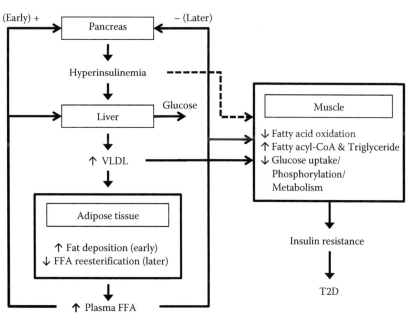

FIGURE 4.1 Hypothetical progression of T2D. (Adapted from McGarry, J. D. 2002. *Diabetes* 51(1): 7–18.)

Chapter 4

insulin becoming obese, hyperglycemic, hyperinsulinemic and insulin resistant within 3–4 months of age. Db/db mice have been used extensively to study various complications of T2D and efficacy of therapeutic intervention.

4.2.1.3 Zucker Diabetic Fatty Rat

The Zucker diabetic fatty (ZDF) rat is a by-product of the obese Zucker fa/fa rats. Obese Zucker fa/fa rats have an autosomal recessive mutation in the fa gene that encodes the leptin receptor. Zucker rats exhibit hyperphagia, obesity, and hyperlipidemia, but only mild elevation of blood glucose levels (Godbole and York 1978). By crossbreeding obese zucker fa/fa rats with Wistar–Kyoto rats, which are insulin-resistant and less tolerant to glucose, the Zucker diabetic fatty (ZDF-fa/fa) rat was established (Ikeda et al. 1981). Overt diabetes was found from an early stage in this model, despite compensatory hypersecretion of insulin, indicating insulin resistance. Consequently, exhaustion of insulin secretion with impaired glucose tolerance promotes overt diabetes

as early as 8 weeks of age. This mimics the pathophysiological profile of human type 2 diabetes. By 12 weeks when placed on Purina chow 5008, ZDF rats are considered diabetic. The ZDF rat has been extensively used in studying the mechanism of insulin resistance, β-cell dysfunction as well as in assessing drug efficacy (Kanda et al. 2009, Larsen et al. 2008, Shoghi et al. 2009).

4.2.2 The Koto–Kakizaki (GK) Rat

The GK rat, a polygenic model of T2D, was developed by selective breeding of Wistar rats with elevated blood glucose levels over several generations (Goto et al. 1976). GK rats are nonobese with moderate but stable fasting hyperglycemia, hypoinsulinemia, and impaired glucose tolerance which appears at 2 weeks of age signifying early onset of T2D (Goto et al. 1976). At birth GK rats have reduced islet count; adult GK rats exhibit a 60% decrease in β-cell mass along with decreased pancreatic insulin stores (Miralles and Portha 2001, Portha 2005).

4.3 Methodological Aspects of Imaging Small Animals

4.3.1 Quantification of Small Animal Pet Images

One of the many advantages of PET imaging is that it offers quantifiable measures of the underlying biology. Yet, in most instances PET images are quantified by normalizing average values in regions of interest (ROIs). This "normalization" scheme takes the form of the "standardized uptake value" (SUV) or related to a reference region presumably devoid of the underlying biological process under investigation. While the application of SUV has been discussed in detail by others (Boellaard 2009, Ito et al. 2009, Lucignani et al. 2004, Ning et al. 2009, Suzuki et al. 2009, Wahl et al. 2009) to name a few, we will discuss quantification of small animal PET images. Quantification in this regards is defined modeling the kinetics of the imaging agents in tissue. Recent examples of kinetic modeling in small animals span many fields, including oncologic imaging (Kim et al. 2008, Pollok et al. 2009), cardiac imaging (Shoghi 2009, Shoghi et al. 2008), and neuroimaging (Guo et al. 2009) among others. By far, the predominant choice of tracer is ^{18}FDG largely owing to its availability. Kinetic modeling offers several advantages in comparison to "normalization"

schemes, in that it offers measures of transport/perfusion rates which can be used to tease out delivery issues from metabolism/binding of ligands. Kinetic analysis also offers means to measure metabolic rates and receptor density, as well as fracture of underlying vasculature which can be used to discern "true" tissue activity especially in situations when one expects changes in vasculature, for example, due to injury. Kinetic analysis also offers means to separate specific from nonspecific signal, for example, when a ligand is found to have a secondary nonspecific-binding site. Finally, with more advanced application of kinetic analysis, spatial information on the kinetics in tissue may be characterized. Quantification of small animal PET images is a process which requires: (1) the derivation of the input function, (2) a notion of a kinetic model for the interaction of the imaging agent in tissue, (3) parameter estimation, and finally (4) validation of the model. To some extent, the latter three steps may be iterated until the appropriate model is confirmed.

4.3.1.1 Derivation of the Input Function in Rodents

Accurate quantification of PET images requires knowledge of the plasma time–activity curve (pTAC) of the

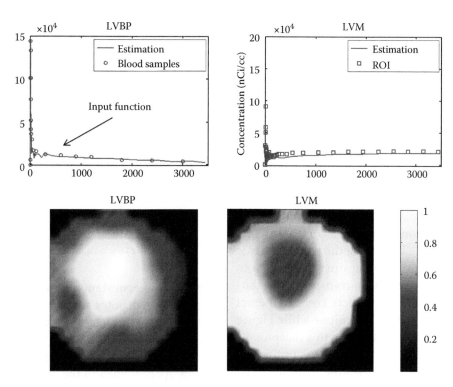

FIGURE 4.2 The application of maximum-likelihood factor analysis for the derivation of the input function. (Adapted from Su, Y., Welch, M. J., and Shoghi, K. I. 2007. *Phys Med Biol* 52(8): 2313–2334.)

radiopharmaceutical (Acton et al. 2004, Wienhard 2002). The pTAC serves as a forcing function to a kinetic model commonly known as the plasma input function (PIF). The gold standard for the determination of PIF is an invasive blood-sampling procedure (Fang and Muzic 2008, Laforest et al. 2005), in which activity concentrations of arterial blood samples are measured directly at select intervals. This procedure is challenging for small animal studies because of the small size of animal blood vessels and the potential perturbation to the physiology due to the loss of blood (Laforest et al. 2005). Therefore, less invasive, image-based PIF estimation techniques are desirable. In Figure 4.2, we depict the application of maximum likelihood factor analysis to derive the input function. In Table 4.1, we survey methods for derivation of the input function in small animals.

4.3.1.2 Kinetic Analysis

A mathematical treatment of kinetic analysis is beyond the scope of this chapter. Instead we provide examples of such analysis in particular in the next section in discussing metabolic targets; for additional detail the reader is referred to excellent reviews by Shoghi (2009) and Watabe et al. (2006). In general, imaging agents can be categorized into four groups,

namely: metabolic agents, "trapped" agents, perfusion agents, and agents targeting receptors. Metabolic agents, as the name implies, are metabolized much like endogenous substrates. These include ^{11}C-labeled glucose or palmitate for example, which are widely

Table 4.1 Survey of Methods to Reconstruct the Input Function in Small Animal Imaging

Method	Description	Refs.
a.	Hybrid image-blood sampling (nonparametric)	Shoghi and Welch (2007)
b.	Application of maximum likelihood factor analysis (MLFA) to heart images	Su et al. (2007)
c.	Hybrid single input dual output (H-SIDO) semi-parametric reconstruction	Su and Shoghi (2010)
d.	Estimation of [18]FDG input function by multitissue kinetic modeling	Ferl et al. (2007)
e.	Spillover and partial volume correction for image-derived input function	Fang and Muzic (2008)
f.	Microfluidic-based blood sampling	Wu et al. (2007)

used to quantify glucose and fatty acid metabolism, respectively. "Trapped" agents are those which have been modified chemically to be trapped within a cell following an enzymatic reaction. The best-known example is ^{18}FDG in which the 2′-OH has been replaced with F-18 thereby blocking further downstream metabolism (although there are reports disputing this assertion (Southworth et al. 2003)). Perfusion agents measure perfusion/blood flow in tissue, for example, myocardial blood flow using [^{15}O]H$_2$O. Finally, the broadest category of imaging agents comprise radiolabeled ligands targeting receptors. Depending on the tracer, a compartmental model is devised for the kinetics of the tracer in tissue and is subsequently optimized against image data to estimate kinetic rates of tracer transport, metabolism, or binding. Figure 4.3 depicts compartmental models for the kinetics of [^{11}C]palmitate while Figure 4.4 depicts a compartmental binding to a receptor.

The compartmental model is then optimized against PET data to estimate the kinetic parameters. Let $\widehat{ROI}(t)$ denote the radioactivity in the ROI based on the kinetic model and let $ROI(t)$ denote the observed radioactivity in the image, parameter estimation amounts to minimization of the weighted residual sum of squares ($WRSS$)

$$WRSS = \sum_{i=1}^{N} w(t_i)\left[\widehat{ROI}(t_i) - ROI(t_i)\right]$$

where N denotes the number of time points (or frames) in the image and $w(t_i)$ denotes the weight assigned to each time point. The weighting is traditionally assumed to be reciprocal to the variance of measurements, that is,

$$w(t_i) = s\sqrt{\frac{\Delta t_i}{ROI(t_i)}}$$

where s being a constant proportional to scanner sensitivity and Δt_i denotes the scan duration at time t_i. In performing parameter estimation, one needs to consider the complexity of the model, number of parameters, as well as visual inspection of fitted model to data. As described in the next section, imaging targets in diabetes encompass a wide range of tracer–target interaction, ranging from metabolic tracer, "trapped" tracers, and tracers targeting receptors.

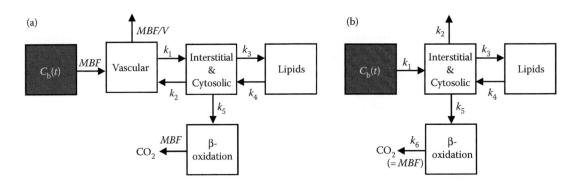

FIGURE 4.3 Compartmental models for the kinetics of [^{11}C]Palmitate. The model depicted in (a) is primarily used in human studies—denoted by "full model." An alternative compartmental model for the kinetics of [^{11}C]Palmitate—denoted as "reduced Palmitate model"— is depicted in (b). The reduced model may be better suited for quantification of rodent images, mice in particular.

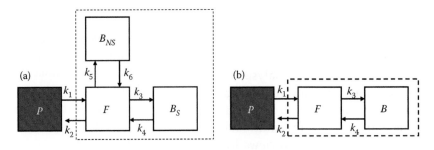

FIGURE 4.4 Compartmental models for receptor binding. In (a) two binding sites are depicted for specific and nonspecific binding while (b) includes a single-specific receptor pool. The input function is denoted by P.

4.4   Imaging Targets in Diabetes

4.4.1   Targets for Imaging β-Cell Mass

T1DM, as discussed earlier, is a result of autoimmune destruction of β-cells, resulting in deficiencies in insulin secretion and hyperglycemia. Generally, clinical manifestation of T1D occurs later in life at which point most of the BCM has been destroyed. T2DM, on the other hand, is a complex disease involving a breakdown in glucose homeostasis generally attributed to deficiencies in insulin signaling in peripheral tissues and increased hepatic glucose production. As the primary role of insulin is enhanced, uptake of glucose into peripheral tissues, deficiencies in insulin signaling results in hyperglycemia. To counteract increased glucose levels, β-cells increase insulin secretions. Overtime, β-cells can no longer meet the increased demand for insulin and BCM declines. With that in mind, developments in earlier diagnosis may assist in reducing the current deleterious effects of reduced BCM as well as assess the efficacy of various experimental treatments to enhance BCM. One of the main challenges in this endeavor is that BCM constitutes 1–2% of the pancreas, thus the imaging target would have to produce sufficient signal-to-background to visualize BCM. One potential imaging biomarker for BCM is the vesicular monoamine transporter type 2 (VMAT2). VMAT2 has been found to be expressed in β-cells and monoaminergic neurons but absent in exocrine pancreas and other abdominal organs (Harris et al. 2008, Maffei et al. 2004). In addition, histochemical data suggest that VMAT2 is found on vesicles that store insulin within β-cells and its presence is independent of insulin synthesis, processing, and secretion (Anlauf et al. 2003, Ichise and Harris 2010).

A specific ligand for VMAT2, [¹¹C]dihydrotetrabenazine ([¹¹C]DTBZ), has been already in clinical use for PET imaging of the central nervous system (CNS) (Vander Borght et al. 1995). Naturally, its utility in quantifying BCM has been attempted by several groups. The first study to demonstrate the utility of [¹¹C]DTBZ is reported by Souza et al. (2006). The authors demonstrate that PET-based quantification of VMAT2 receptors provides a noninvasive measure of BCM. A fluorine-18-labeled analog of DTBZ has been synthesized and is reported to have similar characteristics as the C-11-labeled radiopharmaceutical (Kung et al. 2008, Tsao et al. 2010).

4.4.2   Metabolic Imaging Targets

4.4.2.1   Substrate Metabolism in Liver

Glucose metabolism in the liver can be broken down into four arms: glycolysis (GLY), gluconeogenesis (GNG), glycogenesis (GGN), and glycogenolysis (GGL) as depicted in Figure 4.5. Glucose enters hepatic cells facilitated primarily by GLUT2 transporters. Once in tissue, glucose is converted into glucose-6-phosphate (G6P) by hexokinase (HK). The presence of the enzyme glucose-6-phosphatase in the liver hydrolyzes G6P resulting in the creation of a phosphate group and free glucose which can exit the liver via GLUT2. Unhydrolyzed G6P is directed toward one of three pathways: glycolysis (GLY), glycogenolysis (GGL), or the pentose phosphate pathway (not depicted). In health, insulin inhibits the production and release of glucose in the liver blocking GNG and GGL. In diabetic patients, the contribution of glucose production

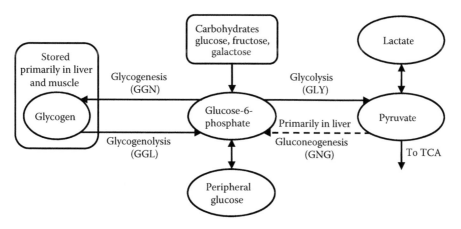

FIGURE 4.5 The fate of carbohydrates in the liver. Glucose metabolism in the liver can be broken down into four pathways: Glycolysis (GLY), gluconeogenesis (GNG), glycogenesis (GGN), and glycogenolysis (GNL).

Chapter 4

through GNG is significantly higher than normal controls (Magnusson et al. 1992) accounting for approximately 90% of total glucose production compared to 70% in controls (Figure 4.6). This occurs through a direct effect of insulin on the liver (Michael et al. 2000) as well as by indirect effects of insulin on substrate availability (Bergman and Ader 2000). On a molecular level, insulin directly controls the activities of a set of metabolic enzymes by phosphorylation or dephosphorylation and also regulates the expression of gene encoding hepatic enzymes of GNG and GLY (Pilkis and Granner 1992). It inhibits the transcription of the gene encoding phosphoenolpyruvate carboxylase, the rate-limiting step in GNG (Sutherland et al. 1996). The hormone also decreases transcription of the genes encoding fructose-1,6-bisphosphatase and glucose-6-phosphatase and increases transcription of glycolytic enzymes such as glucokinase and pyruvate kinase, and lipogenic enzymes such as fatty acid synthase and acetyl-CoA carboxylase. Thus, insulin resistance has profound effects on all carbohydrate-metabolism pathways in the liver. The abovementioned aspects of glucose metabolism in the liver—GLY, GNG, GGN, and GGL—can be potentially quantified by ^{18}FDG, [^{11}C]glucose, and [^{11}C]lactate with the appropriate mathematical model.

The liver takes up nonesterified fatty acids (NEFA) from blood in proportion to their concentration via one of several transporters (including fatty acid transport protein (FATP), fatty acid translocase (FAT), or CD36) diffusion. Within the hepatocytes, NEFA are bound by fatty acid binding protein (FABP) or activated by acyl-CoA synthase (ACS) to form FA-CoA which are transported to intracellular compartments for metabolism or to the nucleus to interact with transcription factors regulating metabolism. Metabolism of NEFA or FA-CoA in the liver takes one of several forms: oxidation, either mitochondrial or peroxisomal; lipid synthesis in the form of triglycerides (TG) and phospholipids; and fatty acid synthesis (i.e., *de novo* lipogenesis) (Canbay et al. 2007, Nguyen et al. 2008). Mitochondrial β-oxidation of fatty acids results in the formation of acetyl-CoA, which can be oxidized completely to CO_2 in the mitochondrial tricarboxylic acid (TCA) cycle. Entry of fatty acids of more than 14-carbons into the mitochondria is facilitated by the activity of CPT-I and CPT-II. FA with 12-carbons or less pass through the mitochondrial membrane, independent of CPT-I, and are activated by ACS within the mitochondria. The activity of CPT-I is indirectly inhibited by insulin to prevent simultaneous oxidation and synthesis of fatty acids. FA oxidation is the predominant source of energy under fasting periods (Canbay et al. 2007). After a carbohydrate-rich meal, high glucose and insulin levels enhance hepatic FA synthesis (Begriche et al. 2006). In addition, pyruvate derived from glycolysis is converted in the mitochondria into acetyl-CoA and then into citrate. In the cytosol, citrate regenerates acetyl-CoA upon which acetyl-CoA carboxylase acts to produce malonyl-CoA. Malonyl-CoA is used as a precursor in FA synthesis catalyzed by fatty acid synthase (FAS). High levels of malonyl-CoA also inhibit CPT-I and decrease FA oxidation (Begriche et al. 2006). Therefore, FAs which are not metabolized are instead shuttled toward the formation of TG or secreted as VLDL. The first committed step in TG synthesis is catalyzed by glycerol-3-phosphate acyltransferase (GPAT). Insulin stimulates fatty acid synthesis in the liver along with the formation of TG. In diabetes, insulin resistance interferes with many of the above processes resulting in enhanced GNG, synthesis of FA, and release of TG as VLDL. Using the imaging markers [^{11}C]palmitate and [^{18}F]FTHA and radiolabeled forms of VLDL or its derivatives (TG), it is potentially feasible to quantify aspects of FA oxidation and storage (as TG) in conjunction with mathematical models of tracer kinetics.

4.4.2.2 Substrate Metabolism in Muscle

Glucose uptake into muscle is insulin dependent via GLUT4 transporter protein. Upon transport across the muscle cell membrane, glucose is rapidly phosphorylated by hexokinase to glucose-6-phosphate

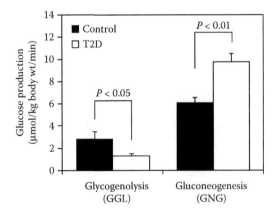

FIGURE 4.6 Glucose production rates from GGL and GNG in control and T2D patients. (Numerical values adapted from Magnusson, I. et al. *J Clin Invest* 90(4): 1323–1327.)

(G6P) which rapidly traps glucose inside the muscle cell. Unlike the liver, skeletal muscle does not contain the enzyme glucose-6-phosphatase. G6P occupies the first branch point in glucose metabolism for oxidative energy and glycogen synthesis by glycogen synthase. Muscle accounts for 60–70% of whole body glucose uptake (Smith 2002). In the fed state, insulin promotes removal of glucose from the periphery which is almost exclusively stored as glycogen (Shulman et al. 1990) via activation of glycogen synthase. This enables energy to be released anaerobically via glycolysis. Muscle does not rely on glucose or glycogen for energy during basal state when insulin levels are low. In insulin resistance, glucose uptake is diminished and muscle glycogen synthesis is impaired largely as a result of reduced intracellular glucose translocation (Hunter and Garvey 1998). For example, it has been shown that glycogen concentrations are 30% lower in T2D patients than normal control (Carey et al. 2003, Shulman et al. 1990). Moreover, the rate of glycogen synthesis in skeletal muscle is approximately 50% lower in diabetic patients compared with normal controls (Shulman et al. 1990). Finally, in a normal physiological response, increased glucose levels promote glucose oxidation and lipid storage, and inhibit fatty acid oxidation. Conversely, increase in fatty acid availability enhances flux of NEFA into the muscle facilitated by FAT/CD36 and FATP-I transport proteins. Increased fatty acid oxidation induces production of both acetyl-CoA and citrate which in turn inhibit pyruvate dehydrogenase and phosphofructokinase activities. Increased supply and/or decreased oxidation of lipids beyond the oxidative capacity leads to accumulation of intracellular lipid metabolites. The content of intramuscular triglycerides (IMTG) has been linked to insulin resistance (Phillips et al. 1996). Falholt et al. showed that the IMTG content in diabetic patients is significantly larger than in healthy individuals and that IMTG levels inversely correlate with insulin levels (Falholt et al. 1988). In addition, a significant inverse relationship between insulin-induced intramuscular activation of glycogen synthase and muscle IMTG has been observed (Phillips et al. 1996).

4.4.2.3 Metabolic Targets in Adipose Tissue

Two types of adipose tissue have been described: white adipose tissue (WAT) and brown adipose tissue (BAT). WAT is specialized to store TG and to release FFA in response to changing energy requirements. In addition, mammals have a second terminally differentiated adipose cell type that is composed of BAT, which is involved in the dissipation of energy via heat generation. BAT is characterized by the presence of multiple lipid droplets and a high number of mitochondria (Nicholls and Locke 1984). While humans have negligible distribution of BAT, both WAT and BAT are insulin target cells, and as such, can be used to study insulin resistance in the progression of diabetes. Intracellular glucose transport into adipocytes is insulin dependent via GLUT4. It is estimated that adipose tissue accounts for about 10% of insulin-stimulated glucose uptake (Smith 2002). Insulin stimulates glucose uptake, promotes lipogenesis while suppressing lipolysis, and enhances free fatty acid flux from the blood stream. Insulin stimulates fatty acid synthesis in adipose tissue along with formation and storage of TG. Fatty acid synthesis is increased by activation and phosphorylation of acetyl-CoA carboxylase while fatty acid oxidation is suppressed by the inhibition of carnitine acyltransferase. TG synthesis is stimulated by esterification of glycerol phosphate while TG breakdown is suppressed by dephosphorylation of hormone-sensitive lipase (Hunter and Garvey 1998). GLUT4 mRNA and protein are downregulated in adipose tissue in the setting of obesity and T2D in both humans and rodents (Shepherd and Khan 1999). Insulin resistance in adipose tissues leads to increased lipolysis, with subsequent release of glycerol and free fatty acids (FFA) into the circulation. Circulating FFA derived from adipocytes are elevated in many insulin-resistant states and have been suggested to contribute the insulin resistance of diabetes and obesity by inhibiting glucose uptake, glycogen synthesis, and glucose oxidation, and increasing hepatic glucose output (Bergman and Ader 2000). Elevated FFAs are also associated with a reduction in insulin-stimulated IRS-1 phosphorylation and IRS-1-associated PI(3)K activity (Shulman 2000). The link between increased circulating FFA and insulin resistance may involve accumulation of TG and FA-derived metabolites (diacylglycerol, fatty acyl-CoA, ceramides) in muscle (i.e., IMTG) and liver (Hwang et al. 2001, Kim et al. 2001). In addition to its role as a storage depot for lipids, the fat cells produce and secrete a number of hormones, referred to as adipokines, which influence metabolism at several levels. Interestingly, PPARγ subtype of PPAR is expressed at high levels in adipose tissue in contrast to a variety of other tissues (Chawla et al. 1994). Activators of PPARγ regulate fatty acid metabolism and can induce adipocyte differentiation and expression of genes involved in the transport and

Chapter 4

sequestration of fatty acids (Martin et al. 1997). Thiazoladinediones (TZDs), which include the drug Rosiglitazone, improve insulin sensitivity by binding to the ligand-binding domain, thus activating PPARγ and increasing transcription of target genes. In imaging adipose tissue, ^{18}FDG can be used to assess glucose uptake; [^{11}C]palmitate and ^{18}FTHA can be used to assess FA oxidation with the former providing also a measure of FA storage as TG.

4.4.2.4 Myocardial Substrate Metabolism

There is increasing evidence suggesting that diabetic patients have a predisposition to heart failure resulting from impairment in heart muscle contraction, particularly abnormalities in diastolic function (Fang et al. 2004). This impairment in muscle contraction, termed diabetic cardiomyopathy, is independent of vascular abnormalities as diastolic dysfunction is evident in both T1D and T2D patients. One prominent hypothesis argues that diabetic cardiomyopathy is a consequence of alterations in myocardial fuel metabolism (Lopaschuk 2002, Lopaschuk et al. 1994). We have quantified myocardial substrate metabolism in an animal model of T2D (Shoghi et al. 2008). Specifically, myocardial glucose utilization (MGU) and fatty acid utilization (MFAU) were quantified in Zucker diabetic fatty (ZDF) rats and their lean littermates using small animal PET with FDG. We showed, noninvasively, that the diabetic heart utilizes significantly less glucose than nondiabetic heart (Figure 4.7a) independent of age (Shoghi et al. 2008). Moreover, we showed that the

diabetic heart utilizes significantly more fatty acids (Figure 4.7b) attributed primarily to enhanced fatty acid oxidation (Figure 4.7c) (Welch et al. 2006).

We validated PET findings by performing gene expression analysis and Western blot analysis. Our data indicated that the decrease in glucose utilization observed in the diabetic heart is partially attributed to decrease in GLUT4 gene expression as shown in Figure 4.8a, which was confirmed by protein expression analysis (Figure 4.8b). Furthermore, we demonstrated that PET measures of MGU$_{UpR}$, a measure which is independent of peripheral glucose levels, significantly correlated with gene expression of GLUT4 (Figure 4.9). The increase in fatty acid utilization and oxidation is attributed to enhanced expression of fatty acid transporters CD36 and FATP-I (Figure 4.10a) and increased expression of MCAD (Figure 4.10b), a key enzyme in the oxidation of fatty acids.

4.4.2.5 Assessing Efficacy of Antidiabetes Therapies

In subsequent work, we assessed the efficacy of antidiabetic therapies metformin (16.6 mg/kg/day) and rosiglitazone (4 mg/kg) on myocardial substrate metabolism. Metformin (MET), a biguanide, is thought to act by decreasing hepatic glucose production through activation of AMP-activated protein kinase (AMPK)—an enzyme that plays an important role in insulin signaling, whole body energy balance, and the metabolism of glucose and fats (Towler and Hardie 2007). Activation of AMPK is required for metformin's

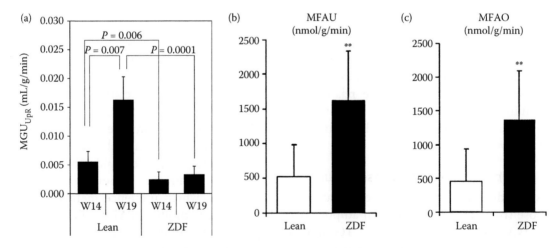

FIGURE 4.7 Alterations in myocardial substrate metabolic. (a) MGU$_{UpR}$ at the age of 14 weeks (W14) and 19 weeks (W19) in lean and ZDF rats. (Adapted from Shoghi, K. I. et al. 2008. *J Nucl Med* 49(8): 1320–1327.) Significance values are denoted above bar plots. (b) Diabetic rats exhibit increased myocardial fatty acid utilization which is attributed to (c) enhanced myocardial fatty acid oxidation. (Adapted from Welch, M. J. et al. 2006. *J Nucl Med* 47(4): 689–697.) $P < 0.05$ was considered significant.

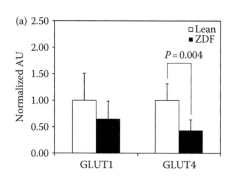
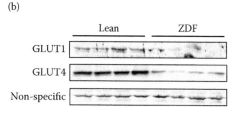

FIGURE 4.8 (a) GLUT1 and GLUT4 gene expression levels in lean and ZDF rats at W19. Gene expression was normalized to levels observed in lean rats. Significance values are denoted above bar plots. $P < 0.05$ was considered significant. (b) Representative autoradiographic results of Western blot analyses performed with antibodies for GLUT-1 and GLUT-4 on whole-cell lysates of hearts of lean and ZDF rats at W19.

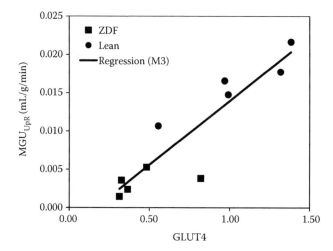

FIGURE 4.9 Regression model to characterize dependence of MGU_{UpR} on gene expression of GLUT-4 with $R^2 = 0.81$, which is significant at $P = 0.0003$.

inhibitory effect on the production of glucose by liver cells (Zhou et al. 2001) resulting in reduced plasma glucose concentration (Johnson et al. 1993). Rosiglitazone (RSG), on the other hand, targets the peroxisome proliferation-activated receptor-gamma (PPARγ) family of receptors. PPARγ agonists have profound effects on glucose and lipid metabolism. This class of drugs has been shown to improve insulin sensitivity in various animal models (Fujita et al. 1983).

In characterizing the therapeutic effects of MET and RSG, each rat underwent a 60 min dynamic PET acquisition with [18]FDG (0.5–0.8 mCi) to characterize glucose utilization; a 20 min dynamic PET acquisition with [11C]acetate (0.6–0.8 mCi) to quantify myocardial blood flow (MBF) and myocardial oxygen consumption (MVO$_2$); followed by a 20 min acquisition with [11C]palmitate (0.6–0.8 mCi) to quantify myocardial fatty acid metabolism. Myocardial glucose utilization was quantified as described earlier (Shoghi et al. 2008) while the kinetics of [11C]palmitate was quantified using the compartmental model depicted in Figure 4.3. We showed,

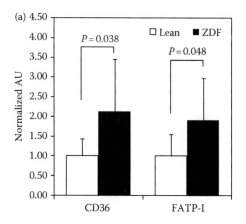
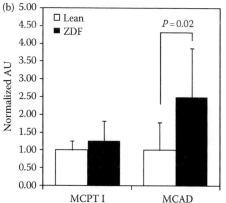

FIGURE 4.10 Expression analysis of fatty acid transporters (a) and gene regulating fatty acid oxidation (b) in lean and ZDF rats.

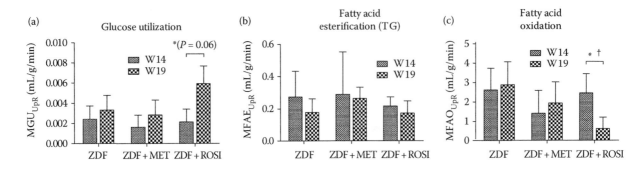

FIGURE 4.11 (a) Myocardial fatty acid utilization uptake rate (MFAU$_{UpR}$), (b) myocardial fatty acid esterification (storage) as TG, and (c) myocardial fatty acid oxidation uptake rate (MFAO$_{UpR}$) in untreated (ZDF), Metformin-treated (ZDF + MET) and rosiglitazone-treated (ZDF + ROSI) ZDF rats at week 14 (W14) and week 19 (W19). MGU$_{UpR}$, MFAE$_{UpR}$, and MFAO$_{UpR}$ represent the *intrinsic* capacity of the heart to utilize glucose and store and oxidize fatty acids, respectively, independent of the concentration of glucose and free fatty acids in plasma. *, Denotes that the treatment is significantly better than no treatment; †, denotes that the rosiglitazone treatment is significantly better than metformin treatment.

noninvasively, that rosiglitazone-treated ZDF rats exhibited significantly higher MGU$_{UpR}$ ($P = 0.0054$) relative to baseline which is only marginally insignificant in comparison to untreated ZDF rats ($P = 0.06$) (Figure 4.11a). In addition, rosiglitazone-treated rats exhibited a significant reduction in MFAO$_{UpR}$ compared to untreated ZDF rats ($P = 0.0189$) and metformin-treated rats ($P = 0.01$) (Figure 4.11c) while TG storage through esterification was unchanged (Figure 4.11b). The combined effect of enhanced glucose uptake rate and reduction in myocardial fatty acid oxidation rate resulted in a significant net gain in fractional glucose utilization (Shoghi 2009).

In order to gain a mechanistic insight on the effects of metformin and rosiglitazone monotherapy, we characterized expression of key genes in glucose and FA metabolism (Figure 4.12). We observed a significant correlation between MGU$_{UpR}$ and gene expression of GLUT4 following treatment with rosiglitazone, confirming that rosiglitazone-enhanced FDG uptake is attributed to increased insulin sensitivity. In order to

validate PET measures of MFAO$_{UpR}$, we characterized the expression profile of genes involved in FA transport and oxidation. Fatty acids enter the myocyte facilitated by several fatty acid transport proteins, including FATP, FATP$_{PM}$, FAT/CD36. Of the fatty acid transport proteins we examined, we found that FATP-I was downregulated by rosiglitazone treatment. In parallel, rosiglitazone significantly inhibited expression of the gene encoding MCAD, a highly regulated step in mitochondrial β-oxidation of fatty acids. Accordingly, cardiac intrinsic measures of myocardial fatty acid oxidation rates, *MFAO*$_{UpR}$, correlated significantly with both FATP-I and MCAD suggesting that PET measures of MFAO$_{UpR}$ provide a noninvasive measure of myocardial fatty acid oxidation (Table 4.2).

4.4.2.6 Gender Differences in Myocardial Fatty Acid Utilization

Having validated our methodology for quantification of myocardial fatty acid metabolism in rodents, we assessed gender differences in fatty acid metabolism.

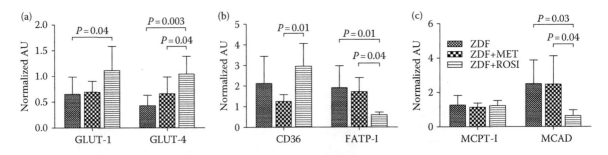

FIGURE 4.12 Gene expression of (a) glucose transporters GLUT1 and GLUT4; (b) Fatty acid transporters FATP-I and CD36; (c) and genes encoding for proteins involved in fatty acid oxidation, namely MCPT-I and MCAD. Legend provided in panel c is applicable to both panels a and b. Significant differences between groups are denoted above bar plots. $P < 0.05$ was considered significant. AU = arbitrary units.

Table 4.2 Pearson Correlation between PET Measures of MGU$_{UpR}$ and MFAO$_{UpR}$ and Gene Expression

		ZDF/ZDF + Met			ZDF/ZDF + Rosi		
		ρ	P	N	ρ	P	N
MGU$_{UpR}$ vs.	GLUT1	0.66	0.02698	9	0.46	0.07531	11
	GLUT4	0.01	0.48850	9	0.71	0.00710*	11
MFAO$_{UpR}$ vs.	CD36	0.61	0.01728	12	−0.08	0.40411	12
	FATP1	0.09	0.39339	11	0.74	0.00441*	11
	mCPT1	0.68	0.00730*	12	0.41	0.09085	12
	MCAD	−0.05	0.43995	11	0.71	0.00741*	11

* Significant following Bonferroni correction for multiple hypothesis testing (Wallenstein et al. 1980); ρ, denotes the Pearson correlation coefficient; P, significance value of correlation; N, number of data points; MGU$_{UpR}$, myocardial glucose utilization uptake rate constant; MFAO$_{UpR}$, myocardial fatty oxidation uptake rate constant.

Age-matched male and female ZDF rats were studied under conditions described earlier. To assess the heart's intrinsic capacity to utilize fatty acids, we quantified measures of myocardial fatty acid oxidation and utilization (MFAO$_{UpR}$ and MFAU$_{UpR}$, respectively). In addition, we quantified absolute measures of MFAO and MFAU. Our data indicate that both lean and ZDF female rats have a higher intrinsic capacity to oxidize fatty acids than male rats ($P < 0.05$) (Figure 4.13). Moreover, absolute oxidation and utilization of fatty acids was significantly ($P < 0.05$) greater in female rats than in male rats, primarily as a result of significant increase in peripheral levels of nonesterified fatty acids (NEFA) more so in female rats than in male rats. Interestingly, female ZDF rats did not become hyperglycemic (i.e., normal levels of glycated hemoglobin, HbA1C) consistent with previous findings (Clark et al. 1983, Tirabassi et al. 2004).

4.4.3 Insulin Cell Signaling Targets

In T2DM and obesity, insulin-stimulated glucose homeostasis is significantly impaired (DeFronzo 1997). The decreased insulin-stimulated glucose

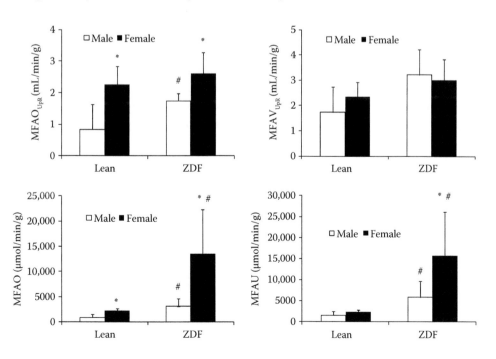

FIGURE 4.13 Gender differences in FA metabolism in ZDF rats. Lean and ZDF female rats have a higher intrinsic capacity to oxidize fatty acids than male rats ($P < 0.05$). Moreover, absolute oxidation and utilization of fatty acids was significantly ($P < 0.05$) greater in female rats than in male rats.

Chapter 4

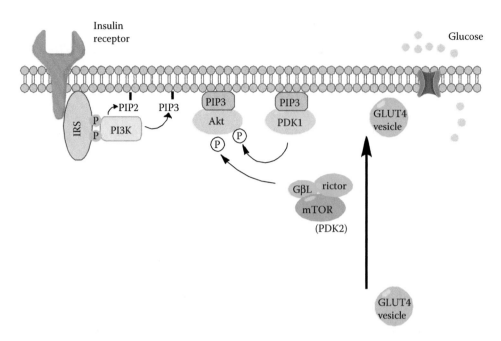

FIGURE 4.14   Insulin signaling pathway. (Adapted from Choi, K. and Kim, Y. B. 2010. *Korean J Intern Med* 25(2): 119–129.)

uptake in peripheral tissues is due to impaired insulin signaling and multiple postreceptor intracellular defects such as impaired glucose transport, reduced glucose oxidation, and glycogen synthesis among others (Bajaj and Defronzo 2003, Bouzakri et al. 2005). Numerous reports suggest that deficiencies in insulin signaling are attributed to increased accumulation of fatty acid and fatty acid metabolites (Bays et al. 2004, Lettner and Roden 2008). Imaging agents which target key pivots of the insulin signaling pathway can provide *in vivo* means of insulin signaling efficiency. The insulin signaling pathway is depicted in Figure 4.14. Binding of insulin to its cell surface receptor activates a cascade of events including receptor autophosphorylation and activation of receptor tyrosine kinases which in turn phosphorylate insulin receptor substrates (IRS) in particular IRS1 and IRS2, IRS3, IRS4 (Taniguchi et al. 2006). Binding of IRSs to regulatory subunits of downstream effectors phosphoinositide 3-kinase (PI3K) (Figure 4.14) activates the phosphoinositide-dependent protein kinase (PDK-1 and PDK-2) resulting in activation of Akt/protein kinase B (PKB) (Farese et al. 2005). Activated Akt, in turn, phosphorylates its 160 kDa substrate (AS160) resulting in translocation of GLUT4 vesicles to the cell surface (Sano et al. 2003). Thus, the IRS-1/PI3K/Akt axis is a viable imaging target to assess the deficiencies in insulin signaling *in vivo*.

References

Acton, P. D., Zhuang, H., and Alavi, A. 2004. Quantification in PET. *Radiol Clin North Am* 42(6): 1055–1062, viii.

Anlauf, M., Eissele, R., Schafer, M. K. et al. 2003. Expression of the two isoforms of the vesicular monoamine transporter (VMAT1 and VMAT2) in the endocrine pancreas and pancreatic endocrine tumors. *J Histochem Cytochem* 51(8): 1027–1040.

Atkinson, M. A. and Leiter, E. H. 1999. The NOD mouse model of type 1 diabetes: As good as it gets? *Nat Med* 5(6): 601–604.

Bajaj, M. and Defronzo, R. A. 2003. Metabolic and molecular basis of insulin resistance. *J Nucl Cardiol* 10(3): 311–323.

Bays, H., Mandarino, L., and DeFronzo, R. A. 2004. Role of the adipocyte, free fatty acids, and ectopic fat in pathogenesis of type 2 diabetes mellitus: Peroxisomal proliferator-activated receptor agonists provide a rational therapeutic approach. *J Clin Endocrinol Metab* 89(2): 463–478.

Begriche, K., Igoudjil, A., Pessayre, D., and Fromenty, B. 2006. Mitochondrial dysfunction in NASH: Causes, consequences and possible means to prevent it. *Mitochondrion* 6(1): 1–28.

Bergman, R. N. and Ader, M. 2000. Free fatty acids and pathogenesis of type 2 diabetes mellitus. *Trends Endocrinol Metab* 11(9): 351–356.

Boellaard, R. 2009. Standards for PET image acquisition and quantitative data analysis. *J Nucl Med* 50(Suppl 1): 11S–20S.

Bolzan, A. D. and Bianchi, M. S. 2002. Genotoxicity of streptozotocin. *Mutat Res* 512(2–3): 121–134.

Bouzakri, K., Koistinen, H. A., and Zierath, J. R. 2005. Molecular mechanisms of skeletal muscle insulin resistance in type 2 diabetes. *Curr Diabetes Rev* 1(2): 167–174.

Brennan, A. M. and Mantzoros, C. S. 2006. Drug insight: The role of leptin in human physiology and pathophysiology—Emerging clinical applications. *Nat Clin Pract Endocrinol Metab* 2(6): 318–327.

Canbay, A., Bechmann, L., and Gerken, G. 2007. Lipid metabolism in the liver. *Z Gastroenterol* 45(1): 35–41.

Carey, P. E., Halliday, J., Snaar, J. E., Morris, P. G., and Taylor, R. 2003. Direct assessment of muscle glycogen storage after mixed meals in normal and type 2 diabetic subjects. *Am J Physiol Endocrinol Metab* 284(4): E688–694.

Chawla, A., Schwarz, E. J., Dimaculangan, D. D., and Lazar, M. A. 1994. Peroxisome proliferator-activated receptor (PPAR) gamma: Adipose-predominant expression and induction early in adipocyte differentiation. *Endocrinology* 135(2): 798–800.

Choi, K. and Kim, Y. B. 2010. Molecular mechanism of insulin resistance in obesity and type 2 diabetes. *Korean J Intern Med* 25(2): 119–129.

Clark, J. B., Palmer, C. J., and Shaw, W. N. 1983. The diabetic Zucker fatty rat. *Proc Soc Exp Biol Med* 173(1): 68–75.

Dahlquist, G. 1998. The aetiology of type 1 diabetes: An epidemiological perspective. *Acta Paediatr Suppl* 425: 5–10.

DeFronzo, R. A. 1997. Insulin resistance: A multifaceted syndrome responsible for NIDDM, obesity, hypertension, dyslipidaemia and atherosclerosis. *Neth J Med* 50(5): 191–197.

Falholt, K., Jensen, I., Lindkaer Jensen, S. et al. 1988. Carbohydrate and lipid metabolism of skeletal muscle in type 2 diabetic patients. *Diabet Med* 5(1): 27–31.

Fang, Y. H. and Muzic, R. F., Jr. 2008. Spillover and partial-volume correction for image-derived input functions for small-animal 18F-FDG PET studies. *J Nucl Med* 49(4): 606–614.

Fang, Z. Y., Prins, J. B., and Marwick, T. H. 2004. Diabetic cardiomyopathy: Evidence, mechanisms, and therapeutic implications. *Endocr Rev* 25(4): 543–567.

Farese, R. V., Sajan, M. P., and Standaert, M. L. 2005. Insulin-sensitive protein kinases (atypical protein kinase C and protein kinase B/Akt): Actions and defects in obesity and type II diabetes. *Exp Biol Med (Maywood)* 230(9): 593–605.

Ferl, G. Z., Zhang, X., Wu, H. M., and Huang, S. C. 2007. Estimation of the 18F-FDG input function in mice by use of dynamic small-animal PET and minimal blood sample data. *J Nucl Med* 48(12): 2037–2045.

Flier, J. S. 2001. Diabetes. The missing link with obesity? *Nature* 409(6818): 292–293.

Fujita, T., Sugiyama, Y., Taketomi, S. et al. 1983. Reduction of insulin resistance in obese and/or diabetic animals by 5-[4-(1-methylcyclohexylmethoxy)benzyl]-thiazolidine-2,4-dione (ADD-3878, U-63,287, ciglitazone), a new antidiabetic agent. *Diabetes* 32(9): 804–810.

Godbole, V. and York, D. A. 1978. Lipogenesis *in situ* in the genetically obese Zucker fatty rat (fa/fa): Role of hyperphagia and hyperinsulinaemia. *Diabetologia* 14(3): 191–197.

Goto, Y., Kakizaki, M., and Masaki, N. 1976. Production of spontaneous diabetic rats by repetition of selective breeding. *Tohoku J Exp Med* 119(1): 85–90.

Guo, Y., Gao, F., Wang, S. et al. 2009. *In vivo* mapping of temporospatial changes in glucose utilization in rat brain during epileptogenesis: An 18F-fluorodeoxyglucose-small animal positron emission tomography study. *Neuroscience* 162(4): 972–979.

Harris, P. E., Ferrara, C., Barba, P. et al. 2008. VMAT2 gene expression and function as it applies to imaging beta-cell mass. *J Mol Med* 86(1): 5–16.

Herberg, L. and Coleman, D. L. 1977. Laboratory animals exhibiting obesity and diabetes syndromes. *Metabolism* 26(1): 59–99.

Hunter, S. J. and Garvey, W. T. 1998. Insulin action and insulin resistance: Diseases involving defects in insulin receptors, signal transduction, and the glucose transport effector system. *Am J Med* 105(4): 331–345.

Hwang, J. H., Pan, J. W., Heydari, S., Hetherington, H. P., and Stein, D. T. 2001. Regional differences in intramyocellular lipids in humans observed by *in vivo* 1H-MR spectroscopic imaging. *J Appl Physiol* 90(4): 1267–1274.

Ichise, M. and Harris, P. E. 2010. Imaging of {beta}-cell mass and function. *J Nucl Med* 51(7): 1001–1004.

Ikeda, H., Shino, A., Matsuo, T., Iwatsuka, H., and Suzuoki, Z. 1981. A new genetically obese-hyperglycemic rat (Wistar fatty). *Diabetes* 30(12): 1045–1050.

Ingalls, A. M., Dickie, M. M., and Snell, G. D. 1950. Obese, a new mutation in the house mouse. *J Hered* 41(12): 317–318.

Ito, K., Kubota, K., Morooka, M. et al. 2009. Diagnostic usefulness of (18)F-FDG PET/CT in the differentiation of pulmonary artery sarcoma and pulmonary embolism. *Ann Nucl Med* 23(7): 671–676.

Johnson, A. B., Webster, J. M., Sum, C. F. et al. 1993. The impact of metformin therapy on hepatic glucose production and skeletal muscle glycogen synthase activity in overweight type II diabetic patients. *Metabolism* 42(9): 1217–1222.

Kahn, B. B. and Flier, J. S. 2000. Obesity and insulin resistance. *J Clin Invest* 106(4): 473–481.

Kanda, S., Nakashima, R., Takahashi, K. et al. 2009. Potent antidiabetic effects of rivoglitazone, a novel peroxisome proliferator-activated receptor-gamma agonist, in obese diabetic rodent models. *J Pharmacol Sci* 111(2): 155–166.

Kim, J. K., Fillmore, J. J., Chen, Y. et al. 2001. Tissue-specific overexpression of lipoprotein lipase causes tissue-specific insulin resistance. *Proc Natl Acad Sci USA* 98(13): 7522–7527.

Kim, S. J., Lee, J. S., Im, K. C. et al. 2008. Kinetic modeling of 3′-deoxy-3′-18F-fluorothymidine for quantitative cell proliferation imaging in subcutaneous tumor models in mice. *J Nucl Med* 49(12): 2057–2066.

Kung, M. P., Hou, C., Lieberman, B. P. et al. 2008. *In vivo* imaging of beta-cell mass in rats using 18F-FP-(+)-DTBZ: A potential PET ligand for studying diabetes mellitus. *J Nucl Med* 49(7): 1171–1176.

Laforest, R., Sharp, T. L., Engelbach, J. A. et al. 2005. Measurement of input functions in rodents: Challenges and solutions. *Nucl Med Biol* 32(7): 679–685.

Larsen, P. J., Wulff, E. M., Gotfredsen, C. F. et al. 2008. Combination of the insulin sensitizer, pioglitazone, and the long-acting GLP-1 human analog, liraglutide, exerts potent synergistic glucose-lowering efficacy in severely diabetic ZDF rats. *Diabetes Obes Metab* 10(4): 301–311.

Lettner, A. and Roden, M. 2008. Ectopic fat and insulin resistance. *Curr Diab Rep* 8(3): 185–191.

Lopaschuk, G. D. 2002. Metabolic abnormalities in the diabetic heart. *Heart Fail Rev* 7(2): 149–159.

Lopaschuk, G. D., Belke, D. D., Gamble, J., Itoi, T., and Schonekess, B. O. 1994. Regulation of fatty acid oxidation in the mammalian heart in health and disease. *Biochim Biophys Acta* 1213(3): 263–276.

Lucignani, G., Paganelli, G., and Bombardieri, E. 2004. The use of standardized uptake values for assessing FDG uptake with PET in oncology: A clinical perspective. *Nucl Med Commun* 25(7): 651–656.

Chapter 4

Maffei, A., Liu, Z., Witkowski, P. et al. 2004. Identification of tissue-restricted transcripts in human islets. *Endocrinology* 145(10): 4513–4521.

Magnusson, I., Rothman, D. L., Katz, L. D., Shulman, R. G., and Shulman, G. I. 1992. Increased rate of gluconeogenesis in type II diabetes mellitus. A 13C nuclear magnetic resonance study. *J Clin Invest* 90(4): 1323–1327.

Makino, S., Kunimoto, K., Muraoka, Y. et al. 1980. Breeding of a non-obese, diabetic strain of mice. *Jikken Dobutsu* 29(1): 1–13.

Makino, S., Muraoka, Y., Kishimoto, Y., and Hayashi, Y. 1985. Genetic analysis for insulitis in NOD mice. *Jikken Dobutsu* 34(4): 425–431.

Martin, G., Schoonjans, K., Lefebvre, A. M., Staels, B., and Auwerx, J. 1997. Coordinate regulation of the expression of the fatty acid transport protein and acyl-CoA synthetase genes by PPARalpha and PPARgamma activators. *J Biol Chem* 272(45): 28210–28217.

McGarry, J. D. 2002. Banting lecture 2001: Dysregulation of fatty acid metabolism in the etiology of type 2 diabetes. *Diabetes* 51(1): 7–18.

Michael, M. D., Kulkarni, R. N., Postic, C. et al. 2000. Loss of insulin signaling in hepatocytes leads to severe insulin resistance and progressive hepatic dysfunction. *Mol Cell* 6(1): 87–97.

Miralles, F. and Portha, B. 2001. Early development of beta-cells is impaired in the GK rat model of type 2 diabetes. *Diabetes* 50(Suppl 1): S84–S88.

Nguyen, P., Leray, V., Diez, M. et al. 2008. Liver lipid metabolism. *J Anim Physiol Anim Nutr (Berl)* 92(3): 272–283.

Nicholls, D. G. and Locke, R. M. 1984. Thermogenic mechanisms in brown fat. *Physiol Rev* 64(1): 1–64.

Ning, X. H., Meng, Q. L., Y. Wang, Z., and Bai, C. M. 2009. Application of 18F-fluorodeoxyglucose positron emission tomography in diagnosis of malignant diseases. *Chin Med Sci J* 24(2): 117–121.

Permutt, M. A., Wasson, J., and Cox, N. 2005. Genetic epidemiology of diabetes. *J Clin Invest* 115(6): 1431–1439.

Phillips, D. I., Caddy, S., Ilic, V. et al. 1996. Intramuscular triglyceride and muscle insulin sensitivity: Evidence for a relationship in nondiabetic subjects. *Metabolism* 45(8): 947–950.

Pilkis, S. J. and Granner, D. K. 1992. Molecular physiology of the regulation of hepatic gluconeogenesis and glycolysis. *Annu Rev Physiol* 54: 885–909.

Pollok, K. E., Lahn, M., Enas, N. et al. 2009. *In Vivo* measurements of tumor metabolism and growth after administration of enzastaurin using small animal FDG positron emission tomography. *J Oncol* 2009: 596560.

Portha, B. 2005. Programmed disorders of beta-cell development and function as one cause for type 2 diabetes? The GK rat paradigm. *Diabetes Metab Res Rev* 21(6): 495–504.

Portha, B., Giroix, M. H., Serradas, P. et al. 2001. beta-cell function and viability in the spontaneously diabetic GK rat: Information from the GK/Par colony. *Diabetes* 50(Suppl 1): S89–S93.

Sano, H., Kane, S., Sano, E. et al. 2003. Insulin-stimulated phosphorylation of a Rab GTPase-activating protein regulates GLUT4 translocation. *J Biol Chem* 278(17): 14599–14602.

Shepherd, P. R. and Kahn, B. B. 1999. Glucose transporters and insulin action–implications for insulin resistance and diabetes mellitus. *N Engl J Med* 341(4): 248–257.

Shoghi, K. I. 2009. Quantitative small animal PET. *Q J Nucl Med Mol Imaging* 53(4): 365–373.

Shoghi, K. I., Finck, B. N., Schechtman, K. B. et al. 2009. *In Vivo* metabolic phenotyping of myocardial substrate metabolism in rodents: Differential efficacy of metformin and rosiglitazone monotherapy. *Circ Cardiovasc Imaging* 2(5): 373–381.

Shoghi, K. I., Gropler, R. J., Sharp, T. et al. 2008. Time course of alterations in myocardial glucose utilization in the Zucker diabetic fatty rat with correlation to gene expression of glucose transporters: A small-animal PET investigation. *J Nucl Med* 49(8): 1320–1327.

Shoghi, K. I. and Welch, M. J. 2007. Hybrid image and blood sampling input function for quantification of small animal dynamic PET data. *Nucl Med Biol* 34(8): 989–994.

Shulman, G. I. 2000. Cellular mechanisms of insulin resistance. *J Clin Invest* 106(2): 171–176.

Shulman, G. I., Rothman, D. L., Jue, T. et al. 1990. Quantitation of muscle glycogen synthesis in normal subjects and subjects with non-insulin-dependent diabetes by 13C nuclear magnetic resonance spectroscopy. *N Engl J Med* 322(4): 223–228.

Smith, U. 2002. Impaired ('diabetic') insulin signaling and action occur in fat cells long before glucose intolerance—is insulin resistance initiated in the adipose tissue? *Int J Obes Relat Metab Disord* 26(7): 897–904.

Southworth, R., Parry, C. R., Parkes, H. G., Medina R. A., and Garlick, P. B. 2003. Tissue-specific differences in 2-fluoro-2-deoxyglucose metabolism beyond FDG-6-P: A 19F NMR spectroscopy study in the rat. *NMR Biomed* 16(8): 494–502.

Souza, F., Simpson, N., Raffo, A. et al. 2006. Longitudinal noninvasive PET-based beta cell mass estimates in a spontaneous diabetes rat model. *J Clin Invest* 116(6): 1506–1513.

Stumvoll, M., Goldstein, B. J., and van Haeften, T. W. 2005. Type 2 diabetes: Principles of pathogenesis and therapy. *Lancet* 365(9467): 1333–1346.

Su, Y. and Shoghi, K. I. 2010. Single-input-dual-output modeling of image-based input function estimation. *Mol Imaging Biol* 12(3): 286–294.

Su, Y., Welch, M. J., and Shoghi, K. I. 2007. The application of maximum likelihood factor analysis (MLFA) with uniqueness constraints on dynamic cardiac microPET data. *Phys Med Biol* 52(8): 2313–2334.

Sutherland, C., O'Brien, R. M., and Granner, D. K. 1996. New connections in the regulation of PEPCK gene expression by insulin. *Philos Trans R Soc Lond B Biol Sci* 351(1336): 191–199.

Suzuki, K., Nishioka, T. Homma, A. et al. 2009. Value of fluorodeoxyglucose positron emission tomography before radiotherapy for head and neck cancer: Does the standardized uptake value predict treatment outcome? *Jpn J Radiol* 27(6): 237–242.

Taniguchi, C. M., Emanuelli, B., and Kahn, C. R. 2006. Critical nodes in signalling pathways: Insights into insulin action. *Nat Rev Mol Cell Biol* 7(2): 85–96.

Tirabassi, R. S., Flanagan, J. F., Wu, T. et al. 2004. The BBZDR/Wor rat model for investigating the complications of type 2 diabetes mellitus. *Ilar J* 45(3): 292–302.

Towler, M. C. and Hardie, D. G. 2007. AMP-activated protein kinase in metabolic control and insulin signaling. *Circ Res* 100(3): 328–341.

Tsao, H. H., Lin, K. J., Juang, J. H. et al. 2010. Binding characteristics of 9-fluoropropyl-(+)-dihydrotetrabenzazine (AV-133) to the vesicular monoamine transporter type 2 in rats. *Nucl Med Biol* 37(4): 413–419.

Vander Borght, T. M., Kilbourn, M. R. Koeppe, R. A. et al. 1995. *In vivo* imaging of the brain vesicular monoamine transporter. *J Nucl Med* 36(12): 2252–2260.

Wahl, R. L., Jacene, H. Kasamon, Y., and Lodge, M. A. 2009. From RECIST to PERCIST: Evolving considerations for PET response criteria in solid tumors. *J Nucl Med* 50(Suppl 1): 122S–150S.

Wallenstein, S., Zucker, C. L., and Fleiss, J. L. 1980. Some statistical methods useful in circulation research. *Circ Res* 47(1): 1–9.

Wang, Z. and Gleichmann, H. 1998. GLUT2 in pancreatic islets: Crucial target molecule in diabetes induced with multiple low doses of streptozotocin in mice. *Diabetes* 47(1): 50–56.

Watabe, H., Ikoma, Y., Kimura, Y., Naganawa, M., and Shidahara, M. 2006. PET kinetic analysis—Compartmental model. *Ann Nucl Med* 20(9): 583–588.

Welch, M. J., Lewis, J. S., Kim, J. et al. 2006. Assessment of myocardial metabolism in diabetic rats using small-animal PET: A feasibility study. *J Nucl Med* 47(4): 689–697.

Wienhard, K. 2002. Measurement of glucose consumption using [18F]fluorodeoxyglucose. *Methods* 27(3): 218–225.

Wu, H. M., Sui, G., Lee, C. C. et al. 2007. *In vivo* quantitation of glucose metabolism in mice using small-animal PET and a microfluidic device. *J Nucl Med* 48(5): 837–845.

Zahner, D. and Malaisse, W. J. 1990. Kinetic behaviour of liver glucokinase in diabetes. I. Alteration in streptozotocin-diabetic rats. *Diabetes Res* 14(3): 101–108.

Zhou, S., Palmeira C. M., and Wallace, K. B. 2001. Doxorubicin-induced persistent oxidative stress to cardiac myocytes. *Toxicol Lett* 121(3): 151–157.

Chapter 4

Radiopharmacology and Radiopharmacy

5. Targeted Radiopharmaceuticals
Advances in Radiopharmacology and Radiopharmacy

Cathy S. Cutler and Sally W. Schwarz

The development goal for most targeted radiopharmaceuticals (RPs) is to translate them from the bench to the clinic, and there are a number of steps that must be taken to accurately navigate the translation successfully. The first and most important step in the process is to determine the clinical end point and match it with a target that is both relevant and capable of accurately assessing the end point. Although many imaging targets have been investigated, none have been validated to the point they are accepted as legitimate end points for clinical trials (CTs).

The push for individualized treatments and more accurate assessment of patients is highlighting the growing need for validated targets. To navigate a target successfully through the milieu, it is necessary to understand the different stages and the role that pharmacology and pharmacy play in validating a target. A subsequent step after selecting a target is evaluating it, often by imaging in animal models. This involves developing targets with different chemical properties, determining how these properties affect the *in vivo* characteristics of the drug to determine targets with optimal properties for initial testing in preclinical studies. The choice of animal model or models is critical to obtain an accurate assessment of the target and data that can be directly applied to humans.

Noninvasive imaging techniques are critical tools for evaluating the pharmacology of targets at different time points in an intact living system. The information gathered from this stage is critical since it will determine how successfully the target will be translated from the research phase to evaluation in the clinical setting. Additionally, advancing the target development will reduce overall costs. Clinical trials require formalized production and quality control (QC) testing that is highly regulated, often changing, and at times governed by a number of differing government agencies. Radiopharmacy can bridge the gap from research and development production to a validated process, which produces a consistent quality product that can be assessed accurately in humans and used in a CT. This chapter summarizes the symposium on pharmacology and pharmacy that was presented at the 18th ISRS meeting in Edmonton, Canada in 2009. It also discusses elements required to integrate pharmacology and pharmacy successfully into target development and evaluation for effective validation of a target as a legitimate end point.

Nuclear imaging is noninvasive and therefore unique in its ability to evaluate biological pathways and function at the molecular level without perturbing the systems under investigation. Other techniques require invasive, destructive sampling of cells and tissues that yields a nonuniform, static snapshot at a single point in time. Molecular imaging is used to elucidate basic biology, diagnose and stage disease, and evaluate the *in vitro* properties of proposed drugs and aid in their translation from the bench to the clinic. This requires evaluating the drugs in appropriate animal models to determine distribution, concentration, and clearance in organs and tissues over time, and this can be done most efficiently and effectively by imaging. The ultimate goal is to develop a data set that can accurately predict what drug dose can be administered safely and effectively to humans.

To gain the optimum insight into how a drug will behave, it is important to understand the differences in

Targeted Molecular Imaging. Edited by Michael J. Welch and William C. Eckelman © 2012 Taylor & Francis Group, LLC. ISBN: 978-1-4398-4195-2

Chapter 5

function in the animal models being tested versus humans. Cecilia Giron started the workshop with a presentation on "Radiopharmaceutical Pharmacokinetics in Animals: Critical Considerations," (Giron 2009) pointing out the importance of interspecies differences and the current lack of knowledge of biochemical pathways that can greatly challenge preclinical evaluations of drugs. When performing preclinical testing in animal models, it is imperative to understand the interspecies differences in ATP-binding cassette transporters, plasma protein binding, and the effects of anesthesia, which can result in data that are not translatable to humans.

Initial studies of a drug involve assessing the compound's distribution *in vivo* and determining its absorption, distribution, metabolism, and excretion referred to as ADME studies. Although there have been major advancements in the development of animal models—including knockout models and spontaneous disease models (animals with spontaneous disease similar to humans)—there is no single model that can predict behavior in man. Some of the best models are those of spontaneous origin, such as metastatic prostate cancer in companion dogs, which has been shown to be similar in function and metastatic spread to humans and serves as a good translational model for humans. However, there are a number of interspecies differences that must be considered, including metabolism, plasma protein binding, biliary excretion, glomerular filtration, and intestinal flora.

In addition, there are a number of changes in biochemical pathways and function to consider when choosing a model. Efflux transporters located in the blood–brain barrier and in the endothelial cells of the kidney and liver serve a role in keeping compounds out or in mediating the secretion or elimination of toxins. Significant variation in these transporters occurs in different species, including nonexpression in certain models, levels of expression, capacity, substrate affinity, and differences in functional response, and can change in response to pathophysiological conditions. The result is that pharmacokinetic (PK) data in animals do not always translate to humans. Some of these effects or differences can be minimized by either saturating the transporter or by administering blockers that inhibit their function. Compounds are being developed and evaluated as selective transporter deactivators to gain a better understanding on these interspecies differences and how they alter or can be altered to influence uptake and excretions. Positron emission tomography (PET) tracers containing

Ga-68, F-18, and C-11 have been developed to evaluate these transporters and the effect of blockers on these substrates. In certain cases, they have been shown to be species dependent, highlighting the differences in P-glycoprotein function, expression, transport, capacity, and substrate affinity. Differences in biological structure have also been shown to alter drug uptake, such as differences in the blood–brain barrier and differences noted in hydrophilic/lipophilic content of the brain.

When a drug is administered, typically by intravenous injection, it comes into contact with a number of components in the blood, such as plasma proteins. The distribution and interaction of the drug with cell membranes and receptors can alter depending on its affinity for the protein. Generally, the bound versus unbound fraction of the drug in plasma is considered the principal parameter to determine a drug's availability to distribute into tissues, interact with receptors, and be eliminated by excretion or metabolism. A drug that is tightly bound can be prevented from diffusing into a tissue, and thus the amount of drug that is transported into a tissue correlates to the concentration of free drug present in the plasma (Collins and Klecker 2002, Singh 2006). Serum albumin, present in the highest concentration in the blood and known to bind a broad spectrum of drugs and compounds, is a key determinant in drug PKs (Ascoli et al. 2006, Kragh-Hansen 1981). Differences in serum albumin are observed among species and the resulting altered drug binding is not readily predictable and therefore not easily translated to humans (Collins and Klecker 2002, Kosa et al. 1997). For example, copper (II) bisthiosemicarbazone (PTSM) has been shown to bind to serum albumin differently in five different species (Colby and Morenka 2004). This example highlights the need to understand these species variations to better predict their behavior in humans.

Anesthesia, commonly used in animal testing to prevent the animal from moving during the scan, has been shown to alter biological functioning of the cardiovascular, respiratory, hepatobiliary, and central nervous system (CNS). Careful monitoring of the animal must be done to prevent induction of pulmonary atelectasis, hypoxia, hypothermia, acidosis, and hepatic toxicity (Colby and Morenka 2004). A number of sedation effects have been observed and shown to alter the *in vivo* PKs of a drug, such as blood perfusion and volume, leading to altered uptake, kinetics, and metabolism. Contradictory results have been observed by merely altering the anesthetic agents (Kilbourn et al. 2007). For example, isoflurane is the most commonly used anesthetic and is administered by inhalation through a vaporizer. In rodents, it has demonstrated increased cerebral blood flow and increased extracellular dopamine concentration, resulting in inhibition of the high-affinity D2 receptor state and uptake and clearance of [123I]-IBZM in mice (Meyer et al. 2008). The use of isoflurane/N_2O in monkeys showed no effect in the uptake of [11C] raclopride (Tsukada et al. 1999). Another example is pentobarbital, a commonly used anesthetic, which has been shown to decrease elimination velocity and altered the *in vivo* behavior of [123I]-2-iodo-1-phenylalanine toward the peripheral compartment, resulting in lower tumor uptake, higher blood pool, and kidney uptake values (Kersemans et al. 2006).

Recently, a study was performed to evaluate the distribution of Fluorodeoxyglucose (FDG) in normal mice with differing anesthetics. The anesthetics included no anesthesia, ketamine and xylazine (Ke/Xy), and varying amounts of isoflurane (0.5%, 1%, and 2%). Isoflurane, which is known to result in decrease in blood pressure, respiration, and glomerular filtration rate, showed an increase in FDG uptake in the lungs, heart, kidneys, and intestine that amplified with the dose of anesthetic (Woo et al. 2008). The mice receiving the Ke/Xy showed higher blood glucose and muscle uptake compared to the others. This illustrates how anesthesia can alter the behavior of drugs *in vivo*, but not always in predictable ways.

To produce meaningful data and minimize detrimental effects that can arise when administering anesthesia long term, it is important to keep the animal warm to avoid hypothermia, to monitor respiratory frequency continuously, and to supplement fluids to ensure homeostasis of fluids, electrolyte balance, and blood glucose levels. Differences in response to anesthesia can arise due to differences in gender, strain, age, stress level, immunity, and tumor genesis (Hildebrandt et al. 2008). Genetically engineered rodents have demonstrated a high susceptibility to morbidity and mortality during anesthesia, mandating that special precautions are required. Anesthetics tend to alter normal neurochemistry and their effects need to be accounted for in such studies.

In summary, this presentation highlighted key factors to consider when evaluating RPs in animal models for the purpose of gathering data for translation to humans. To ensure that data can be used accurately to translate the effects of the drugs in humans, it is necessary to take into account factors such as species differences in transporters, plasma protein binding, and influences of anesthesia.

The second presentation in the pharmacology section of the workshop was given by Kooresh Shoghi. "Modeling and Quantitative Small Animal PET" included an introduction of quantitative PET imaging covering kinetic modeling in quantifying small-animal PET imaging, with an emphasis on cardiac imaging with therapy in mind (Shoghi 2009). Besides its ability to measure function noninvasively, PET can provide quantified measurements of physiological parameters.

The standard uptake value is the normal method used for quantitating PET. It is reached by normalizing counts in a region of interest (ROI). While it is a great tool for evaluating disease progress and comparing different agents, it provides little information about the underlying biological process and can be affected by image noise and low resolution, among other things. In order to avoid these problems, kinetic parameters are often modeled which are more quantitative and minimize errors caused by intrasubject variability. Mathematical models can be developed that describe the distribution of the tracer in tissues versus time, termed tracer kinetic modeling. Quantitation by modeling the kinetics of PET agents in tissues provides spatial information through measurement and discernment of rates of flow/perfusion, metabolism, receptor density, and fraction of underlying vasculature and separation of specific from nonspecific signals.

Performing tracer kinetic modeling requires knowledge of several parameters first; the arterial or plasma input function must be known, as well as having a basic understanding of the mechanism of drug action or how the drug interacts with the target. Next, parameter estimation must be performed by fitting the kinetic model to PET data, followed by assessing the goodness of fit. Finally, the model outcome is validated against a gold standard. The input function is the plasma time–activity curve of the tracer in blood as a function of time. Normally, it is performed by withdrawing and analyzing blood samples at various times; however, this is challenging in small animals as their limited blood volume does not allow for serial withdrawals. A number of less invasive imaging techniques have been developed that allow for determining the input function (Shoghi 2009). Which method is used depends on the agent being evaluated and whether the agent is metabolized. For those agents that do not undergo metabolism, such as FDG, the hybrid method termed as maximum likelihood factor analysis is ideal, as it minimizes the number of blood samples that are needed to reconstruct the input function (Shoghi 2009). Tracers that are short-lived and undergo metabolism can be modeled by use of the hybrid image and blood sampling algorithm that derives the peak of the image from recovery-corrected image left ventricle region of interest (LV ROI), and the tail is derived from 5 to 7 blood samples, which are then linked by a Bezier interpolation algorithm (Shoghi et al. 2007).

Quantitation is dependent on the mechanism of the agent and can be categorized into four main groups: metabolic agents, trapped agents, perfusion agents, and receptor–ligand agents. Metabolic agents are namely tracers that allow for the quantification of *in vivo* metabolic pathways such as [^{11}C]glucose. Trapped agents have been designed so that they will be trapped at the initial point of cell uptake and modification in the process. FDG is the most common example in which an OH group has been replaced by a fluorine atom in the glucose structure and becomes trapped in the first step of glycolytic metabolism. Perfusion agents measure blood flow in tissues. Receptor–ligand agents are tracers designed to target receptors.

As many disease states result in changes in glucose metabolism, PET using [^{18}F]FDG is often used to determine the rate of glucose utilization. [^{18}F]FDG is phosphorylated to FDG-6-PO$_4$, which is unable to diffuse across cell membranes.

A number of factors that can affect quantitation must be considered. The time of analysis postinjection is important; it is most desirable to perform the analysis when a steady state has been reached in the process. Incorrect assumptions regarding how the model is established can lead to faulty data and false conclusions. Other factors that may play a role in quantitation are attenuation, scatter, decay correction, ROI definition (which has been shown to be operator dependent), calibration of the scanner/gamma camera, specific activity of the agent (which, if not optimal, may lead to altered distribution), partial volume effects, reconstruction techniques, and competing transport enzymatic effects. Partial volume effects occur for structures smaller than two times the full width at half-maximum (FWHM) of the imaging system and result in either over- or underestimation of their radioactivity concentration. Partial volume effects are the most important source of errors in quantitation when the size of ROI is small, or there are spill-in or spill-out errors by surrounding regions.

An application of quantitation is assessing the efficacy of antidiabetes therapies on myocardial substrate metabolism. The leading cause of death among diabetic patients is cardiovascular disease (CD). An untested hypothesis is that CD in diabetic patients is a consequence of alterations in myocardial substrate metabolism, causing the heart to utilize less glucose and more fatty acids for energy (Lopaschuk et al. 1994, Lopaschuk 2002). Therapeutic interventions for type 2 diabetes aim to enhance glycemic control; however, their effects on myocardial substrate metabolism are unknown. Two common drugs that are used are Metformin (MET), which is thought to act by decreasing hepatic glucose production through activation of AMP-activated protein kinase and Rosiglitarzone (ROSI), which targets the peroxisome proliferator-activated receptor-gamma nuclear receptor, resulting in improved insulin sensitivity. The effect of treatment with MET and ROSI was assessed by the determination of the rate of myocardial blood flow and oxygen consumption by modeling [^{11}C]acetate uptake. Myocardial fatty acid utilization was assessed by modeling the kinetics of [^{11}C]Palmitate. Finally, myocardial glucose utilization was assessed by modeling [^{18}F]FDG. Results indicated that treatment was better than no treatment and that treatment with ROSI resulted in higher glucose and lower fatty acid utilization and thus a significantly better treatment outcome than MET (Shoghi et al. 2008).

The third talk in the pharmacology section was presented by Jean Luc Vanderheyden, "How Imaging Is Used in Preclinical Drug Development" (Vanderheyden 2009). The author discussed recent developments in molecular imaging and how they were advancing us toward personalized medicine and improved drug selection for human evaluation. The talk highlighted how the future of health care lies in understanding that cellular know-how and converging molecular information will drive more cost-effective choices and improve quality outcomes. In order to market a drug, companies go through multiple phases of the development process to show its quality, efficacy, and safety. Pharmaceutical drug development costs have risen sharply; currently, drug failures represent two-thirds of the cost of drug development and one in three drug projects end due to toxicity and cause 90% of postmarket withdrawals. Molecular imaging offers a view of drug interaction at the cellular level and forms a translational bridge from animals to humans. Compared to biodistribution studies, imaging is noninvasive and can supply meaningful data with fewer animals. It offers high resolution and sensitivity to detect functional pathways and mechanisms of disease. It occurs in real time and gives automated, high throughput to increase research productivity. Imaging provides a regulatory advantage in establishing clinical end points and can markedly decrease the time needed to advance a new drug to market.

Molecular imaging in animals now includes a number of different techniques, such as optical, magnetic imaging, ultrasound, PET, single-photon emission computed tomography (SPECT), and computed tomography (CT) imaging. PET and SPECT, translatable to humans, offer the highest sensitivity with unlimited depth penetration unlike optical imaging. Shown in Table 5.1 are the current specifications for PET and SPECT cameras. Improvements in SPECT imaging are entering the market due to improvements in detector materials such as cadmium, zinc, telluride alloys (CdZnTe), which allow for a higher intrinsic resolution, eliminate dead space at the edges of detectors, and allow for fast dynamic scattering at different resolutions and field of view (FOVs). Avalanche photodiodes technology has delivered high-resolution imaging, dynamic PET imaging, high count rate, and integrated PET and SPECT/PET/CT instruments. Increased computer power affords improved reconstruction algorithms, including modeling of attenuation, scatter, and detector response function. It also permits better resolution, uniformity, and quantitation. These improvements result in overall enhanced imaging allowing for improved diagnosis and staging.

The advantage in high-energy resolution of the new CZT detectors systems now allows the possibility for dual or triple isotope imaging. Tracers labeled with SPECT radionuclides with differing energy emissions can be imaged simultaneously to give information on blood flow, hypoxia, and mapping receptors. The new systems allow for gating for respiratory and cardiac motion in a mouse, which averages 500 cardiac beats/min, requiring high sensitivity as only a few counts per gate will be detected (Vanderheyden 2009). This allows for dynamic imaging in a standard fashion. This ability to correct for motion, through gating, results in improved lesion detection and quantification. These improvements in high resolution and image contrast are allowing for the assessment of new cardiac treatments on rodent models that previously were unobtainable as

it was not possible to differentiate animals with physiological abnormalities. Largest growth has been observed in neurology imaging which requires extremely high resolution and accurate quantitation for studying animals with compromised brain functions and assessing efficacy of therapies. The different imaging modalities are now being offered in one instrument providing maximum flexibility, resulting in lower costs, time, animal constraints, and the benefit that images can be displayed in identical animal configuration.

The increasing utilization of imaging in drug development and research has created a need for the development of technologies that can produce labeled compounds consistently, reliably, and with limited dose to the personnel. This need is being met chemically by the development of labeling techniques using standardized chemical moieties that can be attached simply and quickly to biomarkers of interest. For instance, companies are producing prochelators such as tris-butylated DOTA or NOTA that can be attached to peptides through solid-phase synthesis or the Isothiocyanato (NCS) versions that can be added to an amine group by an increase of pH. Recently, Immunomedics has developed a method for fluorinating peptides by the use of the C-NOTA chelator and the combination of Al^{3+} with F^- (Dijkgraaf et al. 2010). They are planning to launch a kit in the next year. The Tc and Re tricarbonyl approach developed by Alberto et al. (2001) allows for the attachment of these metals to almost any biomarker utilizing a kit commercially available from Covidien. For PET traces, precursors have been developed such as [^{11}C]carbon monoxide and [^{11}C] methylation that can be used to label compounds at different sites and thus evaluate structure–activity relationships. Radiochemistry-automated synthesis units are increasingly being used not only to provide clinical doses but, on the development side, also to provide routine, flexible synthesis at a reduced time and cost. These units allow pharmaceutical and biotech companies as well as hospitals to produce standardized radiolabeled biomarkers that satisfy the Food and Drug Administration (FDA) guidelines for clinical research.

These advancements in imaging are being used to gain information in drug development that answer five basic questions: whether the drug reaches the target tissue, whether the drug binds sufficiently to its primary target (occupancy/inhibition), whether the interaction affects the signaling pathway, whether the drug affects physiology relevant for therapeutic effect, and

Table 5.1 Comparison of PET and SPECT Specifications

	Clinical PET	Clinical SPECT	Preclinical PET	Preclinical SPECT
Sensitivity	1–3%	0.1–0.03%	2–4%	~0.3%
Resolution	~5 mm	~10 mm	~1.5 mm	~1.2 mm
FOV	~50 cm	~50 cm	~7 cm	~8 cm

Chapter 5

whether the drug results in any side effects or toxicity. Additional questions that can be addressed are what route of administration results in sufficient quantities reaching the target. Different analogs can be assessed to determine which is optimal if any to pursue in further clinical evaluation. Furthermore, it can be utilized to determine early response or failure to therapy. This can be assessed at different drug doses and various times to more accurately assess the dosing regimen that will result in highest efficacy and minimize toxicity and in identifying responders versus nonresponders in patient selection for CTs. These measurements can be used to accelerate clinical development or halt drugs early that are either not effective or toxic.

This presentation highlighted how preclinical imaging benefits all areas of molecular imaging research and how recent advancements have greatly extended its use in drug research and development. These advancements have led to standardization of imaging protocols, production of biomarkers, and the ability to perform multimodality imaging on a single instrument utilizing fewer animals, reducing overall costs and personnel. Higher resolution, sensitivity, throughput, and automation are aiding the assessment of new agents in animal models that are leading to drug optimization earlier in the process and reduced costs as well as time to market.

The last talk of the pharmacology section of the workshop was presented by Timothy McCarthy, who spoke on "The Use of Imaging as a Biomarker for Pharmaceutical Development" (McCarthy 2009). From the perspective of a pharmaceutical company, the author discussed how imaging is being used in drug development in CTs as a way of determining early on in the investigation whether the company should proceed with clinical evaluation. Imaging at this stage can answer critical questions and assess the drug distribution and uptake over time. Additionally, the structural and functional changes that occur with intervening treatment can be correlated to molecular and genetic patient profiles. Increasingly, imaging is being used to assess a compound's behavior *in vivo* in appropriate animal models, which aids in the translation and selection of lead compounds for further evaluation in humans. Imaging with biomarkers allows for determination of the drug reaching the target site, assessing changes in function and structure that may occur as a consequence of treatment, and understanding which patient populations will receive the greatest benefit, or those patients who may be negatively affected. Such studies thus allow for rapid decision making with regard to patient stratification, drug regimen selection, toxicity avoidance, and therapeutic monitoring. They also assist with establishing clinical end points and selecting the patient population for clinical evaluation.

McCarthy described the three major categories of biomarkers that have been developed, their use, the need for qualifying the marker, and current challenges for developing novel markers. Imaging is being utilized to examine the genetic and molecular changes that are unique to a patient and his/her disease state, and so treatment options can be matched to the patient's specific disease molecular profile to increase efficacy and minimize toxicity. The three major uses of biomarkers are for proof of target (POT), proof of mechanism (POM), and proof of efficacy (POE) (McCarthy 2009).

Initially, it is important to demonstrate that the drug reaches the intended target *in vivo* and to evaluate the dose and dosing regimen. This is normally performed in a Phase 1 trial of normal volunteers and does not assess clinical benefit. In POT, the aim is to confirm that the drug is interacting with the intended target (McCarthy 2009). Imaging studies are performed to answer the question as to whether the drug is reaching the target, and are used to determine the dose of drug that needs to be delivered to reach 50% receptor occupancy. If sufficient receptor occupancy is not reached, the team might consider stopping the study or begin testing with a different compound. If sufficient receptor occupancy can be reached, then Phase 2 studies can proceed to test proof of concept. Positive results in Phase 2 would allow for moving to Phase 3.

POM studies evaluate whether the drug reaches the target and triggers a change in a biologically relevant parameter such as metabolism, flow, proliferation, or other outcomes. Mechanistic biomarkers can be useful in both normal and patient populations and across disease states, making their use broader than that of target-based markers (McCarthy 2009). While such studies can confirm target expression and aid in patient stratification based on molecular profiling, they do not imply clinical benefit. If a mechanism is established, the team can proceed forward with testing; otherwise, the trial would stop or an alternative compound would be evaluated.

POE markers show that there is a direct link between imaging changes and clinical benefit. These markers allow for imaging a change prior to a clinical outcome, that is, changes that precede clinical benefit. These types of biomarkers are the most valuable;

however, validating them can be very complex and costly. Markers in this category must undergo qualification to establish that the marker is "fit for purpose" and is robust and reproducible enough to be used effectively at multiple centers. Furthermore, the targets must be validated to ensure that they can predict the effect of the treatment on the clinical outcome of interest (Katz 2004). For the imaging to have value, it must have an effective time differential between the time of disease presentation. Initially the biomarker must be analyzed in the natural disease state and then during treatment, and one problem is that it is not always possible to predict how intervention will result in changes in the biomarker.

The three most commonly used imaging techniques are SPECT, PET, and magnetic resonance imaging (MRI). These techniques have been utilized to probe receptor and enzyme distribution *in vivo* to determine the relationship between percent target occupancy and exposure of the test therapeutic. As shown in Figure 5.1, normals are imaged after receiving various doses of the drug to determine the amount of drug required to reach 50% occupancy of the target. Imaging is done to establish that the drug is reaching the desired target and determine how much dose is required to achieve 50% occupancy.

An example PET study to assess *in vivo* occupancy of brain CB-1 receptor (CB1R) after single- and multiple-dose administrations of CBIR receptor agonist, taranabant, is the use of the selective CB1R PET ligand [^{18}F]MK-9470 (Addy et al. 2008). Moreover, the PET study indicated that the taranabant doses evaluated in the 12 week weight-loss efficacy study were associated with brain CB1R occupancies of approximately 10–40%, similar to the range observed in

Proof of mechanism—Endgame

■ Mechanism demonstrated robustly?

    – YES: Proceed to phase 2 POC

    – NO: Avoid phase 2 POC trial (backup compound?)

■ Proof of concept trial?

    – Positive: Proceed to phase 3 with confidence

    – Negative: Question hypothesis, stop program

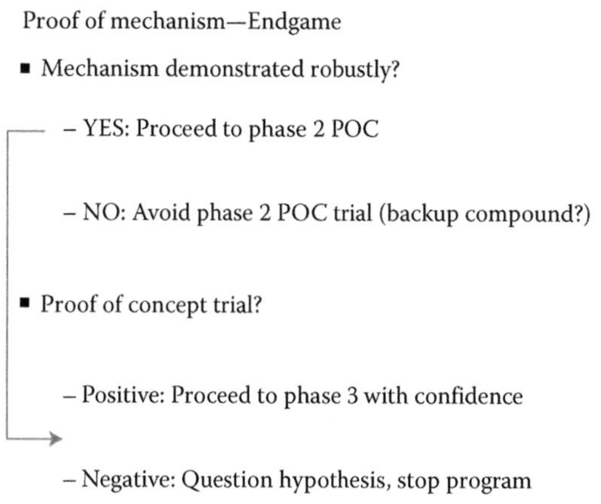

FIGURE 5.2 Proof of target or proof of mechanism endgame scenarios. (Adapted from McCarthy, T.J. 2009. *Q J Nucl Med Mol Imaging*, 53(4), 384.)

preclinical weight-loss efficacy and brain CB1R occupancy studies (Burns et al. 2007).

A second question that can be answered by biomarkers is whether a compound is reaching the target and eliciting the desired biological effect. Imaging studies in oncology are commonly performed to assess if the chemotherapeutic drug is resulting in the desired outcome (i.e., ablation of tumor) by measuring tumor metabolism, perfusion, and permeability. FDG-PET has been approved for many oncology studies to monitor changes in metabolism of tumors following treatment and in CTs to assess novel therapeutics. Figure 5.2 shows how the results of such studies can affect the decisions regarding the testing of the drugs.

The most valuable question a biomarker can address is, does the drug provide clinical benefit? Demonstration of efficacy is particularly important when changes in the biomarker precede clinical changes. Such targets are few and hard to come by due to the complexity and cost required in collecting the data in a longitudinal fashion from numerous clinical sites in a typical Phase 3 CT (McCarthy 2009). Shown in Figure 5.3 is the challenge of utilizing imaging at multiple sites to determine if a drug elicits disease modification. The biomarker must be robust enough to provide statistically significant differences that will delineate whether the test group of patients receiving the novel drug is different from those patients receiving standard of care. The multiple trial sites must utilize similar standards of practice to ensure QC and meaningful correlation of outcomes.

The end point for evaluating therapies of osteoarthritis is their impact on joint space width, which is evaluated by standard x-ray techniques. A clinical

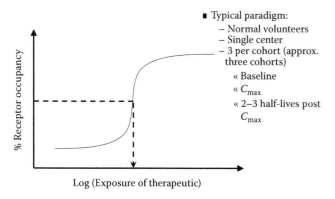

FIGURE 5.1 Goal of receptor occupancy studies: mapping the exposure-to-occupancy relationship. (Adapted from McCarthy, T.J. 2009. *Q J Nucl Med Mol Imaging*, 53(4), 383.)

Chapter 5

- ■ Large datasets from both molecular profiling and imaging are being collected in parallel
  - What is the value proposition for combination?
  - How do we bring data from genomics and imaging together?
- ■ Near term impacts
  - Target selection and understanding of disease
  - Population selection (stratification) for more efficient clinical trials
- ■ Longer term
  - Enabling personalized medicine?
  - How can this be done in a cost effective manner?

FIGURE 5.3 Toward personalized, individualized, and stratified medicine. (Adapted from McCarthy, T.J. 2009. *Q J Nucl Med Mol Imaging*, 53(4), 385.)

study to evaluate the proof of the concept would require following 400 patients for each treatment group, normally three or more groups from 24 to 36 months at 150 clinical sites and have an estimated budget of $15 million (Vignon et al. 2003). This is a very simple method that is highly standardized and illustrates the complexity and cost associated with such studies.

Consortia are stepping up to the challenge of establishing improved methods for imaging and databases that can be used in the qualification and validation of more complex imaging techniques. The hope is that these will result in the development and approval of novel therapies. One such consortium is the Alzheimer's Disease Neuroimaging Initiative. Furthermore, new technologies that are cheaper and easier to deploy are emerging. Contrast-enhanced ultrasound using microbubbles has quantified early changes in tumor perfusion with patients treated with Imatinib, a protein tyrosine kinase inhibitor, at day one and was predictive of future response as measured by progression-free survival. Ultrasound is a technique that can be easily used in a hospital setting at a lower cost than PET or dynamic contrast-enhanced MRI (Hargreaves 2008, Murphy et al. 2008, Willmann et al. 2008).

In summary, this presentation illustrated the critical role imaging biomarkers play in drug development and the types of questions they can address. It discussed the complexities faced to qualify these biomarkers, especially for efficacy, and the challenge in developing novel markers. Biomarker imaging will become increasing more central to the drug development process and in the field of medicine for patient diagnosis, stratification, and treatment monitoring and in the evaluation of regulatory end points.

The second part of the workshop focused on regulatory requirements for drug discovery including application process, production, and QC procedures required for RPs in the United States and the European Union (EU). The first presentation was given by Alan Carpenter, "The Use of the Exploratory IND in the Evaluation and Development of ^{18}F-PET Radiopharmaceuticals for Amyloid Imaging in the Brain: A Review of One Company's Experience" (Carpenter et al. 2009). The report below summarizes Carpenter's presentation, the U.S. Food and Drug Administration (US FDA) development of the Exploratory Investigational New Drug (x-IND) process and Avid Radiopharmaceutical's experience in using the new tool kit.

In 1975, the US FDA required that all RPs, including small molecules and biologics, be regulated as drugs. The FDA also required all human research studies involving RPs to be carried out under an IND application or through a Radioactive Drug Research Committee (RDRC) approval process, overseen by the FDA (FDAGuidanceRDRC 2009). RDRC regulations permit basic research, but do not allow first in human studies. Since RPs are primarily diagnostic agents used as "tracers," they contain a minimal mass quantity and have no expected pharmacological or toxicological effects, yet RPs were required the same safety evaluation as traditional drugs. This requires significant monetary input to develop these agents and will provide much less monetary return when these agents are brought to market.

The FDA launched a plan in 2004 called the Critical Path Initiative, with its report entitled "Innovation or Stagnation: Challenge and Opportunity on the Critical Path to New Medical Products" (CriticalPath 2004). There had been a significant decline in the number of submissions to the FDA for new molecular entities and biologics. There were concerns about the escalating costs of drug discovery and the lengthy time required for the development of new pharmaceuticals which were major motivations behind this initiative. FDA began developing a framework for facilitating drug discovery and development which included ways for manufacturing companies to stop the development of ineffective compounds at an early stage of development.

One of the initiatives the FDA implemented in 2006 which significantly impacted on imaging probe development was the x-IND Guidance issued in January 2006 (FDAGuidanceIND 2006). The x-IND embraces the concept of microdosing, which was initially introduced in the EU (EUMicrodose 2004). The microdose mass was defined as 1/100th of the dose of a test substance calculated to yield a pharmacological effect. If the mass requirement is met, there are reduced pharmacology and toxicology requirements, significantly reducing the cost to bring a new drug to market. This x-IND also allows testing of multiple candidate drugs to identify a lead candidate to move forward to the traditional IND. x-INDs are intended to evaluate the mechanism of action, PKs or pharmacodynamic (PD) properties of the drug candidate. PET RPs for the imaging of molecular targets in human disease are a class of drugs that are very amenable to the use of the x-IND mechanism.

Carpenter described Avid Radiopharmaceutical's experience to demonstrate the utility of the x-IND pathway in the evaluation and eventual selection of an optimal development candidate from a group of PET RP (^{18}F-AV-19, ^{18}F-AV-45, ^{18}F-AV-138, and ^{18}F-AV-144) for imaging amyloid deposits in the human brain. As suggested in the x-IND guidance document, Avid performed appropriate preclinical pharmacology studies, but Carpenter stated that among four F-18 amyloid imaging compounds evaluated, it would have been difficult for AVID to predict, based on preclinical data alone, which compound would have the best overall PET amyloid imaging properties. The use of preclinical models for the identification and optimization of amyloid-targeted molecular imaging agents was not possible due to the lack of good preclinical models of amyloid deposits in the human brain (Bartus 2000, Gotz et al. 2004, Hardy and Selkoe 2002, Watase et al. 2003) and the substantial differences in PK and PD often observed between animals and humans for CNS-targeted molecules (Choi et al. 2009). The x-IND pathway for evaluation of multiple compounds in humans was determined to be the most appropriate strategy for identifying the best compound for further development, which required only a single species (i.e., rat) acute toxicology study to demonstrate adequate safety to justify initial human tracer-dose studies (FDAGuidanceIND 2006). The reduced toxicology requirement in support of x-IND microdose studies saved several months of development time, allowing compounds to progress to the clinic more rapidly than otherwise might have been possible, and allowed multiple compounds to be evaluated simultaneously in a more cost-effective manner.

The x-IND chemistry and manufacturing controls submission is consistent with the traditional IND requirements, and meets the requirements of United States Pharmacopeia (USP) Chapter <823> *Radiopharmaceuticals for Positron Emission Tomography—Compounding*, which was the current regulation for preparation of PET RP (USP Chapter 823 2009).

Following the enrollment of a small number of subjects in the Avid study, Carpenter indicated it was clear that AV-19 did not have the best target to background brain contrast properties and the clinical study of this compound was stopped early. All the three remaining compounds were studied in whole-body imaging protocols in two to three subjects, with subsequent confirmation of appropriate radiotracer dose, prior to enrolling up to 30 subjects (~15 healthy controls and 15 Alzheimer's disease subjects) with dynamic brain PET imaging up to 90–180 min postinjection. ^{18}F-AV-45 was determined to have the best PKs and PDs for imaging brain amyloid deposits in humans versus ^{18}F-AV-144 and ^{18}F-AV-138. ^{18}F-AV-45 had a higher degree of signal-to-background contrast for Alzheimer's versus control subjects in relation to ^{18}F-AV-144. ^{18}F-AV-45 and ^{18}F-AV-138 had similar cortical to cerebellar signal ratios at later time points, but the time frame for achieving the maximal values for ^{18}F-AV-138 was ≥90 min, which was deemed to be suboptimal for the routine clinical imaging, and exhibited higher nonspecific retention in the brain (Carpenter et al. 2009). Therefore, ^{18}F-AV-45 was determined to have the best kinetic balance for brain uptake and background clearance, as well as good amyloid-binding properties and high image contrast (cortical amyloid to white matter ratios) of the ^{18}F-AV RP tested.

The use of the x-IND for the ^{18}F-AV RP allowed a direct comparison of PK, PD, and brain imaging properties for this series of closely related compounds following a common CT design resulting in the selection of ^{18}F-AV-45 for further development under a traditional IND. Prior to human x-IND studies, this compound was not anticipated to be the best compound of the series tested (Carpenter et al. 2009). Therefore, the x-IND mechanism directly resulted in the selection of the best compound in a reduced amount of time at a reduced cost relative to the traditional IND pathway.

In addition to the exploratory IND, the FDA has now moved forward and published the Current Good Manufacturing (CGMP) for PET rule. The FDA Modernization Act (FADAMA) of 1997 had required FDA to develop appropriate CGMP for PET drugs which would be separate from CGMP for traditional drugs. In 2005, the FDA issued a proposed rule (FDADraftGuidancecGMP 2005) which was finalized on December 10, 2009; 21CFR Part 212 CGMP for Positron Emission Tomography Drugs (FDAFinalRulecGMPPET

Chapter 5

2009). Additionally, the agency published a Guidance document, for PET Drugs—CGMP (FDAGuidanccGMP 2009). The Final Rule will become effective on December 12, 2011. The Final Rule still allows a PET drug producer to follow either Part 212 or continue the current practice of using Chapter <823> for the production of investigational PET drugs produced under an IND (Phase 1 and Phase 2) and research drugs produced under the authority of the RDRC. If a PET drug producer intends to seek marketing approval for a PET drug, Phase 3 studies of the drug should be in accordance with the PET CGMP requirements in Part 212.

Regulatory changes are also occurring in the EU as well as the United States. Clemens Decristoforo summarized major EU documents relevant for radiopharmacy practice in Europe and recent developments on the national level especially regarding the small-scale preparation of RPs (including PET) in his presentation "Towards a Harmonized Radiopharmaceutical Regulatory Framework in Europe?" (Decristoforo and Peñuelas 2005). Decristoforo also addressed the EU marketing authorization (MA) required to produce these RP, standards of preparation, quality requirements, education required for production of PET RPs, and CT requirements.

According to Decristoforo, in the EU, the practice of pharmacy in general and of radiopharmacy in particular differs substantially among the EU member countries and is mainly regulated at the national level. The major topics in pharmaceutical legislation are regulated by Directives which are relatively short concise statements which then must be translated into national legislation in detail. Within this translational process, individual national variations occur. This translation reflects national habits and the historic development of radiopharmacy in each country. Standards for the industrial preparation of pharmaceuticals are defined in Directives 2003/94/EC (EUDirective 2003) Article 1 and Directive 2001/83Article 2 and 40 (EUDirective 2001), and the good manufacturing practice (GMP) principle is stated in the GMP text (EUGuidlinesGMP 2005). GMP rules are published in the Eudralex, a web-based compendium European pharmaceutical legislation (EUEudralex 2009). Annex 3 of GMP covering specific GMP regulations related to RPs recently has been revised, March 2009 (EUGMPAnnex3 2008). These rules do not take into account the specific requirements of small scale, especially in a hospital or academic research setting, including involvement of a "Qualified Person" with practical experience in the industrial preparation of pharmaceuticals. EANM has

published several documents based on GMP and to differentiate these guidelines from "conventional" GMP, they were named current "Good Radiopharmaceutical Practice" (cGRPP) (EANMGuidelines 2010). These guidelines specifically address these issues and attempt to harmonize RP preparation across Europe.

The MA is required to use a medicinal product which is defined for patient treatment or diagnosis (and includes RPs), which are produced by an industrial process, and are intended to be sold. Marketing RPs labeled with a short-lived radionuclide is generally not economically feasible since there would be a small potential market, and it may not support the investment required to obtain the MA. Second, RP labeled with an ultra-short-lived radionuclide (e.g., less than 25 min) would require the synthesis (and QC) of many batches per day, boosting the need for equipment and increasing the overall production cost. Additionally, ultra-short-lived RP should need to be produced, analyzed, released, transported, delivered on time, and administered almost immediately to a patient. This makes it impossible to supply remote sites. One exception is ^{18}F-FDG (FDG) which has an MA in most European countries and is shipped by commercial manufacturers to hospitals. In this case, the high production capacity of the F-18 from the current cyclotrons, together with the high yield of the synthetic process and a reasonable half-life of 110 min, allows the production of multi-Curie FDG batches. All these factors combine to allow FDG commercialization. The wide use of FDG in many oncological and neurological diseases has sustained a market worth the investment needed for its production. Currently, there are new ^{18}F-labeled RPs in-process of applying for an MA.

The European Directive 2001/83 (EUDirective 2001) leaves room for exemptions, especially with the "magistral" or "officinal" preparations which are based either on the prescription of a physician or on a country's pharmacopoeial monograph and are produced in a pharmacy. These preparations are outside the EU Directive which requires an MA and therefore becomes the responsibility of the individual EU member states.

In some countries, this exemption is followed more or less straightforward and the hospital preparation of RPs on a small scale is based on the pharmacy status of the radiopharmacy unit, including the requirement for a pharmacist to be responsible for production and release. However, in many countries, laboratories preparing PET RPs have developed in university settings or research laboratories without pharmacy status. Also, local regulations often do not allow a pharmacy to have a licence to handle radioactivity, therefore prohibiting the preparation of RP (e.g., in Germany). In other cases (e.g., in Spain), medicinal products intended for intravenous administration for human are specifically excluded from the possibility of being compounded as a "magistral" preparation.

The production of short-lived RPs can then be based on a specific local regulation where institutions are controlled by a national pharmaceutical inspection scheme, or if the production can be directly under the responsibility of a physician, based on an individual patient's need. Then there is no clear regulatory path. Recent developments indicate that European and National authorities are trying to find specific solutions to allow routine use and application of small-scale preparations without having to apply for an MA or applying for a specific CT to produce these RPs. In Germany, a recent change in regulation allows RP preparation without an MA in a pharmaceutically authorized institution (AMRadV 2007), and in Italy a recent change includes preparation of "experimental RP" without an MA, but requires specific GMP guidance outlined in their National Pharmacopoeia to be followed. In the United Kingdom, preparation of RPs for human use requires GMP conditions under the responsibility of a qualified and authorized person in a facility with a "specials license" authorization. In Spain, a novel legislation specifically dealing with "in house" preparation and use of RPs (mainly based on EANM's cGRPPs (EANMGuidlines 2010)) is likely to be published in 2010. In this case the National Health Authorities will have to inspect each facility and procedure, and authorize each production center for each unregistered RP product prepared. Also, non-EU countries, such as Turkey, have released similar regulations.

The Pharmaceutical Inspection Convention and Pharmaceutical Inspection Co-operation Scheme (jointly referred to as PIC/S) is an international instrument between countries and pharmaceutical inspection authorities, which provides an active and constructive cooperation in the field of GMP. This association recently issued a "PIC/S Guide to Good Practices for the Preparation of Medicinal Products in Healthcare Establishments" (PIC/S 2008) that describes a specific GMP intended for hospital preparation of pharmaceuticals, taking into account the requirements for "magistral" or "officinal" preparations. This document mentions RPs, but does not address their production. PIC/S has also set up PIC/S Expert Circle on the Preparation of Radiopharmaceuticals in Healthcare Establishments with the aim of drafting a specific annex for Radiopharmaceuticals to their general PIC/S guide.

Another initiative begun by the European Pharmacopoeia (EP) is the establishment of a Compounding of Radiopharmaceuticals Working Party for the purpose of including a specific General Chapter to address standards for the extemporaneous preparation of RPs, but it will not be legally binding. This EP Chapter may be similar in concept to the USP Chapter <823>, but differs since the USP Chapter is an enforceable document.

However, the major concern that the authorities have regarding these products remains: who controls and defines environmental and training standards of personnel as well as quality standards of the preparation itself? This issue is also a matter of debate among specialists and authorities and is addressed below.

A major challenge in Europe for radiopharmacy practice has been the change in requirement for conducting a CT in Europe. This began with the so-called Clinical Trial Directive (EUDirective 2003) and a number of guidelines including the Directive defining the preparation of Investigational Medicinal Products according to GMP, which includes the chemical and pharmaceutical documentation required. This formed the basis for issuing a new annex 13 to European GMP for these preparations. The CT legislation was designed almost exclusively for pharmaceutical industry, and has posed excessive hurdles to the development of CT in academic or nonprofit organizations.

Although the promise of new PET imaging agents is great, the process of bringing these agents to commercialization remains in its infancy. Steve S. Zigler of PETNET Solutions stated that the purpose of his presentation, "Instrumentation and Radiopharmaceutical Validation," (Zigler et al. 2009) was to review validation from the perspective of the chemistry, manufacturing, and controls (CMC) for an FDA filing, as well as the validation requirements described in FDA GMP regulations, guidance documents, and general chapters of the USP. His review includes discussion on validation from development to commercial production of PET RP with a special emphasis on equipment and instrumentation used in production and QC testing.

Due to their unique ability to trace *in vivo* biomolecular pathways in health and disease, PET imaging agents play a key role in realizing the promise of molecular medicine. Over the past 20 years, this promise has led to intense interest in the development and commercialization of PET RP. Ziegler echoed Decristoforo's comment that the most important

Chapter 5

PET RP to evolve is FDG, which is employed in more than 1.5 million PET scans per year in the United States (PETMarketSummary 2008). This has encouraged development from commercial companies and universities to move beyond FDG—to determine what will be the PET RP for cardiology, oncology, and neurology.

Thus, the use of PET imaging agents spans a wide variety of stages from discovery to clinical development to routine diagnostic procedures. These stages share many characteristics—the use of sophisticated equipment such as cyclotrons, synthesis modules involving high-pressure liquid chromatography (HPLC) purification, use of laminar air flow enclosures, final product sterilization using membrane filtration, and aseptic process. Additionally, due to the short half-lives of these PET RPs, it is required to perform QC testing on each batch of final products produced. These QC tests must demonstrate that the product has sufficient quality and purity for human-use injection and can involve use of thin-layer chromatography, HPLC, gas chromatography, ionization chambers, gamma-ray spectrometers, and a spectrophotometric device for endotoxin testing.

In order to achieve reliable and reproducible results, each of these production processes and analytical methods must be validated. Process validation and analytical methods validation are also important because they are required for compliance with GMP for PET regulations, as these are part of an overall quality assurance (QA) program (Zigler et al. 2005). These validations are not so important during the initial stages of drug development, but as development moves toward commercialization, key process and analytical methods require validation. Additionally, as a PET RP must be prepared at a number of commercial PET production facilities as part of a distributed manufacturing process, validation is essential at each facility.

Once a process or method is validated, it is important to ensure that the equipment and instrumentation are suitable for use in the routine execution of the validated process or method. When equipment or instrumentation is initially placed into use, it must be qualified to ensure that it can execute the validated process or method. This forms the basis of Installation Qualification, Operational Qualification, and Performance Qualifications (IQ/OQ/PQ). Once qualified, the equipment or instrumentation must remain capable of executing the validated process or method. This forms the basis of system suitability, which is intended to ensure proper performance on an ongoing basis. As in the case of validation, there is often confusion about how much qualification and suitability are necessary, and when it applies during the development process.

The remainder of Zigler's presentation reviewed the regulatory requirements for validation, qualification, and system suitability with an emphasis on equipment and instrumentation (Zigler et al. 2009). But the lines between equipment, process, and chemistry are often blurred in the CMC section of an application or for an MA in European countries.

Before discussing instrumentation, equipment, regulations, and validation, Zigler asked "Why is the PET production environment different?" The characteristics of PET RPs, described by Decristorofo, were reiterated by Zigler, with an added twist.

1. To provide a nationwide supply of PET RPs, companies must rely on numerous small-scale production facilities that serve local communities. This duplicates facilities, equipment, and personnel, and creates challenges inherent in a distributed supply chain.
2. Each site is staffed with a small number of personnel focused solely on production, testing, and shipping. It is impractical to staff each PET facility with the technical expertise that would be expected in a typical vertically integrated central manufacturing facility.
3. Each batch of a PET product yields a limited number of patient doses, which requires the daily production of hundreds of batches to meet the nationwide demand.
4. Each batch of a PET product consists of a single vial with a small volume (e.g., 5–50 mL), which is dramatically different from a traditional pharmaceutical manufacturing process.
5. Each batch of a PET product is subjected to complete QC testing, including pH, visual inspection, radionuclidic identity, radiochemical purity and identity, chemical purity, residual solvent analysis, specific activity (if appropriate), bacterial endotoxins, sterility, and membrane filter integrity test (for intravenously administered products).
6. There is no product inventory at the beginning of each day.
7. A full complement of production and testing equipment is required at each facility.

These conditions create a manufacturing environment that is dramatically different from the traditional pharmaceutical industry (big pharma). More than 120 cyclotron-based manufacturing facilities are required to meet the U.S. demand of FDG of approximately 1.5 million doses (PETMarketSummary 2008) requiring in excess of 50,000 batches. This highly distributed and repetitive model is unprecedented in any other drug manufacturing environment, and demands a unique regulatory approach.

Zigler discussed validation which he stated may be defined as the scientific process of proving that a process is acceptable for its intended use. Additionally, he reviews the associated FDA regulations and guidance ICH guidelines as well as the requirements described in the USP.

FDA Final PET GMP Rule Section 212.30 requires "procedures to ensure that all equipment that could reasonably be expected to adversely affect the strength, quality, or purity of a PET drug product … or give erroneous or invalid test results when improperly used or maintained … is clean, suitable for its intended purposes, properly installed, maintained, and capable of repeatedly producing valid results" (FDAFinalRulecGMPPET 2009). Although the wording in this section lacks detail, the accompanying guidance document (FDAFinalRulecGMPPET 2009) provides examples related to equipment cleaning, system suitability, installation, and maintenance.

Section 212.60 of the FDA PET GMP regulation requires that "testing procedures … be established to ensure that PET drug products conform to appropriate standards … of identity, strength, quality, and purity. Analytical tests must be suitable for their intended purpose and have sufficient sensitivity, specificity, and accuracy" (FDAFinalRulecGMPPET 2009). In addition, the FDA guidance document recommends that "any new analytical test method be validated, through documented data, to show that it will consistently yield results that accurately reflect the quality characteristics of the product tested" (FDAFinalRulecGMPPET 2009). The guidance describes "analytical parameters (e.g., accuracy, precision, linearity, ruggedness) that should be used to validate a new method" according to USP General Chapter <1225> Validation of Compendial Methods and the International Conference on Harmonization of Technical Requirements for Registration of Pharmaceuticals for Human use (ICH) Q2A Text on Validation of Analytical Procedures. The ICH is a unique project that brings together the regulatory authorities of Europe, Japan and the United States and experts from the pharmaceutical industry. The FDA guidance document additionally states that "validation is not required for compendial methods" (FDAFinalRulecGMPPET 2009).

Key elements of PET production processes should be validated. One example of a process element that falls in this category is the use of a laminar air-flow hood in the aseptic assembly of sterile components used in the preparation of the final product assembly. Using media fill process simulations, it is possible to demonstrate that the aseptic assembly process is acceptable for its intended use. Another validation example would include equipment cleaning effectiveness for multibatch production of PET drugs. A validation study is associated with a process or method. As a result of this association, validation should be a one-time process as long as the process or method is not changed. In other words, once a process or method has been validated, it should not be necessary to repeat the validation studies unless there is a major change in the process, formulation, equipment, or method.

Section 212.5 of the PET GMP regulations states that these regulations do not apply to investigational or research PET RPs, but then the question does arise regarding the applicability of GMP for PET to the clinical development phases preceding marketing approval. Section 212.5 also states that investigational or research PET RP must be prepared according to the 32nd edition of USP General Chapter <823>, RP for PET–Compounding (USPChapter823 2009). Chapter <823> addresses (a) the control of components and materials, (b) compounding procedure verification, (c) stability, (d) compounding requirements for human use, (e) quality control, and (f) sterilization and sterility assurance. This chapter does not address any form of validation, per se, but requires "verification studies to ensure that the written compounding procedures, software, equipment and facilities result in a PET RP that meets established acceptance criteria" (USPChapter823 2009). In the area of QC, the chapter requires "verification testing of equipment and procedures used for the QC testing," then goes further to require instrument qualification, system suitability, and maintenance (USPChapter823 2009). Chapter <823> does not address validation, per se, but it uses "verification" in a conceptually similar fashion as "validation" and the chapter includes many practical and specific requirements related to verification. USP General Chapter <1225> is an informational chapter, which describes analytic characteristics that should be addressed in an analytic methods validation: accuracy, precision, specificity, detection limit, quantitation limit, linearity, range, and robustness (EUDirective 2003).

Chapter 5

In 2008, the FDA released a guidance document describing GMP requirements required during Phase 1 CTs (USPChapter1225 2009). Section III.F. of this guidance states that "validation data ... ordinarily need not be submitted at the initial stage of drug development" (USPChapter1225 2009). Instead, "a brief description of the test methods ... should be submitted. Proposed acceptable limits supported by simple analytical data ... of the CTs material should be provided" (USPChapter1225 2009). This same issue is also addressed in the guidance document for analytical methods validation (FDAGuidanceMethodValidation 2000). Section V of this guidance discusses INDs and states that "sufficient information is required in each phase of an investigation to ensure proper identification, quality, purity, strength, and/or potency. The amount of information on analytical procedures and methods validation necessary will vary with the phase of the investigation" (FDAGuidanceMethodValidation 2000). This guidance also states that "analytical procedures should be fully developed and validation completed when the NDA [or] ANDA ... is submitted" (FDAGuidanceMethodValidation 2000).

In addition to validation, the FDA guidance document discusses system suitability testing, stating "prior to each day of its use, the analyst should make sure the ... system is functioning correctly by conducting system suitability testing. At least one injection of the standard preparation ... should be done before the injection of test samples" (FDAFinalRulecGMPPET 2009).

The USP General Chapter <621> on chromatography discusses "system suitability tests as an integral part of gas and liquid chromatographic methods. They are used to verify that the resolution and reproducibility of the chromatographic system are adequate for the analysis to be done. The tests are based on the concept that the equipment, electronics, analytical operations, and samples to be analyzed constitute an integral system that can be evaluated as such" (USPChapter621 2009). The chapter describes several chromatographic parameters that may be used in system suitability tests, including resolution, column efficiency, tailing factors, and replicate injections.

Development and validation are one-time occurrences that precede and define qualification and system suitability. Qualification is used to qualify operators and equipment. Finally, system suitability is performed routinely to ensure that equipment and operators are suitable for use in a given process (Figure 5.4). The manufacturing environment for PET RP is drastically different from big pharma. The development,

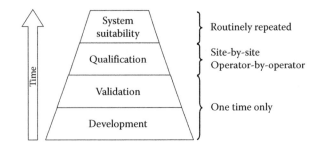

FIGURE 5.4 QA hierarchy depicting the relative relationship between system suitability, qualification, validation, and development. (Adapted from Zigler, S.S. 2009. *Q J Nucl Med Mol Imaging*, 53(4), 409.)

production, and distribution of PET RP is a small industry that is relatively immature by most standards. At this early stage of the evolution, the PET industry has the opportunity to develop practical, scientifically sound practices that will simultaneously maintain our historical track record of safety and allow for commercially viable expansion of the market for PET.

As discussed previously by Ziegler, PET drugs are unique. This includes a short half-life, which is typically minutes to hours. Due to the limited shelf life for this drug, there are unique requirements for completion of the required QC processes as quickly as possible since most QC tests must be performed on each batch released for human injection. In his presentation, Passchier (2009) outlined a new type of "Fast High Performance Liquid Chromatography in Positron Emission Tomography Quality Control and Metabolite Analysis." Passchier discussed how QC and metabolite analysis on compounds labeled with positron-emitting isotopes used for PET studies are constrained by time. For carbon-11-labeled tracers, it is generally accepted that the QC and subsequent release process should take no longer than 20 min to ensure both high product-specific activity and sufficient radioactive product remaining to meet the aims of the studies in which it is used. From a regulatory point of view, PET manufacturing and associated QC testing are also expected to meet GMP standards as discussed in the previous sections.

The challenge for the PET QC laboratory today is to merge the requirement for fast testing with meeting applicable regulations. Passchier's paper aimed to highlight some recent developments that may aid in speeding up both PET QC and metabolite HPLC analysis methods. Since HPLC methods are part of the QC process, these methods must be able to be validated according to accepted standards for validation procedures, depending on the country. There are several method validation guidelines, depending on the country. The ICH guidelines Q2 (R1) cover Validation of Analytical Procedures. Additionally in the United States, USP Chapter <1225> is an informational chapter covering Validation of Procedures (which was also discussed above by Ziegler). In the EU, the *EP Technical Guide*, 4th edition (2005), and the Japanese Pharmacopeia XIV also address these procedures.

Passchier has summarized and described the tests which are generally included as part of methods validation: identification (visual separation of precursor and product), specificity (measurement of the product in the presence of impurities), accuracy (closeness of agreement between a test result and the accepted reference value), precision (agreement between measurements), limits of detection (lowest amount of analyte that can be detected) and quantification (lowest amount of analyte which can be quantitatively determined with precision and accuracy), linearity (ability to obtain results directly proportional to the amount of analyte present), and robustness (capacity to remain unaffected by small but deliberate variation in method parameters) (Passchier 2009). Bearing in mind the requirement for analytical method validation, Passchier has described several ways which can reduce the HPLC analysis time by adjusting the following parameters.

For example, if the observed resolution is good at lower flow rates, increasing the flow rate is a viable option to speed up analysis time. At higher flow rates, system pressure may become an issue with most standard commercial systems having pressure limits of around 400 bar. If pressure is an issue, reduction of column length can be considered. Passchier states that there will be a trade-off with resolution, but this is acceptable as long as $R \leq 1.5$. Reduction of column length can often be combined to good effect with a reduction in particle size. Additionally, smaller particle size (<2 μm) can lead to a reduction in resistance to mass transfer and associated reduction of the theoretical plate height in the Van Deemter equation (i.e., greater resolution) (Mazzeo et al. 2005). Figure 5.5 demonstrates the reduction in analysis time that can be gained by switching from the traditional 5 μm particle size HPLC column to one with 1.7 μm particles (Wu and Clausen 2007).

To address back pressure, consider the use of column ovens set at elevated temperatures. When set at ≤50°C, these will reduce solvent viscosity and can improve analyte diffusion leading to improved HPLC performance (Lestremau et al. 2007), and redesign

system plumbing to use 0.1–0.16 mm inner diameter (id) tubing only where needed while using larger id tubing where possible.

Another way to overcome system pressure is to invest in ultra-high-performance systems that are now provided by most major manufacturers. These systems are designed specifically to cope with the high pressures associated with the use of sub-2 μm particle HPLC columns.

Speeding up analytical HPLC analysis is fairly straightforward if consideration is only given to the analysis of product and precursor(s). In their recent paper, Nakao et al. (2009) presented a 1 min HPLC QC procedure for a wide range of tracers. Although highly interesting and of potential practical use, there are other confounding factors to take into consideration which limit the general utility of this method in a GMP QC setting. In particular, the presence of unknown impurities can lead to inaccurate measurements and this should be taken into account during the method validation process.

Metabolite analysis plays an essential role in PET PK modeling for (1) new tracers whose *in vivo* behavior has not yet been characterized, (2) for tracers that bind to targets for which there is no reference region, and (3) for studies in which it is essential to reduce the bias introduced by the use of simplified analysis methods. Failure to correctly determine the parent tracer fraction in arterial or arterialized blood will lead to an over- or underestimation in the input function which would lead to an under- or overestimation in the binding potential, respectively (Destafano et al. 2008).

The traditional method for metabolite analysis using HPLC involves deproteination of the plasma sample using an organic solvent (e.g., acetonitrile or methanol) or a strong acid (e.g., trifluoroacetic acid or perchloric acid), centrifugation of the resulting denatured protein emulsion, and injection of the supernatant on the HPLC column.

Passchier described a method that is finding increasing use in PET HPLC metabolite analysis—the use of direct plasma injection, either on systems that have been configured for column-switching or through the use of large particle or monolythic columns that can deal with plasma volumes up to 10 mL.

In the column-switching method, a plasma sample is injected directly onto an HPLC system and passed through a suitable precolumn using a low-organic-strength solvent (e.g., 1% acetonitrile in water). The flow is directed to ultraviolet, online radiodetector, and/or fraction collector. Any radioactive peaks

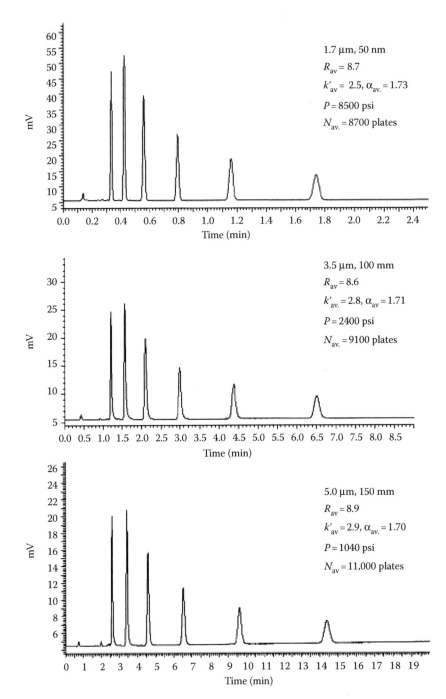

FIGURE 5.5 Proof of target or proof of mechanism endgame scenarios. (Adapted from Passchier, J. 2009. *Q J Nucl Med Mol Imaging*, 53(4), 413.)

coming through this precolumn should reflect polar metabolites. After a set period of time, the precolumn is switched in line with a reversed-phase HPLC column and the nonpolar analytes including the parent are eluted onto the HPLC column using an appropriate higher-organic-strength isocratic or gradient method (Nakao et al. 2009).

One development that was briefly discussed in Passchier's paper was the growing use of liquid chromatography-mass spectrometry in quality control. The big advantage of moving over to such a system is that it allows measurement of radiochemical purity, specific activity, and chemical impurities with a positive confirmation of identity in one analysis (Schou et al. 2009). This usually requires two HPLC injections, and would therefore reduce the time required for analysis by half.

Bacterial endotoxin testing (BET) is another QC test which must be completed prior to intravenous administration of a PET drug to humans. In 1977, the FDA gave initial approval for use of the limulus amebocyte lysate (LAL) for BET testing. LAL is isolated from the horseshoe crab (*Limulus*) and reacts with Gram-negative bacterial endotoxins in nanogram or greater concentrations to form an opaque gel. Gram-negative endotoxins are recognized as the most important source of pyrogen contamination. The BET test is both a rapid and a very sensitive *in vitro* method, and requires 60 min to complete. A 20 min "limit test" was allowed for the release of the PET drug, but the 60 min test also had to be completed (USPChapter823 2009).

The Endosafe Portable Test System (PTS) has been developed and licensed by Charles Rivers Laboratories. It is a software-driven spectrophotometer for measuring and documenting endotoxin, which is approved by the FDA. The system utilizes an assay in which the bacterial endotoxin initiates activation of enzymes in LAL, which finally cleaves a synthetic substrate. Cleavage results in the release of the chromophore, *p*-nitroaniline (PNA). PNA is yellow colored and is measured by the PTS spectrophotometer at 385–410 nm. The system uses a cartridge that contains four channels to which LAL reagent and the chromogenic substrate have been applied. Two channels contain an endotoxin spike and serve as positive product controls. The other two channels are for the test sample. The PTS device has been compared to traditional LAL testing in a multicenter study and showed equivalent results in cell-therapy products (Gee et al. 2008). PTS is particularly suited for PET drugs because the test requires ~20 min, requires <0.1 mL of diluted product (1:100 dilution can be validated), and allows reduced radiation exposure for the QC operator.

In summary, noninvasive imaging techniques are critical tools for evaluating the pharmacology of targeted compounds, and the information gathered can help determine if the compound can be translated from the research phase to the clinical application. Additionally, it can help refine development of a battery of compounds. Subsequent CTs require formalized production and quality control testing that is highly regulated, often changing, and at times governed by a number of differing government agencies. Radiopharmacy can bridge the gap between research and development to formulate a validated production process, with defined quality control specifications that are required for compounds intended for human use.

References

Addy, C., Wright, H., Van Laere, K. et al. 2008. The acyclic CBIR inverse agonist tannabant mediates weight loss by increasing energy expenditure and decreasing caloric intake. *Cell Metab*, 7, 68–78.

Alberto, R., Ortner, K., Wheatley, N., Schibli, R., and Schubiger, R. 2001. Synthesis and properties of boronocarbonate: A convenient *in situ* CO source for the aqueous properties of [99mTc(OH$_2$)$_3$(CO)$_3$]+. *J Am Chem Soc*, 123, 3135–3136.

AMRadV. 2007. Decree on radioactive drugs or drugs treated with ionizing radiation (AmRadV). Ministry of Justice.

Ascoli, G., Domenici, E., and Bertucci, C. 2006. Drug binding to human serum albumin: A bridged review of results obtained with high-performance liquid chromatography and circular dichroism. *Chirality*, 18, 667–679.

Bartus, R. T. 2000. On neurodegenerative diseases, models, and treatment strategies: Lessons learned and lessons forgotten a generation following the cholinergic hypothesis. *Exp Neurol*, 163, 495–529.

Burns, H., Laere, K., Sanabria-Bohorquez, S. et al. 2007. [^{18}F] MK-9470, a positron emission tomography (PET) tracer for *in vivo* human PET brain imaging of the cannabinoid-1 receptor. *PNAS*, 104, 9800–9805.

Carpenter, A. P., Pontecorvo, M. J., Hefti, F. F., and Skovronsky, D. M. 2009. The use of the exploratory IND in the evaluation and development of ^{18}F-PET radiopharmaceuticals for amyloid imaging in the brain: A review of one company's experience. *Q J Nucl Med Mol Imaging*, 53, 387–393.

Choi, S. R., Golding, G., Zhuang, Z. et al. 2009. Preclinical properties of 18F-AV-45: A PET agent for Abeta plaques in the brain. *J Nucl Med*, 50, 1887–1894.

Colby, L. and Morenko, B. 2004. Clinical considerations in rodent bioimaging. *Comp Med*, 54, 623–630.

Collins, J. and Klecker, R. J. 2002. Evaluation of highly bound drugs: Interspecies, intersubject, and related comparisons. *J Clin Pharmacol*, 42, 971–975.

CRITICALPATH. 2004. Innovation or stagnation: Challenge and opportunity on the critical path to new medical products *FDA Challenges and Opportunities Report—March 2004*. Food and Drug Administration.

Decristoforo, A. and Peñuelas, I. 2005. Towards a harmonized radiopharmaceutical regulatory framework in Europe? *J Nucl Med Mol Imaging*, 53, 394–401.

Destafano, J., Langlois, T., and Kirkland, J. 2008. Characteristics of superficially-porous silica particles for fast HPLC: Some performance comparisons with sub-2-micron particles. *J Chromatogr Sci*, 46, 254–260.

Dijkgraaf, I., Franssen, G., Mcbride, W. et al. 2010. Targeting gastrin-releasing peptide receptor-expressing tumors with F-18-labeled bombesin analog. *Eur J Nucl Med Mol Imaging*, 37, S198–S311.

EANMGuidlines. 2010. Guidelines on Current good Radiopharmacy Practice. European Association of Nuclear Medicine.

EUDirective. 2001. Directive 2001. European Union.

EUDirective. 2003. Commission Directive 2003. *Commission Directive 2003/94/EC of 8 October 2003 laying down the*

Chapter 5

principles and guidelines of good manufacturing practice in respect of medicinal products for human use and investigational medicinal products for human use (*Official Journal L 262*, 14/10/2003 pp. 22–26). European Union.

EUEudralex. 2009. EudraLex—Volume 1—Pharmaceutical Legislation Medicinal Products for Human Use. European Union.

EUGMPannex3. 2008. EudraLex The Rules Governing Medicinal Products in the European Union Volume 4 EU Guidelines to Good Manufacturing Practice.

Medicinal Products for Human and Veterinary Use Annex 3 Manufacture of Radiopharmaceuticals. European Union.

EUguidlinesGMP. 2005. Guidelines to Good Manufacturing Practice: Medicinal Products for Human and Veterinary Use. European Union.

EUMicrodose. 2004. Position Paper on Non-Clinical Safety Studies to Support Clinical Trials with a Single Microdose. *Committee for Medicinal Products for Human Use*, ed. London, European Medicines Agency.

FDADraftGuidancecGMP. 2005. Draft guidance on current good manufacturing practice for positron emission tomography drug products; Availability. *Fed Regist*, 70, 55145.

FDAFinalRulecGMPPET. 2009. Food and Drug Administration. *Federal Register Notice: Final Rule—CGMP for PET Drugs Food and Drug Administration*.

FDAGuidancecGMP. 2009. Food and Drug Administration. Guidance: PET Drugs—Current Good Manufacturing Practice (CGMP).

FDAGuidanceIND. 2006. FDA Guidance for Industry, Investigators and Reviewers: Exploratory IND Studies. Rockville, MD, Food and Drug Administration.

FDAGuidanceMethodValidation. 2000. Food and Drug Administration. Guidance for Industry Analytical Procedures and Methods Validation. Food and Drug Administration.

FDAGuidanceRDRC. 2009. Guidance for Industry and Researchers The Radioactive Drug Research Committee: Human Research without an Investigational New Drug Application.

Gee, A., Sumstad, D., Stanson, J. et al. 2008. A multicenter comparison study between the Endosafe PTS rapid-release testing system and traditional methods for detecting endotoxin in cell-therapy products. *Cytotherapy*, 10, 427–435.

Giron, M. C. 2009. Radiopharmaceutical pharmacokinetics in animals: Critical considerations. *Q J Nucl Med Mol Imaging*, 53, 359–364.

Gotz, J., Streffer, J. R., David, D. et al. 2004. Transgenic animal models of Alzheimer's disease and related disorders: Histopathology, behavior and therapy. *Mol Psychiatry*, 9, 664–683.

Hardy, J. and Selkoe, D. J. 2002. The amyloid hypothesis of Alzheimer's disease: Progress and problems on the road to therapeutics. *Science*, 297, 353–356.

Hargreaves, R. 2008. The role of molecular imaging in drug discovery and development. *Clin Pharmacol Ther*, 83, 349–353.

Hildebrandt, I., Su, H., and Weber, W. 2008. Anesthesia and other considerations for *in vivo* imaging of small animals. *ILAR*, 49, 17–26.

Katz, R. 2004. Biomarkers and surrogate markers: An FDA perspective. *NeuroRX*, 1, 189–195.

Kersemans, V., De Spiegeleer, B., Mertens, J., and Slegers, G. 2006. Influence of sedation and data acquisition method on tracer uptake in animal models: [123I]-2-iodo-L-phenylalanine in pentobarbitol-sedated tumor-bearing athymic mice. *Nucl Med Biol*, 33, 119–123.

Kilbourn, M., Ma, B., Butch, E., Quesada, C., and Sherman, P. 2007. Species dependence of [64Cu]Cu-Bis(thiosemicarbazone)

radiopharmaceutical binding to serum albumins. *Nucl Med Biol*, 35, 281–286.

Kosa, T., Maruyama, T., and Otagiri, M. 1997. Species differences of serum albumins: I. Drug binding sites. *Pharm Res*, 14, 1607–1612.

Kragh-Hansen, U. 1981. Molecular aspects of ligand binding to serum albumin. *Pharmacol Rev*, 33, 17–53.

Lestremau, F., De Va, Lyen, F., Cooper, A., Szucs, R., and Sandra, P. 2007. High efficiency liquid chromatography on conventional columns and instrumentation by using temperature as a variable. Kinetic plots and experimental verification. *J Chromatogr A*, 113, 120–131.

Lopaschuk, G. D. 2002. Metabolic abnormalities in the diabetic heart. *Heart Fail Rev*, 7(2), 149–159.

Lopaschuk, G. D., Belke, D. D., Gamble, J., Itoi, T., and Schonekess, B. O. 1994. Regulation of fatty acid oxidation in the mammalian heart in health and disease. *Biochim Biophys Acta*, 1213(3), 263–276.

Mazzeo, J., Neue, U., Kele, M., and Plumb, R. 2005. Advancing LC performance with smaller particles and higher pressure. *Anal Chem*, 77, 460A–467A.

McCarthy, T. J. 2009. The role of imaging in drug development. *Q J Nucl Med Mol Imaging*, 53(4), 382–386.

Meyer, P., Salber, D., Schiefer, J. et al. 2008. Cerebral kinetics of the dopamine D(2) receptor ligand [I123]IBZM in mice. *Nucl Med Biol*, 35, 467–473.

Murphy, P., McCarthy, T., and Dzik-Jurasz, A. 2008. The role of clinical imaging in oncological drug development. *Br J Radiol*, 81, 685–692.

Nakao, R., Ito, T., Hayashi, K., Fukumura, T., Yamaguchi, M., and Suzuki, K. 2009. 1-Minute quality control tests for positron emission tomography radiopharmaceuticals. *J Pharm Biomed Anal*, 50, 245–251.

Passchier, J. 2009. Fast high performance liquid chromatography in PET quality control and metabolite analysis. *Q J Nucl Med Mol Imaging*, 53(4), 411–416.

PETMarketSummary. 2008. PET Imaging Market Summary Report, IMV Medical Information Division. Des Plaines, IL.

PIC/S. 2008. PIC/S Guide to good practices for the preparation of medicinal practices in healthcare. IN SECRETARIAT, P. S. (Ed.) *Pharmaceutical Inspection Convention*.

Schou, M., Zoghbi, S., Shetty, H. et al. 2009. Investigation of the metabolites of (S,S)-[(11)C]MeNER in humans, monkeys and rats. *Mol Imaging Biol*, 11, 23–30.

Shoghi, K., Gropler, R., Sharp, T. et al. 2008. Time course of alterations in myocardial glucose utilization in the Zucker diabetic fatty rat with correlation to gene expression of glucose transporters: A small animal PET investigation. *J Nucl Med*, 49, 1320–1327.

Shoghi, K. I. 2009. Quantitative small animal PET. *Q J Nucl Med Mol Imaging*, 53, 365–373.

Shoghi, K. I. and Welch, M. J. 2007. Hybrid imaging and blood sampling input function for quantification of small animal dynamic PET data. *Nucl Med Biol*, 34, 989–994.

Singh, S. 2006. Preclinical pharmacokinetics: An approach towards safer and efficacious drugs. *Curr Drug Metab*, 7, 165–182.

Tsukada, H., Nishiyama, S., Kakiuchi, T. et al. 1999. Isoflurane anesthesia enhances the inhibitory effects of cocaine and GBR12909 on dopamine transporter: PET studies in combination with microdialysis in the monkey brain. *Brain Res*, 849, 85–96.

USPChapter621. 2009. Chapter <621> Chromatography. *The United States Pharmacopeia and National Formulary* USP 32–NF 27 ed. Rockville, MD.

USPChapter823. 2009. Chapter <823> Radiopharmaceuticals for positron emission tomography: Compounding. *The United States Pharmacopeia and National Formulary* USP 32–NF 27 ed. Rockville, MD, The United States Pharmacopeial Convention, Inc.

USPChapter1225. 2009. Chapter <1225> Validation of compendial procedures. *The United States Pharmacopeia and National Formulary* USP 32–NF 27 ed. Rockville, MD.

Vanderheyden, J. 2009. The use of imaging in preclinical drug development. *Q J Nucl Med Mol Imaging*, 53, 374–381.

Vignon, E., Piperno, M., Le Graverand, M. et al. 2003. Measurement of radiographic joint space width in the tibiofemoral compartment of the osteoarthritic knee: Comparison of standing anterioposterior and Lyon schuss views. *Arthritis Rheum*, 48, 378–384.

Watase, K. and Zoghbi, H. Y. 2003. Modelling brain diseases in mice: The challenges of design and analysis. *Nat Rev Genet*, 4, 296–307.

Willmann, J., Van Brugen, N., Dinkelborg, L., and Gambhir, S. 2008. Molecular imaging in drug development. *Nat Rev Drug Discov*, 7, 591–607.

Woo, S., Lee, T., Kim, K. et al. 2008. Anesthesia condition for [18F]-FDG imaging of lung metastasis tumors using small animal PET. *Nucl Med Biol*, 35, 143–150.

Wu, N. and Clausen, A. 2007. Fundamental and practical aspects of ultrahigh pressure liquid chromatography for fast separations. *J Sep Sci*, 30, 1167–1182.

Zigler, S. S. 2009. Instrumentation and radiopharmaceutical validation. *Q J Nucl Med Mol Imaging*, 53(4), 402–410.

Zigler, S. S., Breslow, K., and Nazerias, M. S. 2005. A quality system for PET: An industry perspective. *Nucl Instrum Methods Phys Res Sec B: Beam Interact Mater Atoms*, 241, 645–648.

Chapter 5

Case Studies on Developing Targeted Nonnuclear Probes

6. Targeted Optical Molecular Imaging with Cytate
From Bench toward Bedside

Yang Liu and Samuel Achilefu

6.1 Introduction

Optical molecular imaging has become a powerful tool for both preclinical research and clinical molecular imaging studies over the last two decades. This is primarily due to the significant advances in the field of molecular science and opto-electronic instrumentation. Unlike the current workhorse of clinical molecular imaging, positron emission tomography (PET) and single photon emission computed tomography (SPECT), optical methods do not involve harmful ionizing radiation. Biomedical optical imaging utilizes diverse reporting mechanisms, such as absorption, scattering, fluorescence, bioluminescence, and photoacoustic techniques to interrogate molecular processes. Moreover, optical imaging also possesses unparalleled high spatial resolution of superficial tissues. For instance, intravital confocal microscopy can resolve structures as tiny as cell organelles (Astner et al. 2008). At the organ level, optical methods are amenable to many applications. Notable examples include the use of optical coherence tomography (OCT) in ophthalmology (Bezerra et al. 2009, Chen and Lee 2007), and endoscopes for imaging the gastrointestinal (GI) system (Zhong et al. 2009). Recently, optical systems have been applied to image thick tissues such as human breasts, where the predominant fatty breast tissue does not attenuate light as much as other organs (Choe et al. 2005, Pakalniskis et al. 2011). The ability of near-infrared (NIR) light to penetrate the human skull with minimal attenuation has enabled its application in noninvasive brain function imaging and spectroscopy (Liao et al. 2010, White et al. 2009). Naturally, the superficial nature of the human skin has favored the use of optical methods to detect, stage, and monitor the treatment of skin-related

Targeted Molecular Imaging. Edited by Michael J. Welch and William C. Eckelman © 2012 Taylor & Francis Group, LLC. ISBN: 978-1-4398-4195-2

Chapter 6

diseases such as melanoma (Chen et al. 2005, Javaheri et al. 2008). Beyond noninvasive imaging applications, a viable path to human translation of optical methods is in the field of image-guided surgical interventions. This approach has been driven by the deficiencies of current methods for surgical guidance. For example, fluoroscopy is widely used in surgical rooms (Hiorns et al. 2006, Rana et al. 2007) but the signal involves the use of ionizing radiation, which is hazardous to human health. Moreover, this method is incapable of detecting molecular processes. Ultrasound imaging provides a viable alternative to fluoroscopy because it is cheap, safe, and simple to use, with an impressive real-time feedback but this method primarily reports the anatomical features of tissues (Dewitt et al. 2006, Rosendahl and Toma 2007). For these reasons, intraoperative optical imaging devices are attractive because they combine the positive attributes of ultrasound with an unparalleled capacity to detect and report molecular targets without jeopardizing the safety of healthcare professionals.

In addition to instrument development, a new generation of optical molecular probes with improved sensitivity, specificity, and signal-to-background ratio (S/N) has been developed. With elegant designs, optical molecular probes are capable of providing information at the molecular, cellular, and tissue levels. Researchers leverage on the fast-growing genomic, proteomic, and nanotechnology data to optimize these probes for tailored applications of interest. The combined knowledge of disease biomarkers, the development of targeted molecular probes, and the availability of highly sensitive optical imaging systems have the potential to significantly improve the detection and assessment of molecular or genetic signatures of diseases, unveil disease molecular mechanisms, detect gene and protein expression, deliver drugs to specific tissue, and monitor disease status. Additional improvement in imaging contrast is readily achieved by utilizing stealth fluorescent probes that can be turned on and off through mechanisms such as fluorescence resonance energy transfer (FRET). This distinct feature is not available to radionuclide imaging.

In this chapter, we describe our efforts toward the development of cytate for clinical use. The rationale for selecting cytate and the intended clinical application are described. We also highlight the advances we have made in the development of a simple imaging device for image-guided surgery, which represents an immediate focus of our clinical application. Considering the numerous reporting strategies available for optical imaging, we briefly review the basis of fluorescence imaging, which is the primary contrast mechanism used in our laboratory at this time.

6.2   Fluorescence Imaging

Currently, there are many optical-based technologies available for molecular imaging, including fluorescence reflectance imaging (Autiero et al. 2009, Waldeck et al. 2008), fluorescence tomography (Nothdurft et al. 2009, Patwardhan et al. 2005), OCT (Applegate and Izatt 2006), and photoacoustic tomography (Pan et al. 2009). Among these modalities, fluorescence imaging has attracted special interest because it offers high detection sensitivity, specificity, and spatial resolution. Fluorescence is a phenomenon initiated by molecular absorption of light at a particular wavelength and the subsequent emission of light at a different wavelength from the excited molecules (Achilefu 2004b). Fluorescence involves electronic transitions between the ground and excitation states, as shown in Figure 6.1 (Achilefu 2004b). In most cases, the emission wavelength is longer than the excitation wavelength, and the energy of the emissive photon is lower than the incident photon. The amplitude of fluorescence emitted from a group of molecules is the fluorescence intensity, and the average time that fluorescent molecules remained in the excitation state before returning to the ground state is termed fluorescence lifetime (Berezin and Achilefu 2010). As fluorescence process is highly dependent on the energy states of molecules, it is highly sensitive to environmental factors, including molecular interactions.

A particular strength of attractive feature of fluorescence for imaging applications is the ability to alter the fluorescence profile and dynamics in response to molecular processes. The somatostatin receptor-specific probe described in this chapter takes advantage of the high detection sensitivity provided by NIR fluorescence, where single molecule detection is possible.

The regenerative nature of fluorescence method facilitates longitudinal imaging of molecular processes without the need for multiple administrations of the imaging agents. In contrast to the radionuclide decay

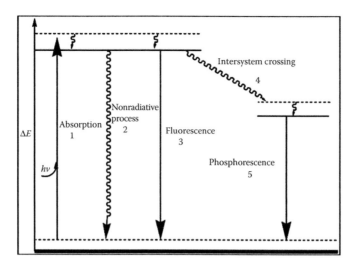

FIGURE 6.1 Illustration of electronic transitions. (Adapted from Achilefu, S. 2004b. *Curr Med Chem:Immunol, Endocrine Metab Agents,* 4(4):253–269.)

process used in nuclear imaging, which depletes the signaling entities over time, each fluorescence emitting species can be excited multiple times to produce the same fluorescence intensity, unless other mitigating factors such as photobleaching interfere with the pro-cess. The above features have favored the use of fluorescence imaging as the lead reporting mechanism for targeted molecular probes used in living organisms, including humans (Frangioni 2003, Rasmussen et al. 2009).

6.3 Near-Infrared (NIR) Fluorophore Development: Cypate

Molecular probes are essential for targeted optical imaging of diseases and the development of optimized probes is critical for the advancement of optical methods. Historically, fluorescence imaging focused on the region of visible spectra, where fluorescent dyes and imaging instrumentations were optimized for detection. Representative classical examples of visible dyes include the use of Fluorescein and Rhodamine used as tracers in the fluorescence microscopy. These dyes are additionally attractive because the colors are visible to the human eye, permitting visual analysis of color and fluorescence intensity changes, albeit with less efficiency when compared with modern detectors. However, fluorescence imaging in the visible range suffers from shallow penetration of light in tissue, confining their use to *in vitro*, *ex vivo*, and superficial tissues. To tackle this challenge, researchers have directed concerted efforts to near-infrared (NIR) fluorescence imaging, which typically spans from 700 to 900 nm. In this region, the absorption, scattering, and autofluorescence in biological tissues are relatively low. This wavelength range is often termed the "transparent window" because it allows for deep tissue imaging with low autofluorescence. Hence, NIR fluorescence is of particular interest for noninvasive *in vivo* or deep tissue imaging.

Among all the NIR molecular probes, carbocyanine dyes are widely used in molecular optical imaging studies. These dyes typically consist of aromatic moieties connected by polymethine groups to form fluorescent compounds. Representative structures are shown in Figure 6.2. A paragon of this class of dyes is Indocyanine Green (ICG), which is approved by US Food and Drug Administration (FDA) for assessing hepatic function, liver blood flow, cardiac output, and ophthalmic angiography (Achilefu 2004a). ICG has excitation and emission spectra centered at 780 and 830 nm, respectively, which are ideal for NIR fluorescence imaging. The emission range is readily detectable by silicon-based photosensors (Achilefu et al. 2000). Unfortunately, ICG is not easily functionalizable because its disulfonate group generally does not react with bioactive moieties under mild conditions. Consequently, applications of ICG are primarily confined to those scenarios where a perfusion agent is needed. However, for molecular imaging, selective targeting of tissues or physiological processes is crucial. To accomplish this goal, a general approach is to introduce a targeting moiety to the fluorescent dye.

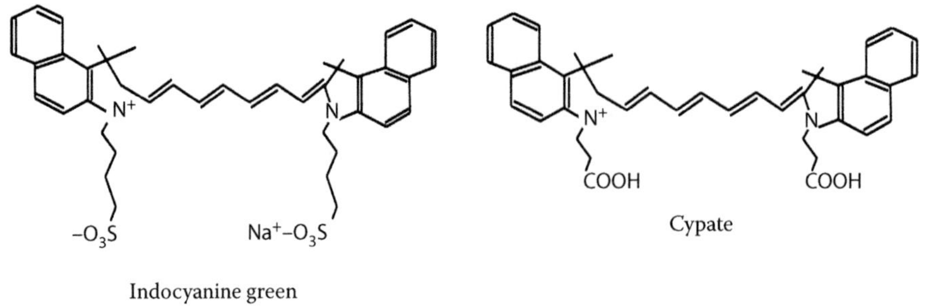

FIGURE 6.2　Chemical stuctures of representative NIR dyes.

Accordingly, an initial primary objective of our program was to develop reactive ICG-analogs for medical application. These efforts led to the development of cypate (Achilefu 2004a, Achilefu et al. 2000, Bugaj et al. 2001). Cypate has similar spectral properties as ICG, but it also has two free carboxyl groups that can

be easily conjugated to targeting moieties to form stable NIR molecular probes (Figure 6.2).

The synthesis of cypate generally begins with preparation of indolenium group using the Fischer indole reaction, as illustrated in Figure 6.3a (Achilefu 2004b). The reactive indolenium compounds generated are

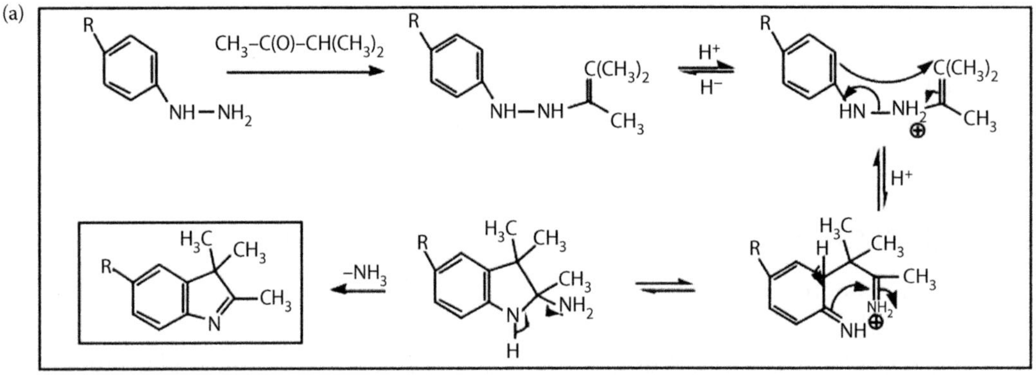

FIGURE 6.3　Synthetic scheme of cypate, compromising the Fischer indole reaction (a) and subsequent coupling with reactive polymethine group (b). (Adapted from Achilefu, S. 2004a. *Technol Cancer Res Treat*, 3(4):393–409.)

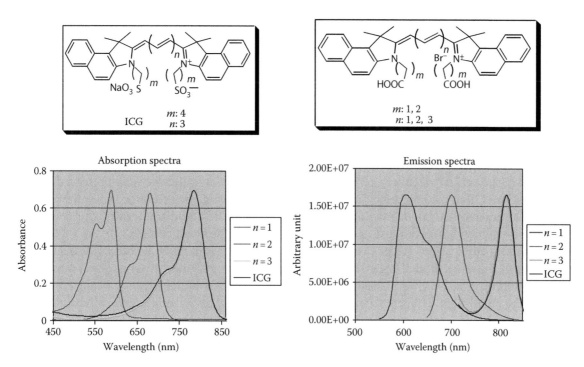

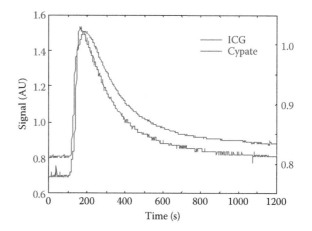

FIGURE 6.4   Spectral properties of NIR carbocyanine dyes in 25% DMSO/water (v/v). Cypate ($m = 2$, $n = 3$) share similar spectral properties as ICG. (Adapted from Achilefu, S. 2004b. *Curr Med Chem:Immunol, Endocrine Metab Agents*, 4(4):253–269.)

subsequently used to synthesize cypate via the route shown in Figure 6.3b (Achilefu 2004b).

During the development of cypate, both the alkyl and polymethine groups are specifically tailored so that the spectral properties of molecule match that of ICG. The absorption and emission spectra of ICG and carbocyanine derivatives are shown in Figure 6.4 (Achilefu 2004a). Since most of the current NIR imaging systems are optimized for ICG, the spectral equivalence of cypate to ICG will facilitate the clinical translation of cypate-based molecular probes.

In addition to the IGC spectral equivalence, cypate also shares similar biodistribution and blood clearance profile as ICG, as shown in Figure 6.5 (Achilefu et al. 2000). In addition, both cypate and ICG are excreted primarily through hepatobiliary route (Figure 6.5) (Achilefu et al. 2000).

Having established the ideal reactive ICG equivalent dye, we then focused on developing a cypate-based

FIGURE 6.5   Pharmacokinetics of cypate and ICG: blood clearance in rats. (Adapted from Achilefu, S. et al. 2000. *Invest Radiol*, 35(8):479–485.)

molecular probe that selectively targets tumors with the goal of using the probe in humans.

6.4   Somatostatin Receptor-Targeted NIR Fluorescent Molecular Probe: Cytate

Targeted molecular probes enhance site-specific uptake in targeted diseased tissues, which can augment specificity, sensitivity, and imaging contrast. A viable approach to accomplish this goal is to conjugate an NIR dye to a molecule that can deliver the signaling species to the site of interest. In general, ligands that

Chapter 6

bind strongly to receptors overexpressed on the diseased cells are used (Achilefu 2004b). Traditionally, nuclear imaging has played the most significant role in targeted molecular imaging, with numerous targeted agents already in clinical trials (Imam 2005). Thus, clinically useful nuclear imaging agents served as models for the initial development of targeted optical molecular probes in our lab.

Our interest narrowed to the overexpression of some receptors on the surface of cancer cells. Earlier studies demonstrated that radiolabeled antibodies could be used for receptor-targeted imaging but the pharmacokinetics and dynamics of these agents were not optimal imaging applications (Cance et al. 1988, Otsuka et al. 1988, Schwarz et al. 1994). This led to interest in developing small peptide-based imaging agents because (a) small-peptide agents can circumvent adverse immunogenic reactions caused by some antibodies; (b) the clearance time and pharmacokinetics of peptide-based agents *in vivo* can be optimized through structural modifications without drastic reduction in the ligand–receptor binding affinity; (c) peptide chemistry is generally straightforward and reproducible; and (d) their relatively smaller size facilitates release of the peptide-based agents into the extravascular space, where diffusion-driven migration to target tissue is essential in some targeted tissues.

At the time of our initial venture into molecular optical imaging, OctreoScan was the first and only available peptide-based nuclear imaging agent approved by the FDA in 1994 (Bruns et al. 1993, Kahaly et al. 1995, Olsen et al. 1995, Ramage et al. 1996). OctreoScan is a radiolabeled peptide used to image neuroendocrine tumors (Fuster et al. 1999) that overexpress somatostatin receptors (STRs). STRs are G protein-coupled seven transmembrane receptors with five known subtypes: STR1, STR2, STR3, STR4, and STR5 (Dalm et al. 2008, Florio 2008). Due to the short half-life of native somatostatin, a stable somatostatin peptide analog (octreotide) that preferentially targets STR2 was used to develop OctreoScan. Another variant of octreotide is octreotate, which has similar peptide sequence but only differs by the replacement of the C-terminal threoninol with the natural threonine. Our interest in octreotate stems from its ninefold higher binding affinity to STR2 than octreotide (Kwekkeboom et al. 2001). Subsequent clinical study demonstrated the feasibility of using [177]Lutetium-DOTA-Octreotate for treatment of neu-

roendocrine tumors in Europe (Kwekkeboom et al. 2005).

Based on OctreoScan's success, we launched a pilot study and successfully synthesized octreotate-cypate-derivatives (OCDs) that preserve the high binding affinities to STR (Achilefu et al. 2000, 2002). The chemical structures of representative OCDs developed are shown in Table 6.1 (Achilefu et al. 2002). Our results showed that the binding affinity peptide and the spectral properties of the dye were preserved. This feat was not obvious at that time because the relatively large size of cypate could undermine the receptor-binding affinity of the small peptide.

Table 6.1 shows that the conjugate of cypate (equivalent to FDA-approved ICG) to octreotate (peptide currently used in human trials) produced cytate, which retained the excellent receptor-binding affinity of the peptide. Spectral studies also showed that the fluorescence properties of the dye were retained. We then examined the ability of cytate to target cells (A427) expressing STR2. The results clearly demonstrated that cytate internalized in the STR2 cells (Figure 6.6) relative to STR-negative cells (not shown).

The tumor targeting ability of cytate was further confirmed by optical imaging *in vivo*. A male Lewis rat model bearing CA20948 tumors that overexpressed STR was used in this study. This same tumor model was used in the preclinical studies used for the approval of OctreoScan for human use. Consequently, the animal served as a good model for comparative study between radiolabeled and fluorescent dye-labeled octreotate. The study showed preferential accumulation and retention of cytate in CA20948 tumors, which was comparable to ^{111}In-DOTA-octreaotate (Figure 6.7) (Achilefu et al. 2000).

After 24 h, the rats were euthanized and representative tissues were removed for *ex vivo* biodistribution study. The fluorescence intensity data showed that the tumor-to-muscle fluorescence intensity ratio was exceptionally high (Figure 6.8) (Achilefu et al. 2000). Beside fluorescence intensity, cytate uptake in human tumor tissue can also be distinguished from normal tissues using polarization optical imaging method. The study further confirmed that the polarization anisotropy of cytate in cancerous tissues was significantly higher than that in normal tissue (Pu et al. 2008).

To the best of our knowledge, this was the first demonstration of using a small peptide as a targeting group for NIR optical imaging.

Table 6.1   STR-Targeting Agents: Octreotate-Cypate-Derivatives (OCDs)

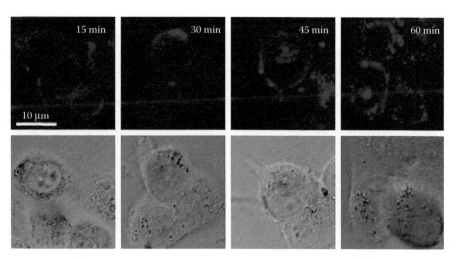

CONHR$_1$ COR$_2$

Probe	Name	R1	R2	n	IC$_{50}$ (nM)	SE ± (nM)
1	Octreotide (Peptide alone)	dF-C-Y-dW-K-T-C-T-ol S————S	–	–	0.4	0.07
2	Octreotate (Peptide alone)	dF-C-Y-dW-K-T-C-T-OH S————S	–	–	0.6	0.10
3	Cytate	-dF-C-Y-dW-K-T-C-T-OH S————S	OH	1	4.6	0.7
4	Cytate-6	-dF-C-Y-dW-K-T-C-T-OH S————S	OH	4	5.4	1.0
5	Cytate-66	-dF-C-Y-dW-K-T-C-T-OH S————S	NHR$_1$	4	1.1	0.1
6	βA-Cytate	-βA-dF-C-Y-dW-K-T-C-T-OH S————S	OH	1	3.4	0.4

Source: Adapted from Achilefu, S. et al. 2002. *J Med Chem*, 45(10):2003–2015.

FIGURE 6.6   Confocal fluorescent (top) and differential interference images (bottom) of STR2-positive A427 cells incubated with 1 μM of cytate as a function of time at 37°C.

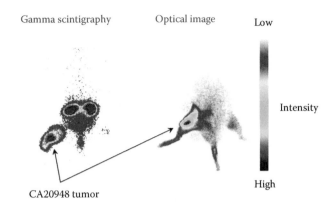

Gamma scintigraphy   Optical image   Low

Intensity

High

CA20948 tumor

FIGURE 6.7 Comparative study of cytate and ^{111}In-DOTA-octreaotate in Lewis rats bearing CA20948 tumors. Left: Gamma scintigraphy; right: optical image. (Adapted from Bugaj, J.E. et al. 2001. *J Biomed Opt*, 6(2):122–133.)

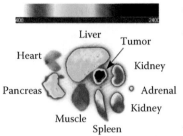

Normal tissue	T/NT
Heart	10
Muscle	7
Kidney	5
Liver	2

FIGURE 6.8 *Ex vivo* biodistribution study of cytate. (a) Biodistribution of cytate 24 h postinjection. (b) Fluorescence intensity ratio of tumor to normal tissues (t/nt). The data were obtained by taking the relative fluorescence intensity averages for each organ 24 h postadministration in six CA20948 tumor bearing rats. (Adapted from Achilefu, S. et al. 2000. *Invest Radiol*, 35(8):479–485.)

6.5   Planned Clinical Application: Intraoperative Detection of Primary Hepatocellular Carcinoma, Positive Nodules, and Liver Metastasis

Although whole-body scan is routinely performed with nuclear imaging modality, optical imaging is depth-limited because of light scattering and absorption by tissue. These physical factors confine optical methods to imaging tissue within a few centimeters from the light source (Culver et al. 2008). Therefore, the selection of medical problems to be addressed with molecular optical imaging must be weighed carefully to accommodate this limitation. Today, studies related to breast, skin, brain, oral, joint, cervical, GI, and intraoperative optical imaging are the primary focus of optical studies in humans. We focused on image-guided HCC surgery as a means to translate optical methods from bench to bedside. This approach was motivated by several factors.

Cancer is one of the most leading causes of death in the world. Among various malignancies, HCC is particularly lethal and it is the most prevalent form of liver cancer (Aguayo and Patt 2001). For example, HCC leads to 600,000 mortalities annually, making it the third most predominate cause of cancer mortality (Parkin et al. 2005). Unfortunately, the current standard of care for HCC is far from satisfactory. As HCC does not respond well to chemotherapy and radiotherapy, the standard of care relies on surgical resection of the tumor. Although the resection of primary HCC is now routinely performed in major hospitals, it is difficult for surgeons to completely remove all the lesions, including small positive nodules that can become aggressive after the removal of the primary tumor. Thus, the surgical outcome is usually poor, with 80–90% of patients having cancer relapse (Tandon and Garcia-Tsao 2009). If the tumors are not completely removed, the prognosis is poor and most patients will die within 3–6 months after surgery. Hence, molecular imaging agents and intraoperative imaging instruments that can accurately detect small nodules and tumor margins are needed, which can help surgeons avoid incomplete resection to achieve complete resection. This strategy involves two steps. First is to develop HCC selective molecular probe, followed by the development of an imaging device that is compatible with intraoperative procedures.

Cytate is ideal for this application because it does not emit radioactivity and the NIR fluorescence can allow visualization of liver tissue within a few centimeters. Furthermore, the low autofluorescence will minimize background signal in a pool of blood during surgery, thereby improving the detection sensitivity of NIR optical imaging devices. Moreover, octreotate, a peptide analog used in the development of cytate, has been used in clinical trials for the treatment of advanced HCC (Kouroumalis et al. 1998). In addition, Fuster et al. 1999 reported that OctreoScan successfully aided in the identification of liver metastasis from primary neuroendocrine tumors. The overexpression of STR in HCC (Kouroumalis et al. 1998) allows us to use cytate as a molecular probe for detecting primary and metastatic liver tumors. Our next effort to accomplish this goal rests in the development of a simple system for use in surgical suites.

6.6 Intraoperative Molecular Imaging: NIR Fluorescence Goggle Device

As discussed in Section 5, we have directed our efforts to HCC management and we have successfully developed a series of STR-targeting agents that are suitable for intraoperative molecular imaging of HCC. From these compounds, we selected cytate as our lead compound for the planned human study. To translate this and other optical molecular probes from small animals to humans, clinical NIR imaging instruments are needed. To date, most intraoperative imaging instruments are limited to ultrasound and nuclear imaging/detection methods (Britten 1999, Kopelman et al. 2005, Perkins and Britten 1999). On the one hand, ultrasound detector is conventionally used to locate the region of the primary tumor based on tissue morphology but it has limited capability to provide molecular information. On the other hand, nuclear imaging uses ionizing radiation that requires stringent handling procedure to prevent the exposure of health professionals to radioactivity. For intraoperative NIR fluorescence imaging, few clinical systems such as SPY imaging system (Kubota et al. 2006, Reuthebuch et al. 2004, Taggart et al. 2003), Hamamatsu Photodynamic Eye (Gotoh et al. 2009), and FLARE™ system (Troyan et al. 2009) are available. However, these systems are bulky and have complex instrumentation. Moreover, all the existing systems need to use external monitors to display the NIR information, which can distract the surgeon with unintended consequences.

To address the need for an accurate and affordable intraoperative imaging instrument, we have developed a new optical molecular imaging device: NIR fluorescence goggle device (Figure 6.9) (Liu et al. 2011a,b). In contrast to conventional imaging instruments, our goggle device does not need any remote monitor (Liu et al. 2011a,b). Instead, it directly displays the fluorescence information in its goggle eyepiece. Hence, the operator can capture functional information from one eye while using the other eye to visualize the anatomical features of the tissue of interest. The goggle system has both NIR and white light source modules capable of offering both probe excitation and illumination of the surgical site at the same time. The goggle device is compact, portable, battery-operated, hands-free, and user-friendly. Furthermore, it is equipped with a wireless communication capability that is capable of sending real-time data to remote sites. The wireless function can enable telemedicine at point-of-care, which reduces the need to have surgeons, radiologists, and pathologists in the same room during the surgery. The fluorescence goggle device is affordable and only cost $1200 for the first prototype. With the development of manufacturing technology, the cost of future models is likely to fall below $1000.

The first prototype of fluorescence goggle was designed to work with cypate dye used in the preparation of cytate with 780/830 nm excitation/emission. It can detect fluorescence from as low as a few nM of ICG. We have applied our goggle device in three types of oncologic surgical procedures. First, we applied it in the sentinel lymph node (SLN) biopsy, which is the standard staging technique for some forms of cancer, such as breast and skin. After intradermal injection of ICG, SLNs were readily found with the assistance of the goggle device (Figure 6.10) (Liu et al. 2011a,b).

Having demonstrated the successful use of the goggle device in fluorescence SLN mapping, it was then

(a)

(b)

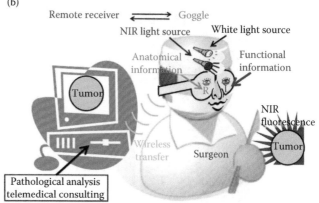

FIGURE 6.9 Wireless NIR goggle device. (a) A prototype device; red arrows indicate NIR light source; white arrows indicate white light source; green arrow indicates the detector. (b) Schematic diagram of the design of goggle device. (Adapted from Liu, Y. et al. 2011a. *Surgery,* 149(5):689–698.)

Chapter 6

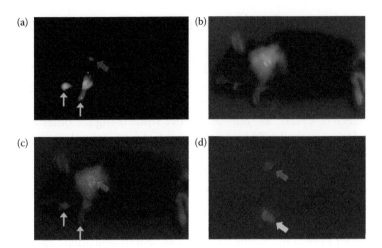

FIGURE 6.10   SLN mapping with ICG and fluorescence goggle. (a) NIR image; green arrows indicate injection points, blue arrow indicates SLN; (b) white light image; (c) NIR-white light composite image, where NIR fluorescence is pseudocolored in red; (d) *ex vivo* examination of SLN and control lymph node. Blue arrow indicates SLN; yellow arrow indicates control node. (Adapted from Liu, Y. et al. 2011a. *Surgery,* 149(5):689–698.)

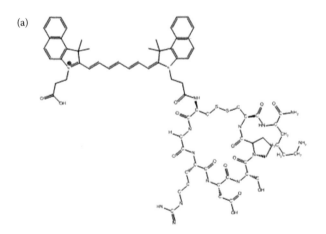

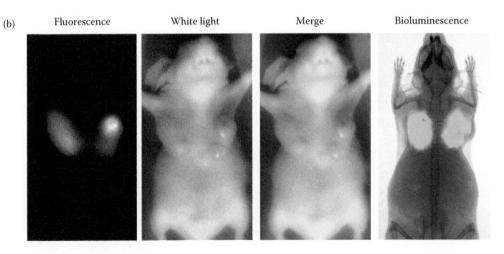

FIGURE 6.11   Targeted molecular probe LS 301 and tumor imaging. (a) Chemical structure of LS 301. (b) *In vivo* imaging of tumor prior to surgery. Fluorescence and white light images were acquired with the goggle device, while bioluminescence and radiograph were acquired with Carestream *In-Vivo* Multispectral FX PRO imaging system. (Adapted from Liu, Y. et al. 2011a. *Surgery,* 149(5):689–698.)

applied to real-time image-guided surgery with a targeted molecular probe. In this study, cypate-derivative LS301 is used (Figure 6.11a) (Liu et al. 2011a). This study was conducted in 4T1-luc tumor-bearing mice. The 4T1-luc cancer cells were transfected with luciferase to track their location by bioluminescence imaging (Figure 6.11b). We imaged the mice prior to surgery and cross-validated the presence of tumors using both NIR fluorescence and bioluminescence imaging.

During surgery, high fluorescence intensity was observed in the tumors (Figure 6.12a). Surgery was then performed under the guidance of the goggle device. The goggle can clearly indicate the tumor margins, as well as small positive nodules and residual tumors that were not obvious to the naked eye (Figure 6.12b). This can help the surgeon to locate the tumors, reduce the size of healthy tissues resected, and minimize the occurrence of incomplete resection.

6.7 Expected Impact on Clinical Care

With the advancements in targeted optical probes and instrument development, optical molecular imaging has emerged as a paradigm-shifting tool for clinical use. This imaging modality is affordable, fast, safe, and provides new imaging option to clinicians. Due to the issue of penetration depth, optical molecular imaging will thrive more in medical applications where limitations of the optical imaging depth are addressed or circumvented. These applications include the imaging of (a) superficial tissues such as skin, tooth, and oral cavity; (b) optically transparent tissues such as the eyes; (c) tissues that have relatively low NIR light attenuation such as breasts and skull; (d) organs that can be easily accessed using an endoscope or a catheter, such as the GI tract and prostate; and (e) intraoperative procedures, where the tissue of interest is exposed. Moreover, unlike most conventional imaging modalities, optical molecular imaging is not limited to traditional hospital settings. Its portability renders it an ideal method for medical applications at point-of-care and military frontiers.

At Washington University Optical Radiology Laboratory, efforts have been directed toward the translation of targeted fluorescence imaging into clinics to

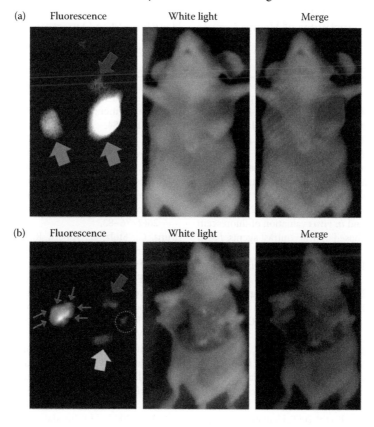

FIGURE 6.12 Image-guided surgery with fluorescence goggle device before (a) and during surgery (b). Green arrows indicate major nodule; purple arrow indicates small nodule; blue arrows indicate tumor margin; green cycle indicates residual tumor; gray arrow indicates liver. (Adapted from Liu, Y. et al. 2011a. *Surgery*, 149(5):689–698.)

Chapter 6

solve challenging medical problems such as HCC. We have developed a library of functionalizable ICG-analogs: cypate and its derivatives. These compounds share similar spectra and biological properties as ICG, with the flexibility of being conjugated to one or multiple targeting moieties. As ICG has been used in clinics for decades, cypate is an ideal ICG-analog for ease of clinical translation. In addition, we have developed various STR-targeting fluorescent molecular probes and animal studies have validated that cytate possesses high targeting capability to tumors expressing this receptor. Since OctreoScan is approved by the FDA for imaging neuroendocrine tumors and [177]Lutetium-DOTA-Octreotate is currently used clinically in Europe for radionuclide treatment of STR-positive tumors, the NIR fluorescent counterpart, cytate, also holds great promise for human use. In addition, the translation of the newly developed goggle device into clinics will be significantly easier. The goggle device is considered as a device of nonsignificant risk because it is noncontact and does not utilize laser. In fact, the first human study of fluorescence goggle device has been initiated.

Coupled with the newly developed hands-free fluorescence goggle system, cytate will improve the management of HCC patients. The poor surgical outcome of HCC will be remedied, as the goggle device can detect nonobvious metastases highlighted by cytate, thereby minimizing incomplete resection and cancer relapse. Furthermore, as the goggle device is affordable and user-friendly, it holds great promise in expanding the reach of fluorescence-guided surgery to currently underserved regions and developing countries, particularly in sub-Saharan Africa, Southeast Asia, and China, where HCC is epidemic. Moreover, as fluorescence goggle is also portable and wireless-capable, it can be used for operations in rural clinics, where inexperienced clinicians can perform life-saving procedures under remote guidance of experts.

Acknowledgments

Liu is supported by US Department of Defense Predoctoral Traineeship Award W81XWH-11-1-0059. Part of the research discussed in this chapter was supported by the US National Institutes of Health Grants R01 EB008111 and EB008458. The content is solely the responsibility of the authors and does not necessarily represent the official views of the National Institutes of Health and the Department of Defense.

References

Achilefu, S. 2004a. Lighting up tumors with receptor-specific optical molecular probes. *Technol Cancer Res Treat*, 3(4):393–409.

Achilefu, S. 2004b. Optical imaging agents and potential application in the assessment of pancreatic beta cells. *Curr Med Chem: Immunol, Endocrine Metab Agents*, 4(4):253–269.

Achilefu, S., Dorshow, R.B., Bugaj, J.E., and Rajagopalan, R. 2000. Novel receptor-targeted fluorescent contrast agents for *in vivo* tumor imaging. *Invest Radiol*, 35(8):479–485.

Achilefu, S., Jimenez, H.N., Dorshow, R.B. et al. 2002. Synthesis, *in vitro* receptor binding, and *in vivo* evaluation of fluorescein and carbocyanine peptide-based optical contrast agents. *J Med Chem*, 45(10):2003–2015.

Aguayo, A. and Patt, Y.Z. 2001. Liver cancer. *Clin Liver Dis*, 5(2):479–507.

Applegate, B.E. and Izatt, J.A. 2006. Molecular imaging of endogenous and exogenous chromophores using ground state recovery pump-probe optical coherence tomography. *Opt Express*, 14(20):9142–9155.

Astner, S., Dietterle, S., Otberg, N., Rowert-Huber, H.J., Stockfleth, E., and Lademann, J. 2008. Clinical applicability of *in vivo* fluorescence confocal microscopy for noninvasive diagnosis and therapeutic monitoring of nonmelanoma skin cancer. *J Biomed Opt*, 13(1):014003.

Autiero, M., Cozzolino, R., Laccetti, P. et al. 2009. Hematoporphyrin-mediated fluorescence reflectance imaging: Application to early tumor detection *in vivo* in small animals. *Lasers Med Sci*, 24(2):284–289.

Berezin, M.Y. and Achilefu, S. 2010. Fluorescence lifetime measurements and biological imaging. *Chem Rev*, 110(5): 2641–2684.

Bezerra, H.G., Costa, M.A., Guagliumi, G., Rollins, A.M., and Simon, D.I. 2009. Intracoronary optical coherence tomography: A comprehensive review clinical and research applications. *JACC Cardiovasc Interv*, 2(11):1035–1046.

Britten, A.J. 1999. A method to evaluate intra-operative gamma probes for sentinel lymph node localisation. *Eur J Nucl Med*, 26(2):76–83.

Bruns, C., Stolz, B., Albert, R., Marbach, P., and Pless, J. 1993. OctreoScan 111 for imaging of a somatostatin receptor-positive islet cell tumor in rat. *Horm Metab Res Suppl*, 27:5–11.

Bugaj, J.E., Achilefu, S., Dorshow, R.B., and Rajagopalan, R. 2001. Novel fluorescent contrast agents for optical imaging of *in vivo* tumors based on a receptor-targeted dye-peptide conjugate platform. *J Biomed Opt*, 6(2):122–133.

Cance, W.G., Otsuka, F.L., Dilley, W.G. et al. 1988. A potential new radiopharmaceutical for parathyroid imaging: Radiolabeled parathyroid-specific monoclonal antibody—I. Evaluation of 125I-labeled antibody in a nude mouse model system. *Int J Rad Appl Instrum B*, 15(3):299–303.

Chen, C.S., Elias, M., Busam, K., Rajadhyaksha, M., and Marghoob, A.A. 2005. Multimodal *in vivo* optical imaging, including confocal microscopy, facilitates presurgical margin mapping for clinically complex lentigo maligna melanoma. *Br J Dermatol*, 153(5):1031–1036.

Chen, J. and Lee, L. 2007. Clinical applications and new developments of optical coherence tomography: An evidence-based review. *Clin Exp Optom*, 90(5):317–335.

Choe, R., Corlu, A., Lee, K. et al. 2005. Diffuse optical tomography of breast cancer during neoadjuvant chemotherapy: A case study with comparison to MRI. *Med Phys*, 32(4): 1128–1139.

Culver, J., Akers, W., and Achilefu, S. 2008. Multimodality molecular imaging with combined optical and SPECT/PET modalities. *J Nucl Med*, 49(2):169–172.

Dalm, V.A., Hofland, L.J., and Lamberts, S.W. 2008. Future clinical prospects in somatostatin/cortistatin/somatostatin receptor field. *Mol Cell Endocrinol*, 286(1–2):262–277.

Dewitt, J., Devereaux, B.M., Lehman, G.A., Sherman, S., and Imperiale, T.F. 2006. Comparison of endoscopic ultrasound and computed tomography for the preoperative evaluation of pancreatic cancer: A systematic review. *Clin Gastroenterol Hepatol*, 4(6):717–725; quiz 664.

Florio, T. 2008. Somatostatin/somatostatin receptor signalling: Phosphotyrosine phosphatases. *Mol Cell Endocrinol*, 286(1–2):40–48.

Frangioni, J.V. 2003. *In vivo* near-infrared fluorescence imaging. *Curr Opin Chem Biol*, 7(5):626–634.

Fuster, D., Navasa, M., Pons, F. et al. 1999. In-111 octreotide scan in a case of a neuroendocrine tumor of unknown origin. *Clin Nucl Med*, 24(12):955–958.

Gotoh, K., Yamada, T., Ishikawa, O. et al. 2009. A novel image-guided surgery of hepatocellular carcinoma by indocyanine green fluorescence imaging navigation. *J Surg Oncol*, 100(1): 75–79.

Hiorns, M.P., Saini, A., and Marsden, P.J. 2006. A review of current local dose-area product levels for paediatric fluoroscopy in a tertiary referral centre compared with national standards. *Why are they so different? Br J Radiol*, 79(940):326–330.

Imam, S.K. 2005. Molecular nuclear imaging: The radiopharmaceuticals (review). *Cancer Biother Radiopharm*, 20(2):163–172.

Javaheri, M., Khurana, R.N., Bhatti, R.A., and Lim, J.I. 2008. Optical coherence tomography findings in paraneoplastic pseudovitelliform lesions in melanoma-associated retinopathy. *Clin Ophthalmol*, 2(2):461–463.

Kahaly, G., Diaz, M., Just, M., Beyer, J., and Lieb, W. 1995. Role of octreoscan and correlation with MR imaging in Graves' ophthalmopathy. *Thyroid*, 5(2):107–111.

Kopelman, D., Blevis, I., Iosilevsky, G. et al. 2005. A newly developed intra-operative gamma camera: Performance characteristics in a laboratory phantom study. *Eur J Nucl Med Mol Imaging*, 32(10):1217–1224.

Kouroumalis, E., Skordilis, P., Thermos, K., Vasilaki, A., Moschandrea, J., and Manousos, O.N. 1998. Treatment of hepatocellular carcinoma with octreotide: A randomised controlled study. *Gut*, 42(3):442–447.

Kubota, K., Kita, J., Shimoda, M. et al. 2006. Intraoperative assessment of reconstructed vessels in living-donor liver transplantation, using a novel fluorescence imaging technique. *J Hepatobiliary Pancreat Surg*, 13(2):100–104.

Kwekkeboom, D.J., Bakker, W.H., Kooij, P.P. et al. 2001. [177Lu-DOTAOTyr3]octreotate: Comparison with [111In-DTPAo]octreotide in patients. *Eur J Nucl Med*, 28(9): 1319–1325.

Kwekkeboom, D.J., Teunissen, J.J., Bakker, W.H. et al. 2005. Radiolabeled somatostatin analog [177Lu-DOTA0,Tyr3] octreotate in patients with endocrine gastroenteropancreatic tumors. *J Clin Oncol*, 23(12):2754–2762.

Liao, S.M., Gregg, N.M., White, B.R. et al. 2010. Neonatal hemodynamic response to visual cortex activity: High-density near-infrared spectroscopy study. *J Biomed Opt*, 15(2):026010.

Liu, Y., Bauer, A., Akers, W. et al. 2011a. Hands-free, wireless goggles for near-infrared fluorescence and real-time image-guided surgery. *Surgery*, 149(5):689–698.

Liu, Y., Bauer, A.Q., Akers, W. et al. 2011b. Compact intraoperative imaging device for sentinel lymph node mapping. *Proc SPIE*, 7910:79100D79101–79107.

Nothdurft, R.E., Patwardhan, S.V., Akers, W., Ye, Y., Achilefu, S., and Culver, J.P. 2009. *In vivo* fluorescence lifetime tomography. *J Biomed Opt*, 14(2):024004.

Olsen, J.O., Pozderac, R.V., Hinkle, G. et al. 1995. Somatostatin receptor imaging of neuroendocrine tumors with indium-111 pentetreotide (Octreoscan). *Semin Nucl Med*, 25(3): 251–261.

Otsuka, F.L., Cance, W.G., Dilley, W.G. et al. 1988. A potential new radiopharmaceutical for parathyroid imaging: Radiolabeled parathyroid-specific monoclonal antibody—II. Comparison of 125I- and 111In-labeled antibodies. *Int J Rad Appl Instrum B*, 15(3):305–311.

Pakalniskis, M.G., Wells, W.A., Schwab, M.C. et al. 2011. Tumor angiogenesis change estimated by using diffuse optical spectroscopic tomography: Demonstrated correlation in women undergoing neoadjuvant chemotherapy for invasive breast cancer? *Radiology* doi: 10.1148/radiol.11100699.

Pan, D., Pramanik, M., Senpan, A. et al. 2009. Molecular photoacoustic tomography with colloidal nanobeacons. *Angew Chem Int Ed Engl*, 48(23):4170–4173.

Parkin, D.M., Bray, F., Ferlay, J., and Pisani, P. 2005. Global cancer statistics, 2002. *CA Cancer J Clin*, 55(2):74–108.

Patwardhan, S., Bloch, S., Achilefu, S., and Culver, J. 2005. Time-dependent whole-body fluorescence tomography of probe biodistributions in mice. *Opt Express*, 13(7):2564–2577.

Perkins, A.C. and Britten, A.J. 1999. Specification and performance of intra-operative gamma probes for sentinel node detection. *Nucl Med Commun*, 20(4):309–315.

Pu, Y., Wang, W.B., Das, B.B., Achilefu, S., and Alfano, R.R. 2008. Time-resolved fluorescence polarization dynamics and optical imaging of cytate: A prostate cancer receptor-targeted contrast agent. *Appl Opt*, 47(13):2281–2289.

Ramage, J.K., Williams, R., and Buxton-Thomas, M. 1996. Imaging secondary neuroendocrine tumours of the liver: Comparison of I123 metaiodobenzylguanidine (MIBG) and In111-labelled octreotide (Octreoscan). *Qjm*, 89(7):539–542.

Rana, A.M., Zaidi, Z., and El-Khalid, S. 2007. Single-center review of fluoroscopy-guided percutaneous nephrostomy performed by urologic surgeons. *J Endourol*, 21(7):688–691.

Rasmussen, J.C., Tan, I.C., Marshall, M.V., Fife, C.E., and Sevick-Muraca, E.M. 2009. Lymphatic imaging in humans with near-infrared fluorescence. *Curr Opin Biotechnol*, 20(1): 74–82.

Reuthebuch, O., Haussler, A., Genoni, M. et al. 2004. Novadaq SPY: Intraoperative quality assessment in off-pump coronary artery bypass grafting. *Chest*, 125(2):418–424.

Rosendahl, K. and Toma, P. 2007. Ultrasound in the diagnosis of developmental dysplasia of the hip in newborns. The European approach. A review of methods, accuracy and clinical validity. *Eur Radiol*, 17(8):1960–1967.

Schwarz, S.W., Connett, J.M., Anderson, C.J. et al. 1994. Evaluation of a direct method for technetium labeling intact and F(ab')2 1A3, an anticolorectal monoclonal antibody. *Nucl Med Biol*, 21(4):619–626.

Chapter 6

Taggart, D.P., Choudhary, B., Anastasiadis, K., Abu-Omar, Y., Balacumaraswami, L., and Pigott, D.W. 2003. Preliminary experience with a novel intraoperative fluorescence imaging technique to evaluate the patency of bypass grafts in total arterial revascularization. *Ann Thorac Surg*, 75(3):870–873.

Tandon, P. and Garcia-Tsao, G. 2009. Prognostic indicators in hepatocellular carcinoma: A systematic review of 72 studies. *Liver Int*, 29(4):502–510.

Troyan, S.L., Kianzad, V., Gibbs-Strauss, S.L. et al. 2009. The FLARE intraoperative near-infrared fluorescence imaging system: A first-in-human clinical trial in breast cancer sentinel lymph node mapping. *Ann Surg Oncol*, 16(10):2943–2952.

Waldeck, J., Hager, F., Holtke, C. et al. 2008. Fluorescence reflectance imaging of macrophage-rich atherosclerotic plaques using an alphavbeta3 integrin-targeted fluorochrome. *J Nucl Med*, 49(11):1845–1851.

White, B.R., Snyder, A.Z., Cohen, A.L. et al. 2009. Resting-state functional connectivity in the human brain revealed with diffuse optical tomography. *Neuroimage*, 47(1):148–156.

Zhong, W., Celli, J.P., Rizvi, I. et al. 2009. *In vivo* high-resolution fluorescence microendoscopy for ovarian cancer detection and treatment monitoring. *Br J Cancer*, 101(12):2015–2022.

7. Molecular Imaging of P-Selectin with Contrast Ultrasound for the Detection of Recent Tissue Ischemia

Brian Davidson and Jonathan R. Lindner

Noninvasive medical imaging techniques such as ultrasound, radionuclide/PET imaging, magnetic resonance imaging, and radiography (computed tomography, angiography) are commonly used in cardiovascular medicine to evaluate the structure and function of the heart, great vessels, and tissue perfusion. Targeted imaging probes for molecular imaging have been developed for each of these modalities in order to improve existing diagnostic capabilities and to expand the impact of imaging in the care of patients with cardiovascular disease. These probes also have potential use as "biologic readouts" in preclinical drug development. Because there are differences in the kinetics of the probes, the capabilities of the detectors, and practical issues such as cost and safety, no single one of these imaging modalities is able to satisfy the diverse needs of biomedical researchers and clinicians. In this chapter, we will focus on molecular imaging utilizing contrast-enhanced ultrasonography (CEU). This technique has practical advantages with regard to portability and the short time required for an imaging protocol to be completed (minutes). Accordingly, it has been proposed that CEU of the heart, or echocardiographic molecular imaging, is an ideal approach for the detection of active or recent myocardial ischemia in patients presenting with recent or active chest pain. This chapter will describe the development and testing of a CEU molecular imaging agent targeted to endothelial selectins which are cell adhesion molecules involved in the inflammatory response that are expressed rapidly during ischemia and persist for hours after resolution of ischemia.

7.1 Contrast-Enhanced Ultrasound

In its current clinical form, CEU relies on the acoustic detection of gas-filled encapsulated microbubbles that are generally 1–5 μm in diameter (Figure 7.1) (Kaufmann et al. 2007b). The concept of using small gas bubbles to

Targeted Molecular Imaging. Edited by Michael J. Welch and William C. Eckelman © 2012 Taylor & Francis Group, LLC. ISBN: 978-1-4398-4195-2

Chapter 7

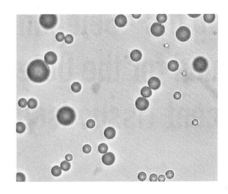

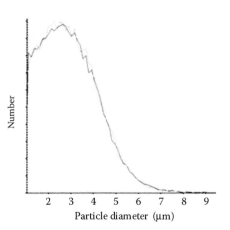

Contrast agent characteristics
1. Size
   – Microbubble (1–5 μm)
   – Nanoparticle (<1 μm)
2. Gas composition
   –Air or nitrogen
   –Sulfur hexafluoride
   –Perfluorocarbons
3. Shell composition
   –Lipid
   –Protein (albumin)
   –Biocompatible polymers

FIGURE 7.1 Microbubble characteristics. The left panel illustrates an example of light-microscopy of a lipid-shelled decafluorobutane gas-filled microbubble agent (diameter range 1–7 μm). The table lists some of the key differentiating factors for nontargeted ultrasound contrast agents; and the graph illustrates a typical size histogram for microbubbles on electrozone sensing.

enhance ultrasound originated in the 1960s when ultrasound signals were seen in the right heart upon rapid injection of indicator-dilution tracers used to evaluate cardiac output (Gramiak and Shah 1968). There has since been an evolution of microbubble technology and ultrasound contrast agents are currently used in patients for cardiovascular applications (e.g., enhancing the blood pool in the ventricles, assessing myocardial or limb perfusion) or for other radiology applications (e.g., assessing liver masses or renal perfusion) (Kaufmann et al. 2007b).

The compressibility of microbubbles in an acoustic field serves as the basis for their use as ultrasound contrast agents (Figure 7.2) (Dayton et al. 1999). During exposure to ultrasound, these microbubble contrast agents vibrate or undergo volumetric oscillation whereby they compress and expand in the pressure fluctuations of acoustic field. This oscillation produces an acoustic signal that is seen as an opacification during an ultrasound examination. If the ultrasound energy is sufficiently strong and within the ideal resonant frequency range, the shape changes of microbubbles can be complex involving nonlinear relation between pressure and microbubble size, unusual nonspherical shape changes, and even compression-only oscillation. These phenomena produce strong signals in the frequency range away from the transmission (fundamental) frequency especially within the harmonic (multiples of the fundamental) and subharmonic (e.g., half the fundamental) portions of the frequency spectra (Figure 7.2) (Dayton et al. 1999, Eckersley et al. 2005; Kaufmann et al. 2007). Nonlinear oscillation can also produce stronger than predicted

(a)

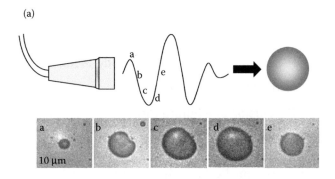

(b)

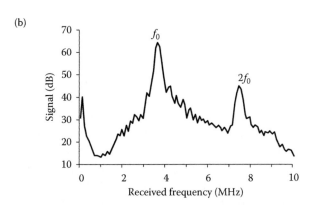

FIGURE 7.2 General principles for contrast-enhanced ultrasound imaging. (a) Signal generation from microbubbles results from the acoustic emissions produced by microbubble cavitation in the ultrasound field. Cavitation can be stable whereby the microbubbles oscillate in the acoustic field without disruption, as shown by the light microscopy images of microbubbles taken at different portions of the pressure wave; or inertial which involves disruption of microbubbles and the brief formation of free gas bubbles. Images are courtesy of N. De Jong, Erasmus University. (b) Example of the frequency–amplitude spectrum from microbubbles exposed to ultrasound at the fundamental (f_0) frequency of 3.7 MHz.

responses within the fundamental frequency range. Filtering for these frequencies and/or multipulse algorithms that eliminate all but the nonlinear signals are currently used to optimize signal-to-noise ratio during CEU imaging.

The *in vivo* stability of microbubbles and their ability to reach the systemic circulation after intravenous administration and reliably produce opacification of the left ventricular cavity, myocardium, liver, and so on, after intravenous contrast administration has been possible with chemical modification of microbubbles. These modifications involve both the shell and gas content. To produce an acoustic signal, microbubbles need to maintain their gas volume. The use of biocompatible gases that are less prone to outward diffusion than air or nitrogen has been one approach. According to Epstein–Plesset models, the outward gas loss for a bubble is dependent upon its size, surface tension, and gas characteristics such as solubility and diffusion capacity (Epstein 1950). Accordingly, the stability of microbubble contrast agents is improved when they contain gases with relatively low diffusion coefficients and low solubility in water or blood (the Ostwald coefficient) (Kabalnov et al. 1998). The most common gases used for this purpose in commercially produced contrast agents approved for use throughout the world include perfluorocarbons such as octafluoropropane [C_3F_8] and decafluorobutane [C_4F_{10}], and sulfur hexafluoride [SF_6]. The encapsulation of the microbubbles with shells composed of protein, lipid, or biopolymers has also been used to enhance *in vivo* stability by reducing outward diffusion and reducing surface tension. The encapsulation of microbubbles has also been an important factor for controlling microbubble size distribution. Ideally, microbubbles should be just small enough to pass freely through the pulmonary and systemic microcirculation without lodging. Yet their signal generation is nonlinearly related to their size. Hence, microbubble agents are generally 1–5 µm in diameter which fortuitously is also in the ideal resonant frequency range for clinical diagnostic ultrasound.

The behavior of microbubble contrast agents in the microcirculation has been extensively studied because of safety considerations and to justify their use as flow tracers. Because of their size, microbubbles remain entirely within the intravascular compartment (Lindner et al. 2002a,c). Microbubble rheology (their behavior in the microcirculation) has been assessed by comparison of the transfer functions of microbubbles and technetium-labeled RBCs through a microvascular bed, and more directly by fluorescent labeling of the microbubble shell and performing intravital microscopy which can be used to visually assess and quantify the *in vivo* behavior of microbubbles in a vessel or microvascular network. These techniques have demonstrated that microbubbles transit the microcirculation of normal muscle beds unimpeded, do not coalesce or aggregate, and have a similar velocity profile as RBCs in arterioles, venules, and capillaries (Jayaweera et al. 1994, Lindner et al. 2002a–c).

For most preparations, a small number (1–2%) of microbubbles are filtered by the pulmonary microcirculation based on size (Lindner et al. 2002). Freely circulating microbubbles are generally cleared from the circulation by the reticuloendothelial organs and almost all the gas volume is exhaled within an hour.

7.2   Microbubbles for Molecular Imaging

Molecular imaging with contrast ultrasound is based on the selective targeting and retention of microbubbles at sites of disease. There are many different determinants of the efficiency for targeting that relate to biologic conditions and tracer properties (Figure 7.3). The intravascular confinement of ultrasound imagining probes limits them to the detection of molecular events occurring with the vascular space. There are nonetheless many important molecular markers or mediators involved in disease states including ischemic injury, inflammation, angiogenesis, and thrombosis that are accessible to ultrasound contrast agents.

Targeting of microbubbles or nanoparticle agents has been accomplished by one of two strategies. A simple approach has been to select certain microbubble shell constituents that mediate or amplify their attachment to certain cells involved in disease related to pathophysiology. The most common example of this strategy has been to use lipid-shelled microbubbles that contain specific phospholipids that promote complement-mediated attachment to activated leukocytes (neutrophils and monocytes) in regions of inflammation (Christiansen et al. 2002, Lindner et al. 2000a,c). Microbubbles with an albumin shell also have the ability to attach to either leukocytes or the endothelium via β1-integrins in the setting of inflammation (Lindner et al. 2000a–c).

A second and more specific strategy has been to attach disease-specific ligands such as monoclonal antibodies, recombinant proteins, glycoproteins, or

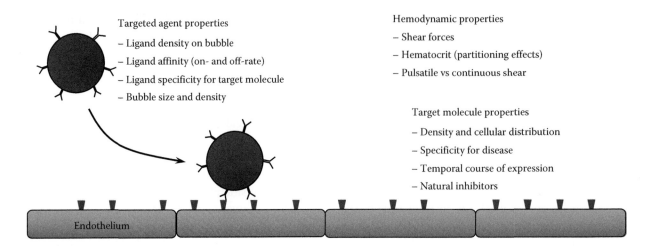

FIGURE 7.3   Schematic illustration of some of the key considerations that determine the success of targeted microbubble design.

small peptides to the microbubble shell surface to produce multivalent acoustically active agents (Lindner 2009). This strategy has generally involved the use of lipid microbubbles that contain bifunctional molecules that have a hydrocarbon domain that allows them to be anchored in the lipid monolayer, and a polyethylene glycol (PEG) spacer arm orient outward (Figure 7.4). Two general strategies exist for this attachment. In the first strategy, the targeting ligand is first coupled to a bifunctional molecule that will be incorporated into the shell surface during formation of microbubbles from an aqueous phase. This approach can work for small organic molecule ligands such as peptides,

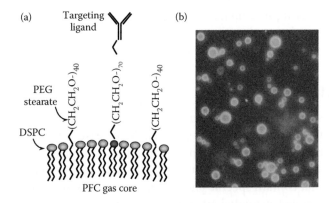

FIGURE 7.4   Example of targeted microbubble design. In the schematic, a monolayer lipid dispersion occurs around a perfluorocarbon (PFC) gas core. The lipid shell contains amphipathic phospholipids, a PEG-stearate component to enhance microbubble stability and reduce nonspecific interactions, and a targeting ligand projected at the end of a longer PEG arm. Successful attachment of targeted ligand on the microbubble surface can be confirmed by fluorescent microscopy (right) obtained after preparing lipid microbubbles bearing the rat antimurine P-selectin mAb RB40.34 at the end of an PEG arm, incubating microbubbles with an FITC-labeled antirat IgG1 antibody, and washing.

carbohydrates, hormones, and vitamins such as biotin. The primary limitation of this technique is that molecules are unstable under conditions of sonication, high shear stress, and high temperatures. An alternative and more commonly used strategy involves ligand conjugation via covalent or noncovalent attachment to preformed microbubbles. With this approach, a reactive moiety is incorporated in the microbubble shell which can then be used to attach the protein ligand to the preformed bubble. Targeting ligands are then linked to the end of the PEG arm via covalent linkage or by biotin–avidin linking strategies.

Often lipid microbubbles will also contain pegylated lipids within their shell to prolong intravascular half-life. In this situation, the ligand is placed at the end of a longer PEG arm creating a "tiered brush border" that projects the ligand beyond the nonconjugated PEG layer. In general, several thousand ligands can be conjugated per square micron of microbubble shell surface area (Lindner et al. 2001, Takalkar et al. 2004). After linking targeting ligands to preformed microbubbles, it is possible to remove free ligand primarily through microbubble flotation strategies. However, for any given application, it is not currently known how important it is to clear free targeting ligands in this manner. To avoid this issue, it has also been possible to control the ratio of targeting ligand to microbubble surface area (e.g., the ratio of biotinylated ligand to PEG–streptavidin) in order to minimize free ligand.

An initial consideration was whether signal from targeted microbubble contrast agents would be affected by attachment to cells within the vasculature thus "dampening" the signal from volumetric oscillation. Early *in vitro* studies of phagocytosed microbubbles by leukocytes showed diminished signal intensity compared to free microbubbles (Dayton et al. 2001). Subsequent studies showed that microbubble attachment to a cell surface without internalization did not significantly affect microbubble signal irrespective of the ultrasound frequency or acoustic pressure (Lankford et al. 2006).

7.3   Targeted Imaging Protocols

The goal of ultrasound molecular imaging is to selectively detect signal from microbubbles that have been retained within the vasculature due to their adherence at disease sites. The way this has been achieved is schematically illustrated in Figure 7.5. After intravenous injection, the majority of signal early after introduction of targeted microbubbles will be from freely circulating bubbles (interval A). Accumulation and retention of targeted microbubbles at the disease site occurs when the concentration remains high (interval B). Late signal (interval C) that occurs after clearance of the freely circulating bubbles therefore represents only that from microbubbles that have been retained within the field of interest through adherence of microbubbles to their target molecule. The relatively short clearance time of freely circulating microbubbles is an advantage over other contrast agents that allows rapid data acquisition and very fast temporal resolution.

The ability to destroy microbubbles with high-power ultrasound within the diagnostic frequency range provides two unique features for CEU-targeted imaging (Chomas et al. 2001, Wei et al. 1997). First, it makes it possible to separate molecular imaging signal from that attributable to freely circulating microbubbles. After

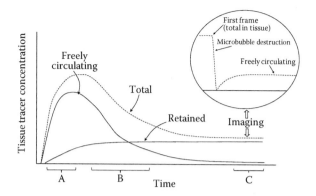

FIGURE 7.5 Schematic illustration of the most common approach for targeted CEU imaging. After bolus intravenous injection of tracer, early signal enhancement on first-pass through tissue (A) is mostly from freely circulating agent. Targeted attachment occurs during first pass and high concentration recirculation phase (A and B). Image acquisition occurs at time when most freely circulating microbubbles have been cleared from the circulation (C), usually 5–10 min after injection. The signal contribution from the few circulating microbubbles can then be established by destroying all microbubbles within the sector imaging and assessing signal after complete replenishment (inset).

microbubble signal in tissue is measured 5–10 min after intravenous injection, all microbubbles within the ultrasound sector are destroyed. Several seconds later the field is replenished with the low concentration of freely circulating microbubbles that remain in the blood pool (Wei et al. 2001). This signal can be measured and digitally subtracted from the initial signal to provide a measure only of retained signal (Lindner et al. 2000a–c). The ability to destroy retained tracer also allows repetition of molecular imaging with the same or a different targeted tracer.

7.4   Selectins in Inflammation and Ischemia

In cardiovascular medicine, the ability to image the immune response is a powerful tool as inflammation is a critical component in the pathophysiology of many disease processes. Activation of the vascular endothelium in response to injury or inflammatory stimuli is a critical event at the onset of inflammation. Activation involves the expression of endothelial cell adhesion molecules that interact with counterligands on the surface of leukocytes (Ley 1996, Springer 1990). The initial step in the process of recruitment of leukocytes to areas of inflammation involves leukocyte capture and rolling along the endothelial surface (Figure 7.6). This process is mediated primarily by the interaction of endothelial selectins (P-selectin, E-selectin) or leukocyte selectins (L-selectin) with their glycoprotein counterligands (Ley 1996, Springer 1990). In muscle tissue, this process occurs primarily in postcapillary venules. Selectin mediate rolling of leukocytes along activated endothelium allows leukocytes to be exposed to pro-inflammatory cytokines that result in the activation of leukocyte heterodimeric integrins (VLA-4, Mac-1, LFA-1, etc.) that interact with endothelial counterligands (ICAMs, VCAM-1). The slow rolling mediated by selectins also allows for these slower on-rate bonds to form and produce firm attachment of the leukocyte to the endothelial surface. Arrested leukocytes may then undergo transendothelial migration according to chemokine signals.

The selectins that are involved in the initial capture and rolling step are a family of adhesion molecules that

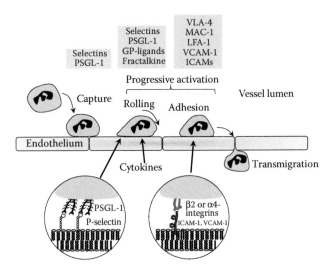

FIGURE 7.6 Intravascular events that occur during the recruitment of leukocytes during the inflammatory response. Key endothelial and leukocyte adhesion molecules that mediate capture, rolling, and adhesion are listed in the shaded blocks.

were originally named to highlight the lectin composition and selective nature of expression of these molecules (Bevilacqua and Nelson 1993, McEver et al. 1989). Each member was then subsequently named according to the cell types on which it was originally identified. Selectins are expressed by platelets (P-selectin), endothelial cells (P-selectin and E-selectin), and leukocytes

(L-selectin). Selectins share a significant degree of sequence homology among themselves and between species (Bevilacqua and Nelson 1993).

P-selectin is a ~140 kDa molecule that is constitutively synthesized and then stored intracellularly within α-granules in platelets and the Weibel–Palade bodies of endothelial cells. In response to an inflammatory stimulus, P-selectin is rapidly mobilized to the endothelial and platelet cell surface with minutes of an inflammatory stimulus allowing it to interact with its primarily counterligand P-selectin glycoprotein ligand-1 (PSGL-1) on leukocytes (Bevilacqua et al. 1993, Chukwuemeka 2005, Ley 1996). The expression of P-selectin, in the absence of a continuous or repeated stimulus, however is transient with peak expression between 60 and 90 min followed by a steady decline as it is returned to its storage bodies. In contrast, E-selectin expression is induced by inflammatory stimuli. Its surface expression is delayed compared to P-selectin with a peak occurring in ~4–6 h, but remains on the endothelial surface for a longer period after stimulation returning to basal levels after 24–48 h. This sequence of selectin response occurs in response to even brief ischemia–reperfusion injury, thereby serving as the basis for using molecular imaging with selectin-targeted microbubbles to detect recent myocardial ischemia in patients with chest pain.

7.5 P-Selectin Imaging with Antibody-Mediated Microbubble Targeting

The term "ischemic memory imaging" was coined to describe molecular imaging of events produced by ischemia–reperfusion injury as means to identify patients subjected to ischemic injury and resulting inflammation but without the functional or anatomic changes conventional imaging can detect. Key properties of selectins that make them attractive as targets of ischemic memory imaging with ultrasound include (1) rapid externalization within minutes from secretory granules in endothelial cells (P-selectin), (2) requirement of only a mild degree of ischemia to stimulate expression, (3) persistent upregulation of synthesis for hours after ischemia (P- and E-selectin), and (4) projection of the lectin-binding domain well away (>40 nm) from the cell surface which is advantageous to microbubble targeting (Bachmann et al. 2006, Bevilacqua and Nelson 1993, Chukwuemeka et al. 2005, McEver et al. 1989). As will be discussed in the remainder of this chapter, the ability to detect and spatially asses recent ischemic injury in the absence of infarction by ultrasound molecular imaging has

now been performed in numerous animal models and in a variety of different organs utilizing two different ligand strategies.

Microbubbles targeting murine P-selectin were initially formulated with a monoclonal antibody (RB 40.34) directed against P-selectin as the targeting ligand attached to the microbubble surface via biotin–streptavidin conjugation (Lindner et al. 2001). Antibody concentration on the microbubble surface was determined to be several thousand per square micron by quantitative fluorescent techniques (Lindner et al. 2001, Takalkar et al. 2004).

Since CEU relies on the attachment of a relatively large multivalent particle within shear flow, parallel plate flow chambers have been used as a high throughput *in vitro* method for characterizing the attachment for selectin-targeted microbubbles under different shear conditions and site densities. In these models, targeted microbubbles are drawn across cell culture dishes coated with an immobilized murine P-selectin/Fc chimera (Takalkar et al. 2004). Illustrated by Figure 7.7,

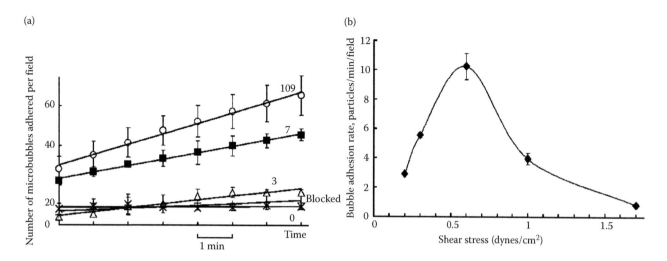

FIGURE 7.7   Flow chamber assessment of P-selectin-targeted microbubble adhesion. (a) Targeted microbubble adhesion over time (*x*-axis) at a shear stress of 0.3 dyne/cm^2 for P-selectin site densities ranging from 0 to 109 molecules/μm^2. Blocking studies were carried out with excess RB40.34 mAb with a site density of 109 molecules/μm^2. (b) P-selectin-targeted microbubble attachment under various shear stresses to a P-selectin-coated plate at a site density of 109 molecules/μm^2. (Reproduced with permission from Takalkar, A.M. et al. 2004. *J Control Release.* 96:473–482.)

attachment of microbubbles targeted via surface conjugation of RB40.34 was influenced by P-selectin site density. It was also affected by shear stress with a maximal "on-rate" at ~0.6 dyn/cm^2 (Figure 7.7) Control experiments on plates lacking P-selectin or blocked with mAb RB40.34 prior to microbubbles demonstrated minimal bubble attachment. These results demonstrate that under physiologic flow conditions, microbubble retention was influenced both by shear stress and surface density of the target receptor. Similarly, detachment rates under high shear have also been shown to be dependent on site density.

As an initial *in vivo* developmental step, the adherence of these same microbubbles bearing a monoclonal antibody against P-selectin (RB40.34) has been tested using intravital microscopy in the mouse cremaster muscle (Kaufmann et al. 2007a, Lindner et al. 2001). P-selectin mobilization to the endothelial surface beyond that which occurs with surgical exteriorization of the cremaster was stimulated by either intrascrotal TNF-α (0.5 μg) or by brief ischemia (10 min) and reperfusion. In both of these situations, rapid surface expression of P-selectin was evidenced by a characteristic increase in the rolling flux fraction and a decrease in rolling velocity for leukocytes in postcapillary venules. Selective retention of the P-selectin targeted microbubbles attached to the endothelial surface confirmed that a P-selectin targeting strategy could be used to detect recent ischemia (Figure 7.8) (Kaufmann et al. 2007a). In the TNF-α model, there was also rare attachment to platelet–leukocyte complexes (Lindner et al. 2001).

Imaging of P-selectin expression in response to ischemia–reperfusion injury was first performed using CEU in a murine model of renal artery ischemia-reperfusion injury (Lindner et al. 2001). The kidney was chosen as the initial target organ because it provided a nonischemic contralateral control organ and because of the minimally invasive approach to renal artery ligation. After ischemia for 30 min and 1 h of reperfusion, CEU imaging was performed 8 min after IV injection of P-selectin-targeted or control microbubbles with either no targeting ligand or with nonbinding isotype control mAb. In the postischemic kidneys, the mean acoustic signal from retained microbubbles was substantially greater for targeted microbubbles bearing RB4.34 than for either of the control agents (Figure 7.9). The specificity of this enhanced signal for P-selectin expression was further confirmed by a reduction in signal enhancement from P-selectintargeted microbubbles in mice genetically deficient for P-selectin (P-/-). A low level of background signal for all microbubbles was noted and was thought to be secondary to nonspecific attachment to leukocytes, which was also seen on intravital microscopy.

The initial observations in the renal ischemiareperfusion model led to the idea that recent myocardial ischemia could be detected even in the absence of infarction and active hypoperfusion. Echocardiographic molecular imaging of P-selectin expression in response to ischemic injury was studied in a murine model of myocardial ischemia without infarction (Figure 7.10) (Kaufmann et al. 2007a). Wildtype mice were subjected

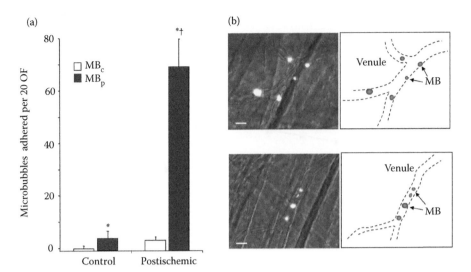

FIGURE 7.8 Adhesion of intravenously injected P-selectin-targeted microbubble (MB$_p$) bearing the mAb RB40.34, and control microbubbles (MB$_c$) by intravital microscopy in postischemic mouse cremaster muscle. The graph illustrates microbubble adhesion in 20 optical fields (OF) after brief ischemia–reperfusion injury and in control uninjured muscle. The images illustrate adhesion of P-selectin-targeted microbubbles to the endothelial surface in postcapillary venules. *$p < 0.05$ versus MB$_c$; †$p < 0.05$ versus control nonischemic conditions. (Reproduced with permission from Kaufmann, B.A. et al. 2007a. *Eur Heart J.* 28:2011–2017.)

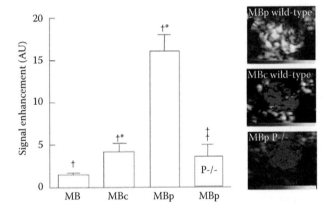

FIGURE 7.9 CEU molecular imaging of P-selectin expression in renal ischemia–reperfusion injury. Signal enhancement from intravenously administered P-selectin-targeted microbubbles (MB$_p$) and control microbubbles bearing nonspecific mAb (MB$_c$) or no antibody (MB) after renal ischemia–reperfusion injury in wildtype and P-selectin-deficient (P-/-) mice. (Reproduced with permission from Lindner, J.R. et al. 2001. *Circulation.* 104:2107–2112.)

to transient myocardial ischemia (10 min) by a suture placed around the left anterior descending artery (LAD). During ischemia, the area at risk was assessed using myocardial contrast echocardiography (MCE) perfusion imaging which was also used to confirm complete reperfusion. Forty-five minutes after reperfusion, molecular imaging with echocardiography was performed after IV administration of microbubbles bearing either a monoclonal antibody against P-selectin (RB40.34) or an isotype control IgG antibody. Selective signal enhancement was seen for targeted microbubbles within the previously ischemic risk area. Interestingly, there was also some signal enhancement in the remote territory which, when comparing data obtained from sham-operated and closed-chest controls, appeared to be a result of selectin expression caused by the surgery and open-chest cardiac exposure. Infarction in these animals was excluded by normal wall motion and lack of necrosis on postmortem tetrazolium tetrachloride staining.

7.6 Nonantibody Selectin Targeting

The results of the initial studies using P-selectin-targeted molecular imaging in renal and myocardial ischemia–reperfusion imaging sparked interest in the development of clinically feasible targeted imaging probes for ischemic memory imaging in patients. As an early step toward the transition of this technology to humans, other P-selectin ligands have been explored as an alternative to mAbs for microbubble targeting. Agents that have been tested have included small molecules such as sialyl-Lewis-X (a carbohydrate component of PSGL-1) and a glycosulfopeptide PSGL-1 mimetic, as well as recombinant dimeric human PSGL-1 glycoprotein fusion protein (Davidson et al. 2010, Rychak et al. 2006, Villanueva et al. 2007). These ligands have been proposed for several reasons. First, there is less immunogenicity, in particular for the small molecule ligands.

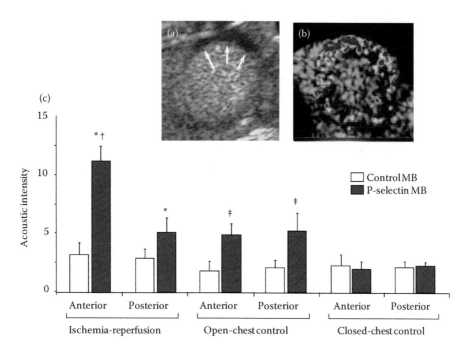

FIGURE 7.10 Ischemic memory imaging with echocardiography and P-selectin-targeted microbubbles in a murine model of myocardial ischemia and reperfusion without infarction. (a) First-pass myocardial contrast echocardiography in the mid-papillary muscle short-axis plane performed during brief left-anterior descending coronary artery ligation illustrating the area at risk in the anterior myocardium (arrows). (b) Molecular imaging performed several minutes after intravenous administration of microbubbles bearing RB40.34 after 8 min of ischemia and 45 min of reperfusion, illustrating signal enhancement in the previously ischemic anterior myocardium. (c) Signal intensity for P-selectin-targeted and control microbubbles in the anterior myocardium and posterior myocardium in animals undergoing LAD (anterior) ischemia–reperfusion, in sham-operated open-chest control animals, and closed chest controls. *$p < 0.05$ versus control MB; †$p < 0.05$ versus control nonischemic posterior region; ‡$p < 0.05$ versus anterior P-selectin-targeted signal in animals undergoing ischemia–reperfusion. (Reproduced with permission from Kaufmann, B.A. et al. 2007a. *Eur Heart J.* 28:2011–2017.)

Lower production costs are also an important consideration. An important clinical issue is that, unlike mAb, all of these alternate ligands can potentially interact not only with P-selectin but also with E-selectin. Extending the binding capabilities for microbubbles to E-selectin could substantially broaden the window for ischemic memory imaging since synthesis-dependent E-selectin expression occurs after the initial P-selectin mobilization and can persist for up to 24 h after ischemic injury.

Similar to the studies using RB40.34 as a targeting ligand, studies with the nonantibody selectin targeting strategies have used intravital microscopy to characterize vascular attachment. This information was particularly important since, in theory, these ligands may support microbubble rolling rather than adhesion. In nature, interaction of selectins with their glycoprotein ligands mediate rolling due to the formation of catch bond kinetics that can mediate leukocyte capture in high shear but also have a fast off-rate. Results from studies using microbubbles targeted with the tetrasaccharide sialyl-Lewis-X in murine cremaster muscle exposed to TNF-α have demonstrated selective retention of these microbubbles on intravital microscopy due primarily to

venular adhesion (Villanueva et al. 2007). There was no report of rolling for this agent. For microbubbles targeted by recombinant dimeric PSGL-1, ischemia–reperfusion injury produced similar retention to RB40.34-bearing microbubbles and again was due to firm attachment to the venular endothelium (Jayaweera et al. 1994). These data suggest that PSGL-1 or mimetics of PSGL-1 that in nature support leukocyte rolling on endothelial selectins, tend to produce adhesion for microbubbles. Reasons for this difference are unclear and may involve site density or distribution of selectin ligands.

Imaging studies of muscle ischemia–reperfusion injury have also been performed. The agent bearing sialyl-Lewis-X was tested in a model of myocardial ischemia produced by 20 min LAD occlusion and reperfusion (Villanueva et al. 2007). These studies demonstrated selective signal enhancement 30 min after reflow in the postischemic bed compared to control nonischemic bed. Signal for the targeted agent was approximately threefold greater for targeted compared to a control nontargeted agent. Selective signal enhancement was still seen at 60–90 min, albeit considerably less than at 30 min. Sham surgery experiments were not performed

Chapter 7

to test the influence of the surgical model on selectin-targeted signal enhancement. However, these studies did demonstrate the important feature that the spatial extent of selectin signal on targeted contrast echocardiography matched well with the territory of recent ischemia on perfusion imaging.

Initial experiments using the dimeric recombinant PSGL-1 fusion protein as a microbubble targeting ligand have focused on how this complex glycoprotein molecule can extend the time window for reproducible opacification. This agent was tested in postischemic leg skeletal muscle due to the high-throughput nature of this model and lack of influence produced by open chest surgery in the myocardial model (Davidson et al. 2010). These studies demonstrated that recombinant PSGL-1 as a targeting moiety gave slightly less signal enhancement than antibody (RB40.34) targeting, but was more specific (less nonspecific enhancement in the contralateral limb) and produced more consistent enhancement over a 6 h period of time after the initial brief ischemic injury. By comparing data for wild-type and P-selectin-deficient mice, it was possible to determine that more consistent enhancement over time was secondary to detection of a late E-selectin expression phase. The safety and efficacy of this agent is now being tested in a nonhuman primate model of closed-chest brief coronary ischemia.

7.7 Conclusion and Future Directions

There has been a practical basis for the development of selectin-targeted microbubbles for contrast ultrasound molecular imaging for ischemic memory. The ability to rapidly provide clinicians with diagnostic information on the presence and extent of recent myocardial ischemia at the bedside makes this the most sensible imaging approach for detecting ischemia with a molecular imaging approach. The future of the technology now depends on pharmacology and toxicology testing of some of the complex multivalent microbubble particles that have been designed using ligand conjugation strategies that are feasible for human use. Although myocardial ischemic memory imaging is the primary indication for these agents, there may be many other inflammatory processes that could potentially be explored since selectins play an integral role in the initial portions of the cellular inflammatory response.

References

Bachmann, C., Klibanov, A.L., Olson, T.S. et al. 2006. Targeting mucosal addressin cellular adhesion molecule (MAdCAM)-1 to noninvasively image experimental Crohn's disease. *Gastroenterology* 130:8–16.

Bevilacqua, M.P. and Nelson, R.M. 1993. Selectins. *J Clin Invest.* 91:379–387.

Chomas, J.E., Dayton, P., Allen, J. et al. 2001. Mechanisms of contrast agent destruction. *IEEE Trans Ultrason Ferroelectr Freq Control.* 48:232–248.

Christiansen, J.P., Leong-Poi, H., Klibanov, A.L. et al. 2002. Noninvasive imaging of myocardial reperfusion injury using leukocyte-targeted contrast echocardiography. *Circulation.* 105:1764–1767.

Chukwuemeka, A.O., Brown, K.A., Venn, G.E. et al. 2005. Changes in P-selectin expression on cardiac microvessels in blood-perfused rat hearts subjected to ischemia–reperfusion. *Ann Thorac Surg.* 79:204–211.

Davidson, B., Kaufmann, B.A., Belcik, T., Yue, Q., and Lindner, J.R. 2010. Molecular imaging using recombinant PSGL-1 as a pan-selectin targeting moiety provides prolonged and selective tissue enhancement for detecting recent ischemia. *Circulation.* 122(Suppl. 21):A12649.

Dayton, P.A., Chomas, J.E., Lum, A.F. et al. 2001. Optical and acoustical dynamics of microbubble contrast agents inside neutrophils. *Biophys J.* 80:1547–1556.

Dayton, P.A., Morgan, K.E., Klibanov, A.L. et al. 1999. Optical and acoustical observations of the effects of ultrasound on contrast agents. *IEEE Trans Ultrason Ferroelectr Freq Control.* 46:220–232.

Eckersley, R.J., Chin, C.T., and Burns, P.N. 2005. Optimising phase and amplitude modulation schemes for imaging microbubble contrast agents at low acoustic power. *Ultrasound Med Biol.* 31:213–219.

Epstein, P.S.P. and Plesset, M.S. 1950. On the stability of gas bubbles in liquid–gas solutions. *J Chem Phys.* 18:1505–1509.

Gramiak, R. and Shah, P.M. 1968. Echocardiography of the aortic root. *Invest Radiol.* 3:356–366.

Jayaweera, A.R., Edwards, N., Glasheen, W.P. et al. 1994. *In vivo* myocardial kinetics of air-filled albumin microbubbles during myocardial contrast echocardiography. Comparison with radiolabeled red blood cells. *Circ Res.* 74:1157–1165.

Kabalnov, A., Klein, D., Pelura, T. et al. 1998. Dissolution of multicomponent microbubbles in the bloodstream: 1. Theory. *Ultrasound Med Biol.* 24:739–749.

Kaufmann, B.A., Lewis, C., Xie, A. et al. 2007a. Detection of recent myocardial ischaemia by molecular imaging of P-selectin with targeted contrast echocardiography. *Eur Heart J.* 28:2011–2017.

Kaufmann, B.A., Wei, K., and Lindner, J.R. 2007b. Contrast echocardiography. *Curr Probl Cardiol.* 32:51–96.

Lankford, M., Behm, C.Z., Yeh, J. et al. 2006. Effect of micro-bubble ligation to cells on ultrasound signal enhancement: Implications for targeted imaging. *Invest Radiol.* 41:721–728.

Ley, K. 1996. Molecular mechanisms of leukocyte recruitment in the inflammatory process. *Cardiovasc Res.* 32:733–742.

Lindner, J.R. 2009. Molecular imaging of cardiovascular disease with contrast-enhanced ultrasonography. *Nat Rev Cardiol.* 6:475–481.

Lindner, J.R., Coggins, M.P., Kaul, S. et al. 2000a. Microbubble persistence in the microcirculation during ischemia/reperfusion and inflammation is caused by integrin- and complement-mediated adherence to activated leukocytes. *Circulation.* 101:668–675.

Lindner, J.R., Dayton, P.A., Coggins, M.P. et al. 2000b. Noninvasive imaging of inflammation by ultrasound detection of phagocytosed microbubbles. *Circulation.* 102:531–538.

Lindner, J.R., Song, J., Christiansen, J. et al. 2001. Ultrasound assessment of inflammation and renal tissue injury with microbubbles targeted to P-selectin. *Circulation.* 104:2107–2112.

Lindner, J.R., Song, J., Jayaweera, A.R. et al. 2002. Microvascular rheology of Definity microbubbles after intra-arterial and intravenous administration. *J Am Soc Echocardiogr.* 15:396–403.

Lindner, J.R., Song, J., Xu, F. et al. 2000c. Noninvasive ultrasound imaging of inflammation using microbubbles targeted to activated leukocytes. *Circulation.* 102:2745–2750.

McEver, R.P., Beckstead, J.H., Moore, K.L. et al. 1989. GMP-140, a platelet alpha-granule membrane protein, is also synthesized by vascular endothelial cells and is localized in Weibel–Palade bodies. *J Clin Invest.* 84:92–99.

Rychak, J.J., Li, B., Acton, S.T. et al. 2006. Selectin ligands promote ultrasound contrast agent adhesion under shear flow. *Mol Pharm.* 3:516–524.

Springer, T.A. 1990. Adhesion receptors of the immune system. *Nature.* 346:425–434.

Takalkar, A.M., Klibanov, A.L., Rychak, J.J. et al. 2004. Binding and detachment dynamics of microbubbles targeted to P-selectin under controlled shear flow. *J Control Release.* 96:473–482.

Villanueva, F.S., Lu, E., Bowry, S. et al. 2007. Myocardial ischemic memory imaging with molecular echocardiography. *Circulation.* 115:345–352.

Wei, K., Le, E., Bin, J.P. et al. 2001. Quantification of renal blood flow with contrast-enhanced ultrasound. *J Am Coll Cardiol.* 37:1135–1140.

Wei, K., Skyba, D.M., Firschke, C. et al. 1997. Interactions between microbubbles and ultrasound: *In vitro* and *in vivo* observations. *J Am Coll Cardiol.* 29:1081–1088.

Chapter 7

8. Evolution of Magnetic Nanoparticles

From Imaging to Therapy

Yali Li and Jason R. McCarthy

8.1   Introduction

Magnetic nanoparticles have been used clinically as contrast agents in magnetic resonance imaging for over 20 years. These particles, synthesized by the simple precipitation of iron salts, have evolved considerably over the past two decades to become both multifunctional and multimodal. This chapter will focus on the synthesis, modification, and application of cross-linked dextran-coated iron oxide (CLIO) nanoparticles in the detection and treatment of atherosclerosis. It will encompass the chemistries involved in the synthesis of the nanoagents, modification with targeting, imaging, and therapeutic moieties, and detailed examples of their utility in the detection and potential treatment of atherosclerosis.

8.1.1   Clinical Applicability of Superparamagnetic Iron Oxides Nanoparticles

Polysaccharide-coated iron oxide nanoparticles have become a mainstay in contrast-enhanced magnetic resonance (MR) imaging. Their utility as T_2 contrast agents, which will be discussed further below, was originally reported by Ohgushi et al. (1978). Approximately one decade later, the first clinical trials were performed to demonstrate the applicability of these particles in the imaging of the liver, the spleen, and their associated tumors (Stark et al. 1988, Weissleder and Stark 1989a, 1989b, Weissleder et al. 1988). The utilized formulations, based on dextran-coated iron oxides, displayed variable

Targeted Molecular Imaging. Edited by Michael J. Welch and William C. Eckelman © 2012 Taylor & Francis Group, LLC. ISBN: 978-1-4398-4195-2

blood half-lives, depending on particle size and surface characteristics, but were localized to the cells of the reticuloendothelial system (RES), particularly in phagocytic macrophages, which allowed for their trafficking to the organs listed above.

This macrophage avidity has been utilized for the detection of clinically occult lymph node metastases in prostate cancer (Harisinghani et al. 2003, Harisinghani and Weissleder 2004). As depicted in Figure 8.1, this methodology is based on the extravasation of the particles from the vasculature into the interstitial space and subsequent localization to the nodes via the lymphatic vessels and uptake by macrophages. In the clinic, patients are initially subject to MR imaging to acquire a precontrast image (i.e., prior to the administration of the nanoagent), in which the lymph nodes appear bright (Figure 8.1a). A second image is acquired 24 h after intravenous injection of the imaging agent. In the healthy nodes, there is a homogeneous decrease in MR signal (Figure 8.1b), an effect known for tissues containing iron oxides, whereas metastatic nodes do not demonstrate this same decrease, or the decrease is inhomogeneous (Figure 8.1d). This can be attributed to portions of the lymph node being occupied by cancerous metastases, thereby preventing the accumulation of iron oxide-containing macrophages. Interestingly, this methodology was able to detect metastases 2 mm in diameter, well below the detection threshold of other imaging modalities.

Iron oxides have also proven useful in the detection of atherosclerotic lesions. Pathologically, one of the main features of atheroma is endothelial cell dysfunction, which promotes disease initiation and the leakage of particles, including low-density lipoproteins, into the interstitium. Additionally, inflamed plaques are composed of a significant number of inflammatory macrophages (up to 20% of the volume), into which the dextran-coated iron oxides readily localize (Jaffer et al. 2007). Schmitz et al. (2001) initially reported the identification of vulnerable atherosclerotic lesions in a prospective study of patients undergoing MR procedures for the identification of lymph node metastases. This was soon followed up in a study by Kooi et al. (2003) on 11 symptomatic patients scheduled for carotid endarterectomy, in which the authors demonstrated histologically and by electron microscopy that iron oxide nanoparticles accumulate within lesions, particularly in the macrophages. Thereafter, Trivedi et al. (2004a,b,

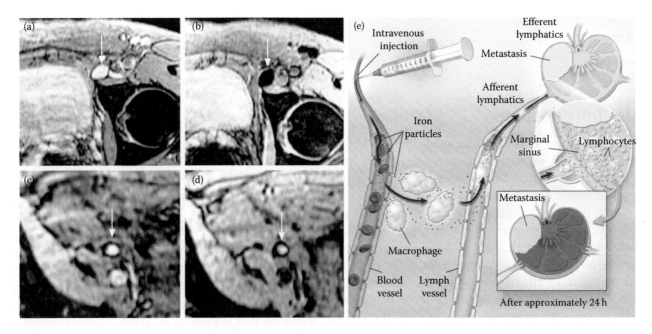

FIGURE 8.1 As compared to conventional MRI (a), MRI obtained 24 h after the administration of lymphotropic superparamagnetic nanoparticles (b) shows a homogeneous decrease in signal intensity due to the accumulation of SPION in a normal lymph node in the left iliac region (arrow). Conventional MRI shows a high signal intensity in an unenlarged iliac lymph node completely replaced by tumor (arrow in c). Nodal signal intensity remains high (arrow in d). (e) The systemically injected long-circulating particles gain access to the interstitium and are drained through lymphatic vessels. Disturbances in lymph flow or in nodal architecture caused by metastases lead to abnormal patterns of accumulation of lymphotropic SPION, which are detectable by MRI. (Reproduced from Harisinghani, M. G. et al. 2003. *N Engl J Med* 348: 2491–2499. With permission.)

2006) demonstrated in a number of patients that high-resolution contrast-enhanced MR imaging is exceedingly capable of detecting inflamed atherosclerotic lesions. Further examination of the capabilities of this class of nanoparticles in the imaging of cardiovascular disease is currently underway (Tang et al. 2009).

8.1.2 Origins of the Utility of Iron Oxide Nanoparticles

As was described earlier, iron oxide nanoparticles are one of the few nanomaterials that have been utilized clinically. This is due to the numerous advantageous properties possessed by this class of nanoparticles, including superparamagnetism, the ability to alter the relaxation rates of adjacent water molecules, and, depending on the surface properties and sizes of the particles, extended blood half-lives. For the most part, they are also biocompatible. Biodegradation of the iron oxide core is assumed to occur within the macrophages of the RES, where the elemental iron is incorporated into the iron pool of the body. This was demonstrated in rats, where 20% of the iron from particles synthesized utilizing iron-59 was found within hemoglobin within 14 days of intravenous administration (Weissleder et al. 1989), which is similar to the biodegradation of radiolabeled ferritin (Hershko et al. 1973). Of course,

overall biocompatibility is based on a number of factors, each of which may alter the toxicity of the nanoagents.

The origin of the utility of magnetic nanoparticles in medical imaging is based on their small size, which leads to properties not observed in bulk materials (Thorek et al. 2006). As opposed to ferromagnetic materials, in which all the unpaired electrons are aligned below a certain temperature, known as the Curie temperature, iron oxide nanoparticles consist of single domains with magnetization that can randomly flip direction, giving an overall average magnetization of ~0. When placed in an external magnetic field, the particles become magnetized, similar to what is observed in paramagnets, yet display significantly greater magnetic susceptibilities. This alignment of spin leads to disturbances in the local magnetic field, thereby dephasing the surrounding protons, a change which can be readily detected by MR imaging. Importantly, when the magnetic field is removed, the particles do not display remnant magnetization. In general, iron oxide nanoparticles have been shown to significantly reduce spin–spin relaxation times (T_2), although they can also alter spin–lattice relaxation times (T_1), albeit to a lesser degree. In T_2-weighted MR images, this decrease in T_2 relaxation results in a decrease in signal intensity, as was observed in Figure 8.1b.

8.2 Biocompatible Iron Oxide Nanoparticles

Nanoparticulate iron oxides are synthesized via two distinct routes, each with its own advantages and disadvantages. Aqueous syntheses result in water-soluble polymer-coated particles, yet give particles with more obtuse size ranges. Organic-based syntheses, on the other hand, result in particles with more homogeneous size distributions and morphology, yet they must be further functionalized with coating materials to make them biocompatible and stable in aqueous environments. This section will focus on the methodologies utilized to synthesize and characterize superparamagnetic iron oxide nanoparticles (SPION) for use in *in vivo* applications.

8.2.1 Synthesis of Superparamagnetic Iron Oxide Nanoparticles

Aqueous SPION syntheses involve the precipitation of iron salts, commonly a mixture of ferric chloride ($FeCl_3$) and ferrous chloride ($FeCl_2$), via the

neutralization of the acidic solution. This process is completed in the presence of an appropriate coating material, in order to affect increased colloidal stability of the resulting particles. Coating materials have included bisphosphonates (Portet et al. 2001), dendrimers (Bulte et al. 2001), lipids (De Cuyper and Joniau 1991, Nitin et al. 2004), liposomes (Bogdanov et al. 1994, Bulte and De Cuyper 2003), polyethylene glycol (Illum et al. 2001), polyacrylamide (Moffat et al. 2003), polysaccharides (Kellar et al. 1999, 2000), and proteins (Bulte et al. 1994, Wilhelm et al. 2003). As described earlier, clinically utilized preparations are commonly based upon dextran or other polysaccharide coating materials. This is due to their intrinsic affinity for iron oxide, as well as their previous use as plasma expanders (Harisinghani et al. 2003, Jung 1995, Stark et al. 1988, Weissleder et al. 1987a,b, 1989a,b, 1990a,b, Weissleder and Stark 1989a,b). While the initial precipitation results in nanoparticulate iron oxides, it is not until the suspension is heated (60–100°C) that it becomes superparamagnetic, and thus useful as an MR imaging agent.

Organic phase syntheses, on the other hand, involve the thermal decomposition of organometallic iron compounds, such as iron pentacarbonyl, in a solution of fatty acid (or amine) surfactants, including tridecanoic acid, oleic acid, and oleylamine in the presence of an oxidizer. These syntheses involve an initial nucleation, followed by crystal growth, which allows for exquisite control over final shape and size via the varying of temperature, reaction time, or other factors (Casula et al. 2006, Hou et al. 2007). Once formed, these organic-soluble particles must be further functionalized to ensure their biocompatibility. This can be accomplished via ligand exchange with tetramethylammonium hydroxide (TMAH) (Guardia et al. 2010), dimercaptosuccinic acid (DMSA) (Jun et al. 2005), and phosphine oxide polymers (Kim et al. 2005), or by the interaction of the fatty acid-coated surface with amphiphilic molecules, such as copolymers, including Pluronic F127 (Qin et al. 2007).

Once synthesized, the nanoparticles can be characterized by a number of means. For size and shape determinations, this can include microscopy techniques such as atomic force microscopy (AFM) and transmission (TEM) and scanning (SEM) electron microscopy. Particle crystallinity is determined by high-resolution TEM, whereas the crystal structure can be determined by x-ray diffraction studies. Nanoparticle hydrodynamic diameter and polydispersity can also be determined by dynamic light scattering. The calculation of particle zeta potential is also highly important as it is an indicator of particle charge. Zeta potential is determined as a function of electrophoretic mobility, and is indicative of colloidal stability, as materials with zeta potentials closer to 0 are more likely to flocculate and settle out of solution than those with far greater positive and negative charges. Charged particles are more likely to form a stable colloid due to electrostatic repulsion. Most important to their utility as MR imaging agents are the magnetic properties of the particles, including relaxivities. These can be determined by using equipment such as a time-domain nuclear magnetic resonance (TD-NMR) instrument or a vibrating sample magnetometer.

Analogous to the clinically utilized preparations described earlier, dextran-coated monocrystalline iron oxide nanoparticles (MION) are synthesized in the aqueous phase. Initially, dextran (T10, ~30 mM) is dissolved in water, along with $FeCl_3 \cdot 6H_2O$ (78 mM), and

the solution is purged of air with nitrogen. After cooling to 2–4°C, a solution of $FeCl_2 \cdot 4H_2O$ (1.5 M) is added, resulting in a 2 : 1 ratio of ferric and ferrous salts (final concentration of ~40 mM $FeCl_2 \cdot 4H_2O$). The solution is rapidly stirred while being neutralized by dropwise addition of ammonium hydroxide (28%) and maintaining the cooling. Once neutralized, the greenish suspension is heated to 75°C over the course of an hour, which is maintained for an additional hour, in order to confer its superparamagnetism. Once cooled, the suspension is purified via ultrafiltration, removing the unused dextran, ammonium chloride, and ammonium hydroxide. This synthesis results in a stable colloid with a core size of around 5 nm, and a 15-nm-thick dextran coat. The particles possess transverse relaxation rates (r_1) around 30 mmol⁻¹ s⁻¹ and longitudinal relaxation rates (r_2) around 125 mmol⁻¹ s⁻¹, making them excellent T_2 contrast agents. The particles also possess zeta potentials around −10 mV.

The aforementioned reaction is run under an inert atmosphere to prevent the oxidation of the ferrous iron to ferric iron, which reduces the magnetic properties of the resulting colloid. In the absence of oxidation, this results in iron oxides with a combination of magnetite (Fe_3O_4) or gamma ferric oxide (γ-Fe_2O_3) structures (($Fe_2O_3)_n(Fe_3O_4)_m$), which possess approximately identical magnetic properties. In order to modulate the overall size of the resulting particles, the ratios of the three reagents and their relative concentrations can be modified.

8.2.2 Cross-Linked Dextran-Coated Iron Oxide Nanoparticles

While MION are useful in MR imaging, there is a serious drawback that is encountered when targeted imaging or functionalization of the particle surface is considered. Since the dextran coating material is noncovalently bound to the particle core, it is possible that it can dissociate. While stored in a vial, this results in an equilibrium between associated and dissociated dextran, yet when injected into an animal, this equilibrium will not exist. Thus, if the surface of the particle is modified with ligands, such as those used in targeting, imaging, or therapy, the dextran may or may not be associated with the particle, giving altered biodistributions, off-target effects, and erroneous results.

In order to circumvent this issue, the dextran surface has been cross-linked (cross-linked iron oxide nanoparticles, Figure 8.2). CLIO is formed by the

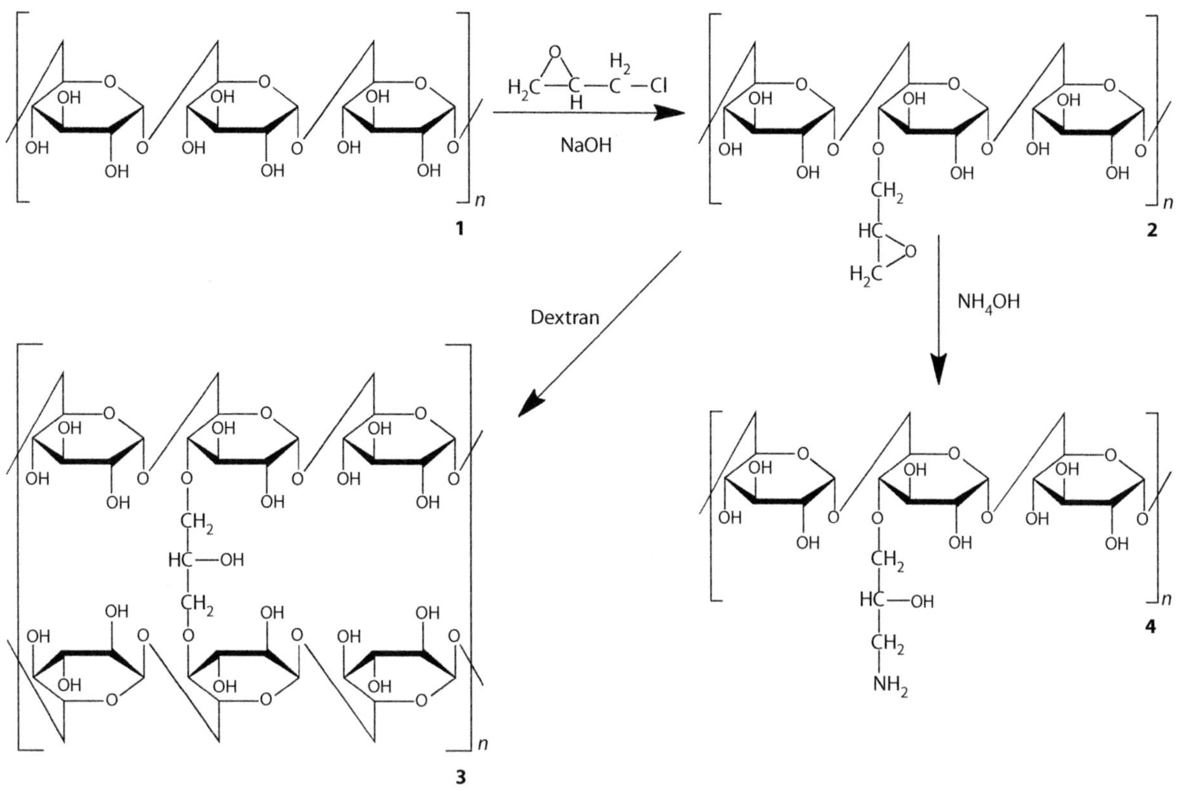

FIGURE 8.2   Epichlorohydrin cross-linking of dextran and its amination. To dextran **1** is added excess epichlorohydrin in 2.5 M NaOH solution. This solution is allowed to stir for 6 h, forming a mixture of epichlorohydrin-modified **2** and cross-linked dextran **3**. To this solution is then added ammonium hydroxide to affect amination **4**.

reaction of MION with a large excess of epichlorohydrin (30-fold) in a 2.5 M NaOH solution. After reacting for 6 h at room temperature, the solution is cooled to 4°C, and excess ammonium hydroxide is added in order to affect amination. At the conclusion of the reaction, the unreacted epichlorohydrin and ammonia are removed by diafiltration. The resulting particles do not demonstrate any appreciable changes in size, shape, or relaxivity, yet display significantly more positive zeta potentials (~10 mV), indicative of the inclusions of amines on the surface. The stability of the coating materials has been tested via a heat stress test, in which the particle solution is heated to 90°C for 30 min, and assayed for free polymer. As compared to the starting MION, the CLIO particles were over sixfold more stable (32% release versus 5% release) (Chen et al. 2009).

8.2.3   Modification with Targeting Ligands

The amination of the CLIO surface facilitates its modification with a plethora of ligands via the formation of amide bonds (Figure 8.3). The simplest examples are reactions of activated esters or isothiocyanates with the particles. This is often used for the inclusion of dyes, such as fluorescein isothiocyanate. Amide bonds can also be formed with carboxylic acid-containing ligands via *in situ* formation of the activated ester with 1-ethyl-3-[3-(dimethylamino)propyl]carbodiimide (EDC) or EDC and *N*-hydroxysulfosuccinimide (Sulfo-NHS). Epoxides or anhydrides can also be reacted with the particle surface under slightly basic conditions (pH 8.6) (Sun et al. 2006b). In order to functionalize the surface with thiol-containing ligands, such as those present in peptides, antibodies, and oligonucleotides, the amines must first be modified with heterobifunctional linkers, such as succinimidyl iodoacetate (SIA), *N*-succinimidyl-3-(2-pyridyldithio)propionate (SPDP), or succinimidyl-4-(*N*-maleimidomethyl)cyclohexane-1-carboxylate (SMCC). SPDP, in particular, possesses several advantages over the other conjugation reagents. When SPDP-modified surfaces are reacted with thiols, pyridine-2-thione is liberated. Since this molecule is a chromophore, it can be used to determine the total number of amines per particle, as well as the number of conjugated species. One drawback of SPDP is that the disulfide bond that is formed between the ligand

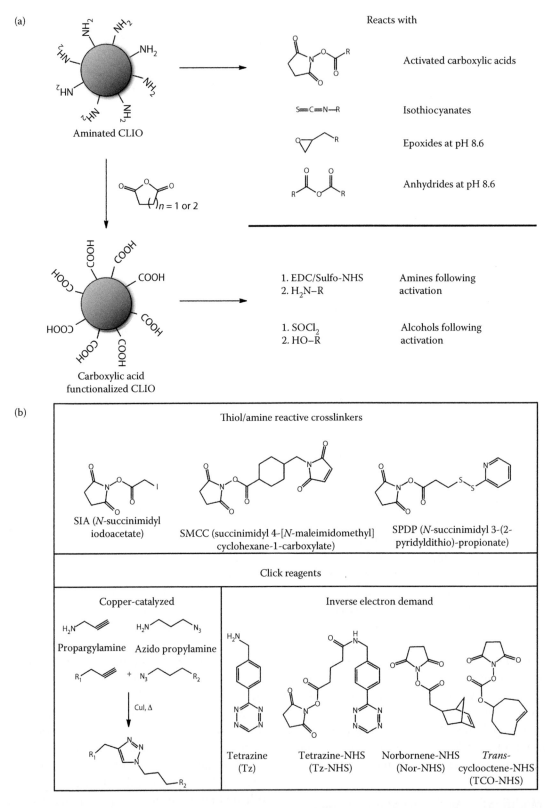

FIGURE 8.3 Chemistries and reagents utilized for the functionalization of CLIO. (a) Reactive moieties relevant to the modification of amine and acid functionalized CLIO; (b) Reagents utilized for the cross-linking of CLIO with thiols, and reagents utilized for the "click" modification of the particles.

and the particle can be exchanged in the presence of other thiols, a problem that does not occur with the thioether bonds formed with SMCC or SIA.

The aminated surface of CLIO can be functionalized with carboxylic acids via reaction with cyclic anhydrides, such as succinic or glutaric anhydride. This modification allows for the reaction of the particle with amine-containing ligands. In order to accomplish this, the particle is initially reacted with the anhydride, followed by a subsequent reaction with EDC and sulfo-NHS in the presence of the amine. Although rarely utilized, the carboxylic acid functionalized particles can also be reacted with alcohols. In order to do this, the particles must first be precipitated from aqueous suspension by addition of acetone, centrifuged to remove the solvent, and washed twice more with acetone to remove the residual water. The particles are then taken up in anhydrous dimethylsulfoxide (DMSO) and reacted with thionyl chloride to form the corresponding acid chloride, which is then reacted with an excess of the alcohol under basic conditions.

Bioorthogonal reactions have also been utilized for the functionalization of the particle surface. These reactions can take place in the presence of a complex biological milieu, and are orthogonal to amide bond formation. The Huisgen 1,3-dipolar cycloaddition involves the copper-catalyzed reaction between an alkyne and an azide to form a triazole. CLIO has been functionalized with both reactive groups via reaction of azido propylamine or propargylamine with carboxylic acid functionalized particles in the presence of EDC and sulfo-NHS (Sun et al. 2006a). Proof-of-principle reactions were then conducted with azide or alkyne modified dyes and biomolecules, including VivoTag680, estradiol, paclitaxel, and Disperse Red 1. Conjugation reactions were accomplished by heating the functionalized particle with the corresponding ligand and CuI to 37°C for 6–8 h, resulting in greater than 90% yield.

Recently, a novel bioorthogonal reaction was reported for the functionalization of CLIO, based upon an inverse electron demand Diels–Alder cycloaddition of olefins with tetrazines (Tz) (Devaraj et al. 2008, 2009, 2010, Haun et al. 2010). In the initial report, the authors described the functionalization of the fluorescent dye VT680 with 3-(4-benzylamino)-1,2,4,5-tetrazine, and its subsequent reaction with a norbornene-modified antibody for live cell labeling (Devaraj et al. 2008). The authors have improved upon this initial report by utilizing a further strained alkene, *trans*-cyclooctenol (TCO), thereby increasing the reaction rate by several orders of magnitude and making the prospect of *in vivo* labeling more realistic (Devaraj et al. 2009).

Haun et al. (2010) have utilized this strategy to increase the sensitivity of cell detection via the amplification of biomarkers (bioorthogonal nanoparticle detection, BOND). In this report, the authors modified CLIO with a succinimidyl ester-functionalized Tz, and an antibody (Her2, EpCAM, or EGFR) with TCO. They then investigated two different methods for cell detection. In one-step BOND (BOND-1), the particles are conjugated to the antibody prior to incubation with the cells, where a two-step BOND (BOND-2) involves the incubation of the cells with the TCO-modified antibody, prior to incubation with the Tz-modified CLIO (Figure 8.4). The advantage of the latter system is that significantly more small-molecule TCO moieties can be conjugated to the antibody without a decrease in binding ability as compared to the nanoparticulate CLIO. When compared *in vitro*, BOND-2 resulted in a 15-fold increase in labeling efficacy versus BOND-1. BOND-2 was also compared to the gold standard avidin/biotin methodology, and proved to be fivefold more efficacious. Additional examples of the utility of these chemistries for the modification of CLIO are forthcoming.

8.3 Utility of CLIO in the Detection of Atherosclerosis

Atherosclerosis is a progressive cardiovascular disease that occurs over decades. Its progression and the subsequent rupture of inflamed plaques can lead to a series of adverse consequences, such as angina, myocardial infarction, embolic stroke, and ischemic limbs, and remains one of the leading causes of death in developed countries (Lloyd-Jones et al. 2010). It is thus critical to identify vulnerable atherosclerotic lesions to prevent the occurrence of traumatic cardiovascular events. Atherosclerosis is asymptomatic in the early

stages; however, narrowing of the blood vessels due to the gradual accumulation of plaque or its acute rupture can occur in the later stages of the disease (Uppal and Caravan 2010). Currently, the majority of atherosclerosis imaging methods are invasive and focus on the characterization of vessel wall stenosis (e.g. coronary angiography, as clinical gold standard), which is a highly useful information, but gives little insight into the identification of the key biological activities that govern the development and progression of the disease

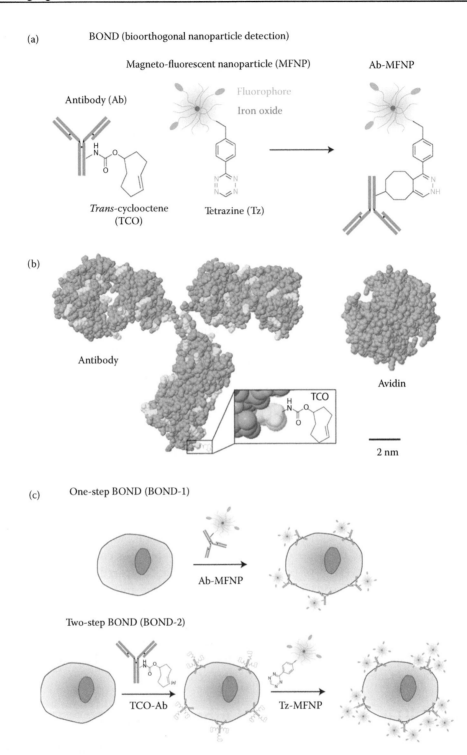

FIGURE 8.4 Bioorthogonal nanoparticle detection. (a) Schematic showing the conjugation chemistry between antibody and nanoparticle. (b) Comparative sizes (to scale) of a representative mouse IgG2a antibody (lysine residues available for TCO modification via amine-reactive chemistry are shown in yellow), a TCO modification, and an avidin protein for comparison. Tetrazine is similar in size to TCO (~200 Da). Protein structures and sizes were obtained from the Protein Data Bank (antibody, 1IGT; avidin, 3FDC). (c) Application of BOND for one-step (direct) and two-step targeting of nanoparticles to cells. Note that the antibody and tetrazine are present in multiple copies per nanoparticle (~2–3 antibodies, Ab; 84 tetrazine, Tz). (Reproduced from Haun, J. B. et al. 2010. *Nat Nanotechnol*. With permission.)

Table 8.1 Inflammatory Targets in Molecular Imaging of Atherosclerosis

Cellular Process	Examples of Imaging Targets	Possible Imaging Methods
Inflammatory cell attachment	VCAM-1	PET, PET-CT
Inflammatory cell accumulation	Labeled monocytes	SPECT, optical
Phagocytosis	Paramagnetic nanoparticle uptake	MRI
Reactive oxidant species production	MPO, hypochlorous acid	Optical, MRI
Protease activity	MMPs, cathepsins	Optical

Source: Reproduced from Libby, P., Nahrendorf, M. and Weissleder, R. 2010. *Tex Heart Inst J* 37: 324–327. With permission.

Note: CT, x-ray computed tomography; MMPs, matrix metalloproteinases; MPO, myeloperoxidase; MRI, magnetic resonance imaging; PET, positron emission tomography; SPECT, single photon-emission computed tomography.

(e.g. inflammation, endothelial dysfunction, thrombus formation, osteogenesis, apoptosis, and angiogenesis). Toward this end, *in vivo* molecular imaging is becoming established as a promising methodology toward the early risk assessment and treatment of atherosclerosis and other cardiovascular diseases. This is due, in part, to its ability to provide additional information about specific molecular targets, fundamental biological processes, and certain cell types in living subjects (Jaffer et al. 2007).

Current approaches used in the molecular imaging of atherosclerosis are based on the specific targets expressed within the inflammatory lesions (Table 8.1). With respect to CLIO-based agents, this has led to the creation of agents targeted to the inflammatory macrophage via the innate affinity of the nanoagent, and the targeted imaging of cellular adhesion molecules on activated endothelium.

8.3.1 Passive Targeting

Macrophages are an essential component of the growth and maturation of atherosclerotic lesions, since they are critically involved in inflammatory response, lesion formation, disease progression, and plaque disruption (Hansson 2005, Hansson and Libby 2006, Jaffer et al. 2007, Libby 2002). Accordingly, the macrophage has emerged as a significant target for *in vivo* atherosclerotic plaque imaging and therapy.

Jaffer et al. (2006) first reported on the applicability of fluorescently labeled CLIO for the multimodal detection of atherosclerotic lesions in a murine atheroma. Initially, the authors incubated activated murine macrophages, endothelial cells, and smooth muscle cells with Cy5.5-labeled CLIO *in vitro*, and demonstrated that the particles were internalized approximately ninefold greater in the macrophages than the

other two cell types, as determined by flow cytometry. The particles were then injected into apolipoprotein E-deficient mice (apoE$^{-/-}$) which spontaneously develop atherosclerotic lesions, especially when fed a high-fat "Western" diet. Apolipoprotein E (ApoE) is a class of lipoproteins which functions to regulate the levels of cholesterol and other fats in the lymphatic and circulatory systems. Therefore, apoE$^{-/-}$ mice have higher levels of serum cholesterol and tend to develop atherosclerotic plaques. An alternate model, not used here, relies upon mice deficient in the low-density lipoprotein reception (LDLR$^{-/-}$) which also has elevated serum cholesterol level of 200–400 mg/dL when on normal diet, and very high levels (>2000 mg/dL) when fed a high-fat diet. Twenty four hours were allowed to pass after injection of the agent in order to allow for localization of the long-circulating particles, and the mice were imaged by MR imaging, revealing accumulation of the particles in the aortic arch and root, locations known to readily develop plaques in this murine model. The mice were subsequently sacrificed and the aortae were resected and imaged by fluorescence reflectance imaging (FRI), demonstrating significant particle accumulation within the atherosclerotic lesion, as compared to control mice receiving saline (plaque-to-background ratio 423% higher in the CLIO group), which was further correlated histologically.

The authors took this strategy one step further by imaging particle accumulation *in vivo* in the carotid arteries of apoE$^{-/-}$ mice using multichannel intravital fluorescence microscopy (IVFM) (Pande et al. 2006). In these experiments, the right carotid artery was exposed via careful dissection of the periadventitial tissues 24 h after nanoagent injection, and was imaged in the Cy5.5-channel (633 nm excitation, bandpass 660–730 nm filter for emission) in order to determine

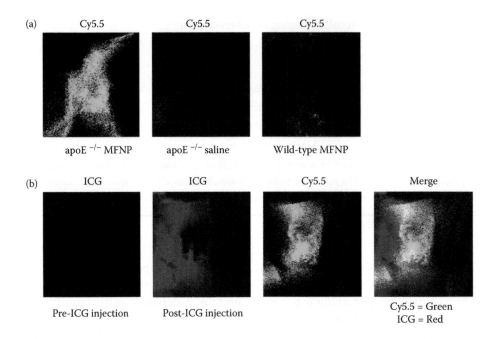

(a)
Cy5.5 | Cy5.5 | Cy5.5

apoE $^{-/-}$ MFNP | apoE $^{-/-}$ saline | Wild-type MFNP

(b)
ICG | ICG | Cy5.5 | Merge

Pre-ICG injection | Post-ICG injection | Cy5.5 = Green
ICG = Red

FIGURE 8.5 (a) Representative intravital laser scanning fluorescence microscopy Cy5.5 channel images from an apoE$^{-/-}$ nanoagent-injected animal, an apoE$^{-/-}$ saline control, and a wild-type nanoparticle control. Strong fluorescence signal in plaques was evident in apoE$^{-/-}$ particle-injected animals but not in controls. The Cy5.5 channel fluorescence images were windowed identically. (b) Additional intravital laser scanning fluorescence microscopy images from a different apoE$^{-/-}$ particle-injected mouse also revealed bright, focal Cy5.5 signal in plaques. Signal in the ICG channel prior to indocyanine green injection was minimal at baseline. Subsequent ICG injection provided a definition of the vascular space and revealed an intravascular filling defect, further confirming the location of plaques detected in the nanoagent channel. (Reproduced from Pande, A. N. et al. 2006. *J Biomed Opt* 11: 021009. With permission.)

agent accumulation. Additionally, indocyanine green (ICG) was injected as an agent to delineate the vasculature (i.e., as a fluorescent angiogram), and imaged in the Cy7 channel (748 nm excitation, long-pass 770 nm filter for emission). As can be seen in Figure 8.5, the agent readily accumulated in the atherosclerotic lesion, as compared to uninjected control or wild-type mice. Interestingly, the accumulation of the particle corresponded to a filling defect in the angiogram. These results were further correlated by *ex vivo* FRI and histology.

More recently, Nahrendorf et al. (2008) have reported the multimodal *in vivo* imaging of plaque macrophages using a trireporter nanoparticle (TNP) based on the combination of MR and optical imaging techniques, and positron emission tomography (PET). The TNP was constructed from CLIO modified with diethylenetriaminepentaacetic acid (DTPA) for the chelation of the PET tracer ^{64}Cu (1 mCi/0.1 mg nanoparticles), and VT680 to allow for fluorescence imaging (Figure 8.6). Initially, the detection threshold of this agent was investigated with regard to its nuclear and MR imaging capabilities. In this experiment, agar phantoms containing dilutions of the radiolabeled TNP were imaged by both modalities, resulting in 50-fold greater sensitivity for the PET imaging (threshold of 0.1 μg Fe/mL for PET vs. 5 μg mL for MR). These nanoagents were then utilized *in vivo* to detect atherosclerotic plaques in the aortae of apoE$^{-/-}$ mice. PET-CT revealed significant signal in the aortic arch and root, with a target-to-background ratio of 5.1 ± 0.9, as compared to surrounding tissues. This was correlated with MR imaging, which demonstrated a significant decrease in signal intensity in the same regions, and *ex vivo* FRI. After sacrifice, the aortae were digested in a cocktail of enzymes (collagenase I, collagenase XI, DNase I, and hyaluronidase) and subjected to flow cytometry in order to investigate the cellular localization of the nanoagent. This experiment revealed that the primary contribution to the *in vivo* imaging signal was accumulation of the agent in monocytes and macrophages (73.9%), while the remaining signal was from neutrophils (17.2%), lymphocytes (4.3%), endothelial cells (4.2%), and smooth muscle cells (0.4%).

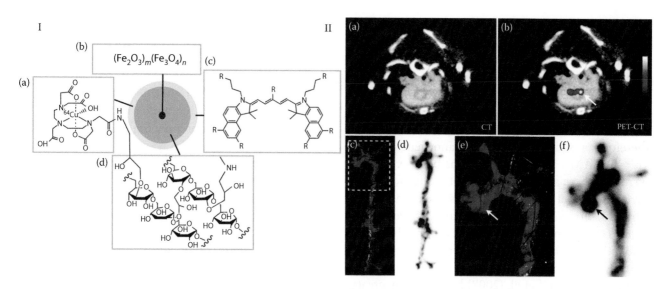

FIGURE 8.6   (I) Schematic view of the trimodality reporter ^{64}Cu-TNP. (a) Derivatization with the chelator DTPA allows attachment of radiotracer ^{64}Cu. (b) Iron oxide core provides contrast in MRI. (c) Fluorophore for fluorescence imaging, including fluorescence microscopy, flow cytometry, and fluorescence-mediated tomography. (d) Cross-linked aminated polysaccharide coating provides biocompatibility, determines blood half-life, and provides linker for attachment of tracers and potential affinity ligands. (II) ^{64}Cu-TNP distributes to atherosclerotic lesions. (a and b) PET-CT shows enhancement of the posterior aortic root (arrow). (c through f) En face Oil Red O staining of the excised aorta depicts plaque-loaded vessel segments, which colocalize with areas of high ^{64}Cu-TNP uptake on autoradiography. (e and f) Zoomed image of the root and arch. Arrows depict a plaque-laden segment of the root with high activity, which corresponds to the *in vivo* signal seen in b. (Reproduced from Nahrendorf, M. et al. 2008. *Circulation* 117: 379–387. With permission.)

8.3.2   Receptor Targeting

Endothelium also plays an important role in the pathogenesis of atherosclerosis. Dysfunctional endothelial cells express leukocyte adhesion molecules, such as vascular cell adhesion molecule-1 (VCAM-1), on the surface, and cause the formation of atheroma. After disease initiation, apoptosis (programmed cell death), angiogenesis, osteogenesis, degradation of extracellular matrix, and thrombus formation will act to cause plaque evolution and complication (Subramanian et al. 2010). Therefore, endothelial VCAM-1 is of particular interest in the *in vivo* imaging of atherosclerotic plaques.

While several approaches have been utilized to identify VCAM-1-expressing endothelial cells using radiolabeled antibodies (O'Brien et al. 1993), the systems suffered from insufficient target-to-background ratios, thereby limiting their applicability *in vivo*. Recently, a new generation of magnetic nanoparticles using antibody as targeting ligands has been developed. Tsourkas et al. (2005) have modified CLIO with VCAM-1-specific antibodies (anti-VCAM-1, ~3 per particle) for receptor targeting and a near-infrared fluorochrome (Cy5.5, 675 nm excitation/694 nm emission, ~10 per particle) with desirable long-wavelength emission for fluorescence imaging. The targeted magnetofluorescent nanoagent (VCAM-NP) was first evaluated *in vitro* against primary cultures of murine cardiac endothelial cells (MCECs) and murine dermal endothelial cells (MDECs), which express high and low levels of VCAM-1, respectively. As expected, VCAM-NP displayed significant binding to MCECs, as shown by flow cytometry. Following this initial experiment, the VCAM-NP was tested in a murine inflammatory ear model, in which the left ear of the mouse was injected with murine tumor necrosis factor-α (TNF-α), inducing expression of VCAM-1, whereas the right ear served as a control. This model was chosen for its simplicity, as well as the superficial nature of the mouse ear vasculature, making it readily imageable noninvasively. As demonstrated by IVFM, maximal levels of fluorescent signal with VCAM-NP were achieved in the TNF-α-treated ear 6 h after administration of the nanoagents, which grew progressively more faint 24 h after administration, likely due to the dissociation of the nanoparticle from the endothelium. A similar VCAM-1 antibody targeted construct, based upon micron-sized iron oxide particles, has also been utilized for the *in vivo* imaging of VCAM-1 expression in the brain (McAteer et al. 2007).

One of the main drawbacks of the aforementioned antibody-functionalized magnetic nanoparticle is that the nanoagent is not internalized by cells; thus, it cannot provide an amplified signal via intracellular trapping. Moreover, as compared to the CLIO, the relatively large size of the antibody causes low surface loading due to steric constraints and colloidal stability issues. Therefore, a second-generation nanoagent modified with a VCAM-1 affinity peptide has been explored to overcome these shortcomings. Peptide ligands can be identified from a number of sources, including natural products, *in silico* data mining, and phage display. *In silico* data mining is a technique that has been used to identify targeting peptide sequences via extracting patterns from existing databases, such as ASPD and UniProt. Schtatland et al. (2007) have recently created a searchable database, PepBank (http://pepbank.mgh. harvard.edu), incorporating data from other database, as well as text searches in Pubmed and PDF files. This database is currently populated with over 21,000 sequences of peptide 20 amino acids or shorter.

Phage display also allows for the rapid identification of prospective peptide sequences from heterogeneous libraries of bacteriophage, viruses that infect bacterial cells (Smith 1985). This technique is utilized in three main ways (Figure 8.7). The simplest is based upon an *in vitro* screen against purified proteins immobilized on plates (Koivunen et al. 1993). While this gives the binding partner of a known protein, the main drawback to this method is that the proteins may not be in their native state; thus, *in vivo* targeting may not be optimal. Alternately, phage display can be done with cells, biasing for cell-bound or cell-internalized peptides (Brown 2000). These screens can be done with a specific biomarker in mind, or with no *a priori* knowledge. For imaging purposes, peptides that are internalized yield agents that are capable of amplification through cellular trapping. Cell-based screens can also be accomplished under flow conditions, allowing for the identification of peptides with higher binding efficiencies, and those that may be better utilized for vascular targets (Kelly et al. 2005). The most advanced methodology involved the screening of phage *in vivo* in disease models (Trepel et al. 2002). This yields phage that are capable of crossing multiple biological barriers to reach their target sites, with higher binding efficiencies, as depicted in the second example, discussed later.

Using cell-based *in vitro* phage display, Kelly et al. (2005) have identified a cyclic peptide sequence (VHSPNKK) that bound to and was internalized by MCECs. Peptide VHSPNKK is a homolog of the α-chain of very late antigen (VLA-4), a known ligand for VCAM-1. To the identified peptide sequence was added a C-terminal GGSKGK peptide extension to allow for fluorescent labeling with fluorescein and subsequent Cy5.5-modified CLIO conjugation. Initially, the authors examined the uptake of the dye-labeled

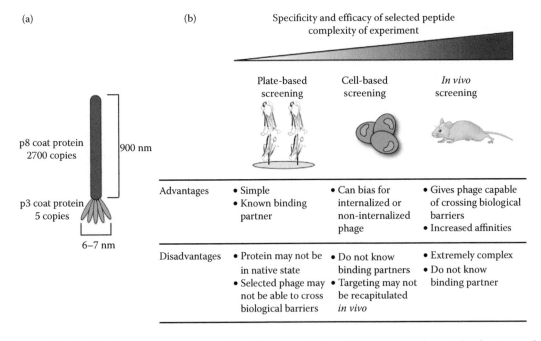

FIGURE 8.7 Phage display. (a) Schematic representation of an M13 bacteriophage. (b) The advantages and disadvantages of the various phage display methodologies.

peptide in MCEC cells. As compared to the control peptide, the targeted peptide demonstrated significant internalization, which could be abrogated by the preincubation of the cells with Fv-VCAM-1. When conjugated to CLIO (approximately four peptides per particle), the particle was also readily internalized by the VCAM-1-expressing MCEC, displaying a 10-fold higher uptake as compared to the control particle conjugated with a scramble peptide. In addition, the authors also demonstrated that VNP exhibited preferred affinity (11-fold higher) to endothelial cells as opposed to primary mouse macrophages. *In vivo* efficacy of the VHSPNKK-modified magnetofluorescent NP (VNP) was tested in TNF-α-induced inflammatory murine model, as had been preformed with VCAM-NP above. As was expected, the inflamed ear displayed elevated vessel labeling at 4 h following injection of VNP, which remained for 24 h due to the long-circulating nature of the nanoagent, whereas no detectable signal was observed in the control ear. Finally, VNP was injected into atherosclerotic lesion-laden apoE$^{-/-}$ mice and its localization examined by MR imaging. Crescent-shaped aortic wall thickening was clearly visualized along with the accumulation of VNP in the lesions, as was corroborated by *ex vivo* MRI and FRI, as well as histologically.

Given that the above peptide was cyclically constrained, its conjugation resulted in low derivatization of the particle surface (four peptides per particle); thus, Kelly et al. (2006) hypothesized that a linear derivative may increase particle loading, thereby increasing the binding affinity of the nanoparticulate VCAM-1 probe via added multivalency. Thus, the authors conducted an *in vivo* phage screen to identify adequate targeting peptides, which resulted in the discovery of a linear peptide sequence VHPKQHR which also had homology to VLA-4, the known ligand for VCAM-1. As was done earlier, a C-terminal linking GGSKGC was appended to the peptide for inclusion of fluorescein and conjugation to CLIO, and the fluorescein-labeled peptide was assayed for binding against MCEC. Once VCAM-1 targeting ability was ascertained, the peptide was conjugated to CLIO, resulting in more than 10 VCAM-1 internalizing peptide$_{28}$ (VINP$_{28}$) per particle. As compared to the aforementioned VNP, the linear derivative VINP$_{28}$-CLIO exhibited a 20-fold increase in MCEC internalization. The *in vivo* efficacy of the nanoagent was next validated in both the TNF-α-induced inflammatory murine model and in atherosclerotic apoE$^{-/-}$ mice (Figure 8.8). As an extension of the initial studies, the authors further demonstrated the ability of VINP$_{28}$-functionalized CLIO to image VCAM-1 expression in human specimens (Nahrendorf et al. 2006). VINP$_{28}$-CLIO conjugates or saline were incubated with freshly resected human endarterectomy specimens for 24 h, and were assessed by MR and fluorescence imaging. Compared with saline, specimen incubated with VINP$_{28}$ presented reduced signal intensity in the areas corresponding to the nanoagent accumulation by MRI, and an elevated fluorescent signal by fluorescence imaging. Furthermore, the authors concluded that based on the early expression of VCAM-1 in the atherosclerotic lesions and excellent avidity of VINP$_{28}$ achieved by several amplification strategies, the novel multifunctional nanoagent might allow for the detection of not only the presence of atherosclerotic lesions but also their degree of inflammatory activation.

8.4 Therapy

The development of therapeutics endowed simultaneously with diagnostic and therapeutic capabilities is an emerging concept in nanomedicine. Compared to conventional therapeutic agents, these integrated "theranostic" nanoagents offer a unique combination of *in situ* feedback mechanisms through noninvasive imaging and enhanced therapeutic efficacy based upon this feedback.

8.4.1 Near-Infrared Light-Activated Therapeutics

As was described earlier, the macrophage is a key component of atherogenesis; thus, it serves as an ideal diagnostic and therapeutic target for the treatment of atherosclerosis. Most therapeutic approaches rely upon the systemic administration of cytotoxic drugs, which can lead to off-target effects or unintended toxicities. To circumvent this and achieve improved therapeutic efficacy, photosensitizers, drugs that are only activated upon exposure to a specific wavelength of light, have been utilized to affect localized photodynamic therapy (PDT) for the treatment of atherosclerosis (Kereiakes et al. 2003, McCarthy et al. 2006, 2010, Rockson et al. 2000a,b). Commonly used photosensitizers are mostly nonpolar aromatic porphyrinic chromophores that generate cytotoxic singlet oxygen upon laser irradiation. The hydrophobic nature of the majority of these

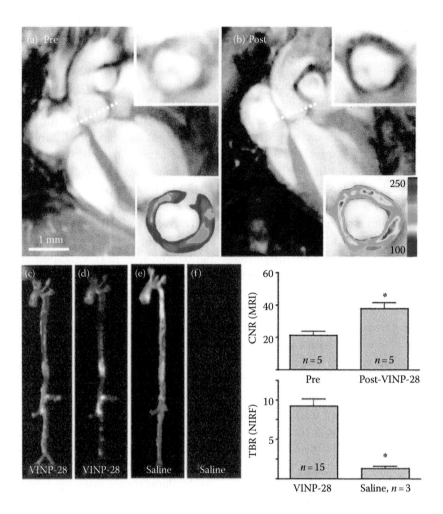

FIGURE 8.8 *In vivo/ex vivo* MR and optical imaging of VCAM-1 expression. (a) MRI before injection of VINP$_{28}$. Dotted line depicts location of short-axis view (insets, lower panel with color-coded signal intensity). (b) Same mouse 48 h after injection of VINP$_{28}$. A marked signal drop in the aortic root wall was noted (insets). The contrast-to-noise ratio (CNR) of the aortic wall was increased significantly after injection of the probe (mean ± SD; *$P < 0.05$ before vs. after injection). (c and e) Light images of excised aortae. (d) NIRF image after VINP$_{28}$ injection demonstrates distribution of the agent to plaque-bearing segments of the aorta, whereas the aorta of the saline-injected apoE$^{-/-}$ shows very little fluorescent signal. (f) Both images were acquired with identical exposure times and were identically windowed. The target-to-background ratio (TBR) was significantly higher in the VINP-28-injected mice (*$P < 0.05$). (Reproduced from Nahrendorf, M. et al. 2006. *Circulation* 114: 1504–1511. With permission.)

compounds allows for their localization to lipid-rich atherosclerotic plaques. However, systemic administration of porphyrinic photosensitizers may also lead to undesired side effects, such as prolonged photosensitivity due to drug accumulation within the skin. For the majority of the reported photosensitizers, plaque localization results in a distribution of the agent between many different cell types, including macrophages, smooth muscle cells, and endothelial cells. Whereas the focal elimination of inflammatory cell types may assist in plaque stabilization and regression, destruction of other cell types, such as endothelial cells, may promote plaque rupture, drastically compounding the nature of the disease. It would thus be advantageous to develop macrophage-specific agents that would overcome these shortcomings and lead to substantially improved therapies for atherosclerotic vascular disease.

A macrophage-targeted theranostic strategy has been reported which utilizes the avidity of inflammatory macrophages for CLIO. In this design, the monocrystalline CLIO nanoparticle was initially modified with Alexa Fluor 750 (AF750) to allow for near-infrared fluorescence imaging and a potent photosensitizer, 5-(4-carboxyphenyl)-10,15,20-triphenyl-2,3-dihydroxychlorin (TPC), with a singlet-oxygen quantum yield of 0.65 when excited at 646 nm (McCarthy et al. 2006). The resulting agent exhibited distinct spectral characteristics of each component, with λ_{max} (iron oxide) <300 nm, λ_{max} (TPC) = 648 nm, and λ_{max} (AF750) = 755 nm, respectively, enabling fluorescent imaging of agent localization at a spectrally distinct wavelength from the photosensitizer, thereby avoiding extraneous cell killing. The cellular uptake of the theranostic agents was subsequently demonstrated by flow cytometry and fluorescence microscopy using murine macrophages (RAW 246.7). The light-induced phototoxicity of the agents was then assessed in both murine and human macrophages (U937 cells) through a cell proliferation assay (MTS) *in vitro*. In the absence of light, the theranostic agent displayed no more toxicity than control particles without TPC. However, once irradiated with light at 650 nm, only 35% of the murine macrophages (0.1 mg Fe/mL, 1 h) and none of transformed human macrophages (0.2 mg Fe/mL, 1 h) remained viable after the light therapy, demonstrating that the agent displayed a highly efficient, dose-dependent phototoxicity. More importantly, when the cells were irradiated at 750 nm with a light dose comparable to intravital fluorescence imaging of atherosclerosis, minimal cell death was observed.

While the TPC-functionalized theranostic agent provided sufficient phototoxicity *in vitro*, there was room to improve the overall phototoxicity of the agent. In a more recent design, a hydrophilic photosensitizer based upon

meso-tetra(*m*-hydroxyphenyl)chlorin (THPC) was utilized, which allowed for the conjugation of at least threefold more dye molecules per AF750-modified nanoparticle (McCarthy et al. 2010). The resulting agent, CLIO-THPC, displayed high macrophage affinity in murine macrophages (RAW246.7) and an exceptional phototoxicity as compared to the commonly used chlorin e_6 (Figure 8.9d). In particular, CLIO-THPC (LD_{50} = 14 nM) demonstrated almost 60 times more phototoxicity than chlorin e_6 (LD_{50} = 800 nM) on a chlorin-per-chlorin basis, and was 1900-fold more toxic on a molecular level (per nanoparticle, LD_{50} of 430 pm). The *in vivo* theranostic efficacy of the agent was next examined in apoE$^{-/-}$ mice. Initially, the mice were injected with the nanoagent 24 h prior the imaging session, in order to allow for the agent to localize to the atherosclerotic lesion. The mice were then anesthetized and the carotid artery was exposed by careful dissection of the periadventitial tissues. The localization of the agent was then determined by IVFM imaging. As can be seen in Figure 8.9e–g, there is significant accumulation of the nanoagent within the plaque. After the initial imaging session, the exposed carotid artery was irradiated with a 650 nm laser in order to elicit a therapeutic effect (150 mW, 3 min). After closing the surgical incisions, the mice were allowed to recover for 1 day before they were

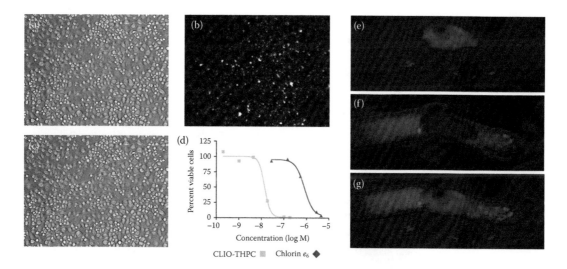

FIGURE 8.9   Cellular uptake and phototoxicity of CLIO-THPC. (a) Brightfield microscopic image of RAW 264.7 murine macrophages (20×). (b) Fluorescent microscopic image of nanoagent uptake in the rhodamine channel (20×). (c) Brightfield and fluorescence channels merged to illustrate localization. (d) Phototoxicity of CLIO-THPC versus chlorin e_6 (650 nm, 50 mW cm^{-2}). *In vivo* localization of the nanoagent, CLIO-THPC, to carotid atheroma, as determined by intravital fluorescence microscopy. (e) Fluorescence image in the AF750 channel demonstrating particle uptake by a carotid plaque. (f) Fluorescence angiogram utilizing fluorescein-labeled dextran to outline the vasculature. (g) Merged image of the two fluorescence channels. (Reproduced from McCarthy, J. R. et al. 2010. *Small* 6: 2041–2049. With permission.)

Chapter 8

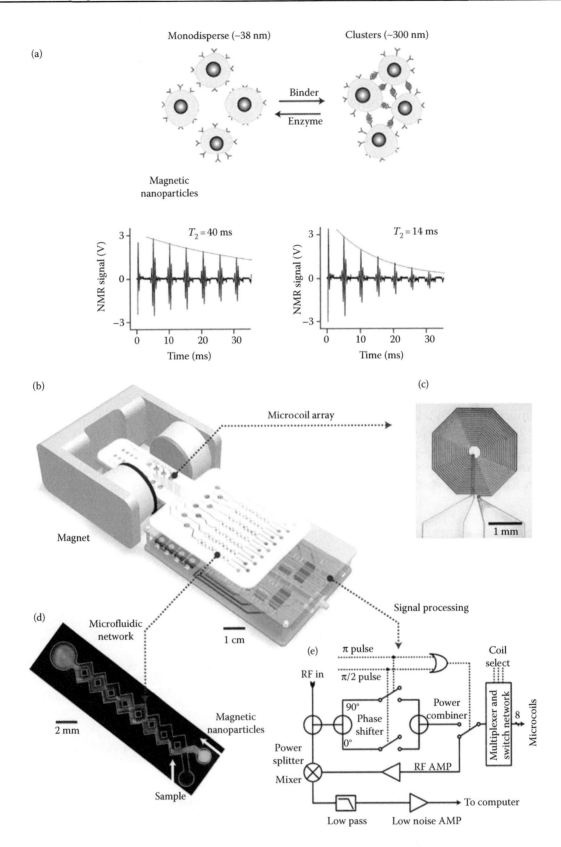

FIGURE 8.10 Principle of the assay and structure of the DMR system. (a) Principle of proximity assay using magnetic particles. When monodisperse magnetic nanoparticles cluster upon binding to targets, the self-assembled clusters become more efficient at dephasing nuclear spins of many surrounding water protons, leading to a decrease in spin–spin relaxation time (T_2). The bottom panel shows an example of the proximity assay measured by the DMR system. Avidin was added to a solution of biotinylated magnetic nanoparticles,

sacrificed for histological assessment. As opposed to mice treated with a control agent, CLIO-AF750, a significant amount of apoptotic cells were observed in the CLIO-THPC-treated mice (0.796% ± 0.10% of the plaque area vs. 55.5 ± 5.1% of the plaque area, respectively). In one final experiment, the skin phototoxicity of the nanoagent was assayed in a mouse hind paw edema model (Peng et al. 2001). In this experiment, wild-type mice were injected with either CLIO-THPC or chlorin e_6, and which was allowed to circulate for either 1 or 24 h. The right paw was then irradiated and measured for the change in thickness 24 h after illumination, while the left paw served as a control. In both time points, CLIO-THPC-treated groups displayed a minimal amount of edema (8 ± 5% and −7 ± 9% change, respectively), while the chlorin e_6-treated groups exhibited much greater change in thickness (32 ± 9% and 14 ± 3% change, respectively), intimating that the nanoagent displays a significantly improved toxicity profile. Further experimentation into the durability of the therapeutic effect is currently underway.

8.5 Impact

Cross-linked dextran-coated iron oxide nanoparticles have been utilized in an astounding number of research applications for over a decade (McCarthy and Weissleder 2008), including the imaging of cancers, diabetes (Denis et al. 2004, Turvey et al. 2005), thrombosis (Chang et al. 2010, McCarthy et al. 2009), transplant rejection (Christen et al. 2009), as well as atherosclerosis. Interestingly, due to the nature of the cross-linking, they may never be clinically viable, as there are questions surrounding the ultimate biodegradability of the nanoagents. Given this, CLIO has been viewed as a reliable platform on which the design and testing of novel imaging and therapeutic paradigms can be tested and then transferred to more clinically relevant platforms. Thus, one of the directions that this research has gone is the search for iron oxide-cored nanoscaffolds comparable to CLIO.

Josephson and coworkers have reported a three-stage screening of approximately two dozen polymers for their ability to form high-relaxivity, high-stability iron oxide nanoparticles (Chen et al. 2009). In the initial screen, tetramethylammonium hydroxide (TMAH)-stabilized particles were plated in a 96-well format, and a solution of the respective polymer was added, effectively exchanging the TMAH for the polymer. Sodium chloride (4 M, 20 μL) was then added, and those particle preparations that were not stable

aggregating, thus failing the screen. The preparations that passed the initial screen (10 polymers) were then assayed as to whether or not they could be used in the synthesis of polymer-coated iron oxide nanoparticles. Several of the polymers from the initial screen were found to form insoluble mixtures under the utilized reaction conditions and were rejected, leaving starch, carboxymethyl dextran (CMD), carboxymethyl poly(vinyl alcohol) (CMPVA), and hydroxyethylstarch (HES) as the only polymers passing the second screen. In the final screen, the remaining particles were tested under the heat stress conditions mentioned earlier. Only one of the preparations, the CMD-coated particles, displayed stability comparable to CLIO; thus, the authors stated that they would be the best choice for further study.

The clinical applicability of CLIO has also moved to another direction. While it cannot be used in patients, it can be used in the clinic for *in vitro* diagnostics. To this end Dr. Hakho Lee and his colleagues have developed a miniaturized diagnostic magnetic resonance (DMR, Figure 8.10b) system that is capable of detecting bacteria or specific cell types (Lee et al. 2008, 2009a,b). In this system, CLIO were modified with ligands for the biological target of interest. Prior to binding the sample, the particles in the system were monodisperse, and thus had a known relaxivity

FIGURE 8.10 (Continued) causing T_2 to decrease from 40 to 14 ms. (b) Schematic diagram of the DMR system. The system consists of an array of microcoils for NMR measurements, microfluidic networks for sample handling and mixing, miniaturized NMR electronics, and a permanent magnet to generate a polarizing magnetic field. The entire setup can be packaged as a handheld device for portable operation. (For clarity, the structure of the magnet is simplified.) (c) High-resolution image of an actual microcoil. The microcoil generates radio frequency (RF) magnetic fields to excite samples and receives the resulting NMR signal. To reduce electrical resistance and hence thermal noise, the metal lines were electroplated with copper. (d) Example of a microfluidic network. Effective mixing between magnetic particles and the target analytes is achieved by generating chaotic advection through the meandering channels. The channel also helps confine the mixture in the most sensitive region of the microcoils, increasing filling factors. (e) Schematic of the NMR electronics. The circuit is designed to perform T_2 and T_1 measurements via CPMG and inversion-recovery pulse sequences, respectively. AMP, amplifier. (Reproduced from Lee, H. et al. 2008. *Nat Med* 14: 869–874. With permission.)

Chapter 8

(Figure 8.10a). When they bind the target, the particles form clusters, drastically decreasing the T_2 value. Initial studies were accomplished using avidin- and biotin-labeled particles, in order to characterize the detection sensitivity of the system, which demonstrated that the system was 80-fold more sensitive than a benchtop relaxometer. CLIO was then modified with vancomycin, which binds to the D-alanyl-D-alanine moieties on the bacterial cell wall. After incubation with *Staphylococcus aureus*, the authors described a concentration-dependent T_2 decrease, which was also observed in a number of different strains. In their initial system, 10 bacteria in 10 µL of solution could be detected. Also, the DMR measurements were markedly faster and simpler than cultivation- or nucleic acid-based analytic methods. Similar experiments were performed targeting epidermal growth factors Her2/neu and EGFR. CLIO was modified with the respective antibody, and the particle was then incubated with a cell line expressing the receptor. T_2 measurements of the washed cells suspended in 10 µL of serum were able to detect both surface markers. Since the initial report, the authors have improved upon the DMR system, and can currently detect as few as two cancer cells in a 1 µL sample volume of unprocessed fine-needle aspirates of tumors (Lee et al. 2009a).

Although CLIO may never be utilized in patients, the establishment of novel techniques and methodologies using this nanoplatform still continues. In particular, CLIO has enabled significant research into the causes of and potential treatments for the sequelae of cardiovascular disease. As such, its continued development is anticipated to produce further innovation.

Acknowledgments

This work was supported in part by NIH grants R21HL093607 and U01-HL080731. We would like to thank Laura Vanderploeg at the UW-Madison Biochemistry Media Lab for providing the clip art mouse depicted in Figure 8.7.

References

Bogdanov, A. A., Martin, C., Weissleder, R. and Brady, T. J. 1994. Trapping of dextran-coated colloids in liposomes by transient binding to aminophospholipid: Preparation of ferrosomes. *Biochim Biophys Acta* 1193: 212–218.

Brown, K. C. 2000. New approaches for cell-specific targeting: Identification of cell-selective peptides from combinatorial libraries. *Curr Opin Chem Biol* 4: 16–21.

Bulte, J. W. and De Cuyper, M. 2003. Magnetoliposomes as contrast agents. *Methods Enzymol* 373: 175–198.

Bulte, J. W., Douglas, T., Mann, S. et al. 1994. Magnetoferritin: Characterization of a novel superparamagnetic MR contrast agent. *J Magn Reson Imaging* 4: 497–505.

Bulte, J. W., Douglas, T., Witwer, B. et al. 2001. Magnetodendrimers allow endosomal magnetic labeling and *in vivo* tracking of stem cells. *Nat Biotechnol* 19: 1141–1147.

Casula, M. F., Jun, Y. W., Zaziski, D. J., Chan, E. M., Corrias, A. and Alivisatos, A. P. 2006. The concept of delayed nucleation in nanocrystal growth demonstrated for the case of iron oxide nanodisks. *J Am Chem Soc* 128: 1675–1682.

Chang, K., Francis, S. A., Aikawa, E. et al. 2010. Pioglitazone suppresses inflammation *in vivo* in murine carotid atherosclerosis: Novel detection by dual-target fluorescence molecular imaging. *Arterioscler Thromb Vasc Biol* 30: 1933–1939.

Chen, S., Reynolds, F., Yu, L., Weissleder, R. and Josephson, L. 2009. A screening paradigm for the design of improved polymer-coated superparamagnetic iron oxide nanoparticles. *J Mater Chem* 19: 6387–6392.

Christen, T., Nahrendorf, M., Wildgruber, M. et al. 2009. Molecular imaging of innate immune cell function in transplant rejection. *Circulation* 119: 1925–1932.

De Cuyper, M. and Joniau, M. 1991. Mechanistic aspects of the adsorption of phospholipids onto lauric acid stabilized magnetite nanocolloids. *Langmuir* 7: 647–652.

Denis, M. C., Mahmood, U., Benoist, C., Mathis, D. and Weissleder, R. 2004. Imaging inflammation of the pancreatic islets in type 1 diabetes. *Proc Natl Acad Sci USA* 101: 12634–12639.

Devaraj, N. K., Hilderbrand, S., Upadhyay, R., Mazitschek, R. and Weissleder, R. 2010. Bioorthogonal turn-on probes for imaging small molecules inside living cells. *Angew Chem Int Ed Engl* 49: 2869–2872.

Devaraj, N. K., Upadhyay, R., Haun, J. B., Hilderbrand, S. A. and Weissleder, R. 2009. Fast and sensitive pretargeted labeling of cancer cells through a tetrazine/trans-cyclooctene cycloaddition. *Angew Chem Int Ed Engl* 48: 7013–7016.

Devaraj, N. K., Weissleder, R. and Hilderbrand, S. A. 2008. Tetrazine-based cycloadditions: Application to pretargeted live cell imaging. *Bioconjug Chem* 19: 2297–2299.

Guardia, P., Perez, N., Labarta, A. and Batlle, X. 2010. Controlled synthesis of iron oxide nanoparticles over a wide size range. *Langmuir* 26: 5843–5847.

Hansson, G. K. 2005. Inflammation, atherosclerosis, and coronary artery disease. *N Engl J Med* 352: 1685–1695.

Hansson, G. K. and Libby, P. 2006. The immune response in atherosclerosis: A double-edged sword. *Nat Rev Immunol* 6: 508–519.

Harisinghani, M. G., Barentsz, J., Hahn, P. F. et al. 2003. Noninvasive detection of clinically occult lymph-node metastases in prostate cancer. *N Engl J Med* 348: 2491–2499.

Harisinghani, M. G. and Weissleder, R. 2004. Sensitive, noninvasive detection of lymph node metastases. *PLoS Med* 1: e66.

Haun, J. B., Devaraj, N. K., Hilderbrand, S. A., Lee, H. and Weissleder, R. 2010. Bioorthogonal chemistry amplifies nanoparticle binding and enhances the sensitivity of cell detection. *Nat Nanotechnol* 5: 660–665.

Hershko, C., Cook, J. D. and Finch, D. A. 1973. Storage iron kinetics. 3. Study of desferrioxamine action by selective radioiron labels of RE and parenchymal cells. *J Lab Clin Med* 81: 876–886.

Hou, Y., Xu, Z. and Sun, S. 2007. Controlled synthesis and chemical conversions of FeO nanoparticles. *Angew Chem Int Ed Engl* 46: 6329–6332.

Illum, L., Church, A. E., Butterworth, M. D., Arien, A., Whetstone, J. and Davis, S. S. 2001. Development of systems for targeting the regional lymph nodes for diagnostic imaging: *In vivo* behaviour of colloidal PEG-coated magnetite nanospheres in the rat following interstitial administration. *Pharm Res* 18: 640–645.

Jaffer, F. A., Libby, P. and Weissleder, R. 2007. Molecular imaging of cardiovascular disease. *Circulation* 116: 1052–1061.

Jaffer, F. A., Nahrendorf, M., Sosnovik, D., Kelly, K. A., Aikawa, E. and Weissleder, R. 2006. Cellular imaging of inflammation in atherosclerosis using magnetofluorescent nanomaterials. *Mol Imaging* 5: 85–92.

Jun, Y. W., Huh, Y. M., Choi, J. S. et al. 2005. Nanoscale size effect of magnetic nanocrystals and their utilization for cancer diagnosis via magnetic resonance imaging. *J Am Chem Soc* 127: 5732–5733.

Jung, C. W. 1995. Surface properties of superparamagnetic iron oxide MR contrast agents: Ferumoxides, ferumoxtran, ferumoxsil. *Magn Reson Imaging* 13: 675–691.

Kellar, K. E., Fujii, D. K., Gunther, W. H. et al. 2000. NC100150 injection, a preparation of optimized iron oxide nanoparticles for positive-contrast MR angiography. *J Magn Reson Imaging* 11: 488–494.

Kellar, K. E., Fujii, D. K., Gunther, W. H., Briley-Saebo, K., Spiller, M. and Koenig, S. H. 1999. "NC100150", a preparation of iron oxide nanoparticles ideal for positive-contrast MR angiography. *Magma* 8: 207–213.

Kelly, K. A., Allport, J. R., Tsourkas, A., Shinde-Patil, V. R., Josephson, L. and Weissleder, R. 2005. Detection of vascular adhesion molecule-1 expression using a novel multimodal nanoparticle. *Circ Res* 96: 327–336.

Kelly, K. A., Nahrendorf, M., Yu, A. M., Reynolds, F. and Weissleder, R. 2006. *In vivo* phage display selection yields atherosclerotic plaque targeted peptides for imaging. *Mol Imaging Biol* 8: 201–207.

Kereiakes, D. J., Szyniszewski, A. M., Wahr, D. et al. 2003. Phase I drug and light dose-escalation trial of motexafin lutetium and far red light activation (phototherapy) in subjects with coronary artery disease undergoing percutaneous coronary intervention and stent deployment: Procedural and long-term results. *Circulation* 108: 1310–1315.

Kim, S. W., Kim, S., Tracy, J. B., Jasanoff, A. and Bawendi, M. G. 2005. Phosphine oxide polymer for water-soluble nanoparticles. *J Am Chem Soc* 127: 4556–4557.

Koivunen, E., Gay, D. A. and Ruoslahti, E. 1993. Selection of peptides binding to the alpha 5 beta 1 integrin from phage display library. *J Biol Chem* 268: 20205–20210.

Kooi, M. E., Cappendijk, V. C., Cleutjens, K. B. et al. 2003. Accumulation of ultrasmall superparamagnetic particles of iron oxide in human atherosclerotic plaques can be detected by *in vivo* magnetic resonance imaging. *Circulation* 107: 2453–2458.

Lee, H., Sun, E., Ham, D. and Weissleder, R. 2008. Chip-NMR biosensor for detection and molecular analysis of cells. *Nat Med* 14: 869–874.

Lee, H., Yoon, T. J., Figueiredo, J. L., Swirski, F. K. and Weissleder, R. 2009a. Rapid detection and profiling of cancer cells in fine-needle aspirates. *Proc Natl Acad Sci USA* 106: 12459–12464.

Lee, H., Yoon, T. J. and Weissleder, R. 2009b. Ultrasensitive detection of bacteria using core-shell nanoparticles and an NMR-filter system. *Angew Chem Int Ed Engl* 48: 5657–5660.

Libby, P. 2002. Inflammation in atherosclerosis. *Nature* 420: 868–874.

Libby, P., Nahrendorf, M. and Weissleder, R. 2010. Molecular imaging of atherosclerosis: A progress report. *Tex Heart Inst J* 37: 324–327.

Lloyd-Jones, D., Adams, R. J., Brown, T. M. et al. 2010. Heart disease and stroke statistics—2010 update: A report from the American Heart Association. *Circulation* 121: e46–e215.

McAteer, M. A., Sibson, N. R., von Zur Muhlen, C. et al. 2007. *In vivo* magnetic resonance imaging of acute brain inflammation using microparticles of iron oxide. *Nat Med* 13: 1253–1258.

McCarthy, J. R., Jaffer, F. A. and Weissleder, R. 2006. A macrophage-targeted theranostic nanoparticle for biomedical applications. *Small* 2: 983–987.

McCarthy, J. R., Kelly, K. A., Sun, E. Y. and Weissleder, R. 2007. Targeted delivery of multifunctional magnetic nanoparticles. *Nanomedicine (Lond)* 2: 153–167.

McCarthy, J. R., Korngold, E., Weissleder, R. and Jaffer, F. A. 2010. A light-activated theranostic nanoagent for targeted macrophage ablation in inflammatory atherosclerosis. *Small* 6: 2041–2049.

McCarthy, J. R., Patel, P., Botnaru, I., Haghayeghi, P., Weissleder, R. and Jaffer, F. A. 2009. Multimodal nanoagents for the detection of intravascular thrombi. *Bioconjug Chem* 20: 1251–1255.

McCarthy, J. R. and Weissleder, R. 2008. Multifunctional magnetic nanoparticles for targeted imaging and therapy. *Adv Drug Deliv Rev* 60: 1241–1251.

Moffat, B. A., Reddy, G. R., McConville, P. et al. 2003. A novel polyacrylamide magnetic nanoparticle contrast agent for molecular imaging using MRI. *Mol Imaging* 2: 324–332.

Nahrendorf, M., Jaffer, F. A., Kelly, K. A. et al. 2006. Noninvasive vascular cell adhesion molecule-1 imaging identifies inflammatory activation of cells in atherosclerosis. *Circulation* 114: 1504–1511.

Nahrendorf, M., Zhang, H., Hembrador, S. et al. 2008. Nanoparticle PET-CT imaging of macrophages in inflammatory atherosclerosis. *Circulation* 117: 379–387.

Nitin, N., LaConte, L. E., Zurkiya, O., Hu, X. and Bao, G. 2004. Functionalization and peptide-based delivery of magnetic nanoparticles as an intracellular MRI contrast agent. *J Biol Inorg Chem* 9: 706–712.

O'Brien, K. D., Allen, M. D., McDonald, T. O. et al. 1993. Vascular cell adhesion molecule-1 is expressed in human coronary atherosclerotic plaques. Implications for the mode of progression of advanced coronary atherosclerosis. *J Clin Invest* 92: 945–951.

Ohgushi, M., Nagayama, K. and Wada, A. 1978. Dextran-magnetite: A new relaxation reagent and its application to T2 measurements in gel systems. *J Magn Reson (1969–1992)* 29: 599–601.

Chapter 8

Pande, A. N., Kohler, R. H., Aikawa, E., Weissleder, R. and Jaffer, F. A. 2006. Detection of macrophage activity in atherosclerosis *in vivo* using multichannel, high-resolution laser scanning fluorescence microscopy. *J Biomed Opt* 11: 021009.

Peng, G., Warloe, T., Moan, J. et al. 2001. Antitumor effect of 5-aminoevulinic acid-mediated photodynamic therapy can be enhanced by the use of a low dose of photofrin in human tumor xenografts. *Cancer Res* 61: 5824–5832.

Portet, D., Denizot, B., Rump, E., Lejeune, J. J. and Jallet, P. 2001. Nonpolymeric coatings of iron oxide colloids for biological use as magnetic resonance imaging contrast agents. *J Colloid Interface Sci* 238: 37–42.

Qin, J., Laurent, S., Jo, Y. S. et al. 2007. A high-performance magnetic resonance imaging T2 contrast agent. *AdvMater* 19: 1874–1878.

Rockson, S. G., Kramer, P., Razavi, M. et al. 2000a. Photoangioplasty for human peripheral atherosclerosis: Results of a phase I trial of photodynamic therapy with motexafin lutetium (Antrin). *Circulation* 102: 2322–2324.

Rockson, S. G., Lorenz, D. P., Cheong, W. F. and Woodburn, K. W. 2000b. Photoangioplasty: An emerging clinical cardiovascular role for photodynamic therapy. *Circulation* 102: 591–596.

Schmitz, S. A., Taupitz, M., Wagner, S., Wolf, K. J., Beyersdorff, D. and Hamm, B. 2001. Magnetic resonance imaging of atherosclerotic plaques using superparamagnetic iron oxide particles. *J Magn Reson Imaging* 14: 355–361.

Shtatland, T., Guettler, D., Kossodo, M., Pivovarov, M. and Weissleder, R. 2007. PepBank—A database of peptides based on sequence text mining and public peptide data sources. *BMC Bioinformatics* 8: 280.

Smith, G. P. 1985. Filamentous fusion phage: Novel expression vectors that display cloned antigens on the virion surface. *Science* 228: 1315–1317.

Stark, D. D., Weissleder, R., Elizondo, G. et al. 1988. Superparamagnetic iron oxide: Clinical application as a contrast agent for MR imaging of the liver. *Radiology* 168: 297–301.

Subramanian, S., Jaffer, F. A. and Tawakol, A. 2010. Optical molecular imaging in atherosclerosis. *J Nucl Cardiol* 17: 135–144.

Sun, E. Y., Josephson, L., Kelly, K. A. and Weissleder, R. 2006b. Development of nanoparticle libraries for biosensing. *Bioconjug Chem* 17: 109–113.

Sun, E. Y., Josephson, L. and Weissleder, R. 2006a. "Clickable" nanoparticles for targeted imaging. *Mol Imaging* 5: 122–128.

Tang, T. Y., Muller, K. H., Graves, M. J. et al. 2009. Iron oxide particles for atheroma imaging. *Arterioscler Thromb Vasc Biol* 29: 1001–1008.

Thorek, D. L., Chen, A. K., Czupryna, J. and Tsourkas, A. 2006. Superparamagnetic iron oxide nanoparticle probes for molecular imaging. *Ann Biomed Eng* 34: 23–38.

Trepel, M., Arap, W. and Pasqualini, R. 2002. *In vivo* phage display and vascular heterogeneity: Implications for targeted medicine. *Curr Opin Chem Biol* 6: 399–404.

Trivedi, R. A., U-King-Im, J. M., Graves, M. J., Kirkpatrick, P. J. and Gillard, J. H. 2004a. Noninvasive imaging of carotid plaque inflammation. *Neurology* 63: 187–188.

Trivedi, R. A., Mallawarachi, C., U-King-Im, J. M. et al. 2006. Identifying inflamed carotid plaques using *in vivo* USPIO-enhanced MR imaging to label plaque macrophages. *Arterioscler Thromb Vasc Biol* 26: 1601–1606.

Trivedi, R. A., U-King-Im, J. M., Graves, M. J. et al. 2004b. *In vivo* detection of macrophages in human carotid atheroma: Temporal dependence of ultrasmall superparamagnetic particles of iron oxide-enhanced MRI. *Stroke* 35: 1631–1635.

Tsourkas, A., Shinde-Patil, V. R., Kelly, K. A. et al. 2005. *In vivo* imaging of activated endothelium using an anti-VCAM-1 magnetooptical probe. *Bioconjug Chem* 16: 576–581.

Turvey, S. E., Swart, E., Denis, M. C. et al. 2005. Noninvasive imaging of pancreatic inflammation and its reversal in type 1 diabetes. *J Clin Invest* 115: 2454–2461.

Uppal, R. and Caravan, P. 2010. Targeted probes for cardiovascular MR imaging. *Future Med Chem* 2: 451–470.

Weissleder, R., Elizondo, G., Stark, D. D. et al. 1989a. The diagnosis of splenic lymphoma by MR imaging: Value of superparamagnetic iron oxide. *AJR Am J Roentgenol* 152: 175–180.

Weissleder, R., Elizondo, G., Wittenberg, J., Rabito, C. A., Bengele, H. H. and Josephson, L. 1990b. Ultrasmall superparamagnetic iron oxide: Characterization of a new class of contrast agents for MR imaging. *Radiology* 175: 489–493.

Weissleder, R., Hahn, P. F., Stark, D. D. et al. 1987a. MR imaging of splenic metastases: Ferrite-enhanced detection in rats. *AJR Am J Roentgenol* 149: 723–726.

Weissleder, R., Hahn, P. F., Stark, D. D. et al. 1988. Superparamagnetic iron oxide: Enhanced detection of focal splenic tumors with MR imaging. *Radiology* 169: 399–403.

Weissleder, R. and Stark, D. D. 1989a. Magnetic resonance imaging of the liver. *Magn Reson Q* 5: 97–121.

Weissleder, R. and Stark, D. D. 1989b. Magnetic resonance imaging of liver tumors. *Semin Ultrasound CT MR* 10: 63–77.

Weissleder, R., Stark, D. D., Compton, C. C., Wittenberg, J. and Ferrucci, J. T. 1987b. Ferrite-enhanced MR imaging of hepatic lymphoma: An experimental study in rats. *AJR Am J Roentgenol* 149: 1161–1165.

Weissleder, R., Elizondo, G., Wittenberg, J., Lee, A. S., Josephson, L. and Brady, T. J. 1990a. Ultrasmall superparamagnetic iron oxide: An intravenous contrast agent for assessing lymph nodes with MR imaging. *Radiology* 175: 494–498.

Weissleder, R., Stark, D. D., Engelstad, B. L. et al. 1989b. Superparamagnetic iron oxide: Pharmacokinetics and toxicity. *Am J Roentgenol* 152: 167–173.

Wilhelm, C., Billotey, C., Roger, J., Pons, J. N., Bacri, J. C. and Gazeau, F. 2003. Intracellular uptake of anionic superparamagnetic nanoparticles as a function of their surface coating. *Biomaterials* 24: 1001–1011.

Case Studies on Developing Targeted Radiotracers

9. Targeting Prostate Cancer with Small Molecule Inhibitors of Prostate–Specific Membrane Antigen

John W. Babich, John L. Joyal, and R. Edward Coleman

9.1 Prostate-Specific Membrane Antigen as a Target for Prostate Cancer

9.1.1 Introduction

Prostate cancer is second only to lung cancer as a leading cause of cancer-related fatalities and is the malignancy with the highest incidence of new cases in men in the United States (American Cancer Society). It was estimated in 2010 that 217,730 new cases of prostate cancer would be diagnosed and that 32,050 men would die of the disease (American Cancer Society; NCI).

Prostate cancer is typically diagnosed by digital rectal exam and prostate-specific antigen (PSA) blood testing. After the introduction of serum PSA screening in 1986, prostate cancer incidence rates increased dramatically, peaking in the early-to-mid-1990s and subsequently declining slightly after the year 2000 (American Cancer Society; Hankey et al. 1999). There is controversy, however, regarding the correlation between the number of new diagnoses that occur as a result of screening and the prostate cancer mortality rate. For example, the prostate, lung, colorectal, and ovarian cancer screening trial demonstrated that the screening group, which received annual digital rectal exam testing for 4 years and PSA testing for 6 years (concurrently

Targeted Molecular Imaging. Edited by Michael J. Welch and William C. Eckelman © 2012 Taylor & Francis Group, LLC. ISBN: 978-1-4398-4195-2

with the first four PSA tests), did not differ significantly from the control group in terms of the mortality rate after an average follow-up of 7–10 years (Andriole et al. 2009). Moreover, serum PSA levels do not always correlate with the extent of disease. For example, it was found that 20–30% of men with prostate cancer had false-negative PSA results (i.e., serum PSA levels within the reference range) (Catalona et al. 1991; Thompson et al. 2004). Additionally, other diseases or conditions, such as benign prostatic hyperplasia, prostatitis, inflammation, trauma, or urinary retention, may lead to elevated serum PSA levels, resulting in false positives (American Cancer Society; Tricoli et al. 2004; Akin and Hricak 2007).

It is clear that there is still a need for tools to aid in the diagnosis and classification of disease stage, the evaluation of treatment success, and the monitoring of recurring disease. The role of imaging in prostate cancer has expanded to address this unmet need, but imaging still has some shortcomings in the diagnosis and staging of prostate cancer. For example, using transrectal ultrasonography (TRUS), the most common imaging technique used in cancer detection, to guide biopsy, needle biopsy obtains only a small sample of tissue (Akin and Hricak 2007). This potential for sampling error can lead to underdetection of disease, as prostate cancer has been identified in only 22–34% of cases on initial TRUS-guided biopsy, as well as the need for repeat biopsy or a perioperative approach such as obtaining more biopsy samples from different regions of the prostate. While TRUS is also frequently used to stage prostate cancer, its role is still limited, as the detection of malignancy relies on the echogenicity of the mass in this modality, and some prostate cancer tumors are isoechoic, limiting their ability to be discovered using TRUS. Furthermore, while prostate cancer masses will usually appear as a hypoechoic area within the peripheral zone on TRUS, this presentation can also be seen in other conditions such as prostatitis or focal atrophy. Computed tomography (CT) also is of limited clinical use in staging, unless in advanced cases, due to poor soft tissue resolution. While magnetic resonance imaging (MRI) has shown promise in the staging of prostate cancer, it has low sensitivity in the assessment of lymph node metastases and can be more expensive than other modalities, such as bone scan, for detecting metastases. Furthermore, for recurring disease, assessing the location and extent of prostate cancer on MRI may be hindered by tissue changes related to radiation therapy.

While imaging has proven to be a useful tool in the diagnosis and staging of prostate cancer, new approaches that work with imaging to more accurately assess the status of the disease and the progress of therapy will allow for the selection of optimal treatment and improve patient outcomes. One such approach is the use of radiopharmaceutical tracers, which localize to a prespecified target on the cell, allowing this area to be visualized.

The design and synthesis of a series of small molecule inhibitors of prostate-specific membrane antigen (PSMA), which can potentially diagnose and classify prostate cancers through common molecular imaging modalities such as single-photon emission computed tomography (SPECT), have been previously described (Maresca et al. 2009). Two of the most potent radioiodinated compounds, ^{123}I-MIP-1072 and ^{123}I-MIP-1095, have been shown to target PSMA and localize to PSMA-expressing tumors in mice and to both bone and soft tissue metastases in patients; these compounds are described in detail later.

9.1.2 Target Protein

PSMA, also known as folate hydrolase I or glutamate carboxypeptidase II, is a transmembrane, 750 amino acid, type II glycoprotein that is primarily expressed in normal human prostate epithelium but is overexpressed in prostate cancer, including metastatic disease (Horoszewicz et al. 1987; Israeli et al. 1993; Fair et al. 1997; Silver et al. 1997).

PSMA is an *N*-acetylated-α-linked acidic dipeptidase (NAALADase) with reactivity toward poly-γ glutamyl folates that has the capability of sequentially removing the poly-γ-glutamyl termini of dipeptides (Horoszewicz et al. 1987; Pinto et al. 1996; Smith-Jones et al. 2000; Hillier et al. 2009). PSMA is an attractive target for prostate cancer therapy and imaging because it is expressed by nearly all prostate cancer types and its expression is augmented in poorly differentiated, metastatic, and hormone-refractory carcinomas (Silver et al. 1997).

PSMA was originally identified as the ligand of the monoclonal antibody 7E11-C5, marketed as Prosta-Scint (capromab pendetide) (Horoszewicz et al. 1987). Although ProstaScint was found in histological studies to have a high degree of specificity for the LNCaP human adenocarcinoma cell line, it was later found to target the intracellular domain of PSMA (amino terminus) and is not believed to bind to viable tumor cells but, rather, to mostly necrotic cells in prostate tumors

(Troyer et al. 1995, 1997; Hillier et al. 2009); as such, the use of ProstaScint is not widespread.

Radiolabeled monoclonal antibodies that bind to the extracellular domain of PSMA have been developed more recently and have been demonstrated to accumulate in PSMA-positive animal prostate tumor models (Smith-Jones et al. 2003; Hillier et al. 2009). In clinical trials, early results have shown PSMA to be a useful diagnostic and therapeutic target (Vallabhajosula et al. 2005; Milowsky et al. 2007). Despite the potential for use of monoclonal antibodies in tumor detection and therapy, because of their long circulating plasma half-lives and low permeability in solid tumors, especially bone metastases, there has been limited clinical success with these agents in other types of cancer besides lymphoma. Lower-molecular-weight small molecules will likely have an advantage over monoclonal antibodies, as they typically have greater permeability in solid tumors and will likely make lesion detection more patent due to generally superior pharmacokinetics in normal tissues compared with intact immunoglobulins.

9.1.3   Chemical Identity and Quantitation of the Probe

Analysis of the crystal structure of PSMA has informed understanding of the critical interactions of potent inhibitors within the active site of the enzyme and has led to the design and synthesis of several classes of NAALADase inhibitors that are substrates or transition-state analogs, including the amide (Figure 9.1a) and urea (Figure 9.1b) substrate analogs, and the phosphinate (Figure 9.1c, where $X = CH_2$), and phosphonate (Figure 9.1c, where $X = O$) transition-state analogs (Speno et al. 1999; Maresca et al. 2009). Herein, we focused on analogs of the urea-linked dipeptide NAALADase inhibitors (Figure 9.1b) (Nan et al. 2000; Kozikowski et al. 2001).

Two amino acids were joined through the amino groups by a urea linkage to form the original structure, as it was believed that the urea group would serve as a replacement for the central phosphinate/phosphonate backbone present in the early lead structures (Kozikowski et al. 2001; Barinka et al. 2002, 2007, 2008). On the basis of the inhibitory potency of the different isomers of the glutamate–urea–glutamate core and subsequent structure–activity relationship studies with other amino acids, information was obtained regarding structural modifications to the pharmacophore and the effect on binding to the enzyme active site (Wang et al. 2007). Two radiolabeled urea-based analogs have been described, N-[N-[(S)-1,3-dicarboxypropyl]carbamoyl]-(S)-[^{11}C]methyl-L-cysteine ([^{11}C]DCMC) and N-[N-[(S)-1,3-dicarboxypropyl]carbamoyl]-(S)-3-iodo-L-tyrosine ([^{125}I]DCIT), which exhibit specific uptake in human prostate cancer xenografts (Pomper et al. 2002; Foss et al. 2005).

The synthesis and *in vitro* evaluation of novel glutamate–urea–lysine heterodimers as PSMA inhibitors, which are prepared by modifying the ε-amino group of the lysine residue of a protected glutamate–urea–lysine heterodimer, accomplished by conversion to halogen-substituted benzylamines, benzamides, and phenylureas, have been described (Maresca et al. 2007). *In vitro* studies investigated the electronic, steric, and regiochemistry properties of the halogen atoms on the phenyl ring, as well as the functionality of the linker bridge from the lysine nitrogen to the aromatic ring, leading to initial insight into the structural prerequisites for PSMA binding of these halogenated glutamate–urea–lysine heterodimers. The results identified several potent inhibitors of PSMA, including two iodine-containing molecules, MIP-1072 and MIP-1095, shown in Figure 9.2, that may be germane to radioimaging or radiotherapy of prostate cancer. MIP-1072 and MIP-1095 are potent inhibitors of the NAALADase enzymatic activity of PSMA with K_i values 4.6 ± 1.6 and 0.24 ± 0.14 nM, respectively. These compounds have been prepared as the radioiodinated (^{123}I and ^{131}I) analogs in high radiochemical yields, high radiochemical purity, and high specific activity and have demonstrated prolonged radiochemical stability at elevated temperature, and have

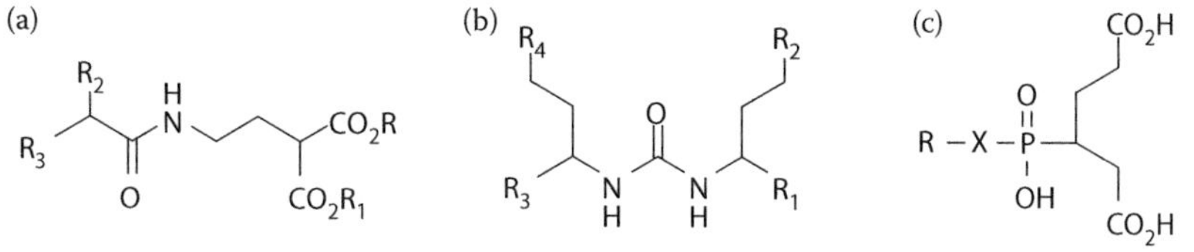

FIGURE 9.1   Substrate (a and b) and transition state (c) analogs of NAALADase based on the structure of N-acetylaspartylglutamate (NAAG).

Chapter 9

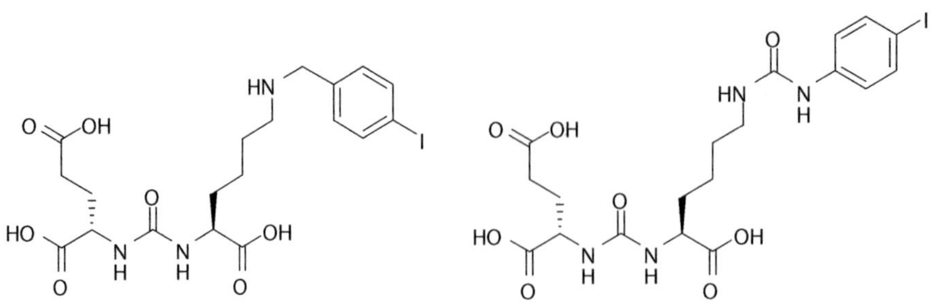

MIP-1072 MIP-1095

FIGURE 9.2 Structures of lead PSMA inhibitors: MIP-1072 and MIP-1095.

progressed through preclinical evaluation to clinical trials in prostate cancer patients.

9.1.3.1 Chemistry and Synthesis of Glutamate–Urea–Lysine

The key intermediate, glutamate-urea–lysine, used to prepare the PSMA inhibitors is prepared by the isocyanate intermediate route as depicted in Scheme 9.1 (Maresca et al. 2009).

Glutamate–urea–lysine is prepared via the isocyanate (**1B**), which is generated *in situ* from (**1**) upon treatment with triphosgene. Subsequent reaction with Cbz–Lys–Ot–Bu, followed by removal of the Cbz protecting group by hydrogenolysis, affords glutamate–urea–lysine in good yield. Compounds are analyzed for purity and confirmation of structure by high-pressure liquid chromatography (HPLC), 1H NMR spectroscopy, and mass spectroscopy.

9.1.3.2 Synthesis of MIP-1072

MIP-1072 is prepared using the route depicted in Scheme 9.2 (Maresca et al. 2009). The key synthetic intermediate, glutamate–urea–lysine, is reacted with the appropriate aldehyde to form the intermediate Schiff base, which

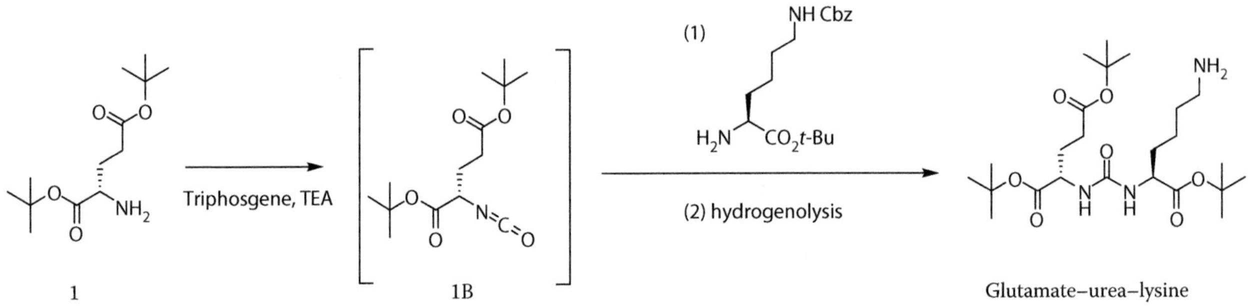

1 1B Glutamate–urea–lysine

SCHEME 9.1 General pathway for the synthesis of the key intermediate (glutamate–urea–lysine) utilized in the synthesis of PSMA inhibitors, via the isocyanate intermediate.

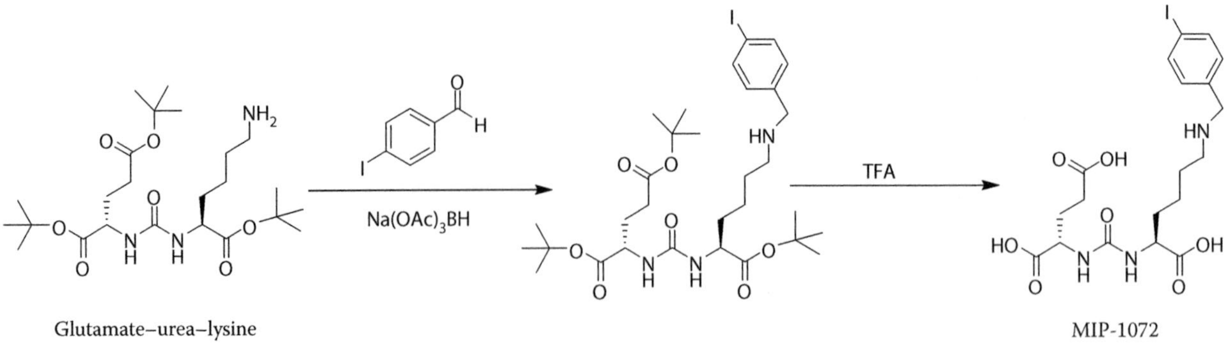

Glutamate–urea–lysine MIP-1072

SCHEME 9.2 General pathway for the synthesis of MIP-1072.

is directly reduced *in situ* with sodium triacetoxyborohydride to produce the desired tri-*tert*-butyl ester-protected benzylamine, MIP-1072. The *tert*-butyl ester-protecting groups are removed using trifluoroacetic acid (TFA), after which the samples are purified by HPLC or recrystallized to form the desired product MIP-1072 in overall yields of 20–90%. Compounds are analyzed for purity and confirmation of structure by HPLC, 1H NMR spectroscopy, and mass spectroscopy.

9.1.3.3 Synthesis of MIP-1095

MIP-1095 was prepared by the route depicted in Scheme 9.3 (Maresca et al. 2009). The key synthetic intermediate, glutamate–urea–lysine, was reacted with the appropriate phenyl isocyanate at room temperature to produce the desired protected intermediate MIP-1095 in good yields. The *tert*-butyl ester-protecting groups were removed with TFA, after which the sample was purified by column chromatography or recrystallized to form the desired products MIP-1095 in

overall yields of 40–90%. Compounds were analyzed for purity and confirmation of structure by HPLC, 1H NMR spectroscopy, and mass spectroscopy.

9.1.3.4 Preparation of ^{123}I-MIP-1072

MIP-1072 and MIP-1095 are readily labeled with ^{123}I or ^{131}I starting with the trimethyl tin precursors as shown in Scheme 9.4. An example of the radiolabeling procedure for MIP-1072 is as follows. Sterile water for injection (SWFI) (100 μL) is added into a 5-cc vial containing ^{123}I-NaI, followed by 305 μL of an acid solution (acetic acid [300 μL] and sulfuric acid [5 μL]), and then 300 μL of an oxidant (acetic acid [0.2 mL] and 30% hydrogen peroxide [0.335 mL] brought to a final volume of 5 mL with SWFI), to which is added 150 μL of Sn-MIP 1072 (1 mg/mL solution in ethanol) (Maresca et al. 2009). The mixture is vortexed for 2 min and incubated at room temperature for an additional 30 min. The reaction is quenched with 200 μL of 0.1 M sodium thiosulfate. The product is

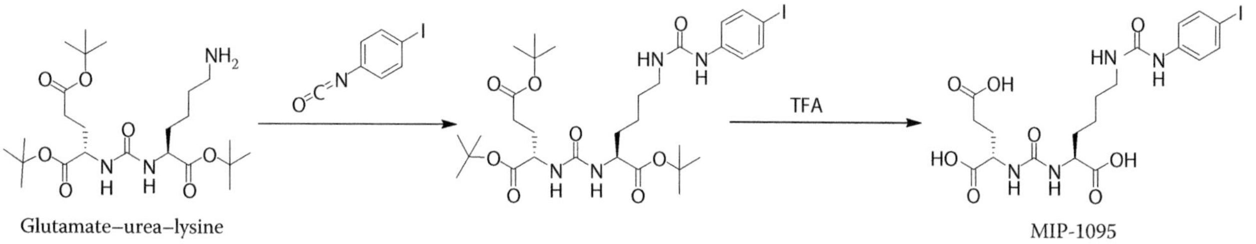

SCHEME 9.3 General pathway for the synthesis of MIP-1095.

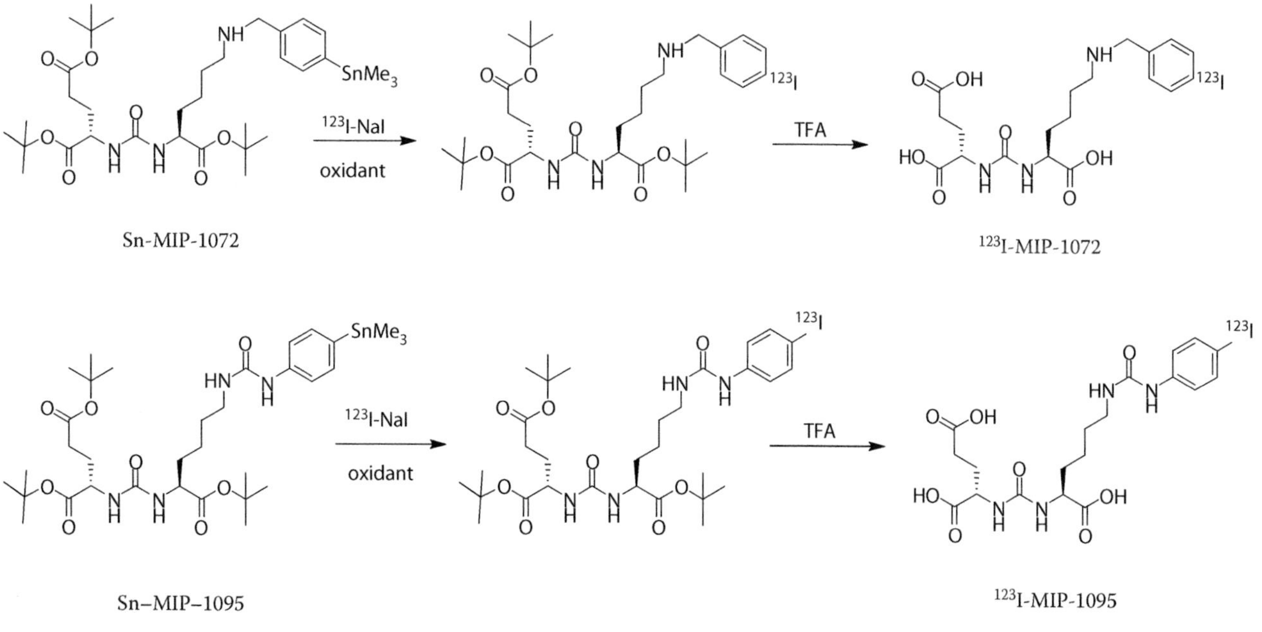

SCHEME 9.4 Iododestannylation to form ^{123}I-MIP-1072 and ^{123}I-MIP-1095.

Chapter 9

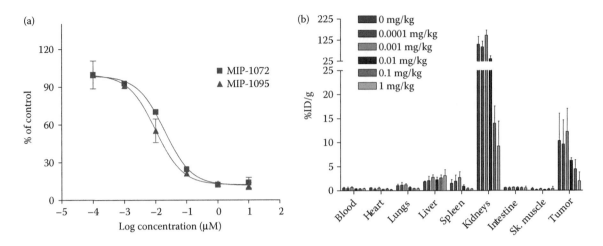

FIGURE 9.3 (a) Competitive binding of MIP-1072 and MIP-1095 on LNCaP cells. (b) Blocking of kidney and tumor uptake with MIP-1072.

diluted in SWFI (18 mL) and loaded onto a C18 Sep Pak Plus column, which is washed with SWFI (60 mL) to remove unreacted radioiodine and inorganic and organic salts. The Sn-MIP-1072 precursor and ^{123}I-MIP-1072 ester are retained on the column. ^{123}I-MIP-1072 ester is eluted from the column using ethanol (6 mL), while the Sn-MIP-1072 precursor is retained on the column. The resulting solution, containing the ^{123}I-MIP-1072 ester, is evaporated to dryness under a stream of nitrogen and the residue dissolved in methylene chloride (0.5 mL) and TFA (2 mL) added to remove the ester-protecting groups. The solution is then incubated at room temperature for 40 min. After deprotection, the solution is evaporated to dryness under a stream of nitrogen and the residue is dissolved in the formulation matrix of gentisate 2% and ascorbate 5% (pH = 5), with radiochemical yields ranging from 50% to 70%. The final product is sterile filtered using a sterile 0.2 μm Millex GV (Millipore, Billerica, MA) (33 mm) filter to yield the final product in >90% radiochemical purity with specific activities ≥4000 mCi/μmol. The radiolabeling procedure is similar in the case of MIP-1095.

9.1.4 Specific Activity

As binding to PSMA on prostate cancer cells is a saturable event, it is imperative that high specific activity radiopharmaceuticals are used to avoid competition with nonradiolabeled or "cold" compound which would result in submaximal signal. In a competitive binding assay *in vitro*, both nonradiolabeled MIP-1072 and MIP-1095 were shown to inhibit the binding of

^{123}I-MIP-1072 to LNCaP cells in a dose-dependent manner with IC-50 values of 21 and 10 nM, respectively (Figure 9.3a). Furthermore, nonradiolabeled MIP-1072 when coadministered with ^{123}I-MIP-1072 inhibited both tumor and kidney uptake in a dose-dependent fashion in mice bearing LNCaP xenografts, at a half-maximal dose of ~0.06 mg/kg (Figure 9.3b).

Nonradiolabeled compound in the radiopharmaceutical preparation can arise from several sources; the presence of active residual precursor ligand if purification is not performed, in this case the trimethyl tin precursor of the glutamate–urea–lysine heterodimers, the presence of nonradiolabeled impurities formed during the radiolabeling process, for example, the deshalogenated analogs of the glutamate–urea–benzyl lysine or glutamate–urea–lysine–phenyl urea, or from the incorporation of low specific activity isotopes. In each of these events, apparent binding of the radiolabeled product to the target may be reduced. In the production of ^{123}I-MIP-1072 and ^{123}I-MIP-1095, we use high specific activity ^{123}I which has a theoretical specific activity of >200 K mCi/μmol. In addition, the radiolabeling process of MIP-1072 and MIP-1095 utilizes Sep-Pak purification after the iododestannylation reaction under conditions that have been shown to remove residual precursor ligand and des-halogenated impurities. As a result, no UV-detectable peak can be seen by HPLC analysis of the final product. While the specific activity can be assumed to be equivalent to the specific activity of the starting ^{123}I, it is customary to use the limit of detection, in this case, 10 ng as the denominator for calculating the worse-case specific activity of the product.

9.1.5 Specificity of Targeting Both *In Vitro* and *In Vivo*

An essential characteristic of a molecular imaging radiopharmaceutical is that it binds with high affinity and specificity to its respective target. Specificity is verified in two ways: via competition with homologous nonradiolabeled compound and by the lack of binding to a cell line that is devoid of the target.

9.1.5.1 Specificity of Targeting *In Vitro*

To examine specificity *in vitro*, LNCaP and PC3 cells were incubated with ^{123}I-MIP-1072 or ^{123}I-MIP-1095 to investigate the specificity of the compounds for PSMA-expressing prostate cancer cells (Hillier et al. 2009). Both ^{123}I-MIP-1072 and ^{123}I-MIP-1095 bound to LNCaP cells but not to the PC3 cells, which were PSMA deficient (Figure 9.4a). Nonradiolabeled compound or PMPA (2-(phosphonomethyl)-pentanedioic acid), a structurally unrelated PSMA inhibitor, inhibited binding to LNCaP cells. To establish the affinity of ^{123}I-MIP-1072 and ^{123}I-MIP-1095 for PSMA expressed on LNCaP cells, saturation binding analysis was conducted. To determine K_d and B_{max} (Figure 9.4b), cells were incubated with 30–300,000 pmol/L ^{123}I-MIP-1072 or ^{123}I-MIP-1095. The K_d and B_{max} were determined by

nonlinear regression analysis. Consistent with the order of potency of the NAALADase inhibition assay, MIP-1095 demonstrated greater affinity for PSMA than MIP-1072 ($K_d = 0.81 \pm 0.39$ and 3.8 ± 1.3 nmol/L, respectively). The B_{max} obtained with MIP-1072 was found to be 1490 ± 60 fmol/10^6 cells (0.9×10^6 sites/cell), and with MIP-1095, 1680 ± 110 fmol/10^6 cells (1×10^6 sites/cell), which was consistent with the value reported for the ProstaScint antibody (Smith-Jones et al. 2003; Hillier et al. 2009).

9.1.5.2 Specificity of Targeting *In Vivo*

To determine if ^{123}I-MIP-1072 and ^{123}I-MIP-1095 bind specifically to PSMA *in vivo*, NCr nu/nu mice bearing either LNCaP or PC3 xenografts were concomitantly injected with ^{123}I-MIP-1072 or ^{123}I-MIP-1095 and 50 mg/kg of the PSMA inhibitor, PMPA. Both ^{123}I-MIP-1072 and ^{123}I-MIP-1095 bound to PSMA-expressing LNCaP tumors but not to the PC3 tumors, which were PSMA deficient. Furthermore, concomitantly injecting the mice with 50 mg/kg PMPA blocked binding to the LNCaP tumor xenografts and kidneys (Figure 9.5).

Consistent with the tissue distribution studies, SPECT/CT imaging at 4 h after injection of ^{123}I-MIP-1072 and ^{123}I-MIP-1095 revealed high uptake and

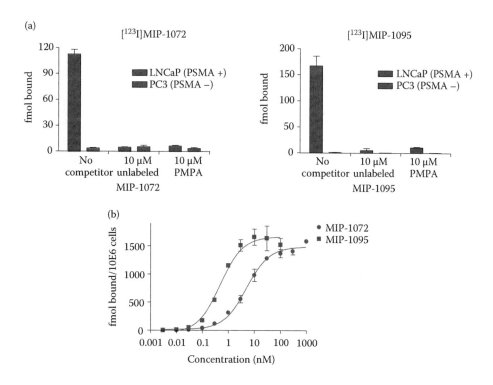

FIGURE 9.4 (a) Binding of 123I-MIP-1072 or 123I-MIP-1095 to LNCaP and PC3 cells. (b) Saturation binding analysis of 123I-MIP-1072 and 123I-MIP-1095.

Chapter 9

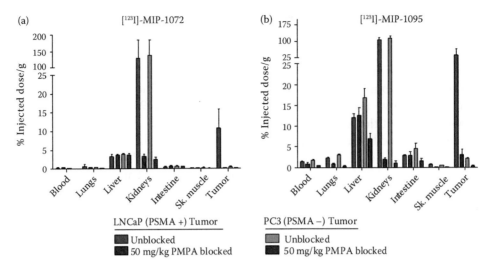

FIGURE 9.5   Specific binding of ^{123}I-MIP-1072 (a) and ^{123}I-MIP-1095 (b) to PSMA *in vivo*. Radiolabeled compound (2 µCi/mouse at >1000 mCi/µmol) was injected alone (LNCaP tumor ■, PC3 tumor ■) or coinjected with 50 mg/kg PMPA (LNCaP tumor ■, or PC3 tumor ■) via the tail vein. Data are expressed as %ID/g.

selectivity for PSMA-expressing tissues: kidney and LNCaP tumor (Figure 9.6, top). Additionally, ^{123}I-MIP-1072 and ^{123}I-MIP-1095 detected PC3 PIP (PSMA +) but not PC3 flu (PSMA −) tumors by SPECT/CT at 2 h after injection (Figure 9.6, bottom), indicating that the uptake is specific to PSMA and is not related to blood flow or permeability differences between cell lines. As anticipated, high uptake was also observed in the kidneys, which express PSMA.

9.1.6   Lipophilicity

Since prostate cancer metastasizes primarily to the pelvic bones and vertebrae, and to abdominal and pelvic lymph nodes, it is critical that the signal to background in these regions is as high as possible to discriminate metastatic lesions. Lipophilic compounds typically demonstrate prolonged blood pool activity, high liver and gut accumulation, and hepatobiliary clearance which confounds their use for molecular imaging of prostate cancer. ^{123}I-MIP-1072 and ^{123}I-MIP-1095 are highly polar molecules containing three carboxyl groups. They exhibit log *P* values of −1.85 and −1.91, respectively, indicating that both are hydrophilic. Both ^{123}I-MIP-1072 and ^{123}I-MIP-1095 display rapid clearance from the blood and nontarget tissues indicating very little nonspecific binding. Little uptake of ^{123}I-MIP-1072 and ^{123}I-MIP-1095 was detected in the brain, which exhibits high NAALADase activity (Blakely et al. 1988), indicating that they do not cross the blood–brain barrier and are unlikely to interfere with the physiologic NAALADase activity of glutama-

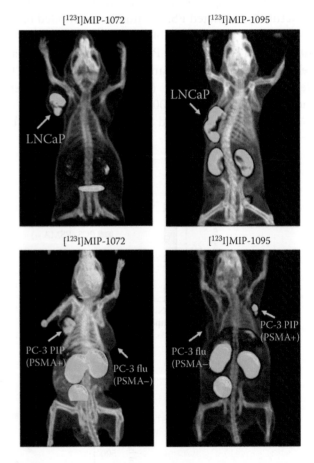

FIGURE 9.6   Selective targeting of PSMA *in vivo* with ^{123}I-MIP-1072 and ^{123}I-MIP-1095. Radiolabeled compound was injected into mice bearing LNCaP xenografts and imaged by SPECT/CT at 4 h (top) or mice bearing PC3 PIP (PSMA +) or PC3 flu (PSMA −) xenografts and imaged by SPECT/CT at 2 h (bottom). Each mouse was injected with approximately 1 mCi of radiolabeled compound at a specific activity >1000 mCi/µmol.

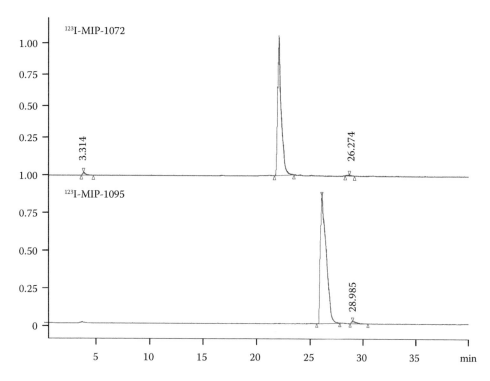

FIGURE 9.7 HPLC chromatograph of ^{123}I-MIP-1072 and ^{123}I-MIP-1095 demonstrating difference in retention time.

tergic neurotransmission. ^{123}I-MIP-1072 is cleared more rapidly from target and nontarget tissues and primarily through urinary excretion, whereas ^{123}I-MIP-1095 is cleared by both urinary and hepatobiliary routes. The differences in the clearance profiles do not seem to be related to metabolism as both compounds are stable in liver microsomes and blood plasma. ^{123}I-MIP-1095 exhibit longer retention (5 min) on reverse phase HPLC (Figure 9.7), compared to ^{123}I-MIP-1072 indicating that it is slightly more hydrophobic, which likely accounts for the dissimilarity in tissue pharmacokinetics.

9.1.7 Kinetics

The pharmacokinetics and tissue distribution of ^{123}I-MIP-1072 and ^{123}I-MIP-1095 were studied in conscious rats at 0.25, 1, 2, 4, 8, and 24 h postinjection (Figure 9.8 and Tables 9.1 and 9.2). Tissues were removed and counted in an automated γ-counter. Radiolabel was detected at varying levels in all tissues examined and decreased readily over time.

Analysis was performed using the mean ± SD for five rats at each time point. To determine the pharmacokinetic parameters, the mean values were modeled using WinNonlin® (St. Louis, Missouri) noncompartmental model.

The clearance from the blood and normal tissues of both agents differed likely due to the higher lipophilicity of MIP-1095. In the blood, after reaching C_{max} at the first time point, blood concentrations of both MIP-1072 and MIP-1095 readily declined with ^{123}I-MIP-1072 being cleared from the vascular compartment three times faster than ^{123}I-MIP-1095. ^{123}I-MIP-1095 was cleared from the vascular compartment and moved into the extravascular space more slowly than ^{123}I-MIP-1072 resulting in a greater total blood exposure. As anticipated, due to the mechanism of action of both compounds, uptake and exposure were greatest in the kidney of the rodent which expresses high levels of PSMA. After reaching C_{max}, 2 h postinjection, kidney concentrations of both compounds declined with ^{123}I-MIP-1072 being five times faster than ^{123}I-MIP-1095. The clearance of ^{123}I-MIP-1072 was renal while the clearance of ^{123}I-MIP-1095 was mixed renal and hepatobiliary.

9.1.8 Sensitivity

It is critical for an effective prostate cancer imaging agent, such as ^{123}I-MIP-1072, to be able to monitor changes in the size of a tumor as they regress in response to therapeutic interventions or continue to grow when they do not respond to therapy. ^{123}I-MIP-1072, in mice-bearing LNCaP xenograft tumors,

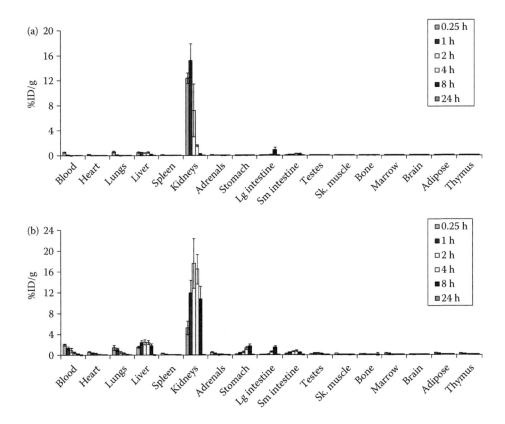

FIGURE 9.8 Tissue distribution of (a) ^{123}I-MIP-1072 and (b) ^{123}I-MIP-1095 in normal Sprague–Dawley rats at 0.25, 1, 2, 4, 8, and 24 h postinjection. Data are expressed as %ID/g.

demonstrates a linear correlation between tumor size and the % injected dose of ^{123}I-MIP-1072 in the tumor, as depicted in Figure 9.9. Based on these data, we examined the ability of ^{123}I-MIP-1072 to track changes in LNCaP tumor mass following treatment with paclitaxel. Figure 9.10 illustrates the mass of individual tumors and uptake of ^{123}I-MIP-1072, expressed as %ID. Tumor mass in the day 2 vehicle group ranged from 54 to 501 mg (mean = 222). As expected, tumors in the day 23 vehicle group, generally exhibited a larger mass as compared to the day 2 vehicle group, ranging from 134 to 890 mg (mean = 415).

After paclitaxel treatment, tumors generally exhibited a reduced mass ranging from 6 to 288 mg (mean = 161). Tumor uptake of ^{123}I-MIP-1072 was proportional to the changes in tumor mass (see Figure 9.11): decreased by paclitaxel treatment and increased in untreated mice, indicating that ^{123}I-MIP-1072 can indeed track changes in tumor size in response to therapeutic interventions (Hillier et al. 2011).

Table 9.1 Pharmacokinetic Analysis of ^{123}I-MIP-1072 in Selected Tissues

	Blood	Kidney	Liver
C_{max} (%ID/g/mL)	0.94	15.2	0.6
AUC_{0-inf} (h*%ID/g/mL)	0.78	38.7	6.2
Clearance (mL/h)	129	2.6	16.2
V_{ss} (mL)	1333	16.7	83
MRT (h)	2.4	2.3	5.7

Table 9.2 Pharmacokinetic Analysis of ^{123}I-MIP-1095 in Selected Tissues

	Blood	Kidney	Liver
C_{max} (%ID/g/mL)	2.1	17.7	2.6
AUC_{0-inf} (h*%ID/g/mL)	9.4	199	33.3
Clearance (mL/h)	10.4	0.1	3.0
V_{ss} (mL)	95.5	2.0	14.7
MRT (h)	6.3	5.9	6.2

Note: Analysis was performed using the mean ± SD for five rats at each time point. To determine the pharmacokinetic parameters, the mean values were modeled using the WinNonlin noncompartmental model.

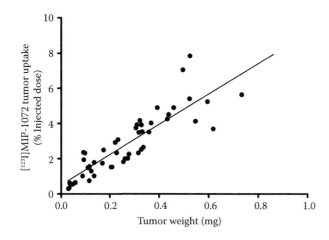

FIGURE 9.9   LNCaP tumor uptake of ^{123}I-MIP-1072 in untreated mice. Mice were administered ^{123}I-MIP-1072 intravenously and sacrificed after 1 h. Tumors were excised, weighed wet, and counted in an automated gamma counter. Data are expressed as %ID.

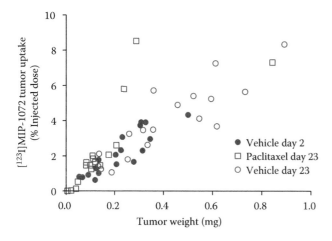

FIGURE 9.10   LNCaP tumor uptake of ^{123}I-MIP-1072 in paclitaxel or vehicle-treated mice. Mice were treated with 6.25 mg/kg paclitaxel or vehicle for 3.5 cycles of 5 days on/2 days off. On day 2 (vehicle) and day 23 (vehicle and paclitaxel), mice were administered ^{123}I-MIP-1072 intravenously and sacrificed after 1 h. Tumors were excised, weighed wet, and counted in an automated gamma counter. Data are expressed as %ID.

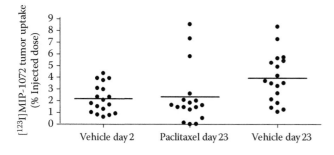

FIGURE 9.11   LNCaP tumor uptake (% injected dose) of ^{123}I-MIP-1072 in paclitaxel (day 23 of treatment) or vehicle-treated mice (day 2 and day 23).

9.1.9 Expected Impact on Clinical Care

Prostate cancer is second only to lung cancer as a leading cause of cancer-related fatalities and is the malignancy with the highest incidence of new cases in men in the United States. Treatment options for prostate cancer include surgery, radiotherapy, hormone therapy, and chemotherapy, as well as combinations of these modalities. National treatment guidelines for prostate cancer have been described and are updated regularly based on current levels of evidence (Mohler 2010). The choice of treatment modality is driven by clinical presentation of the disease or stage as well as the preferences of both physician and patient.

Each therapeutic option has it benefits but also presents the patient with unique risks to their quality of life. Due to the fact that many men survive for many years with the disease, consideration of the impact on quality of life is very important. In some low-risk patients, particularly older men, treatment may be deferred for some time, introducing the concept of watchful waiting or active surveillance. Active surveillance is a strategy of close monitoring for selected patients with very-low-to low-risk prostate cancer. The intent of active surveillance is to avert treatment unless disease progression occurs or a patient chooses treatment, in which case treatment with curative intent is undertaken. This approach has the potential of avoiding possible side effects of definitive therapy that may be unnecessary but the risk of progression or the missed opportunity for cure. A more definitive means of assessing progression, possibly with imaging, may make active surveillance a more robust option.

Accurate assessment of the extent and location of disease is critical as findings will directly influence treatment decision making. In the absence of metastatic disease and when the probability of organ-confined prostate cancer is high, the use of local therapy, such as radical prostatectomy or radiation therapy, is associated with a significant likelihood of cure. However, in the case where prostate cancer has spread to the lymphatic tissue or bone, the opportunity for cure with local therapy is significantly reduced. In both cases, knowing whether disease has spread or not outside the prostate will guide treatment and lead to optimal care and outcomes. It has been reported by Hull et al. (2002) that, at 10 years after radical prostatectomy, the progression-free probability is 92.2% for patients with cancer confined to the prostate, 71.4% for patients with extracapsular extension only, and 37.4%

for patients with seminal vesicle invasion. Therefore, accurate staging is required to select the appropriate treatment option in order to optimize survival while balancing quality of life.

The ability to image lymph node involvement with high sensitivity and specificity could alter patient management. In the case where pelvic lymph node dissection is warranted, usually in the case of patients for whom potentially curative treatments are planned, detecting nodal metastases could spare the patient operative lymphadenectomy. In contrast, in patients who can be deemed node-negative by scanning, nodal staging procedures would be unnecessary before potentially curative treatment.

Imaging plays a fundamental role in the clinical management of all cancer patients including screening, diagnosis, staging, monitoring response to therapy, and continued life-long follow-up. In addition, imaging is increasingly used in image-guided interventions such as biopsy and radiotherapy treatment planning. In the case of prostate cancer, imaging is used in all these applications. However, due to the biochemical nature of prostate cancer and its clinical and anatomical course, current imaging techniques have considerable limitations in accurately defining the presence and extent of disease. Today's clinical imaging modalities are commonly applied to the evaluation of the patient with prostate cancer. These include TRUS for an evaluation of the prostate gland and for guiding biopsy of the gland, CT or MRI for an evaluation of extra-capsular extension, metastatic spread to soft tissue such as lymph nodes and bone marrow, and nuclear medicine bone scans for an evaluation of metastatic spread to the skeleton. Proton magnetic resonance spectroscopy (MRSI) has also been shown to increase the accuracy of MRI in pretreatment prostate cancer detection, staging, and tumor volume estimation (Coakley et al. 2002; Hricak 2005). Overlaying the metabolic data from MRSI directly on top of MR images has been shown to help identify and localize prostate cancer (Mazaheri et al. 2011).

TRUS is a standard imaging technique used primarily for biopsy guidance and brachytherapy seed placement. TRUS is unreliable in differentiating normal prostate gland from cancer tissue, resulting in biopsies not specifically targeted to areas most likely to be malignant. The presence of disease as determined by cross-sectional imaging, such as CT or MRI, is based on size criteria. Such threshold criteria imply suspicion of an abnormal lymph node only, as the cellular (or

molecular) makeup of the lymph node cannot be determined using CT or MRI alone. As metastatic spread to lymph nodes is initially microscopic, the ability to detect early spread using cross-sectional imaging is of little value (Hövels et al. 2008). Bone scans are sensitive for bone metastasis but lack specificity, and changes in disease burden, especially regression, may not be easily derived from bone scans due to the persistent metabolic activity of the bone in cases of inflammation or healing.

PET imaging using ^{18}F-fluorodeoxyglucose (FDG) is an effective tool in the diagnosis of many cancers as it is readily taken up by hypermetabolic tumor cells according to the Warburg effect (Warburg 1931). It has been widely applied in oncology primarily as a staging and restaging tool that can guide patient management. However, because it accurately detects recurrent or residual disease, FDG-PET also has significant potential for assessing therapy response (Kelloff et al. 2005). However, it is essentially ineffective in detecting most prostate cancer due to the low glycolytic rate of the disease (Hofer et al. 1999; Liu et al. 2001). That said the successful application of molecular imaging in many cancers, other than prostate cancer, is exemplified by the use of FDG combined with PET imaging. It has been shown by Hillner et al. (2008) that FDG/PET imaging is associated with a change in the treatment or no-treatment decision in more than one-third of patients scanned. In this study, FDG/PET was associated more commonly with delineation of greater cancer burden or more sites of disease than with downstaging, across several tumor types. When specific changes between nontreatments and the addition or deletion of specific modes of therapy were included, FDG/PET was associated with a change in management in almost three-quarters of patients scanned. This example illustrates the importance of the sensitivity of molecular imaging, over anatomical imaging in a number of cancers. It stands to reason therefore that a molecular imaging agent with particular avidity for prostate cancer would hold the equivalent promise that FDG does in other types of cancer.

In designing an ideal imaging agent for a particular cancer, it is important to select a molecular target that is specific to the cancer in question and one that has consistently high expression levels per tumor cell throughout the natural progression of the disease and ideally does not alter this expression during therapy. In this manner, the signal from tumor cells signifies viable tumor mass and can be used to follow disease burden. This is particularly critical for assessing the presence of tumor and changes in tumor mass and spread, whether it is low-grade localized disease or response to treatment in a late-stage metastatic patient (Moyer and Barrett 2009). Early metastases to bone may be missed with bone scanning because this technique relies on the osteoblastic reaction (mineralization) rather than the actual tumor cells being detected (Toegel et al. 2006). A true and sensitive measure of the presence of disease in bone should lead to earlier detection of metastatic spread and a more robust measure of changes in tumor burden that could guide patient management. Likewise, visualization of the presence of tumor in lymph nodes that are otherwise normal by anatomical criterion would be a significant improvement in the current state of the art and lead to improved staging and therefore better patient management (Kelloff et al. 2009).

As detailed earlier, PSMA was selected as the molecular target since its expression is dramatically upregulated in poorly differentiated, metastatic, and hormone refractory carcinomas (Silver et al. 1997), as well as after androgen deprivation therapy (Wright et al. 1996) and in lymph node metastases (Sweat et al. 1998). Importantly, in normal tissues where expression is found such as in prostate, brain, kidney proximal tubules, and intestinal brush border membranes endogenous expression is lower. The function of PSMA in prostate cancer is unclear, although it is reported to play a role in tumor invasiveness (Yao et al. 2008).

Our results to date suggest that PSMA is a particularly attractive target for the molecular imaging of prostate cancer. We have successfully demonstrated that radiolabeled small molecule inhibitors of PSMA have the potential to localize prostate cancer through the common molecular imaging modalities such as SPECT (Barrett et al. 2010). These novel PSMA targeting tracers should impact clinical management by guiding appropriate patient-specific treatment strategies. It is likely that a sensitive and specific means of imaging prostate cancer and prostate cancer metastases would have a significant impact on the clinical management of the prostate cancer patient by providing greater certainty as to the presence and extent of disease during the course of the patient's life.

References

Akin O and Hricak H. Imaging of prostate cancer. *Radiol Clin N Am* 2007;45:207–222.

American Cancer Society. Cancer facts and figures 2010. Available at: http://www.cancer.org/Research/CancerFactsFigures/index. Accessed November 17, 2010.

Andriole GL, Grubb RL III, Buys SS et al. Mortality results from a randomized prostate-cancer screening trial. *N Engl J Med* 2009; 360:1310–1319.

Barinka C, Hlouchova K, Rovenska M et al. Structural basis of interactions between human glutamate carboxypeptidase II and its substrate analogs. *J Mol Biol* 2008;376:1438–1450.

Barinka C, Rovenska M, Mlcochova P et al. Structural insight into the pharmacophore pocket of human glutamate carboxypeptidase II. *J Med Chem* 2007;50:3267–3273.

Barinka C, Rinnova M, Sacha P et al. Substrate specificity, inhibition and enzymological analysis of recombinant human glutamate carboxypeptidase II. *J Neurochem* 2002;80:477–487.

Barrett JA, LaFrance ND, Armor T et al. Detection of recurrent metastatic prostate cancer in soft tissue, bone and prostate with ^{123}I-MIP-1072: A comparison with ^{111}In-Capromab Pendetide. *Proceedings of the Annual Congress of the European Association of Nuclear Medicine 2010*, Abstract: OP334.

Blakely RD, Robinson MB, Thompson RC, Coyle JT. Hydrolysis of the brain dipeptide N-acetyl-L-aspartyl-L-glutamate: Subcellular and regional distribution, ontogeny, and the effect of lesions on N-acetylated-alpha-linked acidic dipeptidase activity. *J Neurochem* 1988;50(4):1200–1209.

Catalona WJ, Smith DS, Ratliff TL et al. Measurement of prostate-specific antigen in serum as a screening test for prostate cancer. *N Engl J Med* 1991;324:1156–1161.

Coakley FV, Kurhanewicz J, Lu Y et al. Prostate cancer tumor volume: Measurement with endorectal MR and MR spectroscopic imaging. *Radiology* 2002;223:91–97.

Fair WR, Israeli RS, Heston WD. Prostate-specific membrane antigen. *Prostate* 1997;32(2):140–148.

Foss CA, Mease RC, Fan H et al. Radiolabeled small molecule ligands for prostate-specific membrane antigen: *In vivo* imaging in experimental models of prostate cancer. *Clin Cancer Res* 2005;11:4022–4402.

Hankey BF, Feuer EJ, Clegg LX et al. Cancer surveillance series: Interpreting trends in prostate cancer—Part I: Evidence of the effects of screening in recent prostate cancer incidence, mortality, and survival rates. *J Natl Cancer Inst* 1999; 91:1017–1024.

Hillier SM, Maresca KP, Femia FJ et al. Preclinical evaluation of novel glutamate-urea-lysine analogues that target prostate-specific membrane antigen as molecular imaging pharmaceuticals for prostate cancer. *Cancer Res* 2009;69(17):6932–6940.

Hillier SM, Kern AM, Maresca KP et al. [123]I]MIP-1072, A small molecule inhibitor of prostate specific membrane antigen, is effective at monitoring tumor response to taxane therapy. *J Nucl Med* 2011;52:1087–1093.

Hillner BE, Siegel BA, Liu D et al. Impact of positron emission tomography/computed tomography and positron emission tomography (PET) alone on expected management of patients with cancer: Initial results from the National Oncologic PET Registry. *J Clin Oncol* 2008;26:2155–2161.

Hövels AM, Heesakkers RA, Adang EM et al. The diagnostic accuracy of CT and MRI in the staging of pelvic lymph nodes in patients with prostate cancer: A meta-analysis. *Clin Radiol* 2008;63(4):387–395.

Chapter 9

Hofer C, Laubenbacher C, Block T, Breul J, Hartung R, Schwaiger M. Fluorine-18-fluorodeoxyglucose positron emission tomography is useless for the detection of local recurrence after radical prostatectomy. *Eur Urol* 1999;36:31–35.

Horoszewicz JS, Kawinski E, Murphy GP. Monoclonal antibodies to a new antigenic marker in epithelial prostatic cells and serum of prostatic cancer patients. *Anticancer Res* 1987;7:927–936.

Hricak H. MR imaging and MR spectroscopic imaging in the pretreatment evaluation of prostate cancer. *Br J Radiol* 2005; 78(Spec No 2):S103–S111.

Hull GW, Rabbani F, Abbas F, Wheeler TM, Kattan MW, Scardino PT. Cancer control with radical prostatectomy alone in 1,000 consecutive patients. *J Urol* 2002;167:528–534.

Israeli RS, Powell CT, Fair WR, Heston WD. Molecular cloning of a complementary DNA encoding prostate-specific membrane antigen. *Cancer Res* 1993;53:227–230.

Kelloff GJ, Hoffman JM, Johnson B et al. Progress and promise of FDG-PET imaging for cancer patient management and oncologic drug development. *Clin Cancer Res* 2005;11(8): 2785–2808.

Kelloff GJ, Choyke P, Coffey DS et al. Challenges in clinical prostate cancer: Role of imaging. *AJR Am J Roentgenol* 2009;192(6): 1455–1470.

Kozikowski AP, Nan F, Conti P et al. Design of remarkably simple, yet potent urea-based inhibitors of glutamate carboxypeptidase II (NALADase). *J Med Chem* 2001;44:298–301.

Liu IJ, Zafar MB, Lai YH, Segall GM, Terris MK. Fluorodeoxyglucose positron emission tomography studies in diagnosis and staging of clinically organ-confined prostate cancer. *Urology* 2001;57:108–111.

Maresca KP, Hillier SM, Femia FJ et al. Molecular targeting of prostate cancer with small molecule inhibitors of prostate specific membrane antigen (PSMA). *J Nucl Med* Meeting Abstracts, 2007;48:25P.

Maresca KP, Hillier SM, Femia FJ et al. A series of halogenated heterodimeric inhibitors of prostate-specific membrane antigen (PSMA) as radiolabeled probes for targeting prostate cancer. *J Med Chem* 2009;52:347–357.

Mazaheri Y, Shukla-Dave A, Muellner A, Hricak H. MRI of the prostate: Clinical relevance and emerging applications. *J Magn Reson Imaging* 2011;33(2):258–274.

Milowsky MI, Nanus DN, Kostakoglu L et al. Vascular targeted therapy with anti-prostate-specific membrane antigen monoclonal antibody J591 in advanced solid tumors. *J Clin Oncol* 2007;25:540–547.

Mohler JL. The 2010 NCCN clinical practice guidelines in oncology on prostate cancer. *J Natl Compr Canc Netw* 2010;8(2):145.

Moyer BM. and Barrett JA. Biomarkers and imaging: Physics and chemistry for noninvasive analyses. *Bioanalysis* 2009;1(2): 321–356.

Nan F, Bzdega T, Pshenichkin S et al. Dual function glutamate-related ligands: Discovery of a novel, potent inhibitor of glutamate carboxypeptidase II possessing mGluR3 agonist activity. *J Med Chem* 2000;43:772–774.

National Cancer Institute. Cancer advances in focus: Prostate cancer. Available at: http://www.cancer.gov/cancertopics/types/prostate. Accessed November 10, 2010.

Pinto JT, Suffoletto BP, Bergin TM. Prostate-specific membrane antigen: A novel folate hydrolase in human prostatic carcinoma cells. *Clin Cancer Res* 1996;2:1445–1451.

Pomper MG, Musachio JL, Zhang J et al. 11C-MCG: Synthesis, uptake selectivity, and primate PET of a probe for glutamate carboxypeptidase II (NAALADase). *Mol Imaging* 2002;1:96–101.

Silver DA, Pellicer I, Fair WR et al. Prostate-specific membrane antigen expression in normal and malignant human tissues. *Clin Cancer Res* 1997;3:81–85.

Smith-Jones PM, Vallabahajosula S, Goldsmith SJ et al. *In vitro* characterization of radiolabeled monoclonal antibodies specific for the extracellular domain of prostate-specific membrane antigen. *Cancer Res* 2000;60:5237–5243.

Smith-Jones PM, Vallabahajosula S, Navarro V et al. Radiolabeled monoclonal antibodies specific to the extracellular domain of prostate-specific membrane antigen: Preclinical studies in nude mice bearing LNCaP human prostate tumor. *J Nucl Med* 2003;44:610–617.

Speno HS, Luthi-Carter R, Macias WL et al. Site-directed mutagenesis of predicted active site residues in glutamate carboxypeptidase II. *Mol Pharmacol* 1999;55:179–185.

Sweat SD, Pacelli A, Murphy GP, Bostwick DG. Prostate-specific membrane antigen expression is greatest in prostate adenocarcinoma and lymph node metastases. *Urology* 1998; 52(4):637–640.

Thompson IM, Pauler DK, Goodman P et al. Prevalence of prostate cancer among men with a prostate-specific antigen level or 4.0 ng per milliliter. *N Engl J Med* 2004;350:2239–2246.

Toegel S, Hoffmann O, Wadsak W et al. Uptake of bone-seekers is solely associated with mineralization: A study with ^{99m}Tc-MDP, ^{153}Sm-EDTMP and ^{18}F-fluoride on osteoblasts. *Eur J Nucl Med Mol Imaging* 2006;33:491–494.

Tricoli JV, Schoenfeldt M, Conley BA. Detection of prostate cancer and predicting progression: Current and future diagnostic markers. *Clin Cancer Res* 2004;10:3943–3953.

Troyer JK, Beckett ML, Wright GL. Location of prostate-specific membrane antigen in the LNCaP prostate carcinoma cell line. *Prostate* 1997;30:232–242.

Troyer JK, Feng Q, Beckett ML, Wright GL. Biochemical characterization and mapping of the 7E11–5.3 epitope of the prostate-specific membrane antigen. *Urol Oncol* 1995;1:29–37.

Vallabhajosula S, Goldsmith SJ, Kostakoglu L et al. Radioimmunotherapy of prostate cancer using 90Y- and 177Lu-labeled J591 monoclonal antibodies: Effect of multiple treatments on myelotoxicity. *Clin Cancer Res* 2005;11: 7195–7200.

Wang X, Yin L, Rao P et al. Targeted treatment of prostate cancer. *J Cell Biochem* 2007;102:571–579.

Warburg O. *Metabolism of Tumors*. New York, NY: Smith, 1931.

Wright GL Jr, Grob BM, Haley C, Grossman K, Newhall K, Petrylak D, Troyer J, Konchuba A, Schellhammer PF, Moriarty R. Upregulation of prostate-specific membrane antigen after androgen-deprivation therapy. *Urology* 1996;48:326–334.

Yao V, Parwani A, Maier C, Heston WD, Bacich DJ. Moderate expression of prostate specific membrane antigen, a tissue differentiation antigen and folate hydrolase, facilitates prostate carcinogenesis. *Cancer Res* 2008;68(21):9070–9077.

10. Imaging Infection Based on Expression of Thymidine Kinase

Chetan Bettegowda, Youngjoo Byun, Catherine A. Foss, George Sgouros, Richard F. Ambinder, and Martin G. Pomper

Targeted Molecular Imaging. Edited by Michael J. Welch and William C. Eckelman © 2012 Taylor & Francis Group, LLC. ISBN: 978-1-4398-4195-2

Chapter 10

10.1 Introduction: Imaging of Infection

10.1.1 Overview of Imaging Infection: A Clinical Perspective

In this chapter, we focus on the problem of infection in human subjects, as several fine reviews describing infection in preclinical models—using a variety of modalities, including bioluminescence imaging—have recently appeared (Bleeker-Rovers et al., 2004; Doyle et al., 2004). Infection is generally diagnosed on clinical grounds, with fever and often the cardinal signs of inflammation localized to an obvious site, followed by positive bacterial cultures from that site or blood. Imaging is not usually part of the diagnostic algorithm. However, patients may present with fever of unknown origin or with an underlying illness that causes sterile inflammation, such as a collagen vascular disease, rather than infection by a foreign invader. As in the case of cancer and other diseases, imaging becomes important only once spatial data are required to guide management. Such a situation arises most commonly in the context of therapeutic monitoring. For example, the *site* of infection can determine whether a patient remains in the hospital to receive intravenous antibiotics for several weeks or if they may be sent home with an antibiotic taken orally. The *extent* of infection can determine whether surgical evacuation or debridement is necessary rather than more conservative management. Although there are no imaging methods to determine the type of organism involved, or in the case of bacteria, even whether a Gram-positive or Gram-negative species is the intruder, new mechanism-based approaches may enable such noninvasive discrimination.

The treatment of infection remains a major, worldwide public health problem, despite the development of powerful and specific antibiotics. Many infections remain resistant to antibiotic therapy and are becoming increasingly so, in part due to the inappropriate use of antibiotics that are currently available. For example, providing antibiotics up front for patients presenting acutely with presumptive pneumonia may cause errors in the actual diagnosis. Consequently, some patients who were never infected receive antibiotics unnecessarily, contributing further to the pool of resistant organisms (Welker et al., 2008). Antibiotics should be and often are withheld until appropriate cultures are obtained and the organism is identified. Apart from identifying the organism through culture, localization of the infection is also a key issue, as noted above. Occasionally, accurate localization of infection can aid in diagnosis and determination of the underlying pathogen. A combination of anatomic and molecular imaging techniques could synergize toward choosing appropriate antibiotics as soon as possible. Because of the obvious importance of rapid, adequate treatment of infection, there has been a tremendous effort in developing new imaging agents and technologies to localize and quantify the process (Bleeker-Rovers et al., 2004; Palestro et al., 2007; Lankinen et al., 2008; Gemmel et al., 2009). Nevertheless, the current standard remains the radiolabeled leukocyte study (Thakur et al., 1977). That procedure has been in place for over 30 years and is fraught with various problems as discussed below. The emerging standard is positron emission tomography with [^{18}F]fluorodeoxyglucose (FDG-PET). Despite its high sensitivity, however, FDG-PET is of relatively low specificity, and has been proffered as a way not just to image infection but also rheumatoid arthritis as well as its well-known role in imaging cancer (Beckers et al., 2004, 2006). For example, entities with a pressing need for rapid diagnosis and localization include infections involving the musculoskeletal and pulmonary systems. These are also two entities that are well suited to the 2′-fluoro-2′-deoxy-1-β-D-arabinofuranosyl-5-iodouracil (FIAU) imaging method that we discuss in subsequent sections, and are accordingly highlighted below.

10.1.2 Musculoskeletal Infection

We discuss musculoskeletal infections in the realm of infected joint prosthesis, infected fracture-fixation devices, or any chronic suppurative infection after fracture or joint infection of unknown origin. Magnetic resonance imaging is not widely used because prostheses or fixation interferes with the quality of the study due to generation of magnetic susceptibility artifact, and the signals are nonspecific. Various radionuclide scans have not demonstrated sufficient sensitivity or specificity in orthopedic situations (Parvizi et al., 2006). For example, according to one meta-analysis of imaging of potentially infected feet in diabetic patients, ^{99m}Tc-tagged leukocytes demonstrated a sensitivity of 86% and a specificity of only 84% (Capriotti et al., 2006). Managing infections after operative reconstruction and internal fixation is an ongoing dilemma (Rightmire et al., 2008). These are deep infections that generally cannot be cured in the presence of hardware; however, removal of hardware can complicate not only the treatment of the infection, but also the underlying fracture that it is intended to repair. The current state of the art is generally to attempt to suppress the infection until the fracture heals; however, one recent study suggests that this method may not provide optimal results and that it is advisable to remove hardware even with these deep infections (Rightmire et al., 2008). Such infections generally are due to methicillin-resistant *Staphylococcus aureus*. This is an important problem since about 5% of the two million fracture-fixation devices that are placed annually in the United States, a number amounting to about 100,000 such devices, become infected. The average cost to treat such infections is of the order of $15,000 per case. Regarding potentially infected orthopedic prostheses, the situation is equally serious. Approximately 600,000 orthopedic prostheses are placed annually, 2% of which, or 12,000, become infected. Treatment of such cases amounts to about $30,000 per case (Darouiche, 2004). A patient with an orthopedic prosthesis may present with a painful joint. The differential diagnosis for such a patient would be either an infected joint or a loose prosthesis. Obviously, those two entities have vastly different approaches to therapy such that, perhaps more than in any other orthopedic situation, an accurate way to identify infection rather than sterile inflammation becomes critical. Although joint prosthesis and fracture-fixation devices represent two subcategories of surgical implants, notably one half of the two million nosocomial infections that occur each year in the United States are indeed associated with indwelling devices (Darouiche, 2004). Another orthopedic infection is that of the diabetic foot. Fifteen percent of patients with diabetes eventually develop osteomyelitis of the foot (Meller et al., 2007). Although this discussion focuses on orthopedic prosthesis and fracture-fixation

devices, automatic defibrillators, pacemakers, mammary and penile implants, and other implanted devices all become infected with a certain, not insignificant frequency.

10.1.3 Pulmonary Infection

According to the Mayo Clinic, about 60,000 Americans die each year of pneumonia. Acute lower respiratory infections are responsible for more disease and death than any other entities in the United States (Mizgerd, 2008). A differential diagnosis of lung lesions, whether they are malignant or infectious, is critical, because of the obvious therapeutic implications. Once infection is established, monitoring appropriate treatment of the infection can be helpful in patient management. Since the advent of antiretroviral therapy (ART), bacterial pneumonia has replaced *Pneumocystis carinii* pneumonia as the most common HIV-associated opportunistic infection in the United States (Davis et al., 2008). About 8.5 cases of bacterial pneumonia per 100 person years in women infected with HIV are currently seen. In all aspects of HIV-related care, management of opportunistic respiratory infections remains among the most challenging. Localization and therapeutic monitoring of such lesions through targeted imaging could help offset some of the difficulties in managing these patients.

10.1.4 Imaging Infection

Although the tagged white blood cell (WBC) study has been the clinical standard for imaging infection for over 30 years, imaging infection is becoming more sophisticated as targeted agents are beginning to emerge (Bleeker-Rovers et al., 2004; Palestro et al., 2007). There is a shift away from the anatomic techniques of computed tomography (CT) and magnetic resonance (MR) toward functional techniques, such as tagged WBC imaging, but it is the cumbersome nature of the latter that prevents its more widespread use. For example, for osteomyelitis, different imaging modalities have widely differing sensitivities and specificities (Table 10.1; El-Maghraby et al., 2006). Although the functional techniques tend to fare well, with gallium scintigraphy showing a 92% specificity and tagged WBCs showing sensitivities and specificities of the order of 80–90%, these techniques are not used as widely as their purported utility indicates. In the case of tagged WBCs, the method requires autologous retransplantation of radiolabeled cells. About 50 mL of blood

Table 10.1 Imaging Osteomyelitis

Modality	Sensitivity (%)	Specificity (%)
CT	65–75	65–75
MR	82–100	75–96
Bone scan ([99mTc]MDP)	73–100	75
WBC (111In or 99mTc)	80–90	80–90
67Ga-citrate	67–70	92 (with SPECT)
FDG-PET	90–100	70–80

are removed from the patient, the leukocytes are harvested and labeled with 111In, and then are readministered to the patient. Of course, this can be hazardous as misadministration can occur or contamination of the sample with microorganisms can take place, but the technique also fails from the fact that it will localize not only to sites of infection but also those of sterile inflammation, to which leukocytes also migrate. Imaging also occurs at time points somewhat distant from the reinjection of leukocytes, generally out to 24 h after readministration. The technique does not work well in chronic processes or in infections of the axial skeleton. Indium-111 carries a high radiation burden, labeling can be unstable, the cells can become sequestered within the lungs and significant radioactivity is present within the gallbladder and large bowel at >3 h postinjection (El-Maghraby et al., 2006). Nevertheless, a standardized protocol for using tagged WBCs has been in place for many years and results can be compared between institutions. This technique may also be limited in severely immunocompromised patients in whom infection is not always accompanied by inflammation. Consequently, other techniques have evolved in an attempt to overtake the tagged WBC study.

An ideal infection imaging agent should have the following characteristics: (1) efficient accumulation and retention at inflammatory foci; (2) rapid clearance from background; (3) easy, low-hazard preparation, and (4) wide availability at low cost (Bleeker-Rovers et al., 2004). Additionally, it is desirable to have low accumulation in the blood pool or in the bowel. No imaging agent, including those most modern, fulfills those criteria. Newer imaging agents involve sophisticated targets and techniques such as radiolabeling of specific antimicrobial peptides such as IMP-178 (Gratz et al., 2001), which binds to E-selectin, or the use of radiolabeled antibodies, cytokines, or complement factors (Bleeker-Rovers et al., 2004). Most of those agents are still preclinical, due to suboptimal sensitivities and/or specificities (Benitez et al., 2006). Antibiotics

themselves have been radiolabeled, such as ciprofloxacin, which has been commercialized as Infecton™; however, organisms that are resistant to ciprofloxacin may provide false-negative results. In one case in which 25 patients with foot ulcers were studied, six demonstrated false-negative images using that agent (Dutta et al., 2006). Other false negatives, such as in renal abscess (Adams et al., 2006) or in sterile inflammation, have also been reported (Sarda et al., 2003).

One technique worthy of more detailed mention is FDG-PET, which is undergoing trials in a variety of different infectious and inflammatory processes (de Winter et al., 2001; Meller et al., 2002, 2007; El-Haddad et al., 2004; Makinen et al., 2005; Goebel et al., 2007; Keidar et al., 2007). The keyword of the previous sentence is "inflammatory," underscoring the lack of specificity of FDG-PET, the premier metabolic imaging technique for oncology, for imaging infection. As a molecular imaging technique, FDG-PET enables detection of physiological processes, such as infection or inflammation, at a much earlier stage than the standard, anatomic techniques. Furthermore, PET is not affected by metallic implants and does not show the same limitations, for example, in bone infections, for which the three-phase bone scan and tagged WBC studies tend to be limited within the axial skeleton. PET studies can also be performed fairly readily after administration of isotope, in about 2 h. Nevertheless, because FDG uptake is due to an increase in glycolysis in tissues, it can be seen in inflammatory, granulomatous, or frankly infectious processes. In one animal study, 29% of FDG uptake within tumor was actually within associated macrophages and granulation tissue (Beckers et al., 2004). FDG uptake is somewhat dependent on serum glucose level (Meller et al., 2007). FDG-PET has proved utility in chronic osteomyelitis, vasculitis, and in a variety of other infectious and inflammatory entities in the references noted above; however, it is also sequestered by healing infections, particularly within bone (Lankinen et al., 2008). The timing of FDG infection imaging is critical, such that it cannot be performed until several weeks after surgery, after which most of the inflammation should have subsided. Despite those drawbacks, FDG-PET is considered the method of choice for chronic osteomyelitis, particularly of the axial skeleton, and is likely to become the clinical standard (Meller et al., 2007). One may argue that FDG-PET imaging for infection and inflammation is on the right track. It emphasizes use of a low-molecular-weight agent that does not have significant blood pool circulation and localizes to the infectious or the inflammatory foci in a mechanism-based fashion. The FIAU-based method that we discuss can be considered analogous to FDG-PET; however, in part because of the differences in substrate specificity of human versus bacterial TK, FIAU is not as significantly sequestered by mammalian cells and will therefore not be seen in inflammation and will be detected only in the presence of an organism that possesses the appropriate bacterial (or viral or possibly fungal) TK.

10.1.5 New Approaches to Bacterial and Viral Imaging

While using bacterial and viral TK-based approaches to imaging will be discussed in detail below, there are other targets that have not been exploited that are summarized in Bray et al. (2010), particularly in the case of imaging viruses. One must merely review the life cycle of a virus to uncover steps that may be imaged. Those steps include receptor-based fusion of virus to the impending host cell surface; transcription and replication of viral nucleic acids; maturation of virally encoded proteins through cleavage; and, virion assembly and coat protein antigen expression. For example, the protein kinase UL97 of cytomegalovirus, which is the target of a new antiviral drug (mirabavir), could be used to concentrate specific, radiolabeled nucleosides for imaging. In parallel with the development of ART, one could design a radiolabeled HIV protease inhibitor for imaging. Such compounds may serve as suitable imaging agents as the drugs on which they would be based often have subnanomolar affinities for their target, as required for high-sensitivity

imaging of low-capacity targets. Note that many of these targets are enzymes, that is, kinases and proteases. The amplification provided by such species is an important mechanism by which to improve the sensitivity of the technique.

10.1.6 Pressing Needs and Challenges in Imaging Infection

Arguments provided above indicate the necessity of determining the spatial localization and extent of infection for guiding therapy. Perhaps the most important advancement that could occur in this field would be the development of an imaging agent that could distinguish infection from sterile inflammation. FDG is notoriously incapable of doing that, although it is touted as a new agent for this indication, which could replace tagged WBCs. Leveraging targets and mechanisms unique to foreign invaders is the likely way to meet the challenge of specificity. While some have so far failed, for example, targeting DNA gyrase with radiolabeled analogs of ciprofloxacin, others show promise—particularly those that exploit decades of medicinal chemistry for other purposes, namely the development of cytotoxic nucleoside prodrugs for cancer and antiviral therapy. Even in that instance, the application of such well-established reagents must overcome certain hurdles, including those related to specificity, cost, synthetic and radiosynthetic availability, and logistics of dissemination, to achieve widespread acceptance. Currently, the use of various radiolabeled TK substrates for imaging is poised to address some of those challenges.

10.2 Target Protein: Thymidine Kinase and Its Various Families

10.2.1 Classification and Biology of Thymidine Kinases

A readily available pool of deoxyribonucleotides is necessary for the maintenance of genetic stability for all living organisms. There are two pathways by which that pool is generated: the *de novo* pathway through a carefully regulated implementation of ribonucleotide reductase, and the salvage pathway, initiated by deoxyribonucleoside kinases (dNKs), which catalyze the first phosphorylation step of free deoxyribonucleosides *en route* to incorporation into DNA (Welin et al., 2004). Although several steps are involved, such as membrane transport of the deoxyribonucleoside,

and two phosphorylation steps after the action of dNK, it is that first phosphorylation step that is believed to be rate limiting and has been the target of a significant antiviral and, to a lesser extent, anticancer drug development effort. The first phosphorylation step has also more recently been exploited for gene therapy, in which nucleoside analogs behave as prodrugs that upon phosphorylation are transformed to toxic agents that can kill cells harboring the transforming enzyme. The goal and concurrent limitation of such therapy is to assure that only malignant cells express the transforming kinase. As the substrate specificities of the various dNKs are somewhat overlapping, this has been and continues to be a challenge

Chapter 10

for "suicide" gene therapy, which is discussed in more detail in Section 10.4.2.2.

As new and increasingly relevant x-ray structures become available, that is, those with substrate cocrystallized in the active site, a picture of the various dNKs and their relationships begins to emerge. Notably, x-ray structures of the HSV1-TK, human TK1 (hTK1, EC 2.7.1.21), the *drosophila* dNK, and those of various bacterial species, including *Ureaplasma urealyticum* and two *bacillus* species, have contributed significantly to our understanding of the phylogeny, structure, and function of the dNKs (Johansson et al., 2001; Eriksson et al., 2002; Welin et al., 2004). Deoxyribonucleoside kinases can be divided into two families, one including deoxycytidine kinase (dCK), deoxyguanosine kinase (dGK), human TK2 (hTK2), and the herpes virus TKs and the other comprised of hTK1 and homologous TKs from other species, notably bacteria and poxvirus (Sandrini et al., 2006; Segura-Pena et al., 2007). In mammalian cells, dCK and hTK1 are located within the cytoplasm, while dGK and hTK2 are within mitochondria. Human TK1 and hTK2 not only differ in the intracellular location but also have disparate primary sequences, expression levels, and substrate specificities. Human TK1 activity is largely linked to the cell cycle, peaking during S-phase and nearly quiescent within resting cells. It is thought that hTK2 is constitutively expressed throughout the cell cycle and is the predominant TK that is physiologically active in nonproliferating and resting cells. Among endogenous substrates, hTK1 phosphorylates thymidine (dT) and deoxyuridine (dU) exclusively, while hTK2 phosphorylates dT, dU, and dC (Perez-Perez et al., 2008). Several reviews detail the substrate specificities and kinetics of the various TKs, relevant highlights of which are further described below (Johansson et al., 2001; Eriksson et al., 2002; Kosinska et al., 2007; Segura-Pena et al., 2007; Perez-Perez et al., 2008; Deville-Bonne et al., 2010).

Germane to the ensuing discussion on FIAU-based imaging, while both hTK1-like and non-hTK1-like enzymes have shown the capacity to phosphorylate FIAU, hTK2 is thought to mediate the effects seen in humans. While both hTK1 and hTK2 can phosphorylate FIAU, hTK2 is much more efficient. The K_m value—an estimate of binding affinity (Section 10.3.2)—for FIAU with hTK1 is reported to be 140 μM, while that for hTK2 was 4 μM (Wang and Eriksson, 1996). In addition, Lee et al. (2006) discovered that human equilibrative nucleoside transporter 1 (hENT1) is responsible for transporting FIAU into the mitochondria, where

in its triphosphorylated form it can cause mitochondrial DNA depletion and the associated toxicities seen in the clinical trials. Interestingly, rodents do not have a functional equivalent of hENT1, which could explain why no toxicity was seen in preclinical studies, despite doses that exceed 1000-fold the amounts used in the human trials (Lee et al., 2006).

To our knowledge, there are no published reports on the crystal structure of hTK2. Curiously, despite occasionally being grouped together, the amino acid sequence of hTK2 is widely different from that of HSV1-TK, although both have the capacity to phosphorylate FIAU. When combining the observations that both hTK1-like and non-hTK1-like bacterial TKs, HSV1-TK, and hTK2 all are capable of robustly phosphorylating FIAU, it becomes difficult to predict, based on sequence or structure alone, which microbes are capable of incorporating radiolabeled FIAU and being imaged. However, one can begin to hypothesize based on the existing experiences and data *in vivo* such that the potential limitations and opportunities for FIAU (TK)-based bacterial imaging can be evaluated.

Our targets for imaging infection are the TKs of the salvage pathway, which are small, allosteric proteins of between 17 and 45 kDa in size (Deville-Bonne et al., 2010), the primary ones of which, hTK1-like (for bacteria) and the herpes virus TKs, obey Michaelis–Menten kinetics (Section 10.3.2), which has important implications for the sensitivity of detection using external imaging devices. The imaging agents are therefore suitably derivatized nucleoside analogs, such as radiolabeled versions of FIAU, which serve as substrates for these TKs (Figure 10.1). It is often stated that the reason for which HSV1-TK can be used for gene therapy is because its substrate specificity is orthogonal to that of the hTKs, but that is not entirely true. The implication of that is a decreased specificity and higher background signal upon imaging, as only a limited number of nucleoside analogs can be phosphorylated by these somewhat fastidious enzymes. Similarly, a challenge is posed by the fact that bacterial TKs are homologous to hTK1—although several important and exploitable differences exist—in terms of generating specific, TK-based bacterial imaging agents.

10.2.2 hTK1 and Bacterial TKs

Of all the dNKs discussed, hTK1 has the most restricted substrate specificity. As stated above, it is capable only of phosphorylating dT and dU with ATP and dATP,

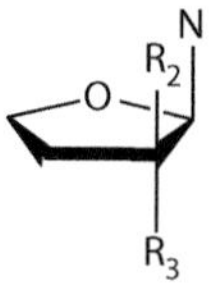

FIGURE 10.1 Natural (dU and dT) and synthetic uracil derivatives that form the basis for imaging agents for thymidine kinase.

Compound	R_1	R_2	R_3
Deoxyuridine (dU)	H	H	H
Thymidine (dT)	CH_3	H	H
FAU	H	F	H
FIAU	I	F	H
FEAU	CH_2CH_3	F	H
FMAU	CH_3	F	H

the preferred phosphate donors. The mechanism for the reaction is shown in Figure 10.2. Although cloned in 1984 (Bradshaw and Deininger, 1984), the crystal structure was not solved until 20 years later (Welin et al., 2004), with additional information about its mechanism deriving from cocrystallization studies of bacterial analogs of the enzyme with dT and deoxythymidine triphosphate (dTTP), respectively, representing quiescent (closed) and active (open) states (Kosinska et al., 2007). Interestingly, in that recent report (Kosinska et al., 2007), the feedback inhibitor dTTP occupies the phosphate donor rather than acceptor site, and causes changes in conformational and tertiary structure, that is, a tetramer is formed. Overall, bacteria have a smaller group of dNKs than do mammalian cells. Most bacteria have a TK that resembles hTK1 with respect to both sequence and structure. Overall, the similarity of bacterial TK to hTK1 varies based on the species, with a range from 13% to 28% homology (Sandrini et al., 2006). Across multiple bacterial genera, TK appears to be fairly well conserved with significant conservation of the catalytic domain in nearly all sequenced bacteria (Bettegowda et al., 2005). There are

a few pathogens that have evolutionarily lost the *tk* gene, namely species in the *pseudomonas* and *mycobacterium* families (Sandrini et al., 2006). There are also one or two non-hTK1-like dNKs in most Grampositive bacteria. Those dNKs are similar to dCK or dGK but likely do not account for a significant amount of the phosphorylation of pyrimidine nucleosides as catalytic efficiencies against such substrates are quite low, often below the limit of detection of the assay system: the k_{cat}/K_m values, which describe the catalytic efficiency of an enzyme for a substrate (Section 10.3.2), for dT to dGK and dCK are $<2 \times 10^2$. Drug development efforts at TK-based bacterial inhibition have focused on differences between hTK1 and the bacterial TKs.

Flexibility of hTK1 has made it difficult to crystallize, and so a truncated form was initially studied in a report that concurrently described the structure of full-length *Ureaplasma* TK, cocrystallized with dTTP (Welin et al., 2004). Each structure is a tetramer with each subunit containing two domains—one α/β domain and a zinc-containing domain, with the active site present between the domains (Figure 10.3a). The

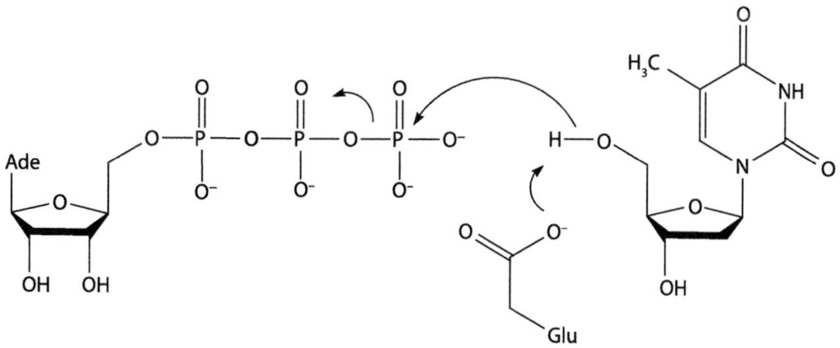

FIGURE 10.2 The mechanism of action for deoxyribonucleoside kinases. Phosphate donor ATP is on the left showing the substrate dT on the right. Glutamate serves as a general base. (Modified from Eriksson, S. et al. 2002, *Cell Mol Life Sci* 59: 1327–46.)

Chapter 10

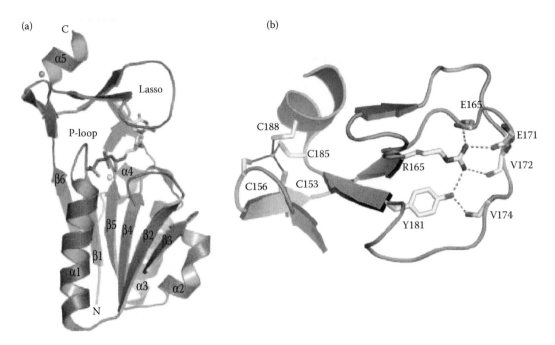

FIGURE 10.3   (a) The subunit structure of human TK1 (hTK1) with dTTP in the active site and color coded according to atom type. Mg²⁺ is in yellow (near the center) with Zn²⁺ in gray (upper left). The full hTK1 is a tetramer. Note the presence of two domains, one α/β domain and a small zinc containing domain (the lasso region). The active site exists between those domains. (b) Detail of the lasso region of hTK1, with coordinating zinc on the left and substrate binding on the right. (From Welin, M. et al., 2004, *Proc Natl Acad Sci U S A* 101: 17970–5. With permission.)

zinc-binding domain contains a lasso-shaped structure that covers the DNK site in hTK1 (Figure 10.3b). Although the hTK1 and *Ureaplasma* TKs share only 30% sequence identity, which is in the range of other degrees of sequence homology between hTK1 and bacterial TKs, their structures are quite similar. With respect to other dNKs, that is, the viral and hTK2-like enzymes, the α/β domain differs in strand order and in the positions of the helices, with the substrate binding site formed from the two helix pairs and the lid structure missing in the non-hTK1 family. Those findings, along with the presence of the lasso-like loop to interact with the deoxynucleoside and a zinc-binding domain, suggest a different evolutionary origin of hTK1 compared to the other TK families. ATP is also capable of generating a conformation change in binding to TK-like enzymes, as discussed in a report that used P¹-(5′-adenosyl)P⁴-(5′-(2′-deoxy-thymidyl)) tetraphosphate (TP4A) to elucidate the ATP-binding mode of both hTK1 and the TK of *Thermotoga maritime* (Segura-Pena et al., 2007). That report describes closed and open conformations of the enzyme, with a more open conformation seen in the *Thermotoga*–TP4A complex. That was presumably in order to provide space for the adenosine moiety. The closed versus open structures of the enzyme are believed to recapitulate the dynamic changes that occur in the tetramer during the catalytic cycle, as it is the weak association of the dimers of the tetramer that comprise the ATP-binding site.

10.2.2.1   Structure–Activity Relationships Surrounding hTK1

Ureaplasma does not have a *de novo* pathway and must rely strictly on salvage to generate precursors for DNA synthesis. Consequently, the *Ureaplasma* TK provides a good target for antibacterial drugs. However, with it established that hTK1 bears striking similarity to bacterial TKs, how can one develop drugs or imaging agents that specifically target the bacterial TKs? Unfortunately, hTK1 has a very narrow substrate specificity. Its active site is much smaller than HSV1-TK, which is far more promiscuous. All polar groups of the base are maintained in tight contact with the protein through hydrogen-bonding interactions. A halogen or ethyl substituent may be tolerated at the 5-position of the base, but bulkier or polar groups are not, as supported by the snug fit of the methyl group of dT. Despite hydrogen bonding of the N3 position of the base to a carbonyl group of the enzyme, bulky substituents have been reported at that location. The aforementioned lasso domain, which covers the nucleoside-binding pocket, may not close completely under such circumstances. As for the sugar moiety, the 2′-position must not be substituted, in keeping with the inability for ribonucleosides to bind, but the 3′-position can be substituted, as in the antiviral drug 3′-azido-3′-deoxythymidine, AZT. Welin et al. (2004) further suggest that the 5-position methyl group may be modified to discriminate between hTK1 and *Ureaplasma* TK, as in the former enzyme that methyl group is in contact with the β-carbon of Thr, while in the latter it is near a Ser residue. However, a subsequent report by the same group that attempted to create compounds designed specifically against *Ureaplasma*, which were 3′-substituted dT analogs based on AZT, showed that although they could generate compounds that were more readily phophorylated by *Ureaplasma* TK than by hTK1, the IC_{50} values of those same compounds were still lower for hTK1, a seemingly paradoxical effect, previously demonstrated (Lin et al., 2010). The somewhat more promiscuous nature of *Ureaplasma* TK still leaves the door open to more rational antibacterial drug design, and could serve as a model for developing agents to image bacterial rather than human versions of TK1. Attempts to exploit the structural divergence of human and bacterial TKs have also involved synthesis of lipophilic dT phosphate-mimicking compounds as new therapeutics for *Bacillus anthracis* (Byun et al., 2008). However, most of those compounds proved inactive, in part due to poor transport into the bacterium.

A recurrent question in imaging of infection—particularly of bacterial infection—is with regard to whether or not specific bacterial species can be detected by a well-designed, specific imaging agent. Although arguably a more daunting task than designing an agent for bacterial rather than hTK, one approach may again be to study differences between the TKs of various bacteria. For instance, Gram-positive bacteria have a TK that resembles hTK1 more closely than do the Gram-negatives (Sandrini et al., 2006). As the dNKs of the non-hTK1-like family are less selective than those similar to hTK1, in principle one could design agents that target various bacteria based on the composition of their cell walls. In a study of the phylogeny of bacterial TKs, Sandrini et al. (2006) arrived at the interesting result that no TKs of Gram-negative bacterial origin were found in the group of Gram-positive TKs investigated, while some Gram-positive species did possess Gram-negative TKs. Although there was some overlap in that small sample ($n = 20$), it does suggest that

structural differences do exist and might be exploited to differentiate various types of bacteria. Perhaps of greatest interest is the fact that certain bacteria have no TK-like gene, indicating that they will not be able to be imaged by the FIAU method. For example, while *Mycobacterium tuberculosis* does not possess a TK, it does have a related kinase that might be leveraged with certain nucleoside analogs (Byun et al., 2008).

10.2.3 Viral TKs

As stated above, although they have been considered as a family unto themselves (Eriksson et al., 2002), the viral TKs are also classified along with the mitochondrial hTKs, such as hTK2, on the basis of sequence homology. These TKs are not cell cycle dependent. The substrate specificity of viral TKs is less restricted than hTK1 and there is no stereospecificity to the phosphorylation of their substrates. Unlike hTK1, these dNKs are homodimers and are structurally much different. There are three conserved sequence motifs of importance: (1) a P-loop motif that binds the donor phosphates (of ATP); (2) a Glu–Arg–Ser sequence in which Glu binds to Mg^{2+} with Arg having a role in catalysis; and (3) the LID domain (The LID domain is the flexible region between the α7 helix and the α8 helix, which contains conserved arginines at the phosphate donor site and undergoes a conformational change upon binding to the phosphate donor group), which closes onto the bound phosphate donor. Because of their use in gene-directed enzyme prodrug therapy (GDEPT), there have been efforts to enhance the efficiency of the first phosphorylation step of these enzymes through site-directed mutagenesis or directed evolution (Deville-Bonne et al., 2010). That has been particularly true of the HSV1-TK, but hTK2 has also been so modified (Gerth and Lutz 2007). That the first phosphorylation step is so inefficient, a point that has come under scrutiny recently (Deville-Bonne et al., 2010), explains why pyrimidine and acycloguanosine analogs are not very potent drugs, which, in turn, suggest why some may be better than others for imaging. Modification of HSV1-TK and hTK2 to provide more selective therapeutic and imaging agents, that is, those with higher affinity to the mutated protein in order to avoid binding to hTK1, for example, is relevant to the use of gene-tagged cells for GDEPT or cell trafficking studies or to enable administration of reporter molecules such as radiolabeled nucleoside analogs to patients already undergoing antiviral therapy. However, insights into substrate selectivity gained from developing those mutated proteins may provide better therapeutic and imaging agents for bacterial or viral kinases. For imaging bacterial and viral infection, we are not imaging exogenously gene-tagged cells; our focus is on the use of agents selective for endogenous viral TK over hTKs, if and when possible.

10.2.3.1 Epstein–Barr Virus and Kaposi's Sarcoma Herpesvirus Thymidine Kinase

Epstein–Barr virus (EBV) is a gammaherpesvirus that is the cause of infectious mononucleosis and is associated with certain malignancies, including endemic Burkitt's lymphoma, nasopharyngeal carcinoma, and 10–20% of gastric cancer (Shibata et al., 1991; Liu et al., 1992). The EBV-TK was first described by Chen et al. (1978). As with the alphaherpesviruses (HSV1-TK), the gammaherpesviruses serve as targets for antiviral and anticancer therapies (Ambinder and Cesarman, 2007).

The EBV-TK has a narrower substrate specificity than does HSV1-TK, with dT analogs serving as the most suitable substrates, and possession of a minor thymidylate kinase activity. EBV-TK can phosphorylate dT but not dC. Acycloguanosines such as ganciclovir (GCV) and acyclovir are substrates for the EBV-TK and an EBV-encoded protein kinase (Moore et al., 2001). Recent evidence suggests that the protein kinase may be more important physiologically (Meng). Studies in tumor cells engineered to express the EBV TK and lacking the remainder of the viral genome, demonstrated that FIAU is phosphorylated directly by the EBV TK. Kaposi's sarcoma herpesvirus (KSHV) also encodes a TK and protein kinase and both phosphorylate GCV, although the protein kinase appears to be more active (Cannon et al., 1999). Further discussion on EBV-TK, with respect to imaging and therapy, is reserved for Section 10.4.

10.3   Imaging Thymidine Kinase: Focus on Bacterial Infection

10.3.1   Enzymes as Imaging Reporters

Molecular-genetic imaging can be accomplished in gene-tagged cells, whereupon a reporter gene is transfected to a cell of interest and a reporter probe is used to image that cell—or tumors derived from it—*in vivo* through radionuclide or optical methods. Occasionally, one may exploit the endogenous expression of a reporter gene, such as a bacterial or viral TK, for imaging. Either way, of the possible imaging targets, that is, enzymes, receptors, transporters, nucleic acids, or other antigens, enzymes tend to make the best imaging reporters due to their ability to provide signal amplification for added sensitivity. For example, attempts to use the dopamine receptor outside of the central nervous system as an imaging reporter have largely been abandoned due to the lack of such amplification, along with the paucity of imaging agents with pharmacokinetics suitable for imaging outside of the brain. Although certain transporters, such as the sodium-iodide symporter or norepinephrine transporter (Mandell et al., 1999; Anton et al., 2004; Dwyer et al., 2005), have been used to good effect, it has been the HSV1-TK and its mutants, reviewed elsewhere, which have been most aggressively pursued (Serganova et al., 2007). That is because the HSV1-TK is a classic non-saturable system, with the potential to provide high sensitivity for detecting transfected cells with an appropriate nucleoside analog (Bray et al., 2010). Mutants of the HSV1-TK, such as the HSV1-sr39TK

and the HSV1-A167Ysr39tk, were developed to enable imaging with substrates that provide higher catalytic efficiency or have better pharmacokinetic properties for imaging, and can be given to patients who are concurrently receiving antiviral nucleoside therapy, as often occurs in immunocompromised patients undergoing bone marrow transplantation (Gambhir et al., 2000; Likar et al., 2008). As mentioned above, other mutants of the HSV1-TK, and more recently of hTK2, have been generated to alter the substrate specificity, perhaps allowing the use of substrates not phosphorylated by mammalian TKs. Recently, workers at the Memorial Sloan-Kettering Cancer Center, who pioneered the use of HSV1-TK for molecular-genetic imaging, have developed nonimmunogenic imaging reporters based on the hTK2 and dCK to avoid this problem, which often confounds HSV1-TK-based imaging and therapy (Ponomarev et al., 2007; Likar et al., 2010). The greatest challenge in developing enzyme-based molecular-genetic imaging systems is the same as for developing selective antibacterial and antiviral agents, or at least those based on nonmammalian TKs (to avoid multidrug resistance): nucleoside substrates for all these TKs, despite insignificant differences at their respective active sites, are similar and overlapping. It is therefore difficult to design an imaging agent that is specific to bacterial or viral TK.

So how can one use radiolabeled nucleosides such as [^{124}I]FIAU for imaging infection when FIAU may serve as a substrate not only for hTK1 (and homologous bacterial TKs) but also hTK2? One of the primary reasons is that we are generally imaging an aggregation of cells into an abscess, rather than target diffusely spread throughout the cytoplasm, as in the case of hTK1. We determined that for preclinical SPECT imaging we could visualize on the order of 2×10^6 cells per cubic centimeter in an animal model of infection (Bettegowda et al., 2005). In other words, approximately, one in 500 cells must express the (endogenous) reporter to be imaged by that modality. Another reason is that the highly polar nucleosides: while they can be transported to the cytoplasm where hTK1 awaits, they cannot easily transgress the inner mitochondrial membrane, the location of hTK2, suggesting that intracellular compartmentalization of target may provide another avenue for the pursuit of selective drugs and imaging agents (Ponomarev et al., 2007). But there is also a difference in the substrate specificity of FIAU for viral and possibly bacterial versus mammalian TKs, in particular for hTK1 which is particularly fastidious. And hTK1, the most accessible of the mammalian TKs, is quiescent unless cells are dividing. In the case of EBV-TK-based imaging and therapy discussed in detail below, which in part forms the basis of selectivity for imaging malignant cells with radiolabeled FIAU—only the viral genome is active during the lytic cycle while hTK1, although in every living mammalian cell in the organism, remains dormant.

10.3.2 Choosing Nucleoside Analogs for Imaging TK: Lessons from Imaging Gene-Tagged Cells

In choosing a radiolabeled nucleoside for imaging, a key parameter on which to base that choice is the catalytic efficiency of the enzyme for the substrate. This can be expressed by the k_{cat}/K_m, under the assumption that the k_{cat} is based on one active site per enzyme monomer (four in the case of hTK1 and bacterial TKs and two for the viral TKs) and that the TK follows Michaelis–Menten kinetics (Eriksson et al., 2002), since k_{cat} and K_m are derived from fitting kinetic data to the Michealis–Menten equation with nonlinear regression analysis. In fact, the goal of mutating the TKs discussed above is to provide the enzyme with a superior K_m, which can be loosely thought of as the affinity of the substrate for the enzyme. More accurately, K_m is the concentration at which the reaction rate is half its maximal value. Intuitively, a lower K_m and a higher k_{cat}, which can also be thought of as the enzyme turnover

number, make for a more efficient enzyme (higher k_{cat}/K_m). For the enzyme to obey Michaelis–Menten kinetics, the enzyme–substrate (ES) complex is in a steady state, without reversion of products to ES. The Michealis–Menten equation states that:

$$V = \frac{V_{max}[S]}{K_m + [S]} = \frac{k_{cat}[E]_t[S]}{k_m + [S]}$$

When considering a radiotracer, $[S] \ll K_m$ such that:

$$V \approx \left\lfloor \frac{k_{cat}}{K_m} \right\rfloor [E]_t [S]$$

and the k_{cat}/K_m is equivalent to the second-order rate constant describing the interaction of E and S. V_{max} is the maximal velocity while $[E]_t$ represents the concentration of enzyme present at time t after beginning the assay. For the usual assay, the limit of detection of k_{cat}/K_m is of the order of 10^2 M^{-1} s^{-1}, providing a rough idea of how to compare and choose the various substrates for imaging based on *in vitro* data. Of course another major criterion is the ability for the substrate to discriminate between mammalian and foreign TK, not to mention pharmacokinetics.

A variety of dT and uracil 2′-deoxy-nucleosides should have been shown to be cytotoxic in cells transfected with HSV1-*tk*. However, such compounds radiolabeled with ^{125}I were unstable *in vivo* due to the rapid deglycosylation by phosphorylase. On the contrary, incorporation of fluorine or a hydroxyl group at the 2′-position generated species resistant to metabolic degradation by nucleoside phosphorylase with the 2′-fluoronucleosides demonstrating a prolonged metabolic half-life in plasma (Fridland et al., 1990). As shown in Figure 10.4, the binding mode of FIAU with HSV1-TK is very similar to that of dT when docked with the known crystal structure of the protein. Fluorine at the 2′-ara-position enables additional hydrogen bonding with the phenolic OH of Tyr 101 and appears to stabilize FIAU in the active site of HSV1-TK. According to Cheng et al. (1981), the binding affinity of nucleosides for TK determines the rate of phosphorylation at low concentration (less than K_m), which obtains under radiotracer conditions, while V_{max} is the dominant factor at higher concentrations, and so the K_m can be used as a surrogate comparison parameter when the efficiency (k_{cat}/K_m) is not available. Binding affinities (K_m) of FIAU toward HSV1-TK, hTK1, and hTK2 can help explain the selectivity of

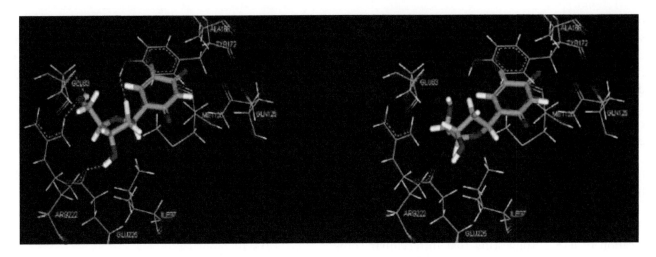

FIGURE 10.4 Docked poses of FIAU (left) and iododeoxyuridine (right) within the HSV1-TK crystal structure.

FIAU for HSV1-TK versus hTK1 or hTK2. Although a K_m value of FIAU for HSV1-TK has not yet been reported, K_i values of FIAU have been determined to be 0.68 µM for HSV1-TK (Cheng et al., 1981) and 1.54 µM for hTK2 (Wang and Errikson, 1996), indicating that FIAU may be slightly more selective for viral TKs than for those that are mammalian derived.

10.3.2.1 Comparison of Various Substrates of Thymidine Kinase for Imaging

There have been several studies that compare the efficiency of phosphorylation of existing nucleosides for various TKs, enabling a rational, if not systematic, structure–activity based method for choosing a radiolabeled nucleoside for imaging in a particular instance (Tjuvajev et al., 2002; Kang et al., 2005; Buursma et al., 2006; Alauddin et al., 2007; Miyagawa et al., 2008). Of course LogD (lipophilicity), metabolism, and the ability of the nucleoside to be transported by specific transporters to the cytoplasm are also important considerations and may underlie the somewhat unpredictable behavior of these substrates for the various TKs. For instance, as discussed above, the 2′-fluorinated derivatives, such as FIAU, were chosen because they are resistant to the operation of nucleoside phosphorylases. When Miyagawa et al. (2008) compared various "reporter probes" for the HSV1-TK, they used sensitivity and selectivity indices relative to the ability of the cells to take up dT and be sensitive to GCV (endogenous TK activity). They found that 1-(2′-deoxy-2′-fluoro-β-D-arabinofuranosyl)-5-ethyluridine (FEAU) demonstrated approximately threefold higher sensitivity and 80-fold higher selectivity for imaging TK-transduced versus wild-type cells. 9-[4-Fluoro-3-(hydroxymethyl)

butyl]guanine (FHBG) demonstrated a similar selectivity to FIAU. At 2 h postinjection, FEAU demonstrated a target-to-nontarget ratio of >10:1 while FIAU demonstrated a ratio of 3:1, *in vivo*. However, a main advantage of imaging with [^{124}I]FIAU is that imaging can be performed many hours after radiotracer injection due to the 4.2 day physical half-life of ^{124}I, with ratios of >60:1 demonstrated at 24 h after injection. This is an important consideration where the kinetics for a particular target—such as bacteria—are unknown. Several studies have shown that acycloguanosines, such as FHBG, demonstrate very high levels of gastrointestinal radioactivity. As these compounds have been used in clinical trials (Yaghoubi et al., 2009), efforts have been made to decrease that confound, but without success (Ruggiero et al., 2010). Despite attempts to compare these imaging agents in carefully controlled *in vitro* and *in vivo* situations, a combination of synthetic accessibility, match of physical to biological half-life, and achievable signal/noise for a particular indication will govern the choice of radionucleoside for TK-based imaging.

10.3.3 Synthesis of Radiolabeled FIAU

The nucleoside we have chosen as the molecular imaging agent for the bacterial and viral TKs is FIAU. Preclinical studies use [^{125}I]FIAU, while those for the clinic have employed [^{124}I]FIAU, mainly due to synthetic accessibility (iodogen method with commercially available iodide) and the unknown pharmacokinetics of the radiotracer for imaging bacteria and herpesviruses, that is, we needed a compound with an extended physical half-life. FIAU has been produced in radiolabeled form for a variety of applications, from testing in cellular

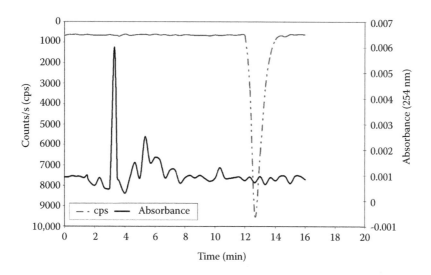

FIGURE 10.5 A typical radiosynthesis of [124I]FIAU. Eluted radioactivity trace is shown in red (top) with mass in black (bottom). Note the lack of mass eluting with the radioactive peak, indicative of material of high-specific radioactivity.

uptake assays to molecular imaging and therapy of cancer. The earliest applications to imaging involved synthesis of the 131I- and 124I-labeled forms by Tjuvajev et al. (1995, 1996, 1998). We have optimized radiolabeling of the 124I-labeled form, which is an adaptation of the earlier method, for higher yield and specific radioactivity (Figure 10.5).

Production of [124I]FIAU according to current Good Manufacturing Practice (cGMP) for our initial clinical investigation resulted in a 90% success rate for the radiochemical synthesis, that is, radiochemical yields producing at least 2 mCi (74 MBq), specific radioactivities of >2000 Ci/mmol (74 GBq/mmol), and radiochemical purities of >95%, which indicate our minimum standards. More recently, we noticed a steep increase in synthesis failure rates. The presence of unreacted radioiodide peaks in the chromatogram prompted us to suspect a contaminant in the radioiodide used for the reaction. As outlined in Zielinski et al. (1977), radioactive iodate (IO_3^-) is frequently present in the 124Te target-generated production of both [123I]NaI and [124I]NaI and reduces radiochemical yields. The radioiodide we receive may contain up to 1 mg/mL of tellurium contaminant, which can oxidize the desirable radioiodide to unreactive IO_3^- and I_3^- over time. To avoid poor radiochemical yields due to unreactive, oxidized radioiodine species, we have implemented a commonly used reduction step of the radioiodine solution prior to reaction with the precursor 1-(2′-deoxy-2′-fluoro-β-D-arabinofuranosyl) uracil (FAU). This has allowed reliable and improved radiochemical yields despite variable tellurium concentrations in each batch we

receive. In test and at-scale reactions, a 5 min reduction of the radioiodine solution with 50 μL of 60 mM $Na_2S_2O_3$ in 50 mM H_2SO_4 prior to addition to FAU reduced the amount of unincorporated radioiodine at the end of synthesis by threefold. It also eliminated the presence of a late-eluting side product, the variable quantity of which would sometimes dominate the yield leading to synthetic failure. All thiosulfate and sulfuric acid are eliminated along with other reaction constituents during purification by high-performance liquid chromatography (HPLC). We have achieved reliable yields of cGMP [124I]FIAU regardless of isotope batch composition with that modification. New at-scale reactions using this modification reliably result in 80–85% radiochemical yield and 99% radiochemical purity.

The theoretical specific radioactivity of [124I]FIAU is 32,000 Ci/mmol (1.2×10^3 GBq/μmol). In practice, we achieve cGMP-specific radioactivities of the order of 10,000 Ci/mmol (370 GBq/μmol). Specific radioactivity measurements are based on UV-HPLC. Standard curves generated using serial dilutions of known concentrations of FIAU were used to determine the concentration of an unknown amount of unlabeled/radioactive FIAU present in the dose. Similarly, standard curves generated for FAU are used to determine the mass of FAU, if present, in the injection dose. For quality assurance between batches, FIAU and FAU standard HPLC runs are always performed prior to the radiopharmaceutical synthesis. Quality of the radiopharmaceutical is based on differences of <10% in area under the curve between the standard curves and the chromatograms obtained during a cGMP synthesis run.

10.3.4   A Bacterial Target for Radiolabeled FIAU

The ability of nucleoside analogs to have an effect on bacteria has been known for nearly 50 years (Hubert-Habart and Cohen 1962). Efforts along those lines were revived in the mid- to late 1980s when several groups began using bacteria as a model system to test candidate antiretroviral agents prior to the advent of robust *in vitro* or *in vivo* systems for HIV infection. For example, Elwell et al. (1987) at Wellcome Research Laboratories found that AZT had potent antibacterial properties at micromolar concentrations in a number of Gram-negative pathogens, including *Escherichia coli*. They purified *E. coli* TK and found that AZT was approximately one-fifth as efficient of a substrate as endogenous dT. By isolating TK from spontaneous colonies that were resistant to *E. coli*, Elwell et al. (1987) found that the TK had a 70- to 300-fold reduction in activity when compared with the susceptible parental clones. This provided some of the early framework proving that not only could nucleoside analogs have antibacterial properties but also that the mechanism of cell kill was mediated through TK.

FIAU garnered significant interest in the early 1990s for demonstrating striking efficacy against the hepatitis B virus (HBV) both in preclinical models and early human trials. In the Oclassan clinical trial R90-001-01, there was a reduction in circulating HBV DNA to undetectable levels and the response was durable over several weeks. Those results were followed up by Eli Lilly and unfortunately resulted in severe toxicity and death in several patients who were administered, in some cases, gram quantities of FIAU over extended periods. That tragic result led to the immediate halting of all clinical activities related to FIAU (Witt, 1993; McKenzie et al., 1995). Simultaneously, there was a burgeoning research interest in viral gene therapies for a number of different medical conditions. During the development of FIAU, it became clear that it was a robust substrate for not only the HBV-TK but also for the HSV1-TK. Given that FIAU was nontoxic in nearly all *in vitro* and animal models tested and the favorable chemistry that allowed its modification with radiolabeled iodine, several groups began using radioactive FIAU for monitoring experimental gene therapies that utilized HSV1-TK as a reporter gene (Tjuvajev et al., 1995). That was later followed up in three small clinical studies that tested the efficacy of viral gene therapies in patients with glioblastoma multiforme. Radioiodinated versions of FIAU were used in those studies to assess noninvasively retrovirally mediated gene transfer in the patients studied (Voges et al., 2003; Jacobs et al., 2001; Dempsey et al., 2006).

10.3.4.1   Preclinical Studies of Infection: Focus on Radiolabeled FIAU

A motivation for imaging bacterial infection that extends beyond assessing for the presence of bacteria that need to be detected and eradicated includes the tracking of therapeutic bacteria. It has been known for many years that bacteria—particularly anaerobes—could be homed to tumors and inhibit their growth (Reilly, 1953). Several bacterial strains, notably *Salmonella typhimurium* and *Clostridium novyi*, have been engineered to enhance their therapeutic efficacy through rendering them nonvirulent or for improved tumor targeting (Bettegowda et al., 2003; Agrawal et al., 2004; Diaz et al., 2005; Soghomonyan et al., 2005). The goal is to have a nontoxic, tumor-specific strain that is therapeutic on its own or by virtue of expression of various therapeutic transgenes. These anaerobes have been shown to accumulate at the tumor transition zone between viable and necrotic tissue where they promote apoptosis after systemic administration or administration of bacterial spores. Recognition that the bacteria tend to accumulate in those transition zones, often avoiding the well-perfused outer shell of the tumor, has led to the investigation of added antivascular agents in preclinical studies in an attempt to avoid that potential confound (Zhou, 2005). Phase I trials using this technology have been performed, but have shown only modestly encouraging results. For example, a small trial ($n = 3$) in which the *Salmonella* strain VNP20009 was used showed that although toxicities were manageable, they were not so low as to permit the use of a strain expressing the HSV1-TK to be used in a stand-alone imaging study (Toso et al., 2002). Significantly, fine-needle biopsy of samples from those patients that failed to identify bacteria initially was detected in the same specimens on total excision. That is where imaging could help—by providing a complete picture of where the vector is located and if it is present in sufficient amounts to be effective.

In a study to assess the utility of radiolabeled FIAU for imaging gene-tagged (HSV1-*tk*) bacteria, Tjuvajev et al. (2001) summarize the advantages of the bacterial tumor treatment method. Some of those advantages include the systemic delivery of replication competent vectors, tumor specificity, ability to deliver an imaging or therapeutic transgene, effectiveness toward a wide variety of tumors, among others. That study demonstrated the direct correlation of radioactivity sequestered within tumor, in the form of [^{14}C]FIAU, and colony-forming units of bacteria setting the stage for an imaging study, which confirmed those findings *in vivo*.

Given the favorable characteristics demonstrated by radiolabled FIAU both in preclinical molecular-genetic imaging studies and in the limited clinical experience, Bettegowda et al. wondered whether FIAU could act as a substrate for bacterial TK similar to AZT and other nucleoside analogs so that the gene-tagging step could be avoided. They demonstrated that FIAU had micromolar MIC_{50} and that bacteria with spontaneous deletions in TK were several hundredfold more resistant to the effects of FIAU, suggesting that FIAU was also acting through TK (Bettegowda et al., 2005). In addition, they found that FIAU had a broader level of activity when compared with other standard nucleoside analogs. For example, AZT has been shown to have antibacterial properties largely restricted to Gram-negative organisms, while gemcitabine acts exclusively on Gram-positive organisms (Bettegowda et al. 2005; Sandrini et al, 2006) demonstrated that [^{125}I]FIAU could be used to image artificial infections caused by both Gram-positive and Gram-negative clinical isolates including *E. coli*, *S. aureus*, *Streptococcus pneumonia*, *Enterococcus faecalis*, *Staphylococcus epidermidis*, and the tumor-targeting anaerobe, *Clostridium novyi-NT*.

Workers at the Memorial Sloan-Kettering Cancer Center have synthesized and tested a wide variety of radiolabeled nucleoside analogs for imaging. The compound that has so far demonstrated arguably the best pharmacokinetics for molecular-genetic imaging has been [^{18}F]FEAU, which was the agent that they used for monitoring the ability for the probiotic strain *E. coli* Nissle 1917 to home to experimental breast tumors (Brader et al., 2008). As for the study in Bettegowda et al. (2005), the rationale behind this study was to enable noninvasive detection of bacteria that may be used to treat tumors, either directly or through delivery of a therapeutic gene. The study by Brader et al. showed a linear correlation of [^{18}F]FEAU and bacterial concentration within tumors, a result not demonstrated for FDG, which was performed in parallel. This group followed this study with one that used a variety of maneuvers, for example, laxatives, to decrease the degree of radioactivity that appeared in the bowel with certain radiolabeled nucleosides, most notably [^{18}F]FHBG (Ruggiero et al., 2010). Such maneuvers did not decrease the amount of background radioactivity suggesting that activity to be due to metabolites rather than radiolabeled parent sequestered within intestinal flora.

As noted above, *M. tuberculosis* does not possess a TK, prompting Davis et al. (2009) to transduce this mycobacterial strain with a bacterial TK to enable detection of this organism in experimentally infected animals. The transduction provided stable integration of one copy of the TK gene into the bacterial genome allowing for consistent comparison of treatment groups. The gene-tagged *M. tuberculosis* proved a useful tool in experimental studies of therapeutic monitoring, with similar antimicrobial susceptibilities between the parental and transduced strains. This study also showed that [^{125}I]FIAU SPECT-CT, for both thigh inoculation and aerosol models, could detect as few as $5\text{--}10 \times 10^6$ bacteria within a granuloma. As there are several orders of magnitude more bacteria within a human lung granuloma ($10^7\text{--}10^9$), this method should prove useful for monitoring bacterial burden in other, more relevant animal models of *M. tuberculosis* infection such as guinea pigs, rabbits, and nonhuman primates.

10.3.4.2 Clinical Studies of Infection: Focus on Radiolabeled FIAU

The study by Bettegowda et al. (2005) was followed up with a small clinical trial that imaged eight patients with presumed bacterial prosthetic joint infection and one control subject with [^{124}I]FIAU (Diaz et al., 2007). Patients were administered 2 mCi (74 MBq) of [^{124}I]FIAU and imaged with PET-CT at 2 h postinjection and in a subset of patients were imaged again at 24 h. Diaz et al. (2007) were able to detect the bacterial nidus accurately in seven of seven patients with infection confirmed by subsequent, intraoperative culture, and correctly identified the two patients who did not have an infection. Radiopharmaceutical uptake in normal tissues was quantified using standardized uptake values, which included values of 2.01, 2.35, and 8.11 in muscle, spleen, and liver, respectively. The images were merely scored as positive or negative based on a visual comparison of uptake in the expected region of infection compared to surrounding tissue (Figure 10.6). The ability to image bacteria in humans has many potential applications. The key to understanding which indications are most appropriate lies, in part, on which cells, both human and bacterial, are capable of phosphorylating FIAU, as covered in the preceding sections. A key element to the FIAU imaging method is that bacteria must provide the TK for imaging. The method should therefore be more specific than existing methods, which may detect sterile inflammation as well as infection as they are based on the presence of coexisting inflammatory cells.

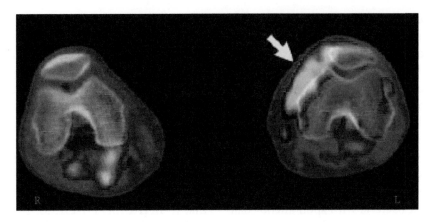

FIGURE 10.6   Necrotizing septic arthritis in the left knee (yellow arrow). Cultures produced methicillin-resistant *Staphlococcus aureus*. (From Diaz, L. A., Jr. et al., 2007, *PLoS One* 2: e1007. With permission.)

10.3.4.3   Toxicity and Metabolism of FIAU

Despite the fact that FIAU proved to be toxic—even lethal—in a small percentage of the patients to whom it was administered, others and we were able to administer it for imaging due to the tracer principle. According to the initial trials, patients who received <200 mg of FIAU for antiviral therapy over a 4 week interval did not demonstrate any toxicity. According to the specific radioactivities we achieved in Diaz et al., we administered 9×10^{-7} mg/kg as a one-time dose to our subjects (Diaz et al., 2007; Pomper, 2009). We also tested liver function before and after imaging in a subset of the patients studied, with stable results between the two sets of tests. Others have also addressed the issue of potential toxicity of administering FIAU for imaging, with the same result that at tracer doses the compound is safe.

FIAU is metabolized and excreted through the enterohepatic system, suggesting that the background signal identified in the liver and gastrointestinal tract could be a concern for quantitative imaging, as noted above (Ruggiero et al., 2010). In addition, it is well known that nucleoside analogs, including FIAU, can cause myopathy due to mitochondrial damage in muscle. However, this damage is usually only induced after high cumulative doses, and so it remains to be seen what will be the baseline uptake in muscle (Perez-Perez et al., 2008). The existing literature also suggests that the background in brain, lungs, and joints is low, perhaps allowing bacterial infections affecting these organs to be visualized clearly with radioactive FIAU (Jacobs et al., 2001; Diaz et al., 2007). While promising data are emerging, further clinical studies need to be performed to elucidate fully the utility of radioactive FIAU for imaging bacterial infections.

10.3.4.4   Dosimetry of Radiolabeled FIAU

The dosimetry for [^{124}I]FIAU has been published; however, as we have proceeded to human studies we have also generated our own estimates. We performed dosimetry calculations using biodistribution data in mice, collected over a 2 h time period (Table 10.2). The time-integrated activity coefficients (cumulated activities) were obtained by fitting single- or double-decreasing exponential expressions to the organ activity concentration (in %ID/g) versus time data. Table 10.3

Table 10.2   Input Biodistribution Data

Time (min)	\multicolumn{10}{c}{Tissue (%ID/g)}
	EBV + Tumor	EBV − Tumor	Blood	Fat	Muscle	Small Intestine	Large Intestine	Liver	Kidney	Spleen
15	0.78	0.34	0.79	0.72	0.46	1.00	0.66	0.73	1.33	1.09
30	0.39	0.30	0.77	0.36	0.23	0.37	0.37	0.40	0.87	0.38
60	0.67	0.40	0.71	0.32	0.18	0.33	0.44	0.39	0.68	0.37
120	0.84	0.14	0.52	0.14	0.09	0.33	0.39	0.21	0.33	0.30
240	1.44	0.07	0.59	0.08	0.04	0.14	0.18	0.08	0.10	0.15

Table 10.3 Fitted Parameter Values

Organ	C1 (%ID/g) ± SD[a]	T1 (h) ± SD	C2 (%ID/g) ± SD	T2 (h) ± SD
Blood	0.76 ± 0.07	7.5 ± 4		
Kidneys	1.31 ± 0.10	1.06 ± 0.06		
Liver	0.61 ± 0.08	1.35 ± 0.2		
Spleen	0.46 ± 0.03	2.6 ± 0.3	100 ± 257	0.03 ± 0.02
Small intestine	0.45 ± 0.06	2.5 ± 0.5	63 ± 256	0.04 ± 0.03
Large intestine	0.48 ± 0.08	3.0 ± 0.8	33 ± 265	0.03 ± 0.05
Muscle	0.3 ± 0.02	1.3 ± 0.1	11 ± 59	0.04 ± 0.06
Fat	0.4 ± 0.1	1.8 ± 0.5	2 ± 3	0.1 ± 0.1

[a] SD reflects the degree to which the fitted value is uniquely determined; if the SD exceeds the fitted value, the value is not uniquely determined and of dubious biological relevance.

lists the fitted parameter values obtained for the expression:

$$C(t) = C1 \cdot e^{-t \cdot \lambda_1} + C2 \cdot e^{-t \cdot \lambda_2}$$

with

$$\lambda_1 = \frac{\ln(2)}{T1}; \quad \lambda_2 = \frac{\ln(2)}{T2}$$

$C(t)$ = activity concentration as a function of time, t. $C1$ and $C2$ represent y-axis intercepts for the rapidly and slowly clearing components of the time–activity curve. $T1$ and $T2$ are the corresponding half-lives. When $C2$ and $T2$ are not listed, an adequate fit was obtained using the single exponential expression (obtained by setting $C2$ to zero).

To convert these mouse tissue %ID/g ($[\%ID/g]_M$) values to human whole-organ %ID/organ ($[\%ID/organ]_H$)

for use in estimating whole-organ residence times, the following equation was used:

$$[\%ID/organ]_H = [\%ID/g]_M \cdot TBM_M \cdot \frac{OM_H}{TBM_H} \quad (10.1)$$

with:
TBM_M = total body mass of the mouse.
OM_H = human organ mass from OLINDA organ mass listing and Reference Man (ICRP 23) (Stabin et al., 2005).
TBM_H = human total body mass (adult male = 70 kg used).

This approach assumes that the concentration of activity in a particular tissue relative to the overall concentration in the whole body is preserved across species (i.e., organ concentration/total body concentration is the same for mouse and man). Hereafter, all %ID/organ will refer to human organ data (Table 10.4). The expressions for human organ time–activity curves $A_H(t)$ were multiplied by an exponential term,

Table 10.4 Human Organ %ID per Organ

Organ	C1 (%ID/organ)	T1 (h)	C2 (%ID/organ)	T2 (h)
Blood	1.5	7.5 ± 4		
Kidneys	0.14	1.06 ± 0.06		
Liver	0.42	1.35 ± 0.2		
Spleen	0.03	2.6 ± 0.3	6.6[a]	0.03 ± 0.02
Small intestine	0.11	2.5 ± 0.5	15[a]	0.04 ± 0.03
Large intestine	0.07	3.0 ± 0.8	4.6[a]	0.03 ± 0.05
Muscle	2.9	1.3 ± 0.1	111[a]	0.04 ± 0.06
Fat	1.9	1.8 ± 0.5	11.2[a]	0.1 ± 0.1

[a] Not uniquely defined.

Chapter 10

corresponding to physical decay of [124]I or ^{131}I and integrated over time to yield the total number of [124]I or ^{131}I disintegrations in each tissue (in units of %ID-h/organ). The results were expressed as residence time. The OLINDA software package was used to obtain the human dose estimates listed in Table 10.5. Note that doses are listed only for those organs that were counted and that were in OLINDA (Stabin et al., 2005).

10.4 Imaging and Therapy of Gammaherpesviruses-Associated Malignancies

10.4.1 Premise

The ability to image and treat gammaherpesvirus-associated malignancies is based on the following premises: (1) lymphomas harboring EBV genomes encode viral enzymes with specificities that differ from those of host cells; (2) these virus-encoded enzymes are not expressed in tumor tissues at high levels at baseline but can be induced pharmacologically; and (3) expression of viral enzymes in even a minority of scattered cells may be sufficient to achieve targeting to tumor sites. As detailed below, radioimmunoconjugates are an important advancement in the targeting of B-cell lymphoma. Cell surface molecules such as CD20 recognized by monoclonal antibodies conjugated to radioisotopes including ^{131}I deliver radiation to CD20-expressing B-cell malignancies (Kaminski et al., 2005). Low-grade B-cell lymphoma, particularly follicular lymphoma, will often respond to unlabeled antibody alone. B-cell lymphomas that are resistant to rituximab in some instances may be effectively treated by radioimmunoconjugates (Bennett et al., 2005). With those nuclear medicine/radiation therapy successes in mind, let us consider EBV-associated lymphoma—noting that while the treatment of some has made strides with the introduction of rituximab (notably posttransplant lymphoma), many remain among the most difficult malignancies to treat (nasal type NK/T-cell lymphoma).

10.4.2 EBV-Associated Lymphomas

B-, T-, and NK-cell lymphomas may harbor the viral genome. Follicular lymphomas and mucosa-associated lymphoid tissue lymphomas rarely, if ever, harbor the virus. In those lymphomas associated with virus, the

Table 10.5 Absorbed Dose Estimates for [[124]I]FIAU

Organ	Residence Time (h)	Absorbed Dose (mSv/MBq)	Absorbed Dose (rem/mCi)
Kidneys	0.0018	0.0017	0.0063
Liver	0.0070	0.0014	0.0050
Spleen	0.0011	0.0017	0.0063
Small intestine	0.0036	0.0020	0.0074
Large intestine[a]	0.0026	0.0022	0.0082
Muscle	0.0493	0.0010	0.0036

[a] Residence time assigned to upper large intestine wall contents.

viral genome is episomal and is readily detected in most or all tumor cells. Three types of EBV-associated tumors deserve comment in the context of this discussion, namely, Hodgkin's lymphoma, nasal type NK/T cell lymphoma, and posttransplant lymphoma.

Hodgkin's lymphoma: Reed–Sternberg cells and their variants harbor the virus in ~30–40% of cases in North America and Europe (Glaser et al., 1997). A substantially higher fraction of cases are associated with EBV in some parts of Latin America, Africa, and Asia. In population-based registry studies, we have found that the presence of EBV in Hodgkin's lymphoma in older patients is an extremely poor prognostic factor—independent of stage or histology (Keegan et al., 2005).

Nasal type NK/T-cell lymphoma: These are almost always associated with EBV (Chiang et al., 1996; Cheung et al., 2003). Many other peripheral T-cell lymphomas, including angiocentric T-cell lymphoma and lymphoma arising in the setting of EBV-associated hemophagocytic syndrome, are commonly EBV associated. Most of these fare poorly with standard therapies including high-dose therapy with stem cell transplant.

Posttransplant lymphoma: These are usually EBV associated, particularly those arising in the first 2 years after transplantation (Chadburn et al., 1998). They may respond to a variety of therapies from withdrawal of immunosuppression when possible, to rituximab, to combination chemotherapy. Nonetheless, a significant fraction does not respond, with progression to death.

10.4.2.1 Targeting EBV in Tumors

Two key elements related to EBV-based tumor targeting include detection and characterization of EBV in clinical specimens and understanding its diagnostic and prognostic import (Wu et al., 1990; Keegan et al., 2005). The second interest is in developing targeted therapies for EBV-associated lymphoma. Targeting viral antigens in lymphoma might appear to be straightforward. O'Riley and colleagues at the Memorial Sloan-Kettering Cancer Center, using donor lymphocyte infusion (Papadopoulos et al., 1994), and Heslop, Rooney, and colleagues at the St. Jude Children's Hospital and more recently at the Baylor School of Medicine, using EBV-activated T-cells, achieved dramatic successes with regression of lymphoma in allogeneic bone marrow transplant recipients or prevention of lymphoma in very high-risk recipients (Savoldo et al., 2007). They have also had some encouraging results but more limited successes in other EBV-associated tumors. Limited viral antigen expression in these other settings is at least part of the explanation. While posttransplant lymphoma may express nine or more viral antigens, only a very few are expressed in most EBV-associated tumors (Thorley-Lawson and Gross, 2004), (Young and Rickinson, 2004). Burkitt's lymphoma shows highly restricted antigen expression. A single EBV antigen, the EB nuclear antigen 1 (EBNA1), is the only viral protein expressed in most of the lymphoma cells. In Hodgkin's lymphoma, in some diffuse large B-cell lymphoma in immunodeficient patients, and in NK/T cell lymphoma, the latency membrane antigens (LMP1 and 2) may also be expressed (Murray et al., 1998).

10.4.2.2   Suicide Gene Therapy

Several investigators have pursued pharmacologic approaches to targeted tumor treatment. As detailed above, when GCV or another nucleoside analog is phosphorylated by a viral kinase, it is trapped in cells because its charge prevents diffusion through the cell membrane. Cellular enzymes carry out further phosphorylation yielding the triphosphate, which inhibits the cellular DNA polymerase and thus kills cycling cells. The HSV1-TK has been used as a suicide gene in a variety of translational clinical applications such as gene therapy of brain tumors (Klatzmann, 1996). There have been many attempts to deliver selectively the gene by retroviral and other vectors. Efficient and selective delivery remains a major obstacle to these approaches. New vectors and other delivery systems are being actively investigated by many groups, as are other suicide genes. Taking advantage of the "natural" presence of EBV in some lymphoma types, butyrate (a histone deacetylase inhibitor known to upregulate some EBV genes in some cell lines) has been administered with

GCV in the hope that activation of gene expression would lead to GCV phosphorylation and cell killing (Westphal et al., 2000). Several groups, including ours, have shown that although sequence homology between the EBV-TK and the HSV1-TK is limited, both phosphorylate GCV (Moore et al., 2001). The approach has met with some encouraging clinical results in a multi-institutional early phase trial (Perrine et al., 2007). Those interested in gene therapy approaches involving the HSV1-TK found that FIAU is readily labeled with various isotopes for imaging, and might be used as a radiotracer in imaging studies to monitor delivery and expression of the HSV1-TK to targeted cells (Tjuvajev et al., 1995, 1998; Blasberg and Tjuvajev, 2003; Koehne et al., 2003), and, by extension, could be used to image tumors derived from cells with an activated EBV-TK.

10.4.2.3   Targeted Radionuclide Therapy

Radiotherapy is effective in a wide range of malignancies, but none are more sensitive than lymphoma. A major limitation of radiation therapy is related to the inability to target radiation selectively to lymphoma cells. Iodine-131-mediated therapy has been in practice for more than 50 years, representing the first and most longstanding example of targeted radionuclide therapy (Benua and Dobyns, 1955). That therapy, which was initially used for thyroid cancer, is effective by virtue of the uptake of radioiodine by the sodium/iodide symporter followed by trapping and organification of the radioactive species. Radioablative doses for thyroid cancer are of the order of 150–175 mCi (5.5–6.5 GBq) (Koral et al., 1986). For the past 20 years, there has been an emphasis on the development of antibody-based strategies for treating specific antigen-expressing tumors (Goldenberg, 2002, 2003; Cardillo et al., 2004; Goldenberg and Sharkey, 2006). This technique has proved particularly useful in the case of lymphoma, culminating in the approval and use of two commercial products, Bexxar™ (labeled with ^{131}I) and Zevalin™ (labeled with ^{90}Y). Both are radiolabeled monoclonal antibodies directed against the CD20 antigen present on both normal and malignant B-cells. Perhaps one of the main advantages of this technique over standard, external beam therapy or chemotherapy is that far fewer side effects have been demonstrated, particularly for non-Hodgkin's lymphoma. Unfortunately, doses proved effective in hematologic malignancies have been shown to be ineffective in solid tumors. Even with doses as high as 3000 cGy to tumor, solid tumors fail to regress. Furthermore, because these antibodies are directed to an antigen present on normal B-cells, severe myelosuppression results, and although that is usually reversible, a significant proportion of patients require supportive care and have even required pretreatment bone marrow harvest for later autologous transplant (Goldenberg and Sharkey, 2006).

Many tumor-targeted, low-molecular-weight radiotherapeutic agents have also been developed, some of which have entered the clinic (Bodei et al., 2003; Oyen et al., 2007). In particular, radiotherapeutic peptides directed against the somatostatin receptor have shown efficacy in early phase II trials (de Visser et al., 2008). Two major benefits of the low-molecular-weight agents, however, include the fact that higher doses can be given because there is less dose-limiting bone marrow toxicity than seen for the antibody-based agents, and also these smaller agents will gain better access to the tumor and will not have to rely on leaky or disorganized tumor vasculature for penetration. Nevertheless, no low-molecular-weight radiotherapeutic agent has been approved for treating cancer, although several, including radioiodinated meta[^{131}I]iodobenzylguanidine ([^{131}I]MIBG), are close.

The choice of radionuclide to be used is dictated by several factors, including ease of introduction to the targeting moiety, size and geometry of the tumor under treatment, biological half-life of the radiolabeled agent, and availability or familiarity of the nuclear medicine department with use of the isotope. There are over a dozen possible radionuclides from which to choose, including the pure β-emitter, ^{90}Y, ^{131}I, which also provides a γ-ray for imaging, ^{177}Lu and ^{213}Bi. Bismuth-213 might be useful for single-cell disease or micrometastisis, as the α-particles emitted have a very high linear energy transfer over a very short range. Its short physical half-life (45.6 min) may also prove useful for labeling of low-molecular-weight agents with rapid pharmacokinetics. An advantage to using ^{90}Y is that patients may leave the hospital immediately after dosing since no γ-rays are emitted. The physical half-life is also amenable to antibody labeling, as is that of ^{131}I. Many radiopharmacies have extensive experience with ^{131}I and it is easily conjugated to antibodies and low-molecular-weight agents alike.

10.4.2.4 Imaging of Gammaherpesviruses-Associated Malignancy with Radiolabeled FIAU

In EBV-associated tumors and KSHV-associated tumors, the viral TKs and protein kinases are not expressed or are expressed only at low levels. Because the viral genome is latent, viral kinase expression must be activated using a pharmacologic inducer. A screen of the Hopkins Drug Library composed principally of FDA-approved compounds, uncovered bortezomib, a proteasome inhibitor, as the most potent compound for lytic induction in gammaherpesvirus-associated tumor cells (Fu et al., 2007). Subsequent work by Shirley et al. showed that bortezomib induces the unfolded protein response, leading to activation of C/EBPβ and induction of EBV lytic gene expression (Shirley et al., 2011). Fu et al. showed that at 96 h after administration of [^{125}I]FIAU, a 22-fold increase in radiopharmaceutical uptake was demonstrated in the animals harboring EBV-associated tumors (Burkitt's lymphoma cell lines Rael and Akata) that were pretreated with bortezomib compared to those which received vehicle. Similar results were found with KSHV-associated primary effusion lymphoma cell lines. Care was taken to prove that it was induction of the viral genome, and not host cells, which accounted for that radiotracer uptake. Furthermore, the effects of bortezomib were shown to be dose dependent. The method proved quite sensitive with as few as only 5% of the cells within tumors needing to express the viral TK for detection by SPECT-CT. These imaging studies set the stage for enzymatic molecular radiotherapy using a therapeutic version of FIAU, [^{131}I]FIAU, in addition to generalization beyond lymphomas to epithelial tumors for which radioimmunotherapy has not yet proved particularly successful.

10.4.3 Molecular Radiotherapy of Gammaherpesvirus-Associated Malignancies

[^{131}I]FIAU has been used to treat tumor cells previously (Schipper et al., 2007). In an effort to demonstrate synergy between molecular radiotherapy and more conventional tumor therapy, in line with other studies, Schipper et al. treated HSV1-*tk*-tagged cells with a combination of [^{131}I]FIAU and GCV. Although each method provided tumor cell growth inhibition in the gene-tagged versus parental cells, no synergy could be demonstrated (Schipper et al., 2007). The authors appropriately note that their *in vitro* result may not reflect the analogous case *in vivo*, where cellular transport and other aspects of pharmacokinetics come into play.

Radioimmunoconjugates can be limited by the degree of expression of target receptor on the cell surface and the availability and affinity of the antibody. A technique that uses a low-molecular-weight agent that could be concentrated within the tumor, in analogy to the treatment of thyroid cancer with [131I]NaI, might be able to deliver the amount of radioactivity needed to treat solid tumors, which is of the order of 0.5–2.0 Gy (Fu et al., 2008). The idea to activate the EBV genome for tumor-specific targeting has been attempted previously with the use of histone deacetylase inhibitors followed by GCV, which provided remissions in certain virus-associated malignancies. Extending the imaging studies discussed in Section 10.4.2.4, we have recently shown in animal models that Bortezomib-induced enzyme-targeted radiotherapy (BETR) was capable of halting the progression of EBV- and KSHV-associated tumors (Fu et al., 2008). An advantage to the use of FIAU rather than antibody-mediated delivery of radioactivity includes that it is metabolically stable (due to the presence of the 2′-fluoro group rendering it insusceptible to phosphorylases). Another advantage is the ability of the [131I]FIAU to be concentrated within cells due to continual turnover by the EBV-TK followed by trapping of the phosphorylated product. The FIAU method shares with the antibody-based method the ability to engender a bystander effect, that is, not all cells need to express the TK to generate tumor growth arrest or eradication (Figure 10.7). Iodine-131 produces a β-particle with an energy of 0.61 MeV, with 90% of its

Table 10.6 Absorbed Dose Estimates for [131I]FIAU

Organ	Residence Time (h)	Absorbed Dose (mSv/MBq)	Absorbed Dose (rem/mCi)
Kidneys	0.0018	0.00086	0.0032
Liver	0.0071	0.00061	0.0022
Spleen	0.0011	0.00085	0.0032
Small intestine	0.0037	0.00064	0.0024
Large intestine[a]	0.0027	0.00086	0.0032
Muscle	0.0497	0.00027	0.0010

[a] Residence time assigned to upper large intestine wall contents.

energy deposited within a 0.7 mm radius of the decay event. This bystander effect, as opposed to the traditional bystander effect which accounts for enhanced therapy of diffusible phosphorylated nucleosides, does not rely on the degree of expression of connexion or gap junctions. Also, the BETR technique is not dependent on the host cell cycle as is treatment with GCV. Clinical studies using BETR to treat gammaherpesvirus-associated tumors are currently under way. Table 10.6 lists the absorbed dose estimates obtained when the pharmacokinetics shown on Table 10.4 are used to calculate absorbed doses for I-131-labeled FIAU.

10.4.4 A Word on Dosimetry for BETR

The possibility of targeting individual cells based on expression of activated EBV-TK and the desire to project the likelihood of tumor control from such an approach has highlighted the need to devise better techniques for cell level and cluster dose calculations that can be reliably translated to tumor control. Calculations of the absorbed dose to each cell making up a spherical cell cluster model on the scale of measurable tumors are not typically possible. Instead, the average absorbed dose to a volume element that includes a large collection of cells, that is, the smallest resolvable element, is calculated and assigned to all cells making up the volume element. For example, in a spherical model with an exponentially decreasing radial dose profile, a single absorbed dose value is calculated for each shell (smallest resolvable element) of the sphere; all cells occupying that particular shell are then assigned this dose value. The frequency distribution of absorbed doses calculated in this manner is then used to calculate the tumor control probability for the spherical tumor cell cluster. This approach overestimates tumor control because it assumes that all cells at a particular

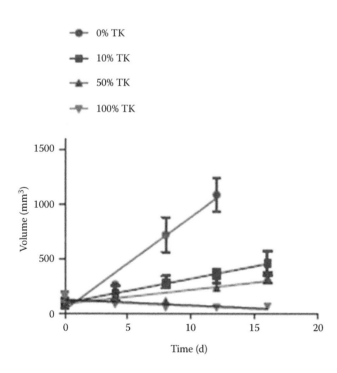

- ● 0% TK
- ■ 10% TK
- ▲ 50% TK
- ▼ 100% TK

FIGURE 10.7 Illustration of the bystander effect. Tumors containing 50% TK+ cells have a similar decrease in volume over 2 weeks as do those with 10% TK+ cells in animals treated with [131I]FIAU. (Modified from Fu, D. X. et al. 2008, *Curr Pharm Des* 14: 3048–65.)

shell receive the same absorbed dose. In reality, there is a distribution of doses in each shell and the lower portion of the distribution will drive the tumor control probability. A methodology has recently been developed that overcomes this problem by assigning a Gaussian dose distribution centered around the single-dose value; the standard deviation of the distribution is obtained from the estimated activity in each cell (Hobbs et al., 2011). The approach is useful when the scale of the calculation precludes Monte Carlo determination of the absorbed dose to each cell (i.e., the smallest resolvable element is made up of a collection of cells). This will occur, for example, when calculating the absorbed dose to a measurable tumor mass in a preclinical model so as to determine the tumor control probability. Details of such a calculation are provided in Hobbs et al. (2011).

10.5 Perspective

Spatial information about the presence of infection often dictates therapy, underscoring the importance of imaging in the management of infectious disease. While there are a variety of experimental imaging techniques, a number of which have been attempted in humans, TK-based imaging is among the most promising due to the near ubiquitous expression of TK or TK-like enzymes in infectious species, the somewhat differing structures of human TKs and those of infectious agents enabling the potential for specificity, and the existing wide array of well-characterized, radiolabeled nucleoside analogs for imaging—and therapy—several of which have been administered to human subjects. Relying on EBV-TK has enabled a new molecular radiotherapy for several intractable tumors, with the associated clinical utility currently under evaluation. Challenges include the structural similarities between human, bacterial, and viral TKs, limiting rational design of agents specific to infectious organisms. While FIAU has proved useful in these early imaging studies, it is still phosphorylated by cellular enzymes—or mitochondrial toxicity would not have been demonstrated in the early human therapy trials—despite the putatively inaccessible intracellular location of the enzyme. As more x-ray structures of these kinases emerge, we may find additional information to direct a rational drug (and imaging agent) discovery effort. Even without that information, agents other than radiolabeled FIAU are emerging, such as [^{18}F] FEAU, which may prove superior in certain instances. The only way to realize the full potential of this technique will be to test the emerging agents in relevant experimental systems and ultimately in humans.

Acknowledgments

We acknowledge Dr. Sridhar Nimmagadda and Dr. Yvette Kasamon for helpful discussions. We also acknowledge CA92871, CA138636, and EB009367 for financial support.

References

Adams, B. K., Youssef, I., and Parkar, S., 2006, Absent Tc-99m ciprofloxacin (infecton) uptake in a renal abscess, *Clin Nucl Med* 31: 211–2.

Agrawal, N., Bettegowda, C., Cheong, I. et al., 2004, Bacteriolytic therapy can generate a potent immune response against experimental tumors, *Proc Natl Acad Sci U S A* 101: 15172–7.

Alauddin, M. M., Shahinian, A., Park, R. et al., 2007, *In vivo* evaluation of 2′-deoxy-2′-[(18)F]fluoro-5-iodo-1-beta-D-arabinofuranosyluracil ([18F]FIAU) and 2′-deoxy-2′-[18F] fluoro-5-ethyl-1-beta-D-arabinofuranosyluracil ([18F]FEAU) as markers for suicide gene expression, *Eur J Nucl Med Mol Imaging* 34: 822–9.

Ambinder, R. F. and Cesarman, E., 2007, Clinical and pathological aspects of EBV and KSHV infection. In *Human Herpesviruses Biology, Therapy and Immunoprophylaxis*, A. Arvin, G. Campadelli-Fiume and P. S. Moore (eds), pp. 885–914. New York: Cambridge University Press.

Anton, M., Wagner, B., Haubner, R. et al., 2004, Use of the norepinephrine transporter as a reporter gene for non-invasive imaging of genetically modified cells, *J Gene Med* 6: 119–26.

Beckers, C., Jeukens, X., Ribbens, C. et al., 2006, (18)F-FDG PET imaging of rheumatoid knee synovitis correlates with dynamic magnetic resonance and sonographic assessments as well as with the serum level of metalloproteinase-3, *Eur J Nucl Med Mol Imaging* 33: 275–80.

Beckers, C., Ribbens, C., Andre, B. et al., 2004, Assessment of disease activity in rheumatoid arthritis with (18)F-FDG PET, *J Nucl Med* 45: 956–64.

Benitez, A., Roca, M., and Martin-Comin, J., 2006, Labeling of antibiotics for infection diagnosis, *Q J Nucl Med Mol Imaging* 50: 147–52.

Bennett, J. M., Kaminski, M. S., Leonard, J. P. et al., 2005, Assessment of treatment-related myelodysplastic syndromes and acute myeloid leukemia in patients with non-Hodgkin's lymphoma treated with Tositumomab and Iodine I 131 Tositumomab (BEXXAR(R)), *Blood* 105(12): 4576–82.

Benua, R. S. and Dobyns, B. M., 1955, Iodinated compounds in the serum, disappearance of radioactive iodine from the thyroid, and clinical response in patients treated with radioactive iodine, *J Clin Endocrinol Metab* 15: 118–30.

Bettegowda, C., Dang, L. H., Abrams, R. et al., 2003, Overcoming the hypoxic barrier to radiation therapy with anaerobic bacteria, *Proc Natl Acad Sci U S A* 100: 15083–8.

Bettegowda, C., Foss, C. A., Cheong, I. et al., 2005, Imaging bacterial infections with radiolabeled 1-(2′-deoxy-2′-fluoro-beta-D-arabinofuranosyl)-5-iodouracil, *Proc Natl Acad Sci U S A* 102: 1145–50.

Blasberg, R. G. and Tjuvajev, J. G., 2003, Molecular-genetic imaging: Current and future perspectives, *J Clin Invest* 111: 1620–9.

Bleeker-Rovers, C. P., Boerman, O. C., Rennen, H. J., Corstens, F. H. and Oyen, W. J., 2004, Radiolabeled compounds in diagnosis of infectious and inflammatory disease, *Curr Pharm Des* 10: 2935–50.

Bodei, L., Cremonesi, M., Zoboli, S. et al., 2003, Receptor-mediated radionuclide therapy with 90Y-DOTATOC in association with amino acid infusion: A phase I study, *Eur J Nucl Med Mol Imaging* 30: 207–16.

Brader, P., Stritzker, J., Riedl, C. C. et al., 2008, *Escherichia coli* Nissle 1917 facilitates tumor detection by positron emission tomography and optical imaging, *Clin Cancer Res* 14: 2295–302.

Bradshaw, H. D., Jr. and Deininger, P. L., 1984, Human thymidine kinase gene: Molecular cloning and nucleotide sequence of a cDNA expressible in mammalian cells, *Mol Cell Biol* 4: 2316–20.

Bray, M., Di Mascio, M., de Kok-Mercado, F., Mollura, D. J., and Jagoda, E., 2010, Radiolabeled antiviral drugs and antibodies as virus-specific imaging probes, *Antiviral Res* 88: 129–42.

Buursma, A. R., Rutgers, V., Hospers, G. A. et al., 2006, 18F-FEAU as a radiotracer for herpes simplex virus thymidine kinase gene expression: In-vitro comparison with other PET tracers, *Nucl Med Commun* 27: 25–30.

Byun, Y., Vogel, S. R., Phipps, A. J. et al., 2008, Synthesis and biological evaluation of inhibitors of thymidine monophosphate kinase from *Bacillus anthracis*, *Nucleosides Nucleotides Nucleic Acids* 27: 244–60.

Cannon, J. S., Hamzeh, F., Moore, S., Nicholas, J., and Ambinder, R. F., 1999, Human herpesvirus 8-encoded thymidine kinase and phosphotransferase homologues confer sensitivity to ganciclovir, *J Virol* 73: 4786–93.

Capriotti, G., Chianelli, M., and Signore, A., 2006, Nuclear medicine imaging of diabetic foot infection: Results of meta-analysis, *Nucl Med Commun* 27: 757–64.

Cardillo, T. M., Karacay, H., Goldenberg, D. M. et al., 2004, Improved targeting of pancreatic cancer: Experimental studies of a new bispecific antibody, pretargeting enhancement system for immunoscintigraphy, *Clin Cancer Res* 10: 3552–61.

Chadburn, A., Chen, J. M., Hsu, D. T. et al., 1998, The morphologic and molecular genetic categories of posttransplantation lymphoproliferative disorders are clinically relevant, *Cancer* 82: 1978–87.

Chen, S. T., Estes, J. E., Huang, E. S., and Pagano, J. S., 1978, Epstein–Barr virus-associated thymidine kinase, *J Virol* 26:203–8.

Cheng, Y. C., Dutschman, G., Fox, J. J., Watanabe, K. A., and Machida, H., 1981, Differential activity of potential antiviral nucleoside analogs on herpes simplex virus-induced and human cellular thymidine kinases, *Antimicrob Agents Chemother* 20: 420–3.

Cheung, M. M., Chan, J. K., and Wong, K. F., 2003, Natural killer cell neoplasms: A distinctive group of highly aggressive lymphomas/leukemias, *Semin Hematol* 40: 221–32.

Chiang, A. K., Tao, Q., Srivastava, G., and Ho, F. C., 1996, Nasal NK- and T-cell lymphomas share the same type of Epstein–Barr virus latency as nasopharyngeal carcinoma and Hodgkin's disease, *Int J Cancer* 68: 285–90.

Darouiche, R. O., 2004, Treatment of infections associated with surgical implants, *N Engl J Med* 350: 1422–9.

Davis, J. L., Fei, M., and Huang, L., 2008, Respiratory infection complicating HIV infection, *Curr Opin Infect Dis* 21: 184–90.

Davis, S. L., Be, N. A., Lamichhane, G. et al., 2009, Bacterial thymidine kinase as a non-invasive imaging reporter for *Mycobacterium tuberculosis* in live animals, *PLoS One* 4: e6297.

de Visser, M., Verwijnen, S. M., and de Jong, M., 2008, Update: Improvement strategies for peptide receptor scintigraphy and radionuclide therapy, *Cancer Biother Radiopharm* 23: 137–57.

de Winter, F., van de Wiele, C., Vogelaers, D. et al., 2001, Fluorine-18 fluorodeoxyglucose-position emission tomography: A highly accurate imaging modality for the diagnosis of chronic musculoskeletal infections, *J Bone Joint Surg Am* 83-A: 651–60.

Dempsey, M. F., Wyper, D., Owens, J. et al., 2006, Assessment of 123I-FIAU imaging of herpes simplex viral gene expression in the treatment of glioma, *Nucl Med Commun* 27: 611–7.

Deville-Bonne, D., El Amri, C., Meyer, P. et al., 2010, Human and viral nucleoside/nucleotide kinases involved in antiviral drug activation: Structural and catalytic properties, *Antiviral Res* 86: 101–20.

Diaz, L. A., Jr., Cheong, I., Foss, C. A. et al., 2005, Pharmacologic and toxicologic evaluation of *C. novyi*-NT spores, *Toxicol Sci* 88: 562–75.

Diaz, L. A., Jr., Foss, C. A., Thornton, K. et al., 2007, Imaging of musculoskeletal bacterial infections by [124I]FIAU-PET/CT, *PLoS One* 2: e1007.

Doyle, T. C., Burns, S. M., and Contag, C. H., 2004, *In vivo* bioluminescence imaging for integrated studies of infection, *Cell Microbiol* 6: 303–17.

Dutta, P., Bhansali, A., Mittal, B. R., Singh, B., and Masoodi, S. R., 2006, Instant 99mTc-ciprofloxacin scintigraphy for the diagnosis of osteomyelitis in the diabetic foot, *Foot Ankle Int* 27: 716–22.

Dwyer, R. M., Bergert, E. R., O'Connor M, K., Gendler, S. J., and Morris, J. C., 2005, *In vivo* radioiodide imaging and treatment of breast cancer xenografts after MUC1-driven expression of the sodium iodide symporter, *Clin Cancer Res* 11: 1483–9.

El-Haddad, G., Zhuang, H., Gupta, N., and Alavi, A., 2004, Evolving role of positron emission tomography in the management of patients with inflammatory and other benign disorders, *Semin Nucl Med* 34: 313–29.

El-Maghraby, T. A., Moustafa, H. M., and Pauwels, E. K., 2006, Nuclear medicine methods for evaluation of skeletal infection among other diagnostic modalities, *Q J Nucl Med Mol Imaging* 50: 167–92.

Elwell, L. P., Ferone, R., Freeman, G. A. et al., 1987, Antibacterial activity and mechanism of action of 3′-azido-3′-deoxythymidine (BW A509U), *Antimicrob Agents Chemother* 31: 274–80.

Eriksson, S., Munch-Petersen, B., Johansson, K., and Eklund, H., 2002, Structure and function of cellular deoxyribonucleoside kinases, *Cell Mol Life Sci* 59: 1327–46.

Fridland, A., Johnson, M. A., Cooney, D. A. et al., 1990, Metabolism in human leukocytes of anti-HIV dideoxypurine nucleosides, *Ann N Y Acad Sci* 616: 205–16.

Fu, D.-X., Tanhehco, Y., Chen, J. et al., 2008, Bortezomib-induced enzyme-targeted radiation therapy in herpesvirus-associated tumors, *Nat Med (NY, United States)* 14: 1118–1122.

Fu, D. X., Foss, C. A., Nimmagadda, S., Ambinder, R. F., and Pomper, M. G., 2008, Imaging virus-associated cancer, *Curr Pharm Des* 14: 3048–65.

Fu, D. X., Tanhehco, Y. C., Chen, J. et al., 2007, Virus-associated tumor imaging by induction of viral gene expression, *Clin Cancer Res* 13: 1453–8.

Chapter 10

Gambhir, S. S., Bauer, E., Black, M. E. et al., 2000, A mutant herpes simplex virus type 1 thymidine kinase reporter gene shows improved sensitivity for imaging reporter gene expression with positron emission tomography, *Proc Natl Acad Sci U S A* 97: 2785–90.

Gemmel, F., Dumarey, N., and Welling, M., 2009, Future diagnostic agents, *Semin Nucl Med* 39: 11–26.

Gerth, M. L. and Lutz, S., 2007, Mutagenesis of non-conserved active site residues improves the activity and narrows the specificity of human thymidine kinase 2, *Biochem Biophys Res Commun* 354: 802–7.

Glaser, S. L., Lin, R. J., Stewart, S. L. et al., 1997, Epstein–Barr virus-associated Hodgkin's disease: Epidemiologic characteristics in international data, *Int J Cancer* 70: 375–82.

Goebel, M., Rosa, F., Tatsch, K. et al., 2007, [Diagnosis of chronic osteitis of the bones in the extremities. Relative value of F-18 FDG-PET], *Unfallchirurg* 110: 859–66.

Goldenberg, D. M., 2003, Advancing role of radiolabeled antibodies in the therapy of cancer, *Cancer Immunol Immunother* 52: 281–96.

Goldenberg, D. M., 2002, Targeted therapy of cancer with radiolabeled antibodies, *J Nucl Med* 43: 693–713.

Goldenberg, D. M. and Sharkey, R. M., 2006, Advances in cancer therapy with radiolabeled monoclonal antibodies, *Q J Nucl Med Mol Imaging* 50: 248–64.

Gratz, S., Behe, M., Boerman, O. C. et al., 2001, (99m)Tc-E-selectin binding peptide for imaging acute osteomyelitis in a novel rat model, *Nucl Med Commun* 22: 1003–13.

Hobbs, R. F., Baechler, S., Fu, D. X., Esaias, C., Pomper, M. G., Ambinder, R. F., and Sgouros, G., 2011, A model of cellular dosimetry for macroscopic tumors in radiopharmaceutical therapy. *Med Phys* 38: 2892–903.

Hubert-Habart, M. and Cohen, S. S., 1962, The toxicity of 9-beta-D-arabinofuranosyladenine to purine-requiring *Escherichia coli*, *Biochim Biophys Acta* 59: 468–71.

Jacobs, A., Voges, J., Reszka, R. et al., 2001, Positron-emission tomography of vector-mediated gene expression in gene therapy for gliomas, *Lancet* 358: 727–9.

Johansson, K., Ramaswamy, S., Ljungcrantz, C. et al., 2001, Structural basis for substrate specificities of cellular deoxyribonucleoside kinases, *Nat Struct Biol* 8: 616–20.

Kaminski, M. S., Tuck, M., Estes, J. et al., 2005, 131I-tositumomab therapy as initial treatment for follicular lymphoma, *N Engl J Med* 352: 441–9.

Kang, K. W., Min, J. J., Chen, X., and Gambhir, S. S., 2005, Comparison of [14C]FMAU, [3H]FEAU, [14C]FIAU, and [3H]PCV for monitoring reporter gene expression of wild type and mutant herpes simplex virus type 1 thymidine kinase in cell culture, *Mol Imaging Biol* 7: 296–303.

Keegan, T. H., Glaser, S. L., Clarke, C. A. et al., 2005, Epstein–Barr virus as a marker of survival after Hodgkin's lymphoma: A population-based study, *J Clin Oncol* 23: 7604–13.

Keidar, Z., Engel, A., Hoffman, A., Israel, O., and Nitecki, S., 2007, Prosthetic vascular graft infection: The role of 18F-FDG PET/CT, *J Nucl Med* 48: 1230–6.

Klatzmann, D., 1996, Gene therapy for metastatic malignant melanoma: Evaluation of tolerance to intratumoral injection of cells producing recombinant retroviruses carrying the herpes simplex virus type 1 thymidine kinase gene, to be followed by ganciclovir administration, *Hum Gene Ther* 7: 255–67.

Koehne, G., Doubrovin, M., Doubrovina, E. et al., 2003, Serial *in vivo* imaging of the targeted migration of human HSV-TK-transduced antigen-specific lymphocytes, *Nat Biotechnol* 21: 405–13.

Koral, K. F., Adler, R. S., Carey, J. E., and Beierwaltes, W. H., 1986, Iodine-131 treatment of thyroid cancer: Absorbed dose calculated from post-therapy scans, *J Nucl Med* 27: 1207–11.

Kosinska, U., Carnrot, C., Sandrini, M. P. et al., 2007, Structural studies of thymidine kinases from *Bacillus anthracis* and *Bacillus cereus* provide insights into quaternary structure and conformational changes upon substrate binding, *FEBS J* 274: 727–37.

Lankinen, P., Makinen, T. J., Poyhonen, T. A. et al., 2008, (68) Ga-DOTAVAP-P1 PET imaging capable of demonstrating the phase of inflammation in healing bones and the progress of infection in osteomyelitic bones, *Eur J Nucl Med Mol Imaging* 35: 352–64.

Lee, E. W., Lai, Y., Zhang, H., and Unadkat, J. D., 2006, Identification of the mitochondrial targeting signal of the human equilibrative nucleoside transporter 1 (hENT1): Implications for interspecies differences in mitochondrial toxicity of fialuridine, *J Biol Chem* 281: 16700–6.

Likar, Y., Dobrenkov, K., Olszewska, M. et al., 2008, A new acycloguanosine-specific supermutant of herpes simplex virus type 1 thymidine kinase suitable for PET imaging and suicide gene therapy for potential use in patients treated with pyrimidine-based cytotoxic drugs, *J Nucl Med* 49: 713–20.

Likar, Y., Zurita, J., Dobrenkov, K. et al., 2010, A new pyrimidine-specific reporter gene: A mutated human deoxycytidine kinase suitable for PET during treatment with acycloguanosine-based cytotoxic drugs, *J Nucl Med* 51: 1395–403.

Lin, J., Roy, V., Wang, L. et al., 2010, 3′-(1,2,3-Triazol-1-yl)-3′-deoxythymidine analogs as substrates for human and *Ureaplasma parvum* thymidine kinase for structure-activity investigations, *Bioorg Med Chem* 18: 3261–9.

Liu, M. Y., Pai, C. Y., Shieh, S. M. et al., 1992, Cloning and expression of a cDNA encoding the Epstein–Barr virus thymidine kinase gene, *J Virol Methods* 40: 107–18.

Makinen, T. J., Lankinen, P., Poyhonen, T. et al., 2005, Comparison of 18F-FDG and 68Ga PET imaging in the assessment of experimental osteomyelitis due to *Staphylococcus aureus*, *Eur J Nucl Med Mol Imaging* 32: 1259–68.

Mandell, R. B., Mandell, L. Z., and Link, C. J., Jr., 1999, Radioisotope concentrator gene therapy using the sodium/iodide symporter gene, *Cancer Res* 59: 661–8.

McKenzie, R., Fried, M. W., Sallie, R. et al., 1995, Hepatic failure and lactic acidosis due to fialuridine (FIAU), an investigational nucleoside analogue for chronic hepatitis B, *N Engl J Med* 333: 1099–105.

Meller, J., Koster, G., Liersch, T. et al., 2002, Chronic bacterial osteomyelitis: Prospective comparison of (18)F-FDG imaging with a dual-head coincidence camera and (111)In-labelled autologous leucocyte scintigraphy, *Eur J Nucl Med Mol Imaging* 29: 53–60.

Meller, J., Sahlmann, C. O., Liersch, T., Hao Tang, P., and Alavi, A., 2007, Nonprosthesis orthopedic applications of (18)F fluoro-2-deoxy-D-glucose PET in the detection of osteomyelitis, *Radiol Clin North Am* 45: 719–33, vii–viii.

Meng, Q., Hagemeier, S. R., Fingeroth, J. D. et al., 2010. The Epstein–Barr virus (EBV)-encoded protein kinase, EBV-PK, but not the thymidine kinase (EBV-TK), is required for ganciclovir and acyclovir inhibition of lytic viral production, *J Virol* 84(9): 4534–42.

Miyagawa, T., Gogiberidze, G., Serganova, I. et al., 2008, Imaging of HSV-tk Reporter gene expression: Comparison between [18F]FEAU, [18F]FFEAU, and other imaging probes, *J Nucl Med* 49: 637–48.

Mizgerd, J. P., 2008, Acute lower respiratory tract infection, *N Engl J Med* 358: 716–27.

Moore, S. M., Cannon, J. S., Tanhehco, Y. C., Hamzeh, F. M., and Ambinder, R. F., 2001, Induction of Epstein–Barr virus kinases to sensitize tumor cells to nucleoside analogues, *Antimicrob Agents Chemother* 45: 2082–91.

Murray, P. G., Constandinou, C. M., Crocker, J., Young, L. S., and Ambinder, R. F., 1998, Analysis of major histocompatibility complex class I, TAP expression, and LMP2 epitope sequence in Epstein–Barr virus-positive Hodgkin's disease, *Blood* 92:2477–83.

Oyen, W. J., Bodei, L., Giammarile, F. et al., 2007, Targeted therapy in nuclear medicine—Current status and future prospects, *Ann Oncol* 18: 1782–92.

Palestro, C. J., Love, C., and Miller, T. T., 2007, Diagnostic imaging tests and microbial infections, *Cell Microbiol* 9: 2323–33.

Papadopoulos, E. B., Ladanyi, M., Emanuel, D. et al., 1994, Infusions of donor leukocytes to treat Epstein–Barr virus-associated lymphoproliferative disorders after allogeneic bone marrow transplantation, *N Engl J Med* 330: 1185–91.

Parvizi, J., Ghanem, E., Menashe, S., Barrack, R. L., and Bauer, T. W., 2006, Periprosthetic infection: What are the diagnostic challenges? *J Bone Joint Surg Am* 88(Suppl 4): 138–47.

Perez-Perez, M. J., Priego, E. M., Hernandez, A. I. et al., 2008, Structure, physiological role, and specific inhibitors of human thymidine kinase 2 (TK2): Present and future, *Med Res Rev* 28: 797–820.

Perrine, S. P., Hermine, O., Small, T. et al., 2007, A phase 1/2 trial of arginine butyrate and ganciclovir in patients with Epstein–Barr virus-associated lymphoid malignancies, *Blood* 109: 2571–8.

Pomper, M. G., 2009, Letter to the editor, *Semin Nucl Med* 39: 354.

Ponomarev, V., Doubrovin, M., Shavrin, A. et al., 2007, A human-derived reporter gene for noninvasive imaging in humans: Mitochondrial thymidine kinase type 2, *J Nucl Med* 48: 819–26.

Reilly, H. C., 1953, Microbiology and cancer therapy: A review, *Cancer Res* 13: 821–34.

Rightmire, E., Zurakowski, D., and Vrahas, M., 2008, Acute infections after fracture repair: Management with hardware in place, *Clin Orthop Relat Res* 466: 466–72.

Ruggiero, A., Brader, P., Serganova, I. et al., 2010, Different strategies for reducing intestinal background radioactivity associated with imaging HSV1-tk expression using established radionucleoside probes, *Mol Imaging* 9: 47–58.

Sandrini, M. P., Clausen, A. R., Munch-Petersen, B., and Piskur, J., 2006, Thymidine kinase diversity in bacteria, *Nucleosides Nucleotides Nucleic Acids* 25: 1153–8.

Sarda, L., Cremieux, A. C., Lebellec, Y. et al., 2003, Inability of 99mTc-ciprofloxacin scintigraphy to discriminate between septic and sterile osteoarticular diseases, *J Nucl Med* 44: 920–6.

Savoldo, B., Rooney, C. M., Di Stasi, A. et al., 2007, Epstein–Barr virus specific cytotoxic T lymphocytes expressing the anti-CD30zeta artificial chimeric T-cell receptor for immunotherapy of Hodgkin disease, *Blood* 110: 2620–30.

Schipper, M. L., Goris, M. L., and Gambhir, S. S., 2007, Evaluation of herpes simplex virus 1 thymidine kinase-mediated trapping of (131)I FIAU and prodrug activation of ganciclovir as a synergistic cancer radio/chemotherapy, *Mol Imaging Biol* 9: 110–6.

Segura-Pena, D., Lutz, S., Monnerjahn, C., Konrad, M., and Lavie, A., 2007, Binding of ATP to TK1-like enzymes is associated with a conformational change in the quaternary structure, *J Mol Biol* 369: 129–41.

Serganova, I., Ponomarev, V., and Blasberg, R., 2007, Human reporter genes: Potential use in clinical studies, *Nucl Med Biol* 34: 791–807.

Shibata, D., Tokunaga, M., Uemura, Y. et al., 1991, Association of Epstein–Barr virus with undifferentiated gastric carcinomas with intense lymphoid infiltration. Lymphoepithelioma-like carcinoma, *Am J Pathol* 139: 469–74.

Shirley, C. M., Chen, J., Shamay, M. et al., 2011, Bortezomib induction of C/EBP{beta} mediates Epstein–Barr virus lytic activation in Burkitt's lymphoma, *Blood* 117(23): 6297–303.

Soghomonyan, S. A., Doubrovin, M., Pike, J. et al., 2005, Positron emission tomography (PET) imaging of tumor-localized Salmonella expressing HSV1-TK, *Cancer Gene Ther* 12: 101–8.

Stabin, M. G., Sparks, R. B., and Crowe, E., 2005, OLINDA/EXM: The second-generation personal computer software for internal dose assessment in nuclear medicine, *J Nucl Med* 46: 1023–7.

Thakur, M. L., Lavender, J. P., Arnot, R. N., Silvester, D. J., and Segal, A. W., 1977, Indium-111-labeled autologous leukocytes in man, *J Nucl Med* 18: 1014–21.

Thorley-Lawson, D. A. and Gross, A., 2004, Persistence of the Epstein–Barr virus and the origins of associated lymphomas, *N Engl J Med* 350: 1328–37.

Tjuvajev, J., Blasberg, R., Luo, X. et al., 2001, Salmonella-based tumor-targeted cancer therapy: Tumor amplified protein expression therapy (TAPET) for diagnostic imaging, *J Control Release* 74: 313–5.

Tjuvajev, J. G., Avril, N., Oku, T. et al., 1998, Imaging herpes virus thymidine kinase gene transfer and expression by positron emission tomography, *Cancer Res* 58: 4333–41.

Tjuvajev, J. G., Doubrovin, M., Akhurst, T. et al., 2002, Comparison of radiolabeled nucleoside probes (FIAU, FHBG, and FHPG) for PET imaging of HSV1-tk gene expression, *J Nucl Med* 43: 1072–83.

Tjuvajev, J. G., Finn, R., Watanabe, K. et al., 1996, Noninvasive imaging of herpes virus thymidine kinase gene transfer and expression: A potential method for monitoring clinical gene therapy, *Cancer Res* 56: 4087–95.

Tjuvajev, J. G., Stockhammer, G., Desai, R. et al., 1995, Imaging the expression of transfected genes *in vivo*, *Cancer Res* 55: 6126–32.

Toso, J. F., Gill, V. J., Hwu, P. et al., 2002, Phase I study of the intravenous administration of attenuated *Salmonella typhimurium* to patients with metastatic melanoma, *J Clin Oncol* 20: 142–52.

Voges, J., Reszka, R., Gossmann, A. et al., 2003, Imaging-guided convection-enhanced delivery and gene therapy of glioblastoma, *Ann Neurol* 54: 479–87.

Wang, J. and Eriksson, S., 1996, Phosphorylation of the anti-hepatitis B nucleoside analog 1-(2′-deoxy-2′-fluoro-1-beta-D-arabinofuranosyl)-5-iodouracil (FIAU) by human cytosolic and mitochondrial thymidine kinase and implications for cytotoxicity, *Antimicrob Agents Chemother* 40: 1555–7.

Welin, M., Kosinska, U., Mikkelsen, N. E. et al., 2004, Structures of thymidine kinase 1 of human and mycoplasmic origin, *Proc Natl Acad Sci U S A* 101: 17970–5.

Welker, J. A., Huston, M. and McCue, J. D., 2008, Antibiotic timing and errors in diagnosing pneumonia, *Arch Intern Med* 168: 351–6.

Westphal, E. M., Blackstock, W., Feng, W., Israel, B., and Kenney, S. C., 2000, Activation of lytic Epstein–Barr virus (EBV) infection by radiation and sodium butyrate *in vitro* and *in vivo*: A potential method for treating EBV-positive malignancies, *Cancer Res* 60: 5781–8.

Chapter 10

Witt, A., Williams, R., and Pierce, R., 1993, *Report of an FDA Task Force. Fialuridine: Hepatic and Pancreatic Toxicity*. Washington, DC: Food and Drug Administration, pp. 1–91.

Wu, T. C., Mann, R. B., Charache, P. et al., 1990, Detection of EBV gene expression in Reed–Sternberg cells of Hodgkin's disease, *Int J Cancer* 46: 801–4.

Yaghoubi, S. S., Jensen, M. C., Satyamurthy, N. et al., 2009, Noninvasive detection of therapeutic cytolytic T cells with 18F-FHBG PET in a patient with glioma, *Nat Clin Pract Oncol* 6: 53–8.

Young, L. S. and Rickinson, A. B., 2004, Epstein–Barr virus: 40 years on, *Nat Rev Cancer* 4: 757–68.

Zhou, S., 2005, Combination therapy with bacteria and angiogenesis inhibitors: Strangling cancer without mercy, *Cancer Biol Ther* 4: 846–7.

Zielinski, F. W., MacDonald, N. S., and Robinson, G. D., Jr., 1977, Production by compact cyclotron of radiochemically pure iodine-123 as iodide for synthesis of radiodiagnostic agents, *J Nucl Med* 18: 67–9.

11. Design Criteria for Targeted Molecules
Muscarinic Cholinergic Systems Biology

William C. Eckelman

11.1 Introduction

The major challenge in radiopharmaceutical design 40 years ago was to develop receptor-targeted probes with the highest affinity possible. Then, the pharmaceutical industry approach involved leads from natural products and endogenous neurotransmitters or from screening organic compounds for a specific action. This usually produced low-affinity, but highly lipophilic, drugs that did not provide high target-to-nontarget ratios *in vivo*, but were considered effective drugs then (Haber 1983). By the 1980s, higher-affinity radiotracers were developed and specific binding to a target protein was clearly detected using external imaging. The challenge shifted to developing a probe at high specific activity that was sensitive to changes in target protein as a function of disease. The initial biodistribution of very high-affinity compounds targeting high-density proteins was heavily influenced by perfusion and permeability and made the determination of changes in target protein less robust even at increased time after injection.

This can be illustrated by the equations discussed by Carson (1996) where K_1 and k_2 are the rate constants for transfer of radiotracer from the plasma to the tissue (K_1) in compartment C_1 and the reverse process (k_2) and k_3

Targeted Molecular Imaging. Edited by Michael J. Welch and William C. Eckelman © 2012 Taylor & Francis Group, LLC. ISBN: 978-1-4398-4195-2

and k_4 are the on-rate and off-rate constants, respectively, for binding to the target protein in compartment 2 (C_2). For the case where k_4 is small or zero, the concentration as a function of time is given by

$$C_1(t) = C_a K_1 \exp\left[-(k_2 + k_3)t\right],$$
$$C_2(t) = C_a \left(K_1 k_3 / (k_2 + k_3)\right)\left(1 - \exp\left[-(k_2 + k_3)t\right]\right),$$

where C_a is the magnitude of an idealized bolus injection.

The rate constant k_4, the off-rate of receptor binding, is often negligible at the chosen imaging interval, but there certainly are examples where k_4 has an impact on biodistribution. The observed tissue concentration is the sum of $C_1(t)$ and $C_2(t)$. For target protein binding in C_2, if k_3 is larger than k_2, then $k_3/(k_2 + k_3)$ is nearly 1, and so the concentration in C_2 is dependent only on $C_a K_1$. This is the case for an "irreversible" ligand such as N-methyl spiperone. Likewise, if k_2 is zero, then the distribution is dependent on $C_a K_1$, that is, influenced by flow and permeability. This is the case for microspheres where the particles are trapped quantitatively in the capillaries (k_3) with no release (k_2).

Nevertheless, if radiotherapy with analogs is the long-term goal, such high-affinity compounds have an advantage because they produce higher target-to-nontarget ratios and therefore higher delivered radiation dose to the target. The caveat is that this applies to high-energy β-particles and not for low-energy emissions resulting from electron capture, α-particles, or toxins delivered by antibodies. Saga et al. (1995) presented evidence for a "binding site barrier" in both bulky tumors and in micrometastases using radiolabeled antibodies. Nonspecific antibodies showed greater penetration into the tumor than specific antibodies that bound predominately on the surface of the tumor. The penetration of the specific antibody could be increased by reducing the specific activity at some cost in specific binding. To put this into perspective, the physical properties of several radiotherapeutic nuclides illustrates the depth of penetration of the β-particle with cell diameter (Carrasquillo and Harbert 1996) (Table 11.1). A similar study has been carried out

Table 11.1 Properties of β- and α-Particle Emissions

Radionuclide	Half-Life (h)	Decay Mode	Maximum Energy (MeV)	Maximum Range in Tissue (μm)
I-125	1440	Electron capture	0.008–0.035	15
Cu-67	62.4	β, γ	0.39–0.58	2200
I-131	192	β, γ	0.6	2400
Re-186	88.8	β, γ	1.02	2000
Re-188	17	β	2.13	11,100
Y-90	64.8	β	2.27	11,900
Bi-212	1	α	6.1	70
At-211	7	α	5.9–7.4	50–90

Source: Carrasquillo JA, Harbert JC, *Nuclear Medicine: Diagnosis and Therapy*, Harbert J, Neumann R, Eckelman W., eds, New York: Thieme Press, 1996, pp. 1125–39 (Reprinted with permission).

using small molecules where the distribution throughout the tumor is more homogenous. One possible explanation for the penetration of a small molecule is that the increased permeability will counteract the binding site barrier effect (Moyes et al. 1989).

For diagnostic purposes, the somewhat lower-affinity compounds produce radiotracers that are more sensitive to changes in target protein density. The shift to probes with equilibrium dissociation constants closer to the target protein density lessened the dependence of the biodistribution on perfusion and permeability. The target-to-nontarget ratio is lower in this case, but the sensitivity to changes in target protein density is increased. The biodistribution of [18F]FP-TZTP produces relatively small target-to-background ratios, but is sensitive to changes in synaptic acetylcholine concentration (see Section 11.9). Although there are several other examples of probes of high and lower affinity for the same target protein (so-called irreversible and reversible, respectively), the high-affinity probe, *RS* 4-[123I]IQNB, and the lower-affinity probe, [18F]FP-TZTP, serve to illustrate the advantages and disadvantages of each approach.

11.2 Choice of Target Protein

There is a hypothesis for specific transmitter system involvement in most diseases, simply because complicated diseases are unlikely to depend on a single biochemical system. (In neurodegenerative disease, these

are called neurotransmitters; in endocrine diseases they are called hormones; and in cancer, they are called signaling pathways.) The cholinergic hypothesis proposes that Alzheimer's disease (AD) is caused by reduced synthesis of the neurotransmitter acetylcholine and is one of the oldest models for drug design (Francis et al. 1999). The cholinergic hypothesis led to the use of inhibitors of acetylcholine esterase (AChE) and muscarinic agonists, but neither have had a dramatic clinical effect. The National Institute of Neurological Disorders and Stroke states:

> Currently there are no medicines that can slow the progression of AD. However, four FDA-approved medications are used to treat AD symptoms. These drugs help individuals carry out the activities of daily living by maintaining thinking, memory, or speaking skills. They can also help with some of the behavioral and personality changes associated with AD. However, they will not stop or reverse AD and appear to help individuals for only a few months to a few years. Donepezil (Aricept), rivastigmine (Exelon), and galantamine (Reminyl) are prescribed to treat mild to moderate AD symptoms. Donepezil was recently approved to treat severe AD as well. The newest AD medication is memantine (Namenda), which is prescribed to treat moderate to severe AD symptoms (www.ninds.nih.gov).

Except for memantine, a moderate affinity N-methyl-D-aspartate (NMDA)-receptor antagonist, these pharmaceuticals all decrease the degradation of acetylcholine by blocking the AChE.

Many neurological and psychiatric diseases have been associated with various neurotransmitter system abnormalities. For example, there is a cholinergic hypothesis of depression, and so the choice of radioligands targeting the mAChR is a reasonable choice in this disease also. Drevets first explored *positron emission tomography* (PET) radiotracers targeted to receptor subtypes and transporters for the serotonin system and, more recently, for the muscarinic system using [^{18}F]FP-TZTP with interesting results (Savitz and Drevets 2009).

In the pregenomic era, the choice of probe to be radiolabeled was often driven by the affinity constant, the ease of synthesis, or autopsy data. Since affinity alone was not sufficient to gauge potential *in vivo* radiotracers, a relationship with the target protein density was introduced as a further refinement although this has been used to exclude candidates more than to vouch for their potential as a radiolabeled probe that could be monitored using external imaging (Eckelman 1982; Katzenellenbogen et al. 1982). At high specific activity, the maximal B/F ratio will be B_{max}/K_D based on the Scatchard transformation of the law of mass action. In the Scatchard transformation ($B/F = B_{max}/K_D - B/K_D$),

the second term is theoretically negligible when the concentration of the bound radiotracer (B) is at high specific activity and therefore much smaller than the total available target protein (B_{max}). Less than 5% occupancy is the usual rule of thumb, but in small animal experiments that can be a challenge. In practice, distribution factors, nonspecific protein binding, metabolism, and other interactions will decrease the maximal B/F ratio when the radiolabeled molecular probe is used *in vivo*. Therefore, this criterion is necessary, but not sufficient, for probes to be used *in vivo*. But this estimation is especially important as targeting of proteins with low nM to pM concentration becomes more prevalent. Only the combination of B_{max}/K_D not K_D alone will accurately estimate the probability of obtaining a reasonable ratio *in vivo*. Of course, if a radiotracer meets this criterion, the influence on flow, permeability, or target protein density will still be controlled by the relationship of k_2 and k_3 as described earlier. A clear demonstration of the use of B_{max}/K_D is the work of Kung who targeted transporters of various neurotransmitters and showed the requirement for the affinity constant as the transporter concentration decreased from a high for the dopamine transporter to a low of the norepinephrine transporter in the brain (Kung et al. 2004; Kung and Kung 2005). The receptor density of muscarinic receptors in the guinea pig heart was quoted as 15 nM and the affinity of both quinuclidinyl benzylate (QNB) and methylquinuclidinyl benzilate (MQNB) were determined to be K_D of 0.30 nM. Using this paradigm, the maximal ratio of target density to equilibrium dissociation constant would be ~50, which is encouraging and would not exclude either QNB or MQNB as potential probes for the muscarinic receptor (Gibson et al. 1979). In fact, tritiated analogs gave heart-to-blood ratios of ~20 in guinea pigs (Gibson et al. 1979). On the other hand, there are examples where the ratios *in vivo* do not approach those obtained using the calculated B_{max}/K_D (Eckelman et al. 2009). For example, it is instructive to consider amyloid-targeted radioligand-binding results (Eckelman et al. 2009). The *in vitro* B_{max} value of Pittsburgh Compound B (PIB)-binding sites in human AD brain tissues is in the range of 1000–2000 nM, and the K_D value of PIB to these sites is about 2 nM. Therefore, the calculated B_{max}/K_D ratio in AD brain is in the range of 500–1000. Yet, the maximum *in vivo* binding potential (BP) value of PIB in AD subjects is only about 2. The reason for this 500-fold difference is likely the result of a very low free fraction in plasma (<0.01 and difficult to accurately quantify), a K_1/k_2 ratio of about 2 and, therefore, a free fraction in tissue of <0.005. Thus, the measured *in vivo* BP values are in the expected range of about 2 in AD subjects. The free fraction in both the plasma and the tissue is often measured under equilibrium conditions; however, the off-rate from plasma may be more critical. Furthermore, there are limited data on how uniform the free fraction in tissue is in various structures of the brain.

For *in vivo* studies, the concept of BP was introduced (Mintun et al. 1984) and led to modifications to account for other factors such as free fraction either in plasma or in the target milieu, nonspecific binding, the exclusion of metabolites, and specific binding to other targets among others (Innis et al. 2007). Mintun et al. attempted to develop an analysis that led to the identification of B_{max} and K_i from *in vivo* studies. As B_{max} and K_i are correlated metrics and difficult to determine independently using a no-carrier-added concentration of radioligand, the BP, which is B_{max}/K_i, could be identified with higher precision. Also, the number of parameters became such that identifiability of unique values was not possible. The challenge of determining B_{max} and K_i independently was a known problem from *in vitro* analysis. The maximal sensitivity for determining B_{max} and K_i in *in vivo* studies was obtained using a concentration of ligand that occupied a significant fraction of the target protein (Vera et al. 1992). Given the physiologic effect of most radioligands at concentrations that would saturate the target protein to that extent, this approach was only possible in a few instances. The group at Orsay used a cold dose following the no-carrier-added radioligand to extract B_{max} densities from studies using radiolabeled β-adrenoceptor ligands (Delforge et al. 2002).

The evolution of drugs for the muscarinic acetylcholine receptor is a prime example of starting with lower-affinity compounds that may be effective drugs, but are not useful radiopharmaceuticals and moving to high-affinity "irreversible" probes and then on to lower-affinity "reversible" probes. Atropine, a muscarinic receptor antagonist, was used as a cosmetic in medieval times to dilate the pupil (hence the name belladonna) and has been used as a pharmaceutical for many years. As tritiated receptor ligands became available, [³H]atropine was tested as a probe of the mAChR receptor. As early as 1973, Farrow and O'Brien (1973) used [³H]atropine to define the mAChR in tissue. Low specific activity and relatively low affinity in combination with a relatively low receptor density led to high nonspecific binding that resulted in minimal specific binding even when studied *in vitro*. Other H-3 compounds were tested, but they also showed little or no specific binding (Eckelman 1982). Not until the high-affinity tritiated 3-QNB was released by the Army for research use, were *in vitro* binding studies of the mAChR possible. For example, receptor distribution could be mapped in isolated tissue using [³H]QNB (Yamamura et al. 1974; Yamamura and Snyder 1974). *In vivo*, the *N*-methyl analog (MQNB) had sufficient affinities such that [³H]MQNB gave high heart-to-blood ratios in guinea pigs (>30 at 1/4 and 2 h) and the uptake was blocked by atropine (Eckelman et al. 1979; Francis et al. 1982a). However, tritiated atropine and *N*-methyl atropine did not bind specifically *in vivo*, which was demonstrated by using co-injected blocking doses of atropine, which was in line with the earlier *in vitro* results. After considering these data, QNB was radioiodinated to give *RS* 4-[¹²⁵I]IQNB and tested in animals. It showed specific binding using blocking studies. *RS* 4-[¹²³I]IQNB was the first ligand to map mAChR in a human using external imaging (Eckelman et al. 1984). The *RS* 4-[¹²³I]IQNB scan was performed on May 11, 1983, and was followed closely by the first PET neuroreceptor study with *N*-[¹¹C]methylspiperone, which mapped the dopamine receptor (Wagner et al. 1983).

11.3   Chemical Identity and Quantitation of the Probes

The radioiodination of QNB in the para-position of the benzylic acid moiety was achieved though a triazene intermediate. The source of iodide and various solvents were tested at equimolar concentrations and then at no-carrier-added concentrations. Trifluoroethanol solvent produced the highest radiochemical yields (~15%) (Francis et al. 1982b). The product, IQNB, contained two chiral centers and four diastereomers. *R*-quinuclidinyl was commercially available but a chiral-substituted benzylic acid was not. The stereochemistry of IQNB was assigned by analogy with the affinity of the α-hydroxy-α-(aryl or alkyl) analog. The ester of the *R*-enantiomer had a higher affinity for the muscarinic receptor than the *S*-enantiomer (Rzeszotarski et al. 1984). This turned out to be an invalid analogy. As a result, before 1997, the stereochemistry of the high-affinity diastereomer at the benzilate center was incorrectly characterized as the *RR* diastereomer. The correct assignment was determined from biological experiments (Zeeberg et al. 1997) and from chemical structure studies (Kiesewetter et al. 1994). Post-1997 publications assigning the stereochemistry of *RS* IQNB as the high-affinity diastereomer are correct. In this review, the correct diastereomer is used independent of the publication date.

The muscarinic agonist approach to the treatment of AD was based on a series of compounds first proposed by Sauerberg et al. (1992) of Novo-Nordisk. These ligands contain a thiadiazolyl moiety attached to various heterocycles, including tetrahydropyridine. Two of these compounds, xanomeline and butylthio-TZTP, demonstrated M1 selectivity and had been labeled with C-11 and studied using PET imaging (Farde et al. 1996). Another compound, (3-(propylthio)-1,2,5-thiadiazol-4-yl)-tetrahydro-1-methylpyridine (P-TZTP), was reported by Novo-Nordisk to be M2 selective. In a NovaScreen assay using brain and heart tissue, it showed specificity for the M2 receptor. The work of Sauerberg et al. was extended at NIH by radiofluorinating P-TZTP at the 3-position of the propyl side chain (Figures 11.1 and 11.2). A more systematic approach would have been to radiolabel P-TZTP with C-11 first and then compare that with FPTZTP. That is what Reid et al. (2008) did after [^{18}F]FP-TZTP was validated as an analog of P-TZTP. They compared the original Sauerberg et al. drug, P-TZTP, with two analogs, namely FP-TZTP and F$_3$P-TZTP, all labeled with C-11 at the methyl pyridine moiety. Values for log *D*, plasma protein binding, and affinity constants were similar for all three. The trifluoro analog was the outlier in measurements of area under the curve and time–activity curves (TACs) for brain regions. P-TZTP and FP-TZTP showed pharmacokinetics (PK) and distribution volumes (DVs) similar to that of [^{18}F]FP-TZTP. Uptake of the three tracers was significantly reduced by coinjection of TZTP indicating saturable binding.

The radiosynthesis and preliminary biodistribution of [^{18}F]FP-TZTP demonstrated M2 selectivity *in vitro*. *In vivo* studies with [^{18}F]FP-TZTP in rat brain showed that

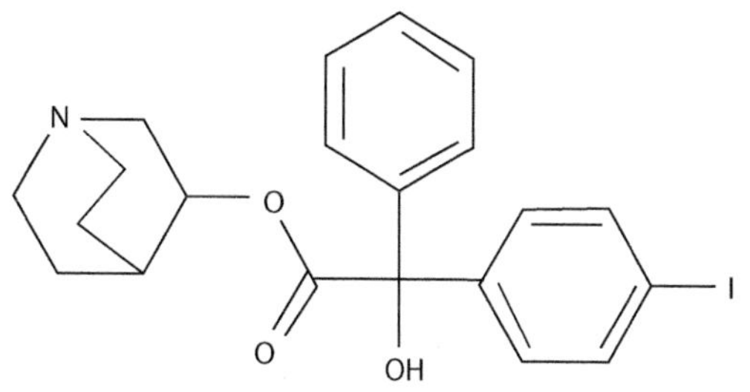

FIGURE 11.1   Structure of 3-quinuclidinyl 4-[^{123}I]iodobenzilate (4-[^{123}I]IQNB). (Adapted from http://pubchem.ncbi.nlm.nih.gov/summary/summary.cgi?cid=122241&loc=ec_rcs)

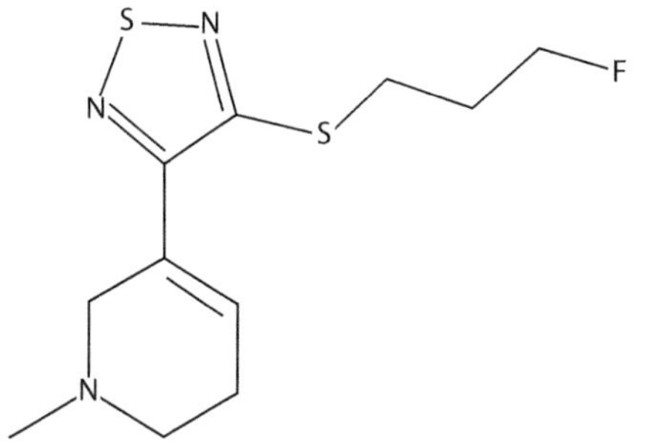

FIGURE 11.2   Structure of 3-(3-[^{18}F]fluoranylpropylsulfanyl)-4-(1-methyl-3,6-dihydro-2H-pyridin-5-yl)-1,2,5-thiadiazole(([^{18}F] FP-TZTP). (Adapted from http://pubchem.ncbi.nlm.nih.gov/summary/summary.cgi?cid=131459&loc=ec_rcs)

the uptake at early times was similar to that obtained with *RS*-[^{18}F]FMeQNB, but the net efflux was faster. Autoradiography using no-carrier-added [^{18}F]FP-TZTP confirmed the uniform distribution of radioactivity throughout the gray matter characteristic of the M2 pattern of localization. This was consistent with M2 specificity, and eliminated M1 subtype receptor as a possible target because its distribution was shown from autopsy data to be concentrated in specific gray matter such as striatum and cortex, but not in cerebellum (Volpicelli and Levey 2004), but it did leave alternate interpretations such as blood flow and so was not a definitive proof.

11.4   Specific Activity

When discussing specific activity it is important to differentiate between specific activity, effective specific activity, and biochemical specific activity (Kilbourn 1990). I-123 iodide produced from a Xe target has a measured specific activity within a factor of 2 of the theoretical specific activity of 237,404 Ci/mmol (8784 TBq/mmol). F-18 fluoride, on the other hand, has been produced on average at a specific activity of 5000 Ci/mmol (185 TBq/mmol), which is several orders of magnitude from the theoretical specific activity for F-18 of 1,712,422 Ci/mmol (63,360 TBq/mmol) (Eckelman et al. 2008). The measured specific activity of the final product is usually less than the specific activity of the starting radionuclide due to isotope dilution from stable isotope in the precursor and reagents used in the preparation. For example, any hydroxide reagent will introduce a small amount of carrier fluoride.

The specific activity of the final product averaged 1100 Ci/mmol (40.7 TBq/mmol) for [125I]IQNB and 552 Ci/mmol (20.4 TBq/mmol) for [123I]IQNB using I-123 from the Crocker Nuclear Laboratory. The radiochemical yield for [123I]IQNB was half of that obtained with [125I]IQNB. I-123 obtained from Brookhaven National Laboratory gave a lower specific activity that averaged 388 Ci/mmol (14.4 TBq/mmol), but a higher yield that averaged 18%. Tailing of the triazene precursor, which eluted first, was one cause of the reduced specific activity. If 1% of the 1 mg of the triazene precursor was present in the collected product volume of 10 mCi, then the specific activity would be limited by that mass and would be of the order of 30 nmol. This would limit the specific activity to <333 Ci/mmol (Rzeszotarski et al. 1984).

The specific activity of [18F]FP-TZTP using the original synthetic scheme was 4377 ± 2011 Ci/mmol (161.9 ± 74.4 TBq/mmol) [end of bombardment (EOB), $n = 100$], whereas the automated procedure by Kiesewetter et al. (2003) gave a specific activity of 4112 ± 2572 Ci/mmol (152.1 ± 25.2 TBq/mmol) (EOB, $n = 25$). A more recent automated synthesis by van Oosten et al. (2009) gave a specific activity of 3732 ± 1109 Ci/mmol (138.7 ± 41.0 TBq/mmol) (EOS, $n = 3$).

11.5 Specificity of Targeting the Muscarinic Receptor Both *In Vitro* and *In Vivo*

The definition of target specificity for a particular radioligand can have various interpretations. Specificity can be determined for target class (muscarinic receptor among all biogenic amines) or subtype specificity of the target class (muscarinic M2 versus muscarinic M1, M3, M4, and M5). The specificity of [123I]IQNB for the muscarinic receptor was determined using muscarinic antagonists, usually QNB or nonradioactive IQNB. Inhibition of *RS* 4-[125I]IQNB and [3H]QNB by cholinergic drugs with similar binding specificity, but different chemical platforms such as atropine, scopalamine, benactazine, oxotremorine, and pilocarpine, produced similar K_i values using rat corpus striatum, which is consistent with mAChR binding (Gibson et al. 1984).

11.5.1 Saturating the Receptor with Nonradioactive Compound by Preinjection, Coinjection or Postinjection

The traditional pharmacologic proof requires (a) specific binding to the target protein in a standard screen containing related receptors and enzymes, (b) subtype specificity demonstrated by competitive binding studies using known subtype-specific ligands, and (c) binding to tissue *in vivo* in proportion to known target protein concentrations. The pharmacologic proof for a subtype-binding radiotracer was unconvincing for [18F]FP-TZTP because the M2 subtype is distributed homogenously throughout the gray matter and there are no high-affinity agonists of a different chemical platform that can be used to block the binding in the brain. Also, in the muscarinic receptor system, antagonists do not block agonist binding and so that approach could not be used (Herschberg et al. 1995). The only blocking studies used the same class of compounds, either P-TZTP or F-TZTP. Coinjection of P-TZTP at 5, 50, and 500 nmol followed by sacrifice at 1 h inhibited [18F]FP-TZTP uptake in a dose-dependent manner (Figure 11.3). The brain distribution of the agonist [18F]FP-TZTP was unaffected by coinjection of 5, 50, or 500 nmol of the antagonist *RS*-IQNB or of QNB itself (Kiesewetter et al. 1995, 1999). Furthermore, inhibition studies, especially with agonists, can produce pharmacologic changes and mask the intended blocking study; therefore, this must be carefully evaluated. Shimoji et al. (2003) showed statistically significant synchronous decreases in both cerebral blood flow (CBF) and mean arterial blood pressure

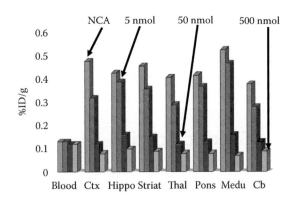

FIGURE 11.3 Uptake of [18F]FP-TZTP in rat brain at 60 min after injection with a co-injection of 0 (NCA), 5, 50, 500 nmol of F-TZTP with [18F]FP-TZTP (cortex (ctx), the hippocampus (Hippo), striatum (Striat), thalamus (Thal), pons, medulla (Med), and cerebellum (Cb). (Adapted from Kiesewetter DO et al. 1995. *J. Med. Chem.* 38:5–8.)

(MABP) within the first minute following administration of inhibiting doses of nonradioactive P-TZTP. If this persisted throughout the biodistribution, a change in biodistribution due to blood flow changes could confound the interpretation of blocking studies. The decreases in both CBF and MABP were prevented by pretreatment with atropine methyl bromide (M-At), a peripheral muscarinic antagonist, and coadministration of M-At with either FP-TZTP or P-TZTP resulted in the same degree of inhibition of cerebral [18F]FP-TZTP uptake at 30 min after administration as observed without M-At. Also, with programmed infusions, which avoid changes in CBF and are designed to produce constant arterial concentrations of [18F]FP-TZTP and FP-TZTP, significant inhibition of [18F]FP-TZTP-binding by FP-TZTP was observed (Shimoji et al. 2003). These results indicate that inhibition of [18F]FP-TZTP-binding in the brain by P-TZTP or FP-TZTP *in vivo* occurs independently of their transient effects on CBF. These methods should be employed for other radiotracers to evaluate physiological effects of blocking agents used to validate other radiopharmaceuticals, especially when the radiotracer is an agonist.

11.5.2 Use of Mice Genetically Manipulated to Suppress the Expression of the Target Protein

However, experiments with knockout (KO) mice led to a convincing proof of the M2 selectivity of [18F] FP-TZTP without the complications of injecting pharmacological doses of drug. Biodistribution of [18F] FP-TZTP was studied in wild-type (WT) mice and genetically engineered mice lacking functional M1, M2, M3, or M4 muscarinic receptors (Jagoda et al.

Wildtype
M1 KO
M3 KO
M4 KO

M2 Knockout

FIGURE 11.4 Gene-manipulated mice with either the M1, the M2, the M3, or the M4 gene deleted, compared to the wild type control at 2 h after injection of [18F]FP-TZTP. On the left hand side is the autoradiograph of a central 20 μm sagittal slice of mouse brain for the control and the M1, M3, and M4 KOs. All are identical indicating that [18F]FP-TZTP binding was not affected. On the right hand side, is the autoradiograph of the M2 KO mouse. There is a clear decrease in binding of [18F]FP-TZTP suggesting M2 specificity for [18F]FP-TZTP. (Adapted from Jagoda EM et al. 2003. *Neuropharmacology* 44(5):653–61.)

2003). Using *ex vivo* autoradiography, the regional brain localization of [18F]FP-TZTP in M2 KO mice was significantly decreased (51.3–61.4%) compared to the WT mice in amygdala, brain stem, caudate putamen, cerebellum, cortex, hippocampus, hypothalamus, superior colliculus, and thalamus. In similar studies with M1 KO, M3 KO, and M4 KO mice compared to WT mice, [18F]FP-TZTP uptakes in the same brain regions were not significantly decreased. Given the fact that large decreases in [18F]FP-TZTP brain uptakes were seen only in M2 KO versus WT mice, [18F] FP-TZTP preferentially labels M2 receptors *in vivo* (Figure 11.4). KO mice are a valuable tool for validating the subtype selectivity and hold great promise for accelerating radioligand development (Eckelman 2003).

11.6 Lipophilicity and Nonspecific Binding in the Brain or Body

As shown in PubChem, 4-IQNB has a molecular weight (MW) of 463.30875, a computed XLogP3-AA of 3.9 (Cheng et al. 2007), one H-bond donor, and four H-bond acceptors. FP-TZTP has an MW of 272.395778, an experimental Log *D* of 2.4 ± 0.1, no H-bond donor, and five H-bond acceptors. Lipinski's rule of five states that an orally active drug can have no more than one violation of the following criteria (Lipinski et al. 1997):

- Not more than five hydrogen bond donors (nitrogen or oxygen atoms with one or more hydrogen atoms)

- Not more than 10 hydrogen bond acceptors (nitrogen or oxygen atoms)
- An MW under 500 Da
- A log of the octanol–water partition coefficient (log P) of <5

There have been refinements of the rule of five (Vistoli et al. 2008), but it is clear from the MWs and the log *P* values, respectively, that FP-TZTP should have the better transport across the blood–brain barrier (BBB) and the lesser nonspecific binding compared to *RS* 4-[123I] IQNB.

Chapter 11

11.7   Mathematical Analysis for Converting ϒ-Ray Emission Rates to Biochemical Parameters

The PK of *RS* 4-[^{123}I]IQNB was studied in rats and humans. The most comprehensive analysis of the kinetics of *RS* 4-[^{123}I]IQNB was carried out by Sawada et al. (1990) in rats. This analysis showed that the uptake in cerebrum is unidirectional during the first 360 min after intravenous administration and that the rate of *RS* [^{123}I]IQNB tissue uptake depends on transport across the BBB as well as the rate of binding to the receptor. However, the data at 24 h is sensitive to receptor concentration. Sawada et al. (1990) using graphical analysis with noniterative linear regression models such as the Patlak plot (Carson 1996) showed significant differences between the *K* for cortex and thalamus, a result not likely to be attributed to flow, but rather receptor density differences. It is difficult to relate these data in rats to equivalent times in human studies, but in rats clearly a receptor-dependent distribution can be obtained at greater than 6 h postinjection.

Since blocking studies cannot be used in humans because of the toxicity of QNB, Hiramatsu et al. (1995) studied the high-affinity diastereomer *RS* 4-[^{123}I]IQNB versus *SS* 4-[^{123}I]IQNB to determine specific binding. In order for the nonbinding stereoisomer to be a control for the mAChR-binding isomer, all properties should be the same, except for the receptor binding. Otherwise, interpretation of the results can be confounded by other differences than mAChR binding. In this case, the low-affinity diastereomer *SS* 4-[^{123}I]IQNB showed much faster clearance from blood than *RS* 4-[^{123}I]IQNB and had a different metabolite profile. Also, the transport kinetics of the enantiomers were different. The estimated BP (proportional to B_{max}/K_D) of *RS* 4-[^{123}I]IQNB was highest in two cortical regions, intermediate in parotid gland, and lowest in cerebellum. As a result, *SS* 4-[^{123}I]IQNB is not an ideal general probe to measure nonspecific binding associated with *RS* 4-[^{123}I]IQNB-specific binding in humans.

The input function for *RS* 4-[^{123}I]IQNB is a key factor. Difference in plasma concentration in four patients injected with RS 4-[^{123}I]IQNB and four patients injected with SS 4-[^{123}I]IQNB as a function of time showed a significant difference (Figure 11.5). Given the rapid clearance of both isomers, the requirement for a delayed image at 24 h to move from a biodistribution dependent on flow and permeability to one dependent on receptor density is not straightforward. But the answer

could be in the lung clearance of the parent compound back into the plasma producing an infusion-like PK. Lung clearance as a function of time in four patients injected with *RS* 4-[^{123}I]IQNB and four patients injected with *SS* 4-[^{123}I]IQNB also showed a difference (Figure 11.6). Although the clearance from plasma is rapid, the %ID/g in brain tissue continues to increase over 24 h. This suggests that the lung is releasing intact *RS* 4-[^{123}I] IQNB that can be taken up by the brain.

In monkeys, control and competition studies for [^{18}F] FP-TZTP were carried out by Carson et al. (1998). [^{18}F] FP-TZTP uptake in the brain was rapid with K_1 values of 0.4–0.6 mL min^{-1} mL^{-1} in gray matter; delivery is flow limited and the shape of the tissue TAC is heavily influenced by flow measured using [^{15}O]H$_2$O flow studies in the same animal. A model with one tissue compartment gave volume of distribution (DV) values that were very similar in cortical regions, basal ganglia, and thalamus, but significantly lower in the cerebellum, consistent with the distribution of M$_2$ cholinergic receptors in monkey. Preblocking studies with unlabeled FP-TZTP reduced *V* by 60–70% in cortical and subcortical regions; approximately one-third of the binding was nonspecific. Physostigmine produced a

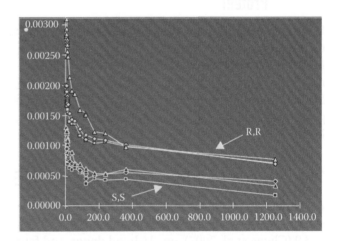

FIGURE 11.5 Difference in plasma concentration in four patients injected with RS 4-[^{123}I]IQNB and four patients injected with SS 4-[^{123}I]IQNB as a function of time. The Y-axis is %ID/g (note the rapid clearance and low concentration at the first measurement point). The X-axis is in seconds. The four upper curves are the RS 4-[^{123}I]IQNB; the four lower curves are the SS 4-[^{123}I] IQNB. The input function for the two isomers is not equal. Therefore, SS 4-[^{123}I]IQNB does not serve as a true tracer for the blood clearance of RS 4-[^{123}I]IQNB. (Adapted from Hiramatsu Y et al. 1995. *Am J Physiol* 268(6 Pt 2):R1491–9).

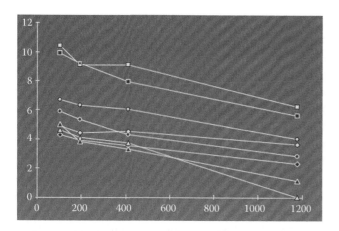

FIGURE 11.6 Difference in lung clearance as a function of time in four patients injected with RS 4-[^{123}I]IQNB (solid symbols) and four patients injected with SS 4-[^{123}I]IQNB (open symbols). The Y-axis is %ID/g (note the high concentration of radioactivity in the lungs). The X-axis is in seconds. Although the clearance from plasma is rapid, the %ID/g in brain tissue continues to increase over 24 hours. This suggests that the lung is releasing intact RS 4-[^{123}I]IQNB. (Adapted from Hiramatsu Y et al. 1995. *Am J Physiol* 268(6 Pt 2):R1491–9.)

35% reduction in cortical specific binding, but the reduction in basal ganglia (12%) was significantly smaller, consistent with its markedly higher AChE activity and reflecting the interaction of synaptic acetylcholine, blocking of AChE by physostigmine, and degradation of acetylcholine by unblocked AChE. Since M2 receptors are distributed throughout gray matter and therefore a receptor-free reference tissue was not available, a metabolite-corrected plasma input function is required to calculate the BP. Recently, Ichise et al. (2008) developed a procedure using a tissue reference model, which eliminated the need for a plasma input function. Without using arterial data or a receptor-free reference region, a receptor parameter called the normalized DV, VT, using a region containing receptors as the input tissue, was developed. Bias for this approach has been evaluated carefully using simulation studies and comparisons with the data obtained using metabolite-corrected plasma input function. This is a practical advantage for the widespread use of this radiotracer.

11.8 Sensitivity of the Detection to a Change in the Target Protein Using Either *In Vitro, Ex Vivo* Biodistribution or *In Vivo* Imaging

The use of receptor-binding radiotracers *in vivo* differs in one important aspect from those used to determine receptor concentration *in vitro*. The biodistribution of the radiotracer *in vivo* must at some point be controlled by the receptor concentration rather than by either blood flow or transport properties. In the case of both *RS* [^{123}I]IQNB and [^{18}F] FP-TZTP, the early biodistribution is influenced by flow and at later times by the systems biology of the muscarinic receptor system. There have been a number of clinical studies using *RS* 4-[^{123}I]IQNB that suggest the ligand is responsive to flow at earlier times and changes in receptor concentration at later imaging times (Weinberger et al. 1990, 1991, 1992) (Figure 11.7 redrawn from Weinberger et al. 1990). Lee et al. (1996) compared the image of the four isomers in humans (Figure 11.8). Likewise, there are examples of changes in [^{18}F]FP-TZTP distribution related to the changes in the systems biology of the muscarinic system in both normal subjects with APOE 4$^+$ allele (Cohen et al. 2003) and patients with bipolar disorder (Cannon et al. 2006).

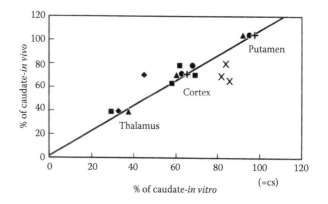

FIGURE 11.7 Data from studies of 4-[^{123}I]IQNB at 24 h after injection in humans show the relationship of %ID/g in various gray matter regions and the B$_{max}$ taken from literature autopsy data. This suggests a sensitivity of 4-[^{123}I]IQNB binding as a function of the receptor density (B$_{max}$). (Adapted from Weinberger et al. 1990. *Adv Neurol* 51:147–50.)

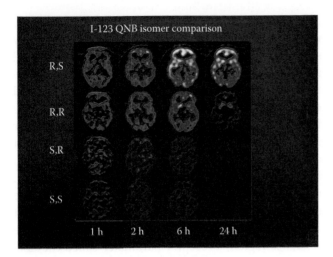

FIGURE 11.8 Injection of the four isomers of 4-[¹²³I]IQNB in humans at 24 h after injection. A well defined difference in the retention of 4-[¹²³I]IQNB is observed among the four isomers. The first designation is for the quinuclidinol moiety and the second designation is for the 4-iodobenzilic acid. RS 4-[¹²³I]IQNB is the diastereomer that has been used in the clinic. (Reprinted by permission of the Society of Nuclear Medicine from Lee KS et al. Improved method for rapid and efficient radioiodination of iodine-123-IQNB. *J Nucl Med.* 1996; 37(12):2021–4. Figure 5.)

11.9 Expected Effect on Clinical Care

Weinberger et al. (1990, 1991, 1992) found clear differences in the image of normals and AD patients performed at 21 h after injection of 5 mCi (185 MBq) of *RS* 4-[¹²³I]IQNB with single-photon emission computed tomography (SPECT). In normal subjects, the images showed a regional pattern that correlated with postmortem studies of the relative distribution of muscarinic receptors in the normal human brain. They observed high radioactivity counts in the basal ganglia, occipital cortex, and insular cortex, low counts in the thalamus, and very low counts in the cerebellum (Figure 11.7). Eight of 12 patients with a clinical diagnosis of AD had obvious focal cortical defects in either frontal or posterior temporal cortex. A region-of-interest statistical analysis of relative regional radioactivity revealed a significant reduction bilaterally in the posterior temporal cortex of the patients with AD compared with normal subjects. In spite of these encouraging data, this radioprobe has not been used extensively because of the flow dependence at early times after injection, which often required two visits to the clinic to obtain the receptor-dependent image. The low sensitivity of SPECT and resolution in the 1980s were additional factors.

The clinical indication most often stated for investigations into radiolabeled probes for the mAChR is suspected AD. A theme throughout the history of radiopharmaceutical development is that many radioligands have been developed without a clear path to clinical impact, that is, creating a radiotracer that binds with high sensitivity to receptor density changes is necessary, but not sufficient, to guarantee its use in the clinic. The reasons for the lack of clinic impact are usually a combination of the lack of drugs to treat the disease or that studies in chronic diseases have progressed to a point where treatment will not be successful. Also, the inability to identify a key target in the progression of the disease in the pregenomic era is a factor. Effort to develop muscarinic receptor-binding radiotracers has been based on deficits found at autopsy studies of AD patients. However, subsequent studies showed that the mAChR density at autopsy varied from downregulated to no change to upregulated and may be related to treatment rather than alterations in the mAChR genome (Teri et al. 2008).

Ng and Hag (2004) recently suggested that pharmaceutical research (and by analogy radiopharmaceutical research) needs to concentrate on finding the right drug against the right target to treat the right disease. In the postgenomic era, one trend in pharmaceutical design is to choose protein expression products that are related to know genetic defects as targets. The current state of molecular genetics of AD focuses on four genes: the β-amyloid precursor protein (β-APP) gene, presenilin 1 gene, presenilin 2 gene, and the apolipoprotein E (APOE) gene (Reitz and Mayeux 2010). The epsilon 4 allele of the (APOE) gene is associated with an increased risk for late-onset AD (>50 years old). Few of the imaging studies to date have been based on these genetic abnormalities. The characteristic neuropathology is neurofibrillary tangles and amyloid plaques. The result of the phenotypic change is a result of an alteration in the processing of β-APP favoring the production of the potentially toxic Abeta₄₂ protein. Pathologists have for many years developed *in vitro* staining reagents for such plaques and radioligands targeted to Abeta₄₂ protein *in vivo* are the most active areas in radiopharmaceutical research (Klunk and Mathis 2008, Kadir and Nordberg 2010). This approach would be considered pregenomic in that the search for radioligands is based on autopsy findings. There has also been a report on the use of [¹⁸F]FDG in a subpopulation of AD patients with the APOE gene (Reiman et al. 2004) although this is an indirect measure of the phenotype. The authors reported on changes in FDG distribution before clinical manifestations.

The primary focus of the PET Department at the Clinical Center of the National Institutes of Health was the development of M2 subtype selective cholinergic ligands, based on the pregenomic observation that this subtype is lost in the cerebral cortex in AD (Quirion et al. 1989; Aubert et al. 1992; Rodriguez-Puertas et al. 1997). In several publications, postmortem quantitation of muscarinic subtypes indicated a selective loss of M2 subtype in cortical regions while the M1 subtype was preserved. Thus, an M2-selective ligand labeled with a positron-emitting radionuclide would allow determination of M2 subtype concentrations in the living human brain. As with *RS* IQNB, most tracers for cholinergic receptors have not demonstrated subtype selectivity. Alternatively, ligands that were developed for *in vitro* applications and are subtype selective typically do not cross the BBB. To develop imaging agents with subtype selectivity, two approaches have been used. In one approach, a ligand that is selective *in vitro*, AF-DX 116 [11-(((2-(diethylamino)-methyl)-1-piperidinyl) acetyl)-5-11-dihydro-6H-pyrido(2,3)(1,4) benzodiazepin-6-one], was modified to increase its BBB permeability while maintaining subtype selectivity (Doods et al. 1993). In the second approach, nonsubtype-selective *RS* IQNB, which crosses into the brain, was modified to analogs that showed increased subtype selectivity (McPherson et al. 1993, 1995). Neither of these approaches has yielded a compound sensitive to changes in a subtype of the muscarinic receptor. *RS*-[¹⁸F]FMeQNB was the most M2 selective of the QNB analog, but defluorination of the benzylfluoride was unpredictable (Lee et al. 1995) (Figure 11.9). Nevertheless, the heart, which contains a high M2 receptor density, was imaged in a monkey, but the uptake in the bone consistent with defluorination was evident (Figure 11.10).

In studies began in normal human volunteers with the eventual goal of studying patients with AD (Podruchny et al. 2003), the initial analysis, based on data from six young control subjects, concentrated on developing the appropriate kinetic model for [¹⁸F]FP-TZTP in humans. In plasma, parent compound represented 68 ± 8, 41 ± 9, and 14 ± 4% of radioactivity at 20, 40, and 120 min, respectively. A model with one tissue compartment produced an excellent fit for the full 120 min of data, so that the additional parameters of a two-compartment model were unidentifiable. *K* values in gray matter regions were high, 0.36–0.56 mL/min mL⁻¹, and showed excellent correlation with cerebral blood flow. *V* values, representing total tissue binding,

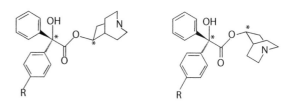

R group	M1 K_D (nM)	M2 K_D (nM)	M3 K_D (nM)
RR fluoromethyl	0.11	0.84	92.2
RS fluoromethyl	0.89	0.13	168
SR fluoromethyl	17.9	11.6	10000
SS fluoromethyl	2.9	16.7	9.5
RR fluoroethyl	0.84	7.6	10000
RS fluoroethyl	0.26	0.31	84.7
RS fluoropropyl	0.20	0.38	59.4
RS IQNB	0.34	4.2	8.1

RS 4-fluoromethylQNB has the greatest M2 selectivity.

FIGURE 11.9 *In vitro* binding data for M2 selective radiotracers based on QNB. (Adapted from Lee JT et al. 1995. *Nucl Med Biol* 22:773–81.)

were very similar in cortical regions, basal ganglia, and thalamus, but were significantly higher (*P* < 0.01) in amygdala. Unlike the results in the monkey, binding in cerebellum was similar to that in the cerebral cortex.

In the first clinical studies, an age-related increase in M2 receptor BP was observed using [¹⁸F]FP-TZTP and PET in normal subjects (Podruchny et al. 2003). There was a significant increase not only in the average DV, but also an increase in variance in the older subjects. This implies that other aspects of the systems biology were coming into play and therefore the data should be modeled by a more complex set of equations. The investigators found that the increase in

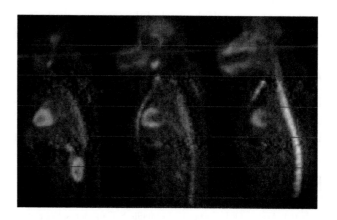

FIGURE 11.10 PET Image (sagittal slices) of R,S [¹⁸F]FMeQNB at 2–3 hours after IV injection of 3 mCi in a baboon. (Adapted from Lee JT et al. 1995. *Nucl Med Biol* 22:773–81.)

Chapter 11

variance for the DV was caused by using only age discrimination. This suggested that the age-related changes might be associated with a specific genotype such as the APOE-epsilon 4+ allele. In the following study, Cohen et al. (2003) found that the gray matter DV for [^{18}F]FP-TZTP was significantly higher in the APOE-epsilon 4+ normal subjects than in the APOE-epsilon-4-normal subjects, whereas there were no differences in global cerebral blood flow. It appears that the agonist [^{18}F]FP-TZTP measures more than just receptor density changes as antagonists do because of possible changes in receptor-binding affinity due to changes in G protein coupling, competition with ACh at the muscarinic receptor, and changes in receptor density. A reasonable hypothesis for the increased DV in elderly normal subjects with APOE-epsilon 4+ is a decreased concentration of ACh in the synapse, which would lead to a higher binding of [^{18}F]FP-TZTP. This type of competition was shown to be possible in the studies in monkeys using physostigmine. However, other factors in the systems biology may come into play as well. Nevertheless, such studies in normal subjects with a genetic predisposition could be key to early detection and the ability to monitor treatment at an early stage of the disease if the preliminary studies by Cohen et al. are supported by larger clinical trials.

What has been the clinical impact of muscarinic receptor-binding radiotracers on patient management? It appears that the use of *RS* 4-[^{123}I]IQNB is now closely tied to the pharmaceutical industry in that imaging has become a major part of the drug development paradigm (Eckelman 2006). Certainly, the pharmaceutical industry can make use of PET and SPECT imaging for preclinical and Phase 0/I studies to identify and quantitate the extent of the drug's interaction with a particular receptor. Examples are in the literature documenting the interaction of olanzapine, risperidone, clozapine, donepezil, and phenserine with the muscarinic receptor using *RS* 4-[^{123}I]IQNB. [^{18}F]FP-TZTP is more likely to be used to identify early disease if clinical studies support the early findings.

11.10   Conclusion

This case study illustrates several points concerning design criteria for targeted molecules. First, the cart should not be before the horse. A clear relationship between the target protein and a disease should be established before the radioligand is chosen, remembering that few diseases are single-gene and single-protein expression product disease. From the technical viewpoint, the ideal affinity and targeted protein density are dependent on the use. If the imaging is to be used as a biomarker for radiotherapy and as a radiotherapeutic with a β-emitting analog, the higher the B_{max}/K_D the better it is to deliver maximal radiation to the target protein. If the goal of the imaging study is to monitor the change in protein density as a function of disease or treatment, then a lower affinity for a particular protein density is optimal. The BP is usually of the order of 2–5 in this case. Radiolabeled agonist (and perhaps antagonist as well) must be evaluated for a confounding physiological effect in saturability studies. Diastereomer pairs of high affinity and low affinity are not always paired in affinity alone and are difficult to interpret to separate specific and nonspecific binding. KO mice are invaluable in validating a tracer's binding specificity and saturability. In general, the right target that is key to a disease process should be chosen first and then the right radioligand for the target should be pursued to increase the probability that the radioligand will play a role in clinical management.

Acknowledgments

I thank my colleagues at George Washington University from 1972 to 1983 and my colleagues in the PET Department/National Institutes of Health from 1983 to 1985 and 1991 to 2004 for their efforts that led to these radiotracers.

References

Aubert I, Araujo DM, Cecyre D, Robitaille Y, Gauthier S, Quirion R 1992, Comparative alterations of nicotinic and muscarinic binding sites in Alzheimer's and Parkinson's diseases. *J Neurochem* 58:529–41.

Cannon DM, Carson RE, Nugent AC et al. 2006, Reduced muscarinic type 2 receptor binding in subjects with bipolar disorder. *Arch Gen Psychiatry* 63:741–7.

Carrasquillo JA, Harbert JC 1996, Radioimmunotherapy. In *Nuclear Medicine: Diagnosis and Therapy* (Harbert J, Neumann R, Eckelman W, eds), pp. 1125–39, New York: Thieme Press.

Carson RE, Kiesewetter DO, Jagoda E, Der MG, Herscovitch P, Eckelman WC 1998, Muscarinic cholinergic receptor measurements with [18F]FP-TZTP: Control and competition studies. *J Cereb Blood Flow Metab* 18:1130–42.

Carson, RE 1996, Mathematical modeling and compartmental analysis. In *Nuclear Medicine: Diagnosis and Therapy* (Harbert J, Neumann R, Eckelman W, eds), pp. 167–94, New York: Thieme Press.

Cheng T, Zhao Y, Li X et al. 2007, Computation of octanol-water partition coefficients by guiding an additive model with knowledge. *J Chem Inf Model* 47:2140–8.

Cohen RM, Podruchny TA, Bokde AL et al. 2003, Higher *in vivo* muscarinic-2 receptor distribution volumes in aging subjects with an apolipoprotein E-epsilon4 allele. *Synapse* 49(3):150–6.

Delforge J, Mesangeau D, Dolle F, Merlet P, Loc'h C, Bottlaender M, Trebossen R, Syrota A 2002, *In vivo* quantification and parametric images of the cardiac beta-adrenergic receptor density. *J Nucl Med* 43(2):215–26.

Doods H, Entzeroth M, Ziegler H, Schiavi G, Engel W, Mihm G, Rudolf K, Eberlein W 1993, Characterization of BIBN 99: A lipophilic and selective muscarinic M2 receptor antagonist. *Eur J Pharmacol* 242:23–30.

Eckelman WC 1982, Radiolabeled adrenergic and muscarinic blockers for *in vivo* studies. In *Receptor Binding Radiotracers* (Eckelman WC, ed.), 1:69–91, Boca Raton, FL: CRC Press.

Eckelman WC 2003, The use of gene-manipulated mice in the validation of receptor binding radiotracer. *Nucl Med Biol* 30(8):851–60.

Eckelman WC 2006, Imaging of muscarinic receptors in the central nervous system. *Curr Pharm Des* 12(30):3901–13.

Eckelman WC, Bonardi M, Volkert WA 2008, True radiotracers: Are we approaching theoretical specific activity with Tc-99 m and I-123? *Nucl Med Biol* 35(5):523–7.

Eckelman WC, Kilbourn MR, Mathis CA 2009, Specific to nonspecific binding in radiopharmaceutical studies: It's not so simple as it seems! *Nucl Med Biol* 36(3):235–7.

Eckelman WC, Reba RC, Gibson RE, Rzeszotarski WJ, Vieras F, Mazaitis JK, Francis B 1979, Receptor-binding radiotracers: A class of potential radiopharmaceuticals. *J Nucl Med* 20(4):350–7.

Eckelman WC, Reba RC, Kelloff GJ 2008, Targeted imaging: An important biomarker for understanding disease progression in the era of personalized medicine. *Drug Discov Today* 13(17–18):748–59.

Eckelman WC, Reba RC, Rzeszotarski WJ, Gibson RE, Hill T, Holman BL, Budinger T, Conklin JI, Eng R, Grissom MP 1984, External imaging of cerebral muscarinic acetylcholine receptors. *Science* 223:291–3.

Farde L, Suhara T, Halldin C et al. 1996, PET study of the M1-agonists [11C]xanomeline and [11C]butylthio-TZTP in monkey and man. *Dementia* 7:187–95.

Farrow JT, O'Brien RD 1973, Binding of atropine and muscarone to rat brain fractions and its relation to the acetylcholine receptor. *Mol Pharmacol* 9(1):33–40.

Francis B, Eckelman WC, Grissom MP, Gibson RE, Reba RC 1982a, The use of tritium labeled compounds to develop gamma-emitting receptor-binding radiotracers. *Int J Nucl Med Biol* 9(3):173–9.

Francis BE, Rzeszotarski WJ, Eckelman WC, Reba RC 1982b, Nucleophilic iodination of 3-quinuclidinyl benzilate (QNB). *J Labelled Cpd Radiopharm* 19:1499–500.

Francis PT, Palmer AM, Snape M, Wilcock GK 1999, The cholinergic hypothesis of Alzheimer's disease: A review of progress. *J Neurol Neurosurg Psychiatr* 66(2):137–47.

Gibson RE, Eckelman WC Vieras F, Reba RC 1979, The distribution of the muscarinic acetylcholine receptor antagonists, quinuclidinyl benzilate and quinuclidinyl benzilate methiodide (both tritiated), in rat, guinea pig, and rabbit. *J Nucl Med* 20(8):865–70.

Gibson RE, Rzeszotarski WJ, Jagoda EM, Francis BE, Reba RC, Eckelman WC 1984, [125I] 3-Quinuclidinyl 4-iodobenzilate: A high affinity, high specific activity radioligand for the M1 and M2-acetylcholine receptors. *Life Sci* 34(23):2287–96.

Haber E 1983, Antibodies as models for rational drug design. *Biochem Pharmacol* 32:1967–77.

Herschberg BT, Mosser VA, Peterson GL, Toumadje A, Vogel WK, Johnson WC Jr, Schimerlik MI 1995, Kinetic and biophysical analysis of the m2 muscarinic receptors. *Life Sci* 56:907–13.

Hiramatsu Y, Eckelman WC, Carrasquillo JA, Miletich RS, Valdez IH, Kurrasch RH, Macynski AA, Paik CH, Neumann RD, Baum BJ 1995, Kinetic analysis of muscarinic receptors in human brain and salivary gland *in vivo*. *Am J Physiol* 268(6 Pt 2):R1491–9.

Ichise M, Cohen RM, Carson RE 2008, Noninvasive estimation of normalized distribution volume: Application to the muscarinic-2 ligand [18F]FP-TZTP. *J Cerebral Blood Flow Metabolism* 28:420–30.

Innis RB, Cunningham VJ, Delforge J et al. 2007, Consensus nomenclature for *in vivo* imaging of reversibly binding radioligands. *J Cerebral Blood Flow Metabolism* 27:1533–9.

Jagoda EM, Kiesewetter DO, Shimoji K, Ravasi L, Yamada M, Gomeza J, Wess J, Eckelman WC 2003, Regional brain uptake of the muscarinic ligand, [18F]FP-TZTP, is greatly decreased in M2 receptor knockout mice but not in M1, M3 and M4 receptor knockout mice. *Neuropharmacology* 44(5):653–61.

Katzenellenbogen JA, Heiman DF, Carlson KE et al. 1982, *In vitro* and *in vivo* steroid receptor assays in the design of estrogen radiopharmaceuticals, In *Receptor Binding Radiotracers* (Eckelman WC, ed.), pp. 93–126, Boca Raton, FL: CRC Press.

Kadir A, Nordberg A 2010, Target-specific PET probes for neurodegenerative disorders related to dementia. *J Nucl Med* 51(9):1418–30.

Kiesewetter DO, Silverton JV, Eckelman WC 1994, Stereoselective synthesis of [R,R]IQNB and fluoroalkyl analogs of QNB. *J Labelled Cpd Radiopharm* 35:419–21.

Kiesewetter DO, Lee J, Lang L, Park SG, Paik CH, Eckelman WC 1995. Preparation of 18F-labeled muscarinic agonist with M2 selectivity. *J Med Chem* 38:5–8.

Kiesewetter DO, Vuong B-K, Channing MA 2003, The automated radiosynthesis of [18F]FP-TZTP. *Nucl Med Biol* 30:73–7.

Kiesewetter, DO, Carson RE, Jagoda EM, Herscovitch P, Eckelman WC 1999, *In vivo* muscarinic binding of 3-(alkylthio)-3-thiadiazolyl tetrahydropyridines. *Synapse* 31:29–40.

Kilbourn MR 1990, *Fluorine-18 Labeling of Radiopharmaceuticals*. pp. 17–8. Prepared for the Committee on Nuclear and Radiochemistry, National Research Council. Washington (DC): National Academy Press.

Chapter 11

Klunk WE, Mathis CA 2008, The future of amyloid-beta imaging: A tale of radionuclides and tracer proliferation. *Curr Opin Neurol* 21(6):683–7.

Kung M-P, Kung HF 2005, Mass effect of injected dose in small rodent imaging by SPECT and PET. *Nucl Med Biol* 32:673–8.

Kung MP, Choi SR, Hou C et al. 2004, Selective binding of 2-[125I]iodo-nisoxetine to norepinephrine transporters in the brain. *Nucl Med Biol* 31(5):533–41.

Lee JT, Paik CH, Kiesewetter DO et al. 1995, Evaluation of stereoisomers of 4-fluoroalkyl analogues of 3-quinuclidinyl benzilate in *in vivo* competition studies for the M1, M2, and M3 muscarinic receptor subtypes in brain. *Nucl Med Biol* 22:773–81.

Lee KS, He X-S, Jones DW, Coppola R, Gorey JG, Knable MB, deCosta BR, Rice KC, Weinberger DW 1996, An improved method for rapid and efficient radioiodination of iodine-123-IQNB. *J Nucl Med* 37:2021–4.

Lipinski CA, Lombardo F, Dominy BW, Feeney PJ 1997, Experimental and computational approaches to estimate solubility and permeability in drug discovery and development settings. *Adv Drug Del Rev* 23:3–25.

McPherson DW, DeHaven-Hudkins DL, Callahan AP, Knapp FF Jr 1993, Synthesis and biodistribution of iodine-125-labeled 1-azabicyclo[2.2.2]oct-3-yl alpha-hydroxy-alpha-(1-iodo-1-propen-3-yl)-alpha-phenylacetate. A new ligand for the potential imaging of muscarinic receptors by single photon emission computed tomography. *J Med Chem* 36:848–54.

McPherson DW, Lambert CR, Jahn K, Sood V, McRee RC, Zeeberg B, Reba RC, Knapp FF Jr 1995, Resolution and *in vitro* and initial *in vivo* evaluation of isomers of iodine-125-labeled 1-azabicyclo[2.2.2]oct-3-yl alpha-hydroxy-alpha-(1-iodo-1-propen-3-yl)-alpha-phenylacetate: A high-affinity ligand for the muscarinic receptor. *J Med Chem* 38: 3908–17.

Mintun MA, Raichle ME, Kilbourn MR, Wooten GF, Welch MJ 1984, A quantitative model for the *in vivo* assessment of drug binding sites with positron emission tomography. *Ann Neurol* 15:217–27.

Moyes JSE, Babich JW, Carter R, Meller ST, Agrawal M, McElwain TJ 1989, Quantitative study of radioiodinated metaiodobenzylguanidine uptake in children with neuroblastoma: Correlation with tumor histopathology. *J Nucl Med* 30:474–80.

Ng JH, Hag LL 2004, Streamlining drug discovery: Finding the right drug against the right target to treat the right disease. *Drug Discov Today* 9:59–60.

Podruchny TA, Connolly C, Bokde A, Herscovitch P, Eckelman WC, Kiesewetter DO, Sunderland T, Carson RE, Cohen RM 2003, *In vivo* muscarinic 2 receptor imaging in cognitively normal young and older volunteers. *Synapse* 48(1):39–44.

Quirion R, Aubert I, Labchak PA et al. 1989, Muscarinic receptor subypes in human neurodegenerative disorders; focus on Alzheimer's disease. *Trends Pharmacol Sci Suppl.* 80–4.

Reid AE, Ding YS, Eckelman WC, Logan J, Alexoff D, Shea C, Xu Y, Fowler JS 2008, Comparison of the pharmacokinetics of different analogs of 11C-labeled TZTP for imaging muscarinic M2 receptors with PET. *Nucl Med Biol* 35(3):287–98.

Reiman EM, Chen K, Alexander GE, Caselli RJ, Bandy D, Osborne D, Saunders AM, Hardy J 2004, Functional brain abnormalities in young adults at genetic risk for late-onset Alzheimer's dementia. *Proc Natl Acad Sci USA* 101(1):284–9.

Reitz C, Mayeux R 2010, Use of genetic variation as biomarkers for mild cognitive impairment and progression of mild cognitive impairment to dementia. *J Alzheimers Dis* 19(1):229–51.

Rodriguez-Puertas R, Pascual J, Vilaro T, Pazos A 1997, Autoradiographic distribution of M1, M2, M3 and M4 muscarinic receptor subtypes in Alzheimer's disease. *Synapse* 26:341–50.

Rzeszotarski WJ, Eckelman WC, Francis BE, Simms DA, Gibson RE, Jagoda EM, Grissom MP, Eng RR, Conklin JJ, Reba RC 1984, Synthesis and evaluation of radioiodinated derivatives of 1-azabicyclo-(2.2.2.)oct-3-Yl alpha-hydroxy-alpha-(4-iodophenyl)phenylacetate as potential radiopharmaceuticals. *J Med Chem* 27:156–9.

Saga T, Neumann RD, Heya T, Sato J, Kinuya S, Le N, Paik CH, Weinstein JN 1995, Targeting cancer micrometastases with monoclonal antibodies: A binding-site barrier. *Proc Natl Acad Sci USA* 92(19):8999–9003.

Sauerberg P, Olesen PH, Nielsen S et al. 1992, Novel functional M1 selective muscarinic agonists. Synthesis and structure-activityrelationshipsof3-(1,2,5-thiadiazoyl)-1,2,5,6-tetrahydro-1-methylpyridines. *J Med Chem* 35:2274–83.

Savitz JB, Drevets WC 2009, Imaging phenotypes of major depressive disorder: Genetic correlates. *Neuroscience* 164:300–30.

Sawada Y, Hiraga S, Francis B et al. 1990, Kinetic analysis of 3-quinuclidinyl 4-[125I]iodobenzilate transport and specific binding to muscarinic acetylcholine receptor in rat brain *in vivo*: Implications for human studies. *J Cereb Blood Flow Metab* 10(6):781–807.

Shimoji K, Esaki T, Itoh Y et al. 2003, Inhibition of [18F]FP-TZTP binding by loading doses of muscarinic agonists P-TZTP or FP-TZTP *in vivo* is not due to agonist-induced reduction in cerebral blood flow. *Synapse* 50(2):151–63.

Manolio TA, Brooks LD, Collins FS 2008, A HapMap harvest of insights into the genetics of common disease. *J Clin Invest* 118:1590–605.

van Oosten EM, Wilson AA, Stephenson KA, Mamo DC, Pollock BG, Mulsant BH, Yudin AK, Houle S, Vasdev N 2009, An improved radiosynthesis of the muscarinic M2 radiopharmaceutical, [18F]FP-TZTP. *Appl Radiat Isot* 67(4):611–6.

Vera DR, Scheibe PO, Krohn KA, Trudeau WL, Stadalnik RC 1992, Goodness-of-fit and local identifiability of a receptor-binding radiopharmacokinetic system. *IEEE Trans Biomed Eng* 39(4):356–67.

Vistoli G, Pedretti A, Testa B 2008, Assessing drug-likeness—What are we missing? *Drug Discov Today* 13(7–8):285–94.

Volpicelli LA, Levey AI 2004, Muscarinic acetylcholine receptor subtypes in cerebral cortex and hippocampus. *Prog Brain Res* 145:59–66.

Wagner HN Jr, Burns HD, Dannals RF et al. 1983, Imaging dopamine receptors in the human brain by positron tomography. *Science* 221(4617):1264–6.

Weinberger DR, Gibson R, Coppola R, Jones DW, Molchan S, Sunderland T, Berman KF, Reba RC 1991, The distribution of cerebral muscarinic acetylcholine receptors *in vivo* in patients with dementia. A controlled study with 123IQNB and single photon emission computed tomography. *Arch Neurol* 48(2):169–76.

Weinberger DR, Jones D, Reba RC, Mann U, Coppola R, Gibson R, Gorey J, Braun A, Chase TN 1992, A comparison of FDG PET and IQNB SPECT in normal subjects and in patients with dementia. *J Neuropsychiatry Clin Neurosci* 4(3):239–48.

Weinberger DR, Mann U, Gibson RE, Coppola R, Jones DW, Braun AR, Berman KF, Sunderland T, Reba RC, Chase TN 1990, Cerebral muscarinic receptors in primary degenerative dementia as evaluated by SPECT with iodine-123-labeled QNB. *Adv Neurol* 51:147–50.

Yamamura HI, Kuhar MJ, Greenberg D et al. 1974, Muscarinic cholinergic receptor binding: Regional distribution in monkey brain. *Brain Res* 66:541–6.

Yamamura HI, Snyder SH 1974, Muscarinic cholinergic binding in rat brain. *Proc Natl Acad Sci USA* 71(5):1725–9.

Zeeberg BR, Boulay SF, Gitler MS, Sood VK, Reba RC 1997, Correction of the seterochemistry assignment of the benzilic acid center in (*R*)-(−)-3-quinuclidinyl (*S*)-(−)-4-iodobenzilate [*R,S*]-4-IQNB]. *Appl Radiat Isot* 48:463–7.

12. Radioligands for the Vesicular Monoamine Transporter Type 2

Michael R. Kilbourn

12.1 Target Protein: The Vesicular Monoamine Transporter Type 2

In vivo imaging studies of the dopaminergic system of the human brain largely developed with radiotracers targeting the enzymatic synthesis of dopamine (e.g., [18F]fluoroDOPA; Garnett et al. 1978) and dopamine receptors (e.g., N-[11C]methylspiperone (Wagner et al. 1983) and numerous subsequent radioligands for D1 and D2/3 receptors). Investigators interested in the disease-related degeneration of the dopamine system, particularly for studies on movement disorders such as Parkinson's disease, sought *in vivo* radioligands for the presynaptic side of the dopaminergic synapse, leading to the development of radioligands for the neuronal dopamine membrane transporter (DAT) (Carroll et al. 1995).

As an alternative to imaging the DAT for an assessment of presynaptic terminal losses, we investigated the second transporter involved in the storage of dopamine and other monoamines in the presynaptic nerve terminals. The vesicular monoamine transporters (VMAT) are proteins specifically located in the vesicular membranes of monoaminergic neurons, including the dopaminergic terminals, and are responsible for the movement of monoamines from the cell cytosol into the storage vesicles, where they are held until released by exocytosis. The vesicular monoamine transporter type 2 (VMAT2: gene SLC18A2) is the transporter responsible for this function in the mammalian brain, with the related VMAT1 functioning in peripheral tissues such

Targeted Molecular Imaging. Edited by Michael J. Welch and William C. Eckelman © 2012 Taylor & Francis Group, LLC. ISBN: 978-1-4398-4195-2

Chapter 12

as the adrenal cortex. The predicted secondary structure of the VMAT2 suggests 12 membrane-spanning segments, a large intravesicular loop, and cytoplasmic tail and head segments. The active transport of substrates is driven by a transmembrane pH and electrochemical gradient generated by a membrane-bound H^+-ATPase (review: Wimalasena et al. 2011).

Why was this protein chosen as the target for *in vivo* radioligand development? In the mammalian brain, the VMAT2 is found only in presynaptic monoaminergic neurons: Although not specific for dopaminergic neurons, the chemical cytoarchitecture of the brain is such that the monoaminergic terminals of the human basal ganglia are predominantly (95%) dopaminergic. Thus, VMAT2 losses in that brain region would largely represent the losses of the dopamine terminals. In other regions of the brain, such as the hypothalamus, the presence of multiple monoaminergic nerve terminal types (serotonin, norepinephrine, and dopamine) means the radiotracer uptake and retention represents a mixture and does not allow assignment of any changes in radiotracer localization as representing any single monoamine type.

When the development of the carbon-11 radioligand for VMAT2 began, there was *in vitro* precedent for the use of VMAT2 radioligands as markers of dopaminergic terminals in the human and animal brain. Several publications had reported on the binding of the tritiated form of dihydrotetrabenazine (abbreviated TBZOH in early publications, now termed DTBZ) in animal and human brain tissue samples (Near 1986; Scherman 1986; Scherman et al. 1988b). The *in vitro* binding of [^{3}H]DTBZ was specific and saturable ($K_d = 2.7$ nM), and the distribution of [^{3}H]TBZOH sites was highly correlated with the regional content of monoamines (Scherman et al. 1986b). The regional distribution of binding sites showed high concentrations throughout the human striatum (750 fmol/mg protein), thus representing a suitable imaging target. *In vitro* radioligand binding in autopsy studies had furthermore demonstrated reductions both with aging and in Parkinson's disease patients (Scherman et al. 1989). Finally, the *in vitro* kinetics of [^{3}H]TBZOH binding to the VMAT2 site was considered highly encouraging, as studies in rat and bovine brain systems

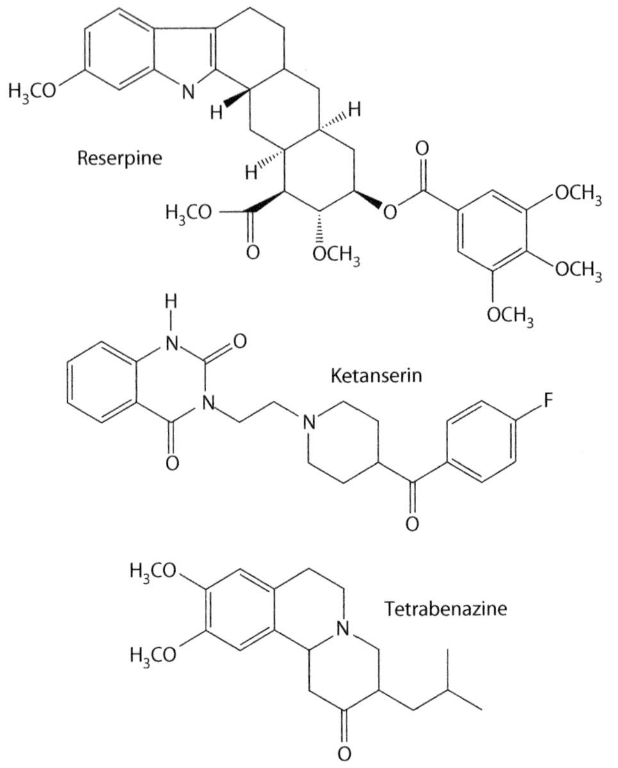

FIGURE 12.1 Structures of high-affinity ligands for the vesicular monoamine transporter type 2 (VMAT2).

showed similar kinetic rates with a clear reversible binding of the radioligand to the VMAT2 site (Near 1986; Darchen et al. 1989b).

The development of radioligands for VMAT2 also benefited from the prior clinical experience with the benzoquinolizine class of pharmaceuticals. Tetrabenazine (3-isobutyl-9,10-dimethoxy-1,3,4,6,7,11b-hexahydropyrido[2,1-a]isoquinolin-2-one, TBZ, Figure 12.1) had been developed in the 1950s as a potential treatment for psychiatric disorders, and had a long history of prior human use (Quinn et al. 1959). The metabolism, distribution, and pharmacology of tetrabenazine had been extensively studied in both animals and humans (Mehvar et al. 1987). Thus, as synthesis of [^{11}C]tetrabenazine was an isotopic substitution, there was no need to perform additional toxicology studies, allowing [^{11}C]tetrabenazine to be rapidly moved from preclinical evaluation to human use without the need to perform toxicology studies, certainly a benefit in advancing this project forward quickly.

12.2 Chemical Structures and Properties of VMAT2 Radioligands

Tetrabenazine is one of only a few chemical structures that exhibits high-affinity binding to the VMAT2 (Figure 12.1). Reserpine, an alkaloid natural product, exhibits very high *in vitro*-binding affinity (reported as low as 30 pM), but although the binding is reversible, the dissociation rate is very slow ($k_{off} = 1.2 \times 10^{-5}$ s^{-1};

k_{on} 4×10^5 M^{-1} s^{-1}) (Darchen et al. 1989b) such that the binding of that compound *in vivo* likely would have been in effect irreversible in the time span of a PET imaging study using carbon-11 or fluorine-18. Reserpine was thus a poor choice for PET radioligand development. Ketanserin has a moderate binding affinity ($K_d = 45$ nM), but since it also has a high affinity for serotonin receptors, it was less appealing as a target for specific *in vivo* radioligand development. Although a radioiodinated ketanserin has been useful for *in vitro* autoradiography studies (Darchen et al. 1989a), the available tritiated dihydrotetrabenazine or a radioiodinated derivative, [^{125}I]iodovinyltetrabenazine (Kung et al. 1994), offers better specificity as *in vitro* radioligands.

12.2.1 Carbon-11–Labeled VMAT2 Radioligands

The development of carbon-11-labeled tetrabenazine ([^{11}C]TBZ), the first in a series of VMAT2 ligands prepared, began with evaluation of procedures to remove one of the two methyl ether substituents (Figure 12.2). Several different *O*-demethylation reactions were evaluated, all of which give complex mixtures containing monodemethylated products (removal of 9- or 10-methoxy groups) and the bisphenol resulting from demethylation of both ether groups. Tedious chromatographic purifications were required to provide pure samples of the individual mono- and the bisdemethylated products. A low-yield demethylation reaction

using boron triiodide was utilized to provide isolated samples of 9-*O*-desmethyl- and 10-*O*-desmethylTBZ, with the chemical structures (9- vs. 10-hydroxyl) initially assigned by ^{1}H-NMR and ^{13}C-NMR spectroscopy. As the metabolism of tetrabenazine had been previously studied (Schwartz et al. 1966), it was determined that the more desired radiochemical ligand should be the (9-[^{11}C]methoxy)tetrabenazine: the normal *O*-demethylation step of metabolism of the radioligand would first occur at that position, releasing the radiolabel as a one-carbon metabolite and more importantly, not produce a radiolabeled and potentially pharmacologically active metabolite (as might arise from labeling in the 10-methoxy position).

The reaction of 9-*O*-desmethylTBZ with [^{11}C]methyl iodide or methyl triflate yielded the isotopically labeled [^{11}C]TBZ with a specific placement of the radiolabel at the 9-methoxy position (DaSilva and Kilbourn 1992), and this was the initial VMAT2 radioligand successfully implemented for human studies (Kilbourn et al. 1993). It did not take long, however, for the realization to hit that although we had accounted for the metabolic route of *O*-demethylation, the very first step in the metabolism of tetrabenazine is actually the reduction of the 2-keto group to form two species termed α- and β-dihydrotetrabenazine (DTBZ), and that these are actually the likely pharmacologically active species in the brain (Mehvar et al. 1987). Both the 2-hydroxy isomers have good *in vitro*-binding affinities (α-DTBZ $K_i = 3$ nM, β-DTBZ $K_i = 20$ nM: Scherman et al. 1988a). Metabolite studies in the rat using [^{11}C]tetrabenazine clearly showed rapid formation of [^{11}C]DTBZ

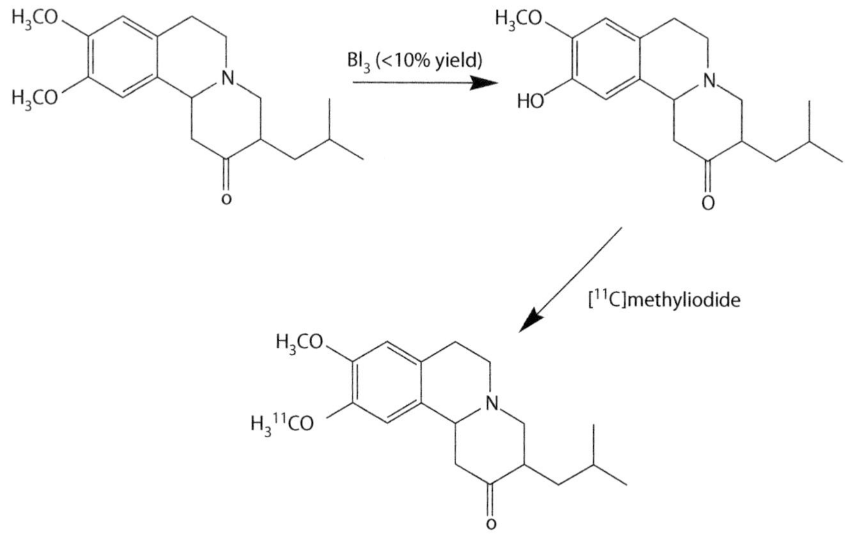

FIGURE 12.2   Radiochemical synthesis of carbon-11-labeled tetrabenazine ([^{11}C]TBZ).

Chapter 12

in the blood (DaSilva et al. 1994). The synthesis of the carbon-11 form of the higher-affinity isomer, α-DTBZ, thus required a revisiting of the precursor preparation to prepare 9-*O*-desmethyl-α-DTBZ. A higher yield (30%) and more regiospecific demethylation reaction using sodium hydride/*N*-methylaniline/HMPA was employed to obtain the desired precursor for radiolabeling (Figure 12.3). The reaction of 9-*O*-desmethyl-α-DTBZ with [¹¹C]methyl triflate thus yielded the desired 9-*O*-[¹¹C]methoxy-α-DTBZ. As it was well known that DTBZ is formed rapidly and almost completely in the human body after a therapeutic administration of TBZ (Mehvar et al. 1987), it was successfully argued that there were no additional toxicology concerns with this chemical species, and in fact in retrospect the prior imaging studies with [¹¹C]tetrabenazine

most likely truly represented largely the imaging of [¹¹C]DTBZ, albeit a mixture of α- and β-isomers.

Although α-[¹¹C]DTBZ was a satisfactory imaging agent, some attempts were then made to further improve the properties by alterations of the chemical structure. A brief synthetic sojourn was taken into the preparation of (2-[¹¹C]methoxy)-α-dihydrotetrabenazine (MTBZ; Figure 12.4) in an attempt to block metabolism (blocking the conjugation of the 2-hydroxyl function and excretion) affecting the biodistribution of [¹¹C]DTBZ (DaSilva et al. 1993b). Labeling of [¹¹C]MTBZ was accomplished using the reaction of [¹¹C]methyl triflate with α-DTBZ under stronger base conditions (sodium hydride). Although human studies were completed (Vander Borght et al. 1995a), it was determined that [¹¹C]MTBZ presented additional challenges in the

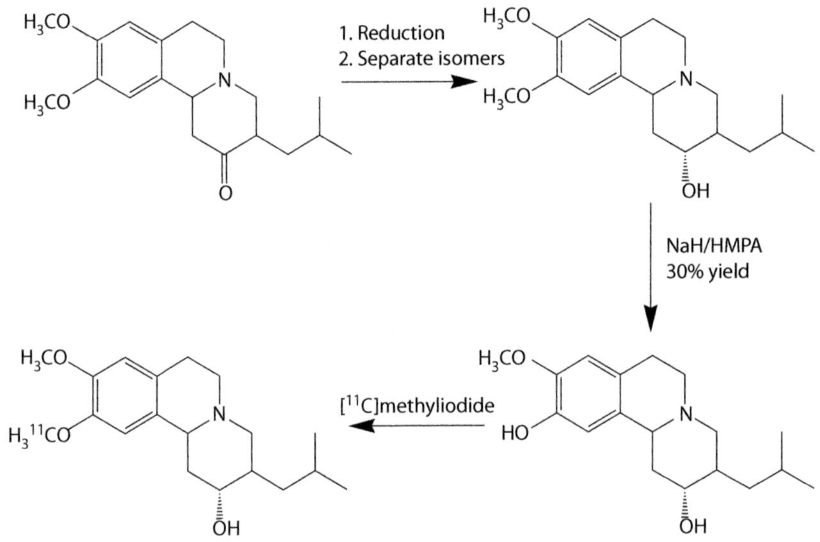

FIGURE 12.3 Radiochemical synthesis of carbon-11-labeled dihydrotetrabenazine ([¹¹C]DTBZ).

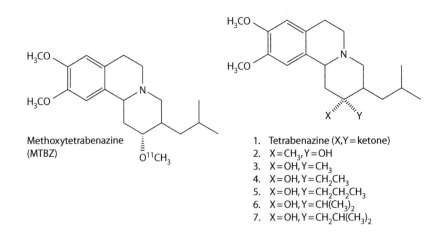

FIGURE 12.4 Chemical structures of [¹¹C]methoxytetrabenazine (MTBZ) and 2-alkyl-substituted dihydrotetrabenazine derivatives.

radiochemical synthesis without clear advantages for *in vivo* imaging. However, the corresponding tritiated molecule, [³H]MTBZ, was an excellent *in vitro* radioligand for binding (K_d = 3.9 nM) and autoradiography studies (Vander Borght et al. 1995c) with a longer shelf life than [³H]DTBZ before undergoing radiolytic decomposition. The finding of good affinity for the 2-methyl ether of DTBZ was followed some years later by the preparation by others of [¹⁸F]fluoroalkyl ethers of DTBZ, attaching the fluoroalkyl groups to the 2-hydroxyl position to yield high-affinity compounds of potential interest as imaging agents (Amarasinghe et al. 2009).

Similarly, the preparation of a series of DTBZ analogs bearing 2-alkyl substituents ranging in size from methyl to isobutyl (Figure 12.4) demonstrated tolerance of bulky groups at that ring position but did not produce derivatives of higher affinity for the VMAT2 (Lee et al. 1996), and attention was returned to perhaps improving the α-[¹¹C]DTBZ itself.

Tetrabenazine, the clinically used drug, contains two chiral centers (3- and 11b-positions) and was (and continues to be) marketed as the racemic mixture of isomers. The reduction of TBZ to DTBZ formed an additional chiral center at the 2-position, producing a potential for eight stereoisomers. The two isomers formed from the ketone reduction (termed as the α- and β-isomers) are easily separated by simple column chromatography (DaSilva et al. 1993a); both have reasonably high affinity for the VMAT2 (3 and 20 nM, respectively: Scherman et al. 1988a) and were shown to block radioligand binding *in vivo* (DaSilva et al. 1994). However, nothing was known of the binding affinities or pharmacological activities of the stereoisomers arising from chirality at the 3- or 11b-positions; in fact, even the presence of the other potential stereoisomers in preparations of TBZ was unknown. We thus began a study on the α-DTBZ structure (the higher-affinity 2-hydroxy compound) using chiral HPLC and enzymatic separation methods. Commercially available TBZ yielded, after reduction and simple column chromatographic isolation of α-DTBZ, a pair of diastereomers separable by preparative chiral chromatography (termed (+)- and (−)-isomers based on optical rotation measurements). *In vitro*-binding assays demonstrated that only one of the isomers, the (+)-isomer, had a high affinity for the VMAT2 (K_i = 0.97 nM), with the mirror image (−)-isomer having essentially no affinity for the transporter (K_i = 2.2. μmol) (Kilbourn et al. 1995a).

Application of the chiral HPLC separation to racemic 9-*O*-desmethyl-α-DTBZ followed by isolation and crystallization of one of the two isomers provided

material suitable for x-ray crystallographic determination of the absolute configuration and the assignment of the structure for the two isolated products from the chiral column: The high-affinity (+)-α-isomer was then assigned the (2R,3R,11bR) absolute configuration (Figure 12.5) (Lee et al. 1996). The x-ray crystallographic data also confirmed that we had, all along, obtained the desired 9-*O*-desmethyl compound. The radiolabeling of each isolated isomer with carbon-11 methyl iodide provided the individually labeled single isomers, termed as (+)- and (−)-α-[¹¹C]DTBZ, which were then evaluated for stereospecificity of *in vivo* binding to the VMAT2 (Kilbourn et al. 1995a). As would be consistent with the *in vitro*-binding affinities, the (+)-[¹¹C]DTBZ showed high and specific binding in the rodent striatum, whereas the opposite isomer (−)-[¹¹C]DTBZ showed a uniform distribution in the brain representing only nonspecific distribution.

More than a decade later, the syntheses of all eight isomers of DTBZ (Figure 12.5) were accomplished (Tridgett et al. 2008; Tridgett and Filloux, 2008); it turned out that the isomer chosen for development as a radioligand, 2R,3R,11bR-DTBZ, is the highest-affinity isomer. Most recently, entirely new and much more efficient chiral preparations of (+)-TBZ and (+)-α-DTBZ

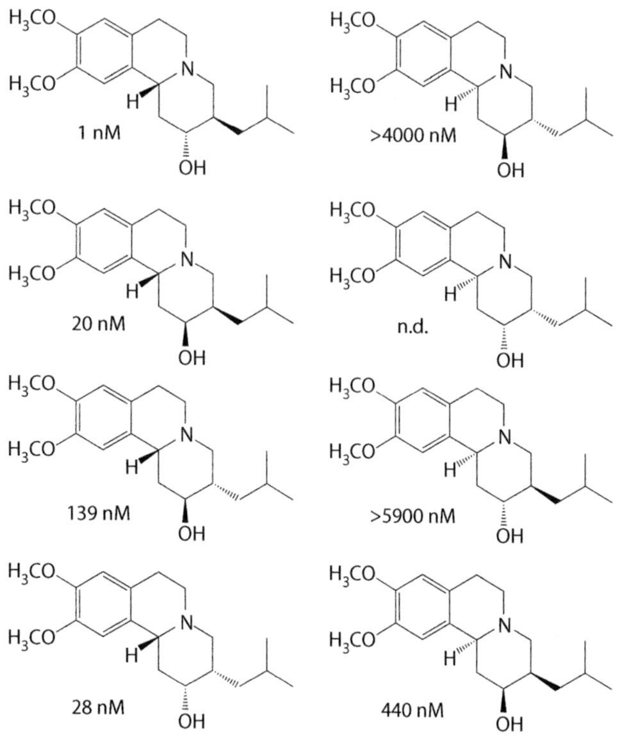

FIGURE 12.5 Configurations and *in vitro*-binding affinities for the eight stereoisomers of dihydrotetrabenazine (DTBZ).

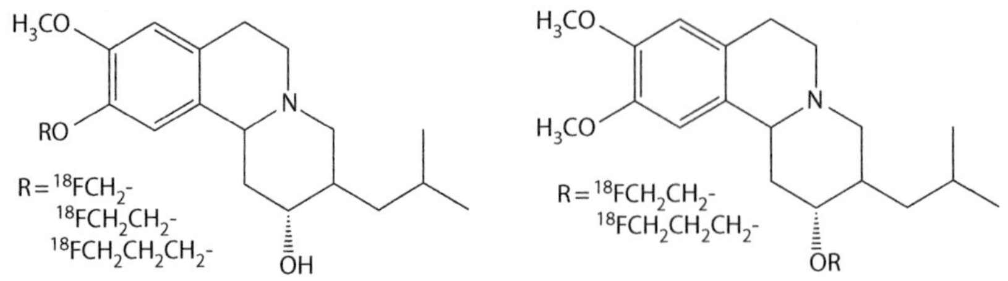

FIGURE 12.6 Chemical structures of 9-*O*-([^{18}F]fluoroalkyl)- and 2-*O*-([^{18}F]fluoroalkyl)-derivatives of dihydrotetrabenazine (DTBZ).

have been published (Boldt et al. 2008, 2009; Rishel et al. 2009; Paek et al. 2010; Yu et al. 2010) and have independently confirmed the original assignments of absolute stereochemical structure for (+)-α-DTBZ.

12.2.2 Fluorine-18-Labeled VMAT2 Radioligands

Although a successful radioligand, widespread clinical applicability of (+)-α-[^{11}C]DTBZ was limited by the short half-life (20.4 min) of the radionuclide, prompting efforts to prepare radioligands with the radionuclide fluorine-18. As fluorine-18 has a significantly longer half-life (110 min) and radiopharmaceuticals labeled with fluorine-18 are readily transported to sites distant from accelerators (cyclotrons) used for generation of the radionuclide, it was an obvious choice for development after the demonstration of the utility of the carbon-11-labeled compound.

The approach to fluorine-18 labeling, that of substituting a [^{18}F]fluoroalkyl substituent in the place of the carbon-11-labeled methyl ether, was quite simple and surprisingly effective. Analogs labeled with [^{18}F]fluoromethyl, [^{18}F]fluoroethyl, and [^{18}F]fluoropropyl groups attached to the 9-phenolic position (Figure 12.6) were prepared either by alkylation with a [^{18}F]fluoroalkylating group or by reaction of [^{18}F]fluoride ion with an appropriately constructed sulfonic acid ester precursor. Consistent with the parent compounds, higher affinity was found for the (+)-isomer; furthermore, and somewhat surprisingly, the fluoropropyl derivative has higher *in vitro*-binding affinity (K_i = 0.1 nM) that (+)-α-DTBZ itself. The synthesis of 9-*O*-(3-[^{18}F]fluoropropyl)-(+)-α-DTBZ has been improved by simplification and automation to provide a method with excellent yields (up to 41% isolated), purities (95%), and specific activity.

As an alternative to the 9-fluoroalkyl DTBZ derivatives, efforts at attaching alkyl groups to the 2-hydroxyl positions have produced 2-*O*-fluroethyl and 2-*O*- fluoropropyl derivatives (Figure 12.6) with high binding affinities (K_i values of 3.2 and 6.4 nM, respectively).

12.3 Specific Activity of VMAT2 Radioligands

Carbon-11-labeled DTBZ was originally prepared by the reaction of the 9-*O*-desmethylDTBZ precursor with high specific activity, [^{11}C]methyl iodide or [^{11}C]methyl triflate, in the presence of a small amount of base (DaSilva et al. 1993a). Syntheses have been simplified, and the reaction can be done in solution or using thin-film solid supported chemistry captive-solvent chemistry (Jewett et al. 2000) or so-called loop chemistry (Quincoces et al. 2008). Specific activity of the product is thus completely dependent on the quality of the carbon-11 labeling systems of each institution. At the University of Michigan, (+)-α-[^{11}C]DTBZ has been routinely synthesized with an end-of-synthesis-specific activity averaging close to 370 GBq/mol (10,000 Ci/mmol) with specific activity determined by reverse-phase HPLC analysis.

Similarly, the final isolated specific activity of a fluorinated VMAT2 ligand such as 9-*O*-(3-[^{18}F] fluoropropyl)-(+)-α-DTBZ is dependent on the quality of the [^{18}F]fluoride ion available from the cyclotron target, and the care taken in excluding trace [^{19}F]fluoride ion from chemicals and solvents used in the synthesis and the operation (including cleaning) of any automated apparatus. With appropriate care, specific activities of 185–555 GBq/mol are easily achievable. It should be noted that trace amounts of the hydroxyl compound (which would result from hydrolysis of the sulfonic acid ester precursor), if not completely removed from the preparation, do not compete for the *in vivo* binding of the radiotracer (Zhu et al. 2010).

12.4 Specificity of Radioligands for the VMAT2

As the VMAT2 is a protein resident in all monoaminergic nerve terminal vesicles, radiotracers binding to the VMAT2 do not strictly target dopaminergic neurons. Both *in vitro* and *in vivo*, radioligands such as [^{11}C or ^{3}H]DTBZ/MTBZ show high-affinity binding to the VMAT2 of dopaminergic, noradrenergic, and serotonergic nerve terminals and cell bodies, with little affinity for any other binding sites.

12.4.1 *In Vitro* Assay of DTBZ Binding to Other Receptors

Benzoisoquinolizine ligands for the VMAT2 show very little affinity for any other binding sites. To evaluate this *in vitro*, racemic α-DTBZ was submitted to the NIMH-sponsored NovaScreen receptor assay screen. At 10^{-5} M concentration (10 µM), the only interactions picked up by the screen were with dopamine D2 receptors (TBZ $K_i = 2$ µM; Reches et al. 1983) and serotonin S2. No affinity was observed at 10^{-5} M for receptors for adenosine, dopamine D1, GABA$_A$, GABA$_B$, serotonin1, NMDA, kainate, quisqualate, benzodiazepine, glycine, PCP, MK-801, angiotensin II, arg-vasopressin, bombesin, central CCK, peripheral CCK, substance P, substance K, neuropeptide Y, neurotensin, somatostatin, VIP, EGF, or NGF receptors: there was no affinity for calcium, chloride, or potassium channels, and there was no affinity for second messenger systems such as forskolin, phorbol ester, or inositol phosphate.

12.4.2 *In Vitro* Competition Assays for VMAT2 Radioligand Binding

The abilities of a wide range of drugs and neurotransmitters to compete for radioligand binding to the VMAT2 have been examined using *in vitro*-binding assays or autoradiography using the tritiated ligands (+/–)-[^{3}H]DTBZ or (+/–)-[^{3}H]MTBZ. The normal neurotransmitter substrates for the VMAT2, as well as a variety of drugs targeting transporters and receptors of the monoaminergic systems, all show micomolar or lower *in vitro* affinities for the VMAT2 (Table 12.1) with the exception of the known VMAT2 ligands: reserpine, TBZ, and ketanserin.

12.4.3 *In Vivo* Competition Assays for VMAT2 Radioligand Binding

In animal studies (rodents, primates) conducted in our laboratories, we have been repeatedly unable to

demonstrate any effect of acute or chronic-competing drugs for dopamine receptors or neuronal monoamine transporters on the *in vivo* regional distribution and binding of [^{11}C]DTBZ in the brain. Compounds tested for abilities to affect *in vivo* radioligand binding in the rodent or primate brain include acute GBR 12935 (dopamine transporter), scopolamine, DOPA, methylphenidate, haloperidol, pargyline, and chronic or repeated administrations of pargyline, deprenyl, or L-DOPA/benserazide. The only compounds that have been found to block *in vivo* VMAT2 radioligand binding have been TBZ, reserpine, and ketanserin, all known inhibitors of VMAT2.

Table 12.1 Drugs and Neurotransmitters Tested for *In Vitro* Competition against VMT2 Radioligand Binding

Drug	K_i(µM)
Reserpine	0.31[a], 0.073[b]
TBZ	0.0065[a], 0.0023[b]
DTBZ	0.008[a]
Serotonin	2200[a], 3.4[b]
Dopamine	5500[a], 2800[b], >1[c]
EPI	8800[a]
Norepinephrine	3100[b]
Histamine	3200[b]
Spiperone	>10[a]
N-Methylspiperone	>1[c]
Bromocriptine	>1[c]
Chlorpromazine	8[a]
Haloperidol	4[a], 1.86[b], >1[c]
WIN 35,065	>30[a]
RTI-30	>30[a]
GBR 12909	2.24[b]
Mazindol	23[b]
Deprenyl	>1[c]
Amphetamine	>20[a], 301[b]
Nomifensine	>20[a], 416[b], >1[c]
MPP+	1.6[b]
6-Hydroxydopamine	>1[c]

[a] [^{3}H]DTBZ, native and cloned animal transporters (Scherman et al. 1988a).
[b] [^{3}H]DTBZ, cloned human transporters (Gonzalez et al. 1994).
[c] [^{3}H]MTBZ, rat brain autoradiography (Vander Borght et al. 1995c).

Chapter 12

12.5 Lipophilicity and Nonspecific Binding for VMAT2 Radioligands

Dihydrotetrabenazine (DTBZ), with a measured log *P* value of 2.1 (Scherman et al. 1988a), has a very advantageous lipophilicity for human PET studies. Brain uptake across the blood–brain barrier is very good, with the delivery (influx constant, k_1) useful as a surrogate measure of blood flow (Albin et al. 2010).

Studies in the rat brain demonstrated that nonspecific distribution, as indicated either by (a) complete pharmacological block of the specific VMAT2 binding of (+)-α-DTBZ or (b) distribution of the inactive (−)-α-[¹¹C]DTBZ isomer, was ~20–25% of the total brain radioactivity in target regions (e.g., striatum). In human subjects, the nonspecific distribution of (+)-α-[¹¹C]DTBZ contributes a similar 20–25% of the total signal in the striatal regions .

For 9-*O*-(3-[¹⁸F]fluoropropyl)-(+)-α-DTBZ, the fluorinated DTBZ derivative in current human studies, simple consideration of substitution of the larger propyl chain (three carbons) for a methyl group (one carbon) would suggest a significant increase in lipophilicity, and perhaps a corresponding increase in nonspecific distribution. The log *P* of 9-*O*-(3-[¹⁸F] fluoropropyl)-(+)-α-DTBZ has, however, been determined experimentally as 2.6, a value in the optimal range for blood–brain barrier permeability (Dischino et al. 1983). Animal studies (rat and monkey) have consistently demonstrated that specific-binding indices (tissue concentration ratios or binding potentials) are greater for 9-*O*-(3-[¹⁸F]fluoropropyl)-(+)-α-DTBZ than for (+)-α-[¹¹C]DTBZ, indicative of no significant increase in nonspecific binding of the compound due to slightly increased lipophilicity resulting from the longer alkyl group.

12.6 *In Vitro* and *In Vivo* Kinetics of VMAT2 Radioligand Binding

The reversibility of DTBZ binding to the VMAT2 had been demonstrated early using the tritiated ligand and *in vitro* assays. Both the association and dissociation rates (Table 12.2) were highly encouraging that the radioligand would behave as a reversibly binding radiotracer for *in vivo* imaging studies.

Kinetic studies performed using bolus injections of radioligand and PET imaging, initially with [¹¹C] TBZ and later with (+)-α-[¹¹C]DTBZ, showed high initial brain uptake (consistent with the good log *P*) and reversible-binding kinetics in the basal ganglia of the human brain, representing regions of high concentrations of predominantly dopaminergic neurons.

The kinetic analysis of (+)-α-[¹¹C]DTBZ PET imaging data has been extensively studied in both animals and humans. As a reversibly binding radiotracer, time–tissue activity data from bolus injections of (+)-α-[¹¹C] DTBZ (Figure 12.7) can be modeled using both compartmental pharmacokinetic analysis (using appropriately determined metabolite-corrected blood input data) or graphical analysis (e.g., Logan plots) (Koeppe et al. 1997). Alternatively, the radioligand is very suitable in both animal (Kilbourn and Sherman 1997) and human studies (Koeppe et al. 1997, 1999) for use in an equilibrium infusion approach, whereby the regional distributions at equilibrium directly provide

Table 12.2 *In Vitro*-Binding Kinetics of (+/−)-α-[³H] DTBZ to the VMAT2 of Bovine or Rat Striatum

	Bovine Striatum[a]	Rat Brain[b]
k_{on}	$1.2 \times 10^5\ M^{-1}\,s^{-1}$	$1.3 \times 10^5\ M^{-1}\,s^{-1}$
k_{off}	$6.5 \times 10^{-4}\,s^{-1}$	$6.6 \times 10^{-4}\,s^{-1}$

[a] Near (1986)

[b] Darchen et al. (1989a)

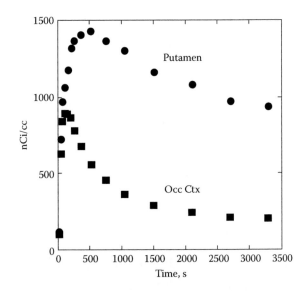

FIGURE 12.7 Regional human brain tissue time–activity curves following bolus intravenous injection of (+)-α-[¹¹C] dihydrotetrabenazine.

good estimates of the distribution volume ratios (DVR). Both the Logan plot and equilibrium approaches do not require obtaining and analyzing serial blood samples, simplifying the imaging protocol.

As noted earlier, 9-*O*-(3-fluoropropyl)-(+)- α-DTBZ has a higher binding affinity than (+)-α-DTBZ, yet the kinetics in both animal and human brain are very similar for the carbon-11 and fluorine-18 compounds. Despite the higher affinity, the [^{18}F]fluoropropyl compound is also clearly reversibly bound, as demonstrated by the rapid egress of radioactivity from the high binding region (striatum) of the monkey brain following administration of cold tetrabenazine at a time point 40 min after bolus radiotracer injection (Figure 12.8). In a very similar fashion to the methods used for [^{11}C]DTBZ, human PET studies with 9-*O*-(3-fluoropropyl)-(+)-α-DTBZ have been also successfully analyzed using Logan plots with the primary visual cortex and the reference region, to yield estimates of the binding potential (Okamura et al. 2010).

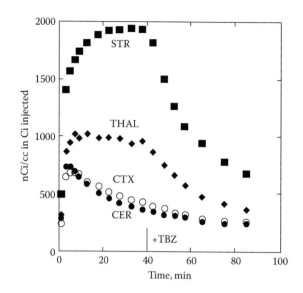

FIGURE 12.8 Regional monkey brain tissue time–activity curves following bolus injection of (+)-α-9-*O*-(3-[^{18}F]fluoropropyl) dihydrotetrabenazine.

12.7 Sensitivity of *In Vivo* Measures of VMAT2 Radioligand Binding

The sensitivity of *in vitro* binding for VMAT2 radioligands has been demonstrated in a number of studies. *In vitro* studies of [^{3}H]DTBZ and [^{3}H]MTBZ binding in 6-hydroxydopamine-lesioned rats demonstrated good correlations between radioligand binding and independent measures of dopaminergic nerve terminal densities (e.g., tyrosine hydroxylase activity) (Masuo et al. 1990b; Vander Borght et al. 1995b,c). In both rats and baboons, neurotoxic methamphetamine treatment produced losses of *in vitro* [^{3}H]DTBZ binding which correlated with losses of other *in vitro* dopaminergic markers (dopamine levels, [^{3}H]WIN-35,428 binding) (Frey et al. 1997; Villemagne et al. 1998). Finally, a loss of [^{3}H]DTBZ binding *in vitro* was reported after lesioning rat brains with ibotenic acid, a neurotoxin that does not specifically target dopaminergic nerve terminals (Masuo et al. 1990a).

The sensitivity of [^{11}C]DTBZ binding to changes of VMAT2-binding sites has also been demonstrated by a variety of different *in vivo* studies in animal models. Following a pharmacological dose of tetrabenazine, which results in a complete block of [^{11}C]DTBZ binding, the steady recovery of VMAT2-binding sites as the cold drug was cleared from the brain was followed by repeated *ex vivo* assays over a subsequent 24 h period (Kilbourn 1997). *In vivo* studies using PET imaging and unilateral 6-hydroxydopamine lesions showed that changes in [^{11}C]DTBZ binding were due to changes in

the number of binding sites (B_{max}) but not the affinity (apparent K_d) (Sossi et al. 2007). Losses of [^{11}C]DTBZ binding to the VMAT2 have been demonstrated *in vivo* using MPTP lesions in mice, including a demonstration of the differential rates of loss and recovery of *in vivo* binding of [^{11}C]DTBZ and [^{11}C]methylphenidate, a radioligand for the neuronal membrane dopamine transporter (Kilbourn et al. 2000). In monkeys, unilateral MPTP administration produced a clear unilateral striatal loss of *in vivo* [^{11}C]DTBZ binding (Dasilva et al. 1993a; Doudet et al. 2006), with losses of binding of the VMAT2 radioligand described as more sensitive for detection of MPTP-induced dopaminergic losses than *in vivo* measures of the DAT or using [^{18}F]fluoroDOPA to mark dopamine synthesis (Doudet et al. 2006). Following a chronic low-dose systemic administration of MPTP to monkeys, striatal losses of [^{11}C]DTBZ binding were shown in presyptomatic animals and preceded evidence for changes in the DAT or dopamine D2/D3 receptors (Chen et al. 2008). The progressive loss of *in vivo* [^{11}C]DTBZ binding in the striatum was also demonstrated following low-dose repeated administration of MPTP in monkeys (Blesa et al. 2010). The ability to measure the opposite—an increase in the numbers of VMAT2-binding sites—was demonstrated in tottering mice, a genetic mouse model of monoaminergic hyperinnervation (Kilbourn et al. 1995b). Thus, in preclinical studies, *in vivo* radioligand binding proved to be

Chapter 12

a sensitive measure of both reductions and increases in VMAT2-binding sites.

In human autopsy samples, *in vitro* studies of [³H] DTBZ binding had demonstrated reductions of VMAT2 binding with increasing age (Scherman et al. 1989). Studies of the age dependence of *in vivo* PET imaging of VMAT2 using the radioligand [¹¹C]DTBZ have produced mixed results, with studies showing an age-dependent decrease in DTBZ binding in normal controls (Frey et al. 1996; Bohnen et al. 2006), and a more recent report by Troiano et al. (2010) not showing an age-related decline of radioligand binding.

Finally, PET imaging with [¹¹C]DTBZ has been undertaken in a variety of neurological diseases where losses of monoaminergic nerve terminals might be involved, including Parkinson's disease, Lewy body disease, Huntington's disease, Alzheimer's disease, Tourette's disease, DOPA-responsive dystonia, and multiple system atrophy. In many diseases, losses of [¹¹C]DTBZ binding could be clearly demonstrated. In Parkinson's disease, the most studied condition, the sensitivity of the radioligand-binding method to the severity and duration of the disease has been demonstrated. Losses of [¹¹C]DTBZ binding are evident in early, untreated Parkinson's disease patients (Martin et al. 2008), including presymptomatic disease (Bohnen et al. 2006) and losses of [¹¹C]DTBZ binding correlate with clinical and behavioral measures (Bohnen et al. 2006). The sensitivity of [¹¹C]DTBZ binding *in vivo* to changes in the numbers of binding sites has thus been repeatedly demonstrated both in preclinical animal studies, and in human studies of disease, supporting a proposal that DTBZ might be used as a stable biomarker of monoaminergic nerve terminal densities in the human brain. Such application is bolstered by the inability to alter the number of VMAT2-binding sites, *in vitro* or *in vivo*, by repeated or chronic administration of drugs expected to affect the dopamine system (Vander Borght et al. 1995a;

Kilbourn et al. 1996). However, the potential competition for radioligand binding by endogenous neurotransmitter has been significantly examined only recently. These studies were in response to the observations made in human studies of DOPA-responsive dystonia, a genetic disease arising from the complete lack of a crucial enzyme in the biosynthesis of dopamine, where *in vivo* binding of [¹¹C]DTBZ was reported to be significantly higher than in normal subjects (De La Fuentes-Fernandez et al. 2003); these findings suggest that the lack of vesicular dopamine removes competing molecules and allows increased radioligand binding. Despite the low affinity of the natural substrates (dopamine, norepinephrine, and serotonin) for the binding site of DTBZ (Table 12.1), the concentration of dopamine in vesicles is very high, suggesting a potential for competition for [¹¹C]DTBZ binding. In animal studies, near-complete depletion of dopamine in the brain using the synthesis inhibitor alpha-methyl *p*-tyrosine (AMPT) resulted in enhanced *in vivo* binding of [¹¹C]DTBZ (Kilbourn et al. 2008, 2010; Tong et al. 2008), although numbers of VMAT2 determined *in vitro* were not changed; the *in vivo* increase of [¹¹C] DTBZ binding could be blocked by restoration of dopamine levels by administration of L-DOPA, which is converted to dopamine by the second step of biosynthesis that is not blocked by AMPT (Kilbourn et al. 2008, 2010). These animal studies support the fact that the very high concentrations of intravesicular dopamine can in some instances compete for *in vivo* radioligand binding to the VMAT2. The potential for endogenous dopamine levels to affect *in vivo* imaging has begun to be explored in human subjects: L-DOPA administration to advanced Parkinson's disease patients resulted in lowered [¹¹C]DTBZ binding in the striatum (De la Fuentes-Fernandez et al. 2009), but administration of low dose of amphetamine (a stimulant that can deplete vesicular dopamine stores) failed to alter *in vivo* [¹¹C]DTBZ binding (Boileau et al. 2010).

12.8 Impact of VMAT2 Radioligands on Clinical Care

12.8.1 Studies of VMAT2 in the Brain

Carbon-11-labeled dihydrotetrabenazine has become a very useful research radiopharmaceutical, with applications in studies of a wide variety of neurodegenerative and psychiatric diseases, and for studies relating to possible losses of neuronal densities in drug abuse studies. The utility of the radiopharmaceutical is only limited by

the short half-life of the radionuclide, carbon-11 ($t_{1/2} = 20.4$ min), which requires its use at locations closely placed to a cyclotron facility. Fortunately, imaging of VMAT2 has now been extended to sites more remote from a cyclotron due to the development of fluorine-18 ($t_{1/2} = 110$ min)-labeled VMAT2 radioligands. Substitution of a [¹⁸F]fluoroalkyl group (fluoromethyl, fluoroethyl, or fluoropropyl) for the 9-methoxy of dihydrotetrabenzine produced radioligands of as good or better affinity than [¹¹C]DTBZ with excellent pharmacokinetics (Goswami et al. 2006; Kung et al. 2007; Kilbourn et al. 2007). Clinical studies in humans, including evaluation in Parkinson's disease patients, has already been reported for one of these new radioligands, 9-[¹⁸F]fluoropropyl-α-(+)-DTBZ (Frey et al. 2008; Okamura et al. 2010). As an alternative, fluoroalkyl ether derivatives of DTBZ with the ether group at the 2-hydroxyl position (analogous to [¹¹C]methoxytetrabenazine) have been recently reported in the patent literature (Amarasinghe et al. 2009).

VMAT2 imaging using PET (Bohnen and Frey 2007) and one of the fluorine-18 radioligands may have clinical applications in the differential diagnosis of neurodegenerative diseases, such as distinguishing idiopathic Parkinson's disease from atypical parkinsonian disorders, or the differential diagnosis of Alzheimer's disease from dementia with Lewy bodies (Koeppe et al. 2008). PET imaging may also provide biomarkers to follow the progression of disease, or evaluation of therapeutic interventions.

12.8.2   Studies of VMAT2 in the Pancreas

Although studies of VMAT2 radioligands have been predominant in the brain, in recent years, applications of VMAT2 radioligand imaging to imaging beta cell densities in the pancreas have been pursued. Based on reports of the presence of VMAT2 in beta cells, the clear uptake of VMAT2 ligands into the pancreas and the ability to image that organ have been established, and PET studies of [¹¹C]DTBZ uptake and retention have been performed in animal models of beta cell losses and in several small studies of patients with diabetes (Ichise and Harris 2010). To date, results have been mixed (Ichise and Harris 2010; Kilbourn et al. 2010; Tsao et al. 2010) and have not yet fully validated VMAT2 imaging in the pancreas as a reliable, stable marker of beta cell densities. The potential clinical application of such an imaging biomarker is highly significant, due to the increasing prevalence of diabetes, and continued development of newer VMAT2 radioligands that might have improved imaging characteristics in the pancreas is being pursued.

Acknowledgments

The author thanks Kirk Frey, Robert Koeppe, Jean DaSilva, Lihsueh Lee, Phillip Sherman, Thierry Vander Borght, Doug Jewett, Roger Albin, Hank Kung, and David Kuhl for their many contributions to the development and applications of VMAT2 radioligands. This work was supported by grants from the National Institutes of Health (NS15655 and MH47611) and the Office of Science, U.S. Department of Energy (DE-FG02–87ER60561).

References

Albin RL, Koeppe RA, Burke JF, Giordani B, Kilbourn MR, Gilman S, and Frey KA. 2010. Comparing fludeoxyglucose F18-PET assessment of regional cerebral glucose metabolism and [¹¹C]dihydrotetrabenazine-PET in evaluation of early dementia and mild cognitive impairment. *Arch Neurol* 67:440–446.

Amarasinghe K, Rishel M, Dinn, S, and Johnson B. 2009. Fluorinated dihydrotetrabenazine ether imaging agents and probes. World Patent WO 2009/05520 A2.

Blesa J, Juri C, Collantes M, Peñuelas I, Prieto E, Iglesias E et al. 2010. Progression of dopaminergic depletion in a model of MPTP-induced Parkinsonism in non-human primates. An [¹⁸F]F-DOPA and [¹¹C]-DTBZ PET study. *Neurobiol Dis* 38:456–463.

Bohnen NI, Albin RA, Koeppe RA, Wernette K, Kilbourn MR, Minoshima S, and Frey KA. 2006. Positron emission tomography of monoaminergic vesicular binding in aging and Parkinson's disease. *J Cerebral Blood Flow Metab* 26:1198–1212.

Bohnen NI and Frey KA. 2007. Imaging of cholinergic and monoaminergic neurochemical changes in neurodegenerative disorders. *Mol Imaging Biol* 9:243–257.

Boldt KG, Biggers MS, Phifer SS, Brine GA, and Rehder KS. 2009. Synthesis of (+)- and (−)-tetrabenazine from the resolution of α-dihydrotetrabenazine. *Synth Comm* 20:3574–3585.

Boldt KG, Brine GA, and Rehder K. 2008. Synthesis of (+)-9-O-desmethyl-dihydrotetrabenazine, precursor for the high affinity VMAT2 imaging PET radioligand [¹¹C]-(+)-dihydrotetrabenazine. *Org Prep Proc Int* 40:379–384.

Carroll FI, Sheffel U, Dannals RF, Boja JW, and Kuhar MJ. 1995. Development of imaging agents for the dopamine transporter. *Med Res Rev* 15:419–444.

Chen MK, Kuwabara H, Zhou Y, Adams RJ, Brasić JR, McGlothan JL et al. 2008. VMAT2 and dopamine neuron loss in a primate model of Parkinson's disease. *J Neurochem* 105:78–90.

Darchen F, Masuo Y, Vial M, Rostene W, and Scherman D. 1989a. Quantitative autoradiography of the rat brain vesicular monoamine transporter using the binding of [³H]dihydrotetrabenazine and 7-amino-8-[¹²⁵I]iodoketanserin. *Neuroscience* 33: 341–349.

Darchen F, Scherman D, and Henry JP. 1989b. Reserpine binding to chromaffin granules suggests the existence of two conformations of the monoamine transporter. *Biochemistry* 28:1692–1697.

DaSilva JN and Kilbourn MR. 1992. *In vivo* binding of [¹¹C]tetrabenazine to vesicular monoamine transporters in mouse brain. *Life Sci* 51:593–600.

DaSilva JN, Kilbourn MR, Carey JE, Sherman, and Pisani T. 1994. Characterization of [¹¹C]tetrabenazine as an *in vivo* radioligand for the vesicular monoamine transporter. *Nucl Med Biol* 21:151–156.

DaSilva JN, Kilbourn MR, and Domino EF. 1993a. *In vivo* imaging of monoaminergic nerve terminals in normal and MPTP-lesioned primate brain using positron emission tomography (PET) and [¹¹C]tetrabenazine. *Synapse* 14:128–131.

DaSilva JN, Kilbourn MR, and Mangner TJ. 1993b. Synthesis of a [¹¹C]methoxy derivative of α-dihydrotetrabenazine: A radioligand for studying the vesicular monoamine transporter. *Appl Radiat Isot* 44:1487–1489.

De la Fuente-Fernandez R, Furtado S, Guttman M, Furukawa Y, Lee CS, Calne DB, Ruth TJ, and Stoessl AJ. 2003. VMAT2 binding is elevated in dopa-responsive dystonia: Visualizing empty vesicles by PET. *Synapse* 49:20–28.

De La Fuentes-Fernandez R, Sossi V, McCormick S, Schulzer M, Ruth TJ, and Stoessl AJ. 2009. Visualizing vesicular dopamine dynamics in Parkinson's disease. *Synapse* 63:713–716.

Dischino DD, Welch MJ, Kilbourn MR, and Raichle ME. 1983. The relationship between lipophilicity and the brain extraction of carbon-11 radiopharmaceuticals. *J Nucl Med* 24:1030–1038.

Doudet DJ, Rosa-Neto P, Munk OL, Ruth TJ, Jivan S, and Cumming P. 2006. Effect of age on markers for monoaminergic neurons of normal and MPTP-lesioned rhesus monkeys: A multi-tracer PET study. *Neuroimage* 30:26–35.

Frey K, Kilbourn M, and Robinson T. 1997. Reduced striatal vesicular monoamine transporters after neurotoxic but not after behaviorally-sensitizing doses of methamphetamine. *Eur J Pharm* 334:273–279.

Frey KA, Koeppe RA, Kilbourn MR, Manchanda R, Pontecorvo MJ, and Skovronsky DM. 2008. Imaging VMAT2 in Parkinson's disease with [F-18]AV-133. *J Nucl Med* 49:5P.

Frey KA, Koeppe RA, Kilbourn MR, Vander Borght TM, Albin RL, Gilman S, and Kuhl DE. 1996. Presynaptic monoamine vesicles in Parkinson's disease and normal aging. *Ann Neurol* 40:873–884.

Garnett ES, Firnau G, Chan PK, Sood S, and Belbeck LW. 1978. [18F]Fluoro-dopa, an analogue of dopa, and its use in direct external measurements of storage, degradation, and turnover of intracerebral dopamine. *Proc Natl Acad Sci* 75:464–467.

Gonzalez AM, Walther D, Pazos A, and Uhl GR. 1994. Synaptic vesicular monoamine transporter expression: Distribution and pharmacologic profile. *Mol Brain Res* 22:219–226.

Goswami R, Kung M-P, Ponde D, Hou C, Kilbourn MR, and Kung HF. 2006. Fluoroalkyl derivatives of dihydrotetrabenazine as PET imaging agents targeting vesicular monoamine transporters. *Nucl Med Biol* 33:685–694.

Ichise M and Harris PE. 2010. Imaging of β-cell mass and function. *J Nucl Med* 51:1001–1004.

Jewett DM, Kilbourn MR, and Lee LC. 2000. A simple synthesis of [¹¹C]dihydrotetrabenazine (DTBZ). *Nucl Med Biol* 27:529–532.

Kilbourn MR. 1997. Time-dependent recovery of *in vivo* binding sites after drug dosing: A method for radiotracer evaluation. *Nucl Med Biol* 24:115–118.

Kilbourn MR, Butch ER, Desmond T, Sherman P, and Frey KA. 2008. Dopamine depletion increases *in vivo* [¹¹C]DTBZ binding in awake rat brain. *NeuroImage* 41:Suppl. 2; T54.

Kilbourn MR, Butch ER, Desmond T, Sherman P, Harris P, and Frey KA. 2010. Dopamine depletion increases *in vivo* [¹¹C]dihydrotetrabenazine ([¹¹C]DTBZ) binding in rat striatum. *Nucl Med Biol* 37:3–8.

Kilbourn MR, DaSilva JN, Frey KA, Koeppe RA, and Kuhl DE. 1993. *In vivo* imaging of vesicular monoamine transporters in human brain using [¹¹C]tetrabenazine and positron emission tomography. *J Neurochem* 60:2315–2318.

Kilbourn MR, Frey KA, Vander Borght T, and Sherman PS. 1996. Effects of dopaminergic drug treatments on *in vivo* radioligand binding to brain vesicular monoamine transporters. *Nucl Med Biol* 23:467–471.

Kilbourn MR, Hockley B, Lee L, Hou C, Goswami R, Ponde DE, Kung M-P, and Kung HF. 2007. Pharmacokinetics of [¹⁸F]fluoroalkyl derivatives of dihydrotetrabenazine (DTBZ) in rat and monkey brain. *Nucl Med Biol* 34:233–237.

Kilbourn MR, Kuszpit K, and Sherman P. 2000. Rapid and differential losses of *in vivo* dopamine transporter (DAT) and vesicular monoamine transporter (VMAT2) radioligand binding in MPTP-treated mice. *Synapse* 35:250–255.

Kilbourn MR, Lee L, Vander Borght T, Jewett D, and Frey K. 1995a. Binding of α-dihydrotetrabenazine to the vesicular monoamine transporter is stereospecific. *Eur J Pharmacol* 278:249–252.

Kilbourn M and Sherman P. 1997. *In vivo* binding of (+)-α-[³H] dihydrotetrabenazine to the vesicular monoamine transporter of rat brain: Bolus vs. equilibrium studies. *Eur J Pharm* 331:161–168.

Kilbourn MR, Sherman PS, and Abbott LC. 1995b. Mutant mouse strains as models for *in vivo* radiotracer evaluations: [¹¹C] Methoxytetrabenazines ([¹¹C]MTBZ) in tottering mice. *Nucl Med Biol* 22:565–567.

Koeppe RA, Frey KA, Kuhl DE, and Kilbourn MR. 1999. Assessment of extrastriatal vesicular monoamine transporter binding site density using stereoisomers of [¹¹C]dihydrotetrabenazine. *J Cerebral Blood Flow Metab* 19:1376–1384.

Koeppe RA, Frey KA, Kume A, Albin R, Kilbourn MR, and Kuhl DE. 1997. Equilibrium versus compartmental analysis for assessment of the vesicular monoamine transporter using (+)-α-[¹¹C]dihydrotetrabenazine (DTBZ) and PET. *J Cerebral Blood Flow Metab* 17:919–931.

Koeppe RA, Gilman S, Junck L, Wernette K, and Frey KA. 2008. Differentiating Alzheimer's disease with Lewy bodies and Parkinson's disease with (+)- [¹¹C]dihydrotetrabenazine positron emission tomography. *Alzheimers Dement* 4 (1 Suppl 1):567–76.

Kung M-P, Canney DJ, Frederick D, Zhuang Z, Billings JJ, and Kung HF. 1994. Binding of 125I-iodovinyltetrabenazine to CNS vesicular monoamine transport sites. *Synapse* 18: 225–232.

Kung M-P, Hou C, Goswami R, Ponde DE, Kilbourn MR, and Kung HF. 2007. Characterization of optically resolved 9-fluoropropyl-dihydrotetrabenazine as a potential PET imaging agent targeting vesicular monoamine transporters. *Nucl Med Biol* 34:239–246.

Lee LC, Vander Borght T, Sherman PS, Frey KA, and Kilbourn MR. 1996. *In vitro* and *in vivo* studies of benzoisoquinoline ligands for the brain synaptic vesicle monoamine transporter. *J Med Chem* 3:191–196.

Martin WR, Wieler M, Stoessl AJ, and Schulzer M. 2008. Dihydrotetrabenazine positron emission tomography imaging in early, untreated Parkinson's disease. *Ann Neurol* 63: 388–394.

Masuo Y, Montagne MN, Pélaprat D, Scherman D, and Rostène W. 1990a. Regulation of neurotensin-containing neurons in the rat striatum. Effects of unilateral striatal lesions with quinolinic acid and ibotenic acid on neurotensin content and its binding site density. *Brain Res* 520:6–13.

Masuo Y, Pelaprat D, Scherman D, and Rostene W. 1990b. [³H] Dihydrotetrabenazine, a new marker for the visualization of dopaminergic denervation in the rat striatum. *Neurosci Lett* 114:45–50.

Mehvar R, Jamali F, Watson MWB, and Skelton D. 1987. Pharmacokinetics of tetrabenazine and its major metabolite in man and rat. *Drug Metab Disp* 12:250–255.

Near JA. 1986. [³H]Dihydrotetrabenazine binding to bovine striatal synaptic vesicles. *Mol Pharmacol* 30:252–257.

Okamura N, Villemagne VL, Drago J, Pejoska S, Dhamija RK, Mulligan RS et al. 2010. *In vivo* measurement of vesicular monoamine transporter type 2 density in Parkinson disease with (18)F-AV-133. *J Nucl Med* 51:223–228.

Quinn GP, Shore PA, and Brodie BB. 1959. Biochemical and pharmacological studies of RO 1–9568 (tetrabenazine), a non-indole tranquilizing agent with reserpine-like effects. *J Pharmacol Exp Ther* 127:103–109.

Quincoces G, Collantes M, Catalán R, Ecay M, Prieto E, Martino E et al. 2008. [Quick and simple synthesis of (¹¹)C-(+)-alpha-dihydrotetrabenazine to be used as a PET radioligand of vesicular monoamine transporters]. *Rev Esp Med Nucl* 27:13–21.

Paek S-M, Kim N-M, Jung J-K, Jung J-W, Chang D-J, Moon H, and Suh Y-G. 2010. A concise total synthesis of (+)-tetrabenazine and (+)-α-dihydrotetrabenazine. *Eur J Chem* 16:4623–4628.

Reches A, Burke RE, Kuhn CM, Hassan MN, Jackson VR, and Fahn S. 1983. Tetrabenazine, an amine-depleting drug, also blocks dopamine receptors in rat brain. *J Pharmacol Exp Ther* 225:515–521.

Rishel M, Amarasinghe KKD, Dinn SR, and Johnson BF. 2009. Asymmetric synthesis of tetrabenazine and dihydrotetrabenazine. *J Org Chem* 74:4001–4004.

Schwartz D, Bruderer H, Rieder J, and Brossi A. 1966. Metabolic studies of tetrabenazine, a psychotropic drug in animals and man. *Biochem Pharmacol* 15:645–665.

Scherman D. 1986. Dihydrotetrabenazine binding and monoamine uptake in mouse brain regions. *J Neurochem* 47:331–339.

Scherman D, Boschi G, Rips R, and Henry J-P. 1986b. The regionalization of [³H]dihydrotetrabenazine binding sites in the mouse brain and its relationship to the distribution of monoamines and their metabolites. *Brain Res* 370:176–181.

Scherman D, Desnos C, Darchen F, Pollak P, Javoy-Agid F, and Agid Y. 1989. Striatal dopamine deficiency in Parkinson's disease: Role of aging. *Ann Neurol* 26:551–557.

Scherman D, Gasnier B, Jaudon P, and Henry J-P. 1988a. Hydrophobicity of the tetrabenazine-binding site of the chromaffin granule monoamine transporter. *Mol Pharmacol* 33:72–77.

Scherman D, Raisman R, Ploska A, and Agid Y. 1988b. [³H] Dihydrotetrabenazine, a new *in vitro* monoaminegic probe for human brain. *J Neurochem* 50:1131–1136.

Sossi V, Holden JE, Topping GJ, Camborde ML, Kornelsen RA, McCormick SE, Greene J, Studenov AR, Ruth TJ, and Doudet DJ. 2007. *In vivo* measurement of density and affinity of the monoamine vesicular transporter in a unilateral 6-hydroxydopmaine rat model of PD. *J Cereb Blood Flow Metab* 27:1407–1415.

Tong J, Wilson AA, Boileau I, Houle S, and Kish SJ. 2008. Dopamine modulating drugs influence striatal (+)-[¹¹C]DTBZ binding in rats: VMAT2 binding is sensitive to changes in vesicular dopamine concentration. *Synapse* 62:873–876.

Tridgett R, Clarke I, Turtle R, and Johnston G. 2008. Dihydrotetrabenazines and pharmaceutical compositions containing them. US Patent 2008/0108645A1.

Tridgett R and Filloux T. 2008. Use of 3,11B-cis-dihydrotetrabenazine for the treatment of symptoms of Huntington's disease. US Patent 2008/0319000A1.

Troianao AR, Schulzer M, De La Fuentes-Fernandez R, Mak E, McKenzie J, Sossi V, McCormick S, Ruth TJ, and Stoessl AJ. 2010. Dopamine transporter PET in normal aging: Dopamine transporter decline and its possible role in preservation of motor function. *Synapse* 64:146–151.

Tsao HH, Lin KJ, Juang JH, Skovronsky DM, Yen TC, Wey SP, and Kung MP. 2010. Binding characteristics of 9-fluoropropyl-(+)-dihydrotetrabenzazine (AV-133) to the vesicular monoamine transporter type 2 in rats. *Nucl Med Biol* 37:413–419.

Vander Borght TM, Kilbourn MR, Desmond TJ, Kuhl DE, and Frey KA. 1995a. The vesicular monoamine transporter is not regulated by dopaminergic drug treatments. *Eur J Pharmacol* 294:577–583.

Vander Borght TM, Kilbourn MR, Koeppe RA, DaSilva JN, Carey JE, Kuhl DE, and Frey KA. 1995b. *In vivo* imaging of the brain vesicular monoamine transporter. *J Nucl Med* 36:2252–2260.

Vander Borght TM, Sima AAF, Kilbourn MR, Desmond TJ, and Frey KA. 1995c. [³H]Methoxytetrabenazine: A high specific activity ligand for estimating monoaminergic neuronal integrity. *Neuroscience* 68:955–962.

Villemagne V, Yuan J, Wong DF, Dannals RF, Hatzidimitriou G, Mathews WB, Ravert HT, Musachio J, McCann UD, and Ricaurte GA. 1998. Brain dopamine neurotoxicity in baboons treated with doses of methamphetamine comparable to those recreationally abused by humans: Evidence from [¹¹C]WIN-35,428 positron emission tomography studies and direct *in vitro* determinations. *J Neurosci* 18:419–427.

Wagner HN Jr, Burns HD, Dannals RF, Wong DF, Langstrom B, Duelfer T et al. 1983. Imaging dopamine receptors in the human brain by positron tomography. *Science* 23:1264–1266.

Wimalasena K. 2011. Vesicular monoamine transporters: Structure-function, pharmacology, and medicinal chemistry. *Med Res Rev* 31:483–519, doi 10.1002/med.20187.

Yu Q, Luo W, Deschamps J, Holloway HW, Kopajtic T, Katz JL, Brossi A, and Greig NH. 2010. Preparation and characterization of tetrabenazine enantiomers against vesicular monoamine transporter 2. *Med Chem Lett* 1:105–109.

Zhu L, Liu Y, Plössl K, Lieberman B, Liu J, and Kung HF. 2010. An improved radiosynthesis of [¹⁸F]AV-133: A PET imaging agent for vesicular monoamine transporter 2. *Nucl Med Biol* 37:133–141.

Chapter 12

13. Radiopharmaceuticals for Imaging Proliferation

Kenneth A. Krohn, John R. Grierson, Mark Muzi, and Jeffrey L. Schwartz

13.1   The Imaging Objective

Unregulated cellular growth is one of the hallmarks of cancer. Because of its central role, quantifying cellular proliferation has long been a goal for imaging. The standard imaging protocol for measuring the growth of a tumor is called RECIST and is basically an anatomic image of the tumor mass quantified as a physical dimension in two orthogonal planes. These measurements fail to appreciate the dynamic nature of cancer, with new cells being produced and damaged cells being destroyed by programmed cell death and other death mechanisms.

In this chapter, cellular growth is considered from a chemist's perspective. When a cell divides to produce

Targeted Molecular Imaging. Edited by Michael J. Welch and William C. Eckelman © 2012 Taylor & Francis Group, LLC. ISBN: 978-1-4398-4195-2

Chapter 13

two new cells, it requires a number of biomolecules for building the new cells: nucleosides to produce new DNA and RNA, lipids and phospholipids to build the cell membrane, energy substrates to support the demand for ATP, and amino acids. While cellular growth is often inferred from FDG-PET, which reflects the energy needs of cells, there are many reasons for the increased uptake of FDG, including a shift in metabolism from oxidative phosphorylation to glycolysis as might be upregulated as a consequence of hypoxia. In fact, a "killed" cancer cell, one that can no longer replicate, will still need energy for maintaining ion gradients, producing new proteins and supporting ATP-binding cassette transporters. Thus, there are many examples where FDG uptake increases after therapy. Labeled amino acids are also widely used to infer growth but cells continue to produce new proteins after their machinery for cell division has been irreparably damaged, so labeled amino acids are only an indirect measure of growth. Acetate labeled with ^{11}C has been touted as a proliferation agent but it is also involved in energy production. Yet another approach involves choline, which can be labeled with either ^{11}C or ^{18}F. The approach still needs considerable validation but may be useful where FDG uptake is low and background is high—for example, in tumors of the prostate.

The most unambiguous measure of growth comes from imaging the process of cell division and replication where DNA unwinds and is replicated during the S-phase of the cell cycle. During the gap that follows, G2-phase, the cell will continue to grow and produce enzymes, and checkpoint processes test that everything is ready for cell division. At some point, cell growth stops and all the energy is used for an orderly division of the cytoplasm into two daughter cells in M-phase, mitosis. An unscheduled parallel process takes place in intracellular mitochondria to produce more mtDNA. Thus, the measurement of cell division is best monitored by the replication of DNA during S-phase and it would be advantageous if the production of nuclear DNA could be distinguished from the production of mtDNA.

DNA and RNA are each chains of repeating units of a sugar 2′-deoxyribose and 3′,5′-phosphates linked to different nitrogen bases. Adenine, guanine, and cytosine are common to both DNA and RNA but only DNA incorporates thymidine; RNA incorporates uracil. Thus, 50 years ago, cellular biologists appreciated the fact that the incorporation of thymi-

dine into cells was a useful measure of cell division for cells in culture and in small animals (Mendelsohn 1962). The technique has even been applied to plant growth (Wimber and Quastler 1963). In these early experiments, the thymidine was labeled with tritium and the labeling index (LI) was measured as the fraction of cells with labeled nuclei as determined autoradiographically. Numerous reports showed that the *in vitro* LI of tumors corresponded closely with LI determined by injection of [^{3}H]-thymidine *in vivo*. The technique with tritium has been extended to the measurement of cellular proliferation in human tissues (Fabrikant 1970) and even for measuring the rate of DNA synthesis in carrot suspension cultures (Slabas et al. 1980).

With the advent of flow cytometry, a new and conveniently automated technology replaced the tritiated thymidine-labeling index (TLI). In flow cytometry, disaggregated cells pass in a single file through an optical detection cell illuminated with a laser beam so that large numbers of cells can be analyzed, one at a time. Initially, cells were analyzed for their size but new fluorochromes allow individual cells to be painted to reflect specific properties and each of these dyes fluoresces with unique wavelengths so that with multiple lasers and wavelength analysis of the fluorescent emissions, multiple properties of the cell can be evaluated. Several fluorochromes bind directly to DNA and are used to estimate the amount of DNA in a cell to determine if it has entered the cell cycle. The results of flow cytometry have demonstrated a strong correlation with TLI although the technique can overestimate %S when TLI is low, apparently because of DNA debris in mechanically disaggregated samples (Meyer and Coplin 1988).

Because of the importance of a noninvasive assay for DNA synthesis, chemistry was developed to incorporate ^{11}C into the thymidine molecule (Christman et al. 1972). Cyclotron-produced ^{11}CO$_2$ was reduced to formaldehyde, which was used for the enzymatic conversion of the uridylate nucleotide to the thymidylate, and alkaline phosphatase was used to yield the nucleoside, [^{11}C]-thymidine, which was purified by anion-exchange resin. The synthesis introduced the label at the methyl position and took about five half-lives of ^{11}C ($T_{1/2} = 20.4$ min) and the product was used in mice for periods up to 3 h using the cut-and-count method. Because imaging instrumentation for positron-emitters was not yet widely available, there was very little use of [^{11}C]-thymidine for more than a decade.

In the meantime, alternative assays for measuring cell proliferation were being developed. Monoclonal antibodies and immunohistochemical (IHC) methods were developed for detecting bromodeoxyuridine, BrdU, a pyrimidine analog of thymidine that is incorporated into DNA during S-phase (Dolbeare et al. 1983). Antibodies to BrdU can be labeled with FITC or detected by a secondary immunoperoxidase assay and the resulting LI is analogous to the TLI calculation. In mice, this experiment can be as simple as including BrdU in the animal's drinking water for a few days; in humans, the thymidine analog is delivered by intravenous injection at the time of surgery. When BrdU is supplied continually for a period lasting at least as long as the doubling time of a tumor, the assay will provide a reliable measure of growth fraction (Yoshii et al. 1986). Noncycling G0 cells will fail to incorporate BrdU during that extended time interval. TLI by microautoradiography and BrdU LI by immunoperoxidase have been compared and gave equivalent results (Richter et al. 1992). Because, BrdU can replace thymidine during DNA replication, it is mutagenic and a potential health hazard, and it sensitizes skin to UV damage.

13.2 Thymidine Metabolism Reflects Proliferation

If imaging proliferation means measuring the rate of DNA synthesis, the concept is similar, whether the probe/detection system is tritiated thymidine and microautoradiography, BrdU and IHC, or [^{11}C]-thymidine and positron emission tomography (PET). However, the pathway to the nucleotide and to the DNA is not a simple linear sequence—thymidine can be metabolized before it gets into the cell and there is a secondary pathway for making the nucleotide intermediate. The route from extracellular thymidine is called the salvage pathway and the route by which deoxyuridine monophosphate (dUMP) is methylated in the cytosols is called the *de novo* pathway (Figure 13.1). Thymidine phosphorylase (TP) is the enzyme that catalyzes the reversible phosphorolysis of thymidine to thymine and 2-deoxy-D-ribose-1-phosphate. TP is also known as platelet-derived endothelial cell growth factor and is an angiogenic factor that stimulates the chemotaxis of endothelial cells and confers resistance to apoptosis. TP helps regulate the level of nucleosides in cells and is particularly important in maintaining the appropriate amount of thymidine in the mitochondria. Deficiencies or mutations in TP can lead to elevated plasma thymidine levels that potentially impair mtDNA replication (Spinazzola et al. 2002). Thymidylate synthase (TS) is the enzyme that catalyzes the addition of single-carbon units at the pyrimidine-5 position of dUMP to form thymidine monophosphate (dTMP). Formate and serine C3 are the source for these single-carbon units. Thymidine kinase (TK) activity is controlled by the cell cycle and it catalyzes phosphate transfer from ATP to thymidine to form the 5′-monophosphate, dTMP. Once either pathway produces dTMP, another nucleoside kinase adds the second and third phosphate groups to eventually produce dTTP, which is the substrate for DNA polymerase.

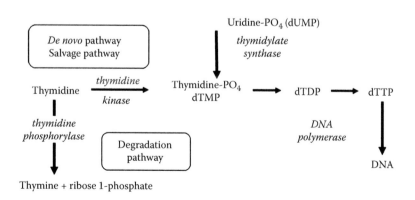

FIGURE 13.1 Biochemical pathways for thymidine metabolism and synthesis of DNA. There are two routes to thymidine monophosphate (dTMP). The imaging studies described in this chapter focus on the more prevalent salvage pathway by which exogenous thymidine enters cells and is sequentially phosphorylated before DNA polymerase adds the triphosphate to a strand of DNA. The *de novo* pathway produces the same intermediate dTMP through methylation of uridine monophosphate (dUMP).

Chapter 13

13.3 Development and Evaluation of [^{11}C]-Thymidine

13.3.1 [Methyl-^{11}C]-Thymidine

The initial enzymatic synthesis of [^{11}C]-thymidine (Christman et al. 1972) was described above. This resulted in a single pioneering report with images from a VX2 rabbit tumor model that clearly showed the tumor (Crawford et al. 1978). An alternative organic synthesis using ^{11}CH$_3$I and trimethyl derivatives of the Li salt of BrdU was developed by Langstrom et al. (1981) but was superceded by a reaction using ^{11}CH$_3$I and a tetrahydropyranyl (THP)-protected derivative of the Li salt of BrdU to provide [methyl-^{11}C]-thymidine in high yield and purity (Sundoro-Wu et al. 1984). The THP-protecting group proved stable in the strong base of *n*-BuLi and it could be readily prepared and stored for months in crystalline form. The labeling involved 2 eq of *n*-BuLi and then ^{11}CH$_3$I to yield [methyl-^{11}C]-thymidine and 2′-deoxyuridine via the pyrimidyl dianion (Scheme 13.1). The desired product was purified by reversed-phase HPLC and the product was verified by coelution with an authentic standard and by its mass spectra. This radiosynthesis resulted in 17–25% radiochemical yield based on starting ^{11}CH$_3$I in 45 min with radiochemical purity >99%.

Only a few PET imaging groups have used this radiosynthesis of [methyl-^{11}C]-thymidine for studies of tumor growth. One study evaluated 10 patients with non-Hodgkin's lymphoma and found a mean tumor/muscle ratio of 11.8 ± 1.7 in intermediate-grade tumors (Martiat et al. 1988). The tumor uptake reached a plateau level of 0.42 ± 0.22%/100 cc by 10 min after injection. Some studies in mice and dogs, including three dogs with spontaneous tumors, compared thymidine labeled with tritium and ^{14}C as well as ^{11}C, all in the methyl position, and demonstrated that a minor fraction of the activity at 60 min after administration was in DNA, with the majority in metabolites (Shields et al. 1990). The percentages in DNA for lymphoma masses were 33, 40, and 37 in three dogs whereas the DNA percentage in normal spleen was 49 and 41 in two dogs. Considerably less of the ^{14}C was in the DNA fraction in slowly proliferating normal tissues like liver, heart, and kidney. There was also evidence supporting the process of local reutilization of the radiocarbon from the methyl group as evidenced by the presence of radiocarbon in proteins. These experiments also showed that ^{14}C and ^{11}C tracked each other but tritium in the methyl position should not be used as it was partially lost as water. This report described an HPLC separation that distinguished thymine from thymidine and showed that after only 2 min most of the thymidine had been converted to thymine and a few minutes later the label showed up as low MW metabolites.

13.3.2 Thymidine Metabolites

Thymidine is rapidly degraded by TP to produce deoxyribose and thymine, the fragment that carries the ^{11}C label. Thymine is reduced stepwise to dihydrothymine and then β-ureidoisobutyric acid and β-aminoisobutyric acid plus CO$_2$ and NH$_3$ (Scheme 13.2).

The rapid time course of metabolite formation measured by HPLC analysis of a time series of blood samples is shown in Figure 13.2. In this assay, blood samples were mixed immediately with three volumes of MeOH/MeCN (3:1 v:v) and centrifuged. The supernatant was separated by reverse phase using a mobile phase of 2% MeCN in water. This provided a baseline separation between thymidine (triangles and solid

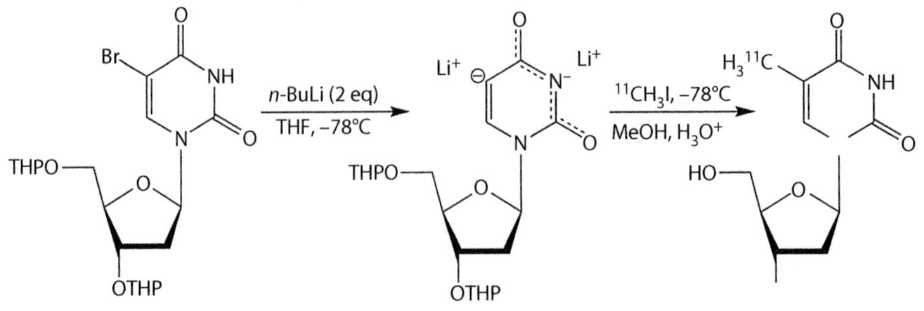

SCHEME 13.1 Synthesis of [methyl-^{11}C]-thymidine.

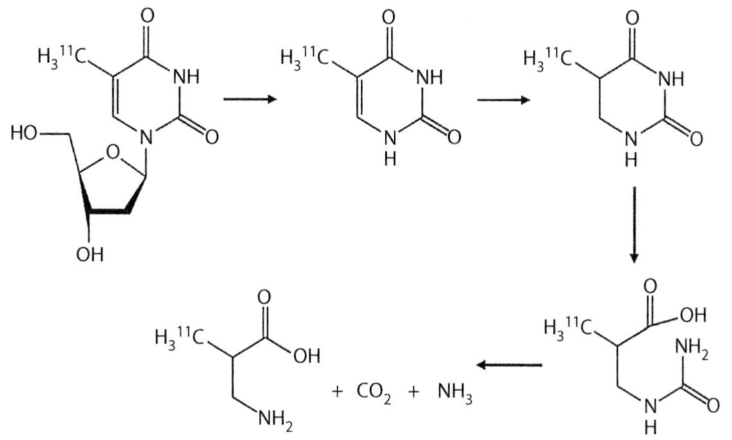

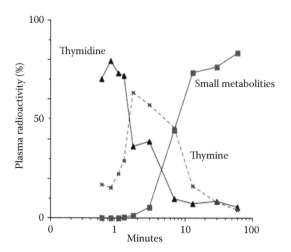

SCHEME 13.2   Degradation of thymidine.

line) and thymine (crosses and dashed line) but did not separate thymine from dihydrothymine. The isobutyric acid metabolite is charged and does not cross the blood–brain barrier.

The time course of metabolite production in plasma samples (Shields et al. 1990, Conti et al. 1994) is an essential component of the input data for modeling cellular growth from PET imaging. Other groups have confirmed that kinetic differences in various ^{11}C-metabolites are small (Goethals et al. 1995). The complex metabolic profile of thymidine, especially with the label in the pyrimidine methyl position, is

probably the main reason why more research with this compound has not occurred.

13.3.3   [2-^{11}C]-Thymidine

The advantage of thymidine labeled in the 2-position of the pyrimidine ring is that its ultimate metabolic fate is [11]CO_2, which washes out of tissues and can be more easily quantified and modeled than β-ureidoisobutyric acid. This advantage was first proposed by scientists from the University of Louvain and their synthesis used ^{11}C]-urea as the precursor in a cyclocondensation reaction with a dialkyl β-methylmalate (Vander Borght et al. 1991). The solvent for cyclocondensation is fuming sulfuric acid (30% SO_3), which provides a challenge for the subsequent conversion of the thymine to the nucleoside using an enzyme. This synthesis is now used in several PET laboratories although there are two routes to the ^{11}C]-urea precursor. The common route is described in the synthetic Scheme 13.3 (Link et al. 1995), but an alternative route derives the urea from ^{11}C]-phosgene (Steel et al. 1999). The simple radiometabolite profile following injection of [2-^{11}C]-thymidine has been demonstrated by measuring metabolites in a series of 17 studies in 14 patients. The results of blood metabolism were fit with a compartmental model and showed that the percentage of intact thymidine could be inferred with as few as three measured points with a root mean square error in the integrated blood curve of only 2% (Shields et al. 1996b), although five samples are commonly assayed to provide the input data for thymidine modeling.

FIGURE 13.2   The time course of metabolite formation after injection of [2-^{11}C]-thymidine. Note that the time axis is a logarithmic scale. Thymidine is rapidly converted to thymine by action of thymidine phosphorylase and then is sequentially degraded to dihydrothymine and β-ureidoisobutyric acid and the label eventually ends up as ^{11}C]CO_2. This graph shows that thymine (dashed line) is in circulation for only a few minutes and most of the activity is small molecules, principally CO_2.

Chapter 13

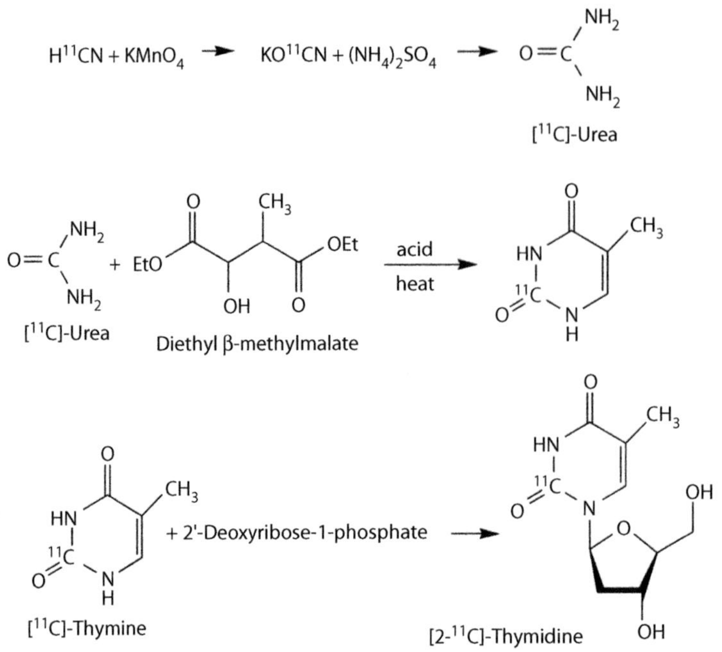

SCHEME 13.3 Synthesis of [2-¹¹C] thymidine.

13.4   Quantification of Thymidine Metabolism

The early tritium work showed that thymidine cleared from the circulation very rapidly (Hughes et al. 1958). Thus, the blood flow dependency of tracer distribution kinetics was the first consideration with the interpretation of [¹¹C]-thymidine images. This led us to examine the time course of [methyl-³H]-thymidine distribution and relative blood flow using [¹⁴C]-iodoantipyrine in mice and dogs, and collecting data in tumors as well as several normal tissues (Shields et al. 1984). The results at 20 s after injection of thymidine into the tail vein of normal AKR mice showed a remarkable correlation with the flow tracer. Only the brain took up significantly less thymidine than expected from its relative perfusion. By 1 min after injection, the correlation was lost and at 1 h there appeared to be an inverse correlation with relative flow that was accounted for by the proliferation characteristics of the organs—brain, heart, and lung had high flow but low thymidine retention whereas spleen, duodenum, and thymus had relatively low flow but high thymidine uptake. These results were verified in two dogs with spontaneous tumors and, together with the studies in mice, provide convincing evidence that there is little influence of blood flow on the ultimate distribution of labeled thymidine. While the tracer must get to a tissue before it can be retained, the intracellular metabolism of thymidine is relatively more important than blood flow.

In order to interpret [¹¹C]-thymidine PET studies of the salvage pathway as a measure of proliferation, it was essential to show that the relative use of endogenous and exogenous thymidine was equivalent. This was established by growing cells in bromodeoxyuridine (BrUdR) and then analyzing the density of DNA using cesium sulfate gradients to separate authentic DNA from that containing the heavier bromine (Shields et al. 1987). Studies were done in HeLa M-1 human cells, SL2 mouse cells, and Cf2Th canine cells and showed an excellent correlation between the percentage exogenous nucleoside, BrUdR, and the micromolar concentration of the BrUdR (over three orders of magnitude) in which the cells were grown. Local reutilization is another issue that could limit the interpretation of thymidine imaging; evidence of negligible local reutilization would simplify the interpretation of [¹¹C]-thymidine images. In order to study this process, both [¹⁴C]-thymidine and tritiated iododeoxyuridine, IUdR, were used to quantify nucleoside reutilization *in vivo* in normal mice (Quackenbush and Shields 1988). Negligible reutilization would be implied if the results showed a close correlation between the two nucleosides. No measurable differences were found in thymus, spleen, or bone marrow although some local reutilization was found in duodenum. There were, however, substantial differences between [³H]-IUdR and $[^{125}I]$-IUdR, suggesting an aggressive deiodination of the nucleoside analog. These two studies provide an important example of the rigorous validation required for new imaging agents.

With validation that thymidine uptake and retention is not limited by blood flow and experimental evidence that there is very little endogenous synthesis and negligible reutilization, a model describing the time course of the ^{11}C imaging agent can be constructed. The experimental data for modeling analysis include the time–activity curve from tissue regions in the image, the blood clearance curve from a combination of imaging data and calibration by a few blood samples, and metabolite analysis to correct the blood input function. In addition to PET imaging after administration of $[^{11}C]$-thymidine, an injection of $[^{11}C]CO_2$ can be used to independently measure the values for K1c, k2c, and k3c in the model (Figures 13.3 and 13.4). This dual injection protocol takes more time but has proved useful in imaging brain tumors (Wells et al. 2002a).

$[^{11}C]$-Thymidine PET with modeling has proven useful for measuring proliferation and pharmacodynamic response to chemotherapy. An innovative example shows that when patients with advanced gastrointestinal cancer are treated with a thymidylate synthase inhibitor to shut down the *de novo* pathway, the tumor compensates by upregulating the salvage pathway as demonstrated by increased uptake of exogenous $[^{11}C]$-thymidine (Wells et al. 2003).

When a five-compartment model was validated for incorporation of $[^{11}C]$-thymidine into DNA in somatic tissues, simulations show that it is difficult to estimate individual rate parameters, but the flux constant for incorporation into DNA can be accurately estimated (Mankoff et al. 1998). In a same-day comparison of FDG and thymidine imaging as measures of response to cytotoxic chemotherapy, the thymidine flux constant declined much more than the FDG SUV after 1 week of treatment in patients who responded and the thymidine flux was unchanged in nonresponders (Shields et al. 1998b). This initial report of patients with non-small-cell lung cancer or sarcoma involved only six patients but it was encouraging in that thymidine flux was shut down in response to successful therapy, whereas FDG SUV was still above normal. Validation studies in animals were perhaps more convincing.

Uptake studies in rats measured the incorporation of ^{11}C into DNA and showed that the model was able to fit the time course of uptake and the fractional incorporation into DNA (Mankoff et al. 1999). This study also reported results from sequential injections of $[2-^{11}C]$-thymidine and $[2-^{11}C]$-thymine, the first metabolite. Thymine uptake in normal volunteers showed only a minimal amount of trapping in bone marrow

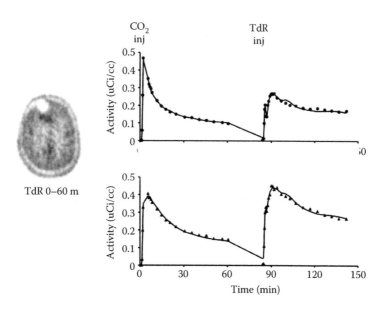

FIGURE 13.3 Results from a sequential imaging protocol with the injection of $[^{11}C]CO_2$, followed 90 min later by the injection of $[2-^{11}C]$-thymidine. The line graphs show activity in normal brain and tumor. The patient had a previously treated glioblastoma multiforme that was contrast enhancing by MRI but showed negligible FDG uptake. The integrated image after thymidine injection shows a concentration of activity along the rim of the surgical resection cavity. The background throughout the brain is from $[^{11}C]CO_2$, which washes out slowly.

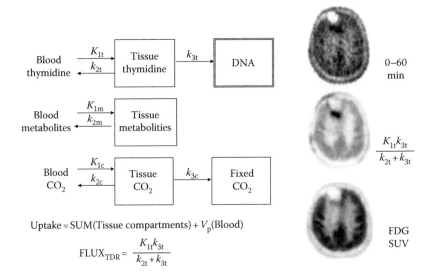

FIGURE 13.4 Compartmental model for the estimation of thymidine flux from a PET study of [2-^{11}C]-thymidine. The input for this model includes a blood clearance curve for the major metabolite, CO_2, and the blood clearance curve for thymidine with HPLC metabolite correction. The images on the right are for the same patient shown in Figure 13.3. The flux parametric image shows very little normal brain background and tumor uptake that is much more conspicuous than in the FDG-PET study.

that was accounted for by a small amount of fixation of labeled CO_2, a process included in the model.

Potential limitations imposed by the blood–brain barrier (BBB) necessitated further model validation prior to measuring proliferation in brain tumors. Simulation studies were used to determine whether kinetic analysis could distinguish between transport and retention (Wells et al. 2002a). These studies showed that non-CO_2 metabolite parameters could be fixed without compromising thymidine parameter estimates. Both K_{1t} (tracer delivery) and thymidine flux could be estimated accurately with standard errors <15% and the model was able to distinguish increased transport associated with BBB breakdown from increased retention associated with tumor proliferation. This report was followed by a study of 20 patients with brain tumors who underwent the sequential

protocol of [^{11}C]CO_2 and [2-^{11}C]-thymidine images for estimation of transport and flux into brain tumors and normal brain (Wells et al. 2002b). With arterial sampling and metabolite analysis, the data could be fit with the five-compartment model described above. As expected, K_{1t} was higher in MR contrast-enhancing tumors. Flux increased with grade and was lower after treatment in high-grade tumors. The modeling analysis was able to distinguish increased transport from increased proliferation and showed good agreement with the clinical and pathological features of a wide range of brain tumors. Figure 13.4 shows images from one of these PET studies with thymidine and also FDG. It involved applying the five-compartment model to a PET study with [^{11}C]CO_2 followed by a dynamic data acquisition with [2-^{11}C]-thymidine to allow generation of a parametric map of thymidine flux.

13.5 Sensitivity and Clinical Impact

These complex imaging protocol and analysis are time consuming and one has to ask whether or not it is worth the effort. Is a parametric flux image more informative than an integrated sum image? It is clear from the images in Figure 13.4 that the contrast is greater in the flux map but one might question whether a threshold to reduce the presumably homogeneous $^{11}CO_2$ background in the 0–60 min image could yield a similar level of contrast. Some anecdotal data to answer this question come from a series of four studies on a

single patient (Figure 13.5). This individual first presented with a low-grade astrocytoma that was not contrast enhancing. After 15 months, the patient's tumor showed MR enhancement and its volume was increasing. At surgery, it was diagnosed by pathology as an anaplastic mixed glioma and was subsequently treated with chemotherapy and radiation. Another imaging study was done after 1 year that showed reduced transport and flux for thymidine but an increased FDG SUV. At this time, the patient was clinically stable.

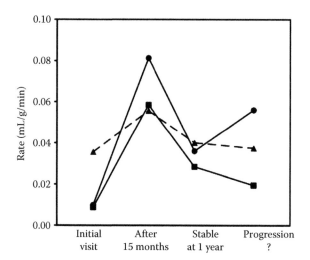

FIGURE 13.5 A time series of studies for a patient who initially presented with a low-grade astrocytoma that was contrast enhancing (image time 1). Fifteen months later, the patient's MRI showed increased volume of enhancement and he was operated for an anaplastic mixed glioma (image time 2). The patient was treated with chemotherapy and radiation. Eleven months later, the patient was stable by clinical exam and MRI and was continued on chemotherapy (image time 3). Note that the SUV (dashed line) and the parametric images for transport (K_1, solid squares) and proliferation ($FLUX_{TdR}$, solid circles) tracked together for these three PET studies. Thirteen months later, the patient showed increased enhancement by MRI and the neuro-oncologist was faced with the question of whether this was progression or radionecrosis. This time the transport and proliferation parameters diverged, showing a substantial increase in transport but a continuing decline in flux. The flux image reflected accurately the absence of progression, which was confirmed by his clinical course. Follow-up MRIs showed resolution of the enhancement.

After another year, the patient showed increased enhancement by MRI and had another PET study with [^{11}C]-thymidine and FDG to differentiate between tumor progression and radionecrosis. The value of modeling the data was revealed in the last study, which showed that transport was increased but flux into DNA was decreased, showing clearly that the disease was not progressing. At this time, the FDG-PET was inconclusive. The patient's clinical course was consistent with the parametric flux image of no proliferation and subsequent MRIs showed resolution of the enhancement.

13.6   Search for Thymidine Analogs That Are Not Rapidly Metabolized

There are several impediments to the use of [2-^{11}C]-thymidine for imaging proliferation. It is a challenging radiosynthesis involving a radionuclide with a short half-life. It requires metabolite analysis by HPLC and the data analysis works much better with a parallel imaging study, $^{11}CO_2$ for brain studies and $H_2{}^{15}O$ for somatic tumors. Lastly, the study requires extensive modeling effort for data analysis. In spite of these limitations, the protocol works and has provided useful insight into the biology of tumors. However, the demands of the method motivated the search for a nonmetabolized analog, preferably labeled with ^{18}F, which would have a clear practical advantage.

With the advent of HIV/AIDS, medicinal chemists began a search for a nucleoside analog reverse transcriptase inhibitor as a type of antiretroviral drug for the treatment of the disease. Reverse transcriptase is necessary for the production of the viral double-stranded DNA that is integrated into the genetic material of virally infected cells. Azidothymidine (AZT) was the first approved treatment for HIV and was heralded as a major breakthrough (Mitsuya et al. 1985). As it relates to development of better imaging agents, it had three critical advantages: (i) the azido group increased its lipophilicity, allowing it to cross cell membranes, including the blood–brain barrier, more easily; (ii) it was not a substrate for TP, so it stayed in the plasma; and (iii) even though it formed the requisite 5′-triphosphate substrate for DNA polymerase, the azido group at the 3′ position led to premature termination of DNA chains (Furman et al. 1986).

The Seattle group developed [^{11}C]-pseudothymidine as an alternative to address the shortcomings of [^{11}C]-thymidine (Grierson et al. 1995). In this analog, the pyrimidine atom involved in the glycosidic linkage was C rather than N and the atom at position 5 was N rather than C so the methyl was an N-Me substitution. An earlier strategy from Edmonton involved difluoro derivatives at the 2′ position. These difluoro compounds were transported into cells and activated by monophosphorylation to molecules that inhibited the critical enzyme in the *de novo* pathway (Mercer et al. 1987).

Armed with these precedents, we surveyed a series of potential imaging agents for their ability to withstand enzymatic degradation by TP (Shields et al. 1996). The molecules were incubated at 37°C in dog plasma and analyzed by HPLC. Thymidine was totally degraded within 5 h, pseudothymidine was <10% degraded in this time, and the molecules listed in Table 13.1 were comparably stable for 24 h.

Table 13.1 Alternative Nucleosides for Imaging

Nucleoside	Small Molecules (%)	RNA + DNA (%)	Protein (%)
FFUdR	90	3.5	6.3
FFaraU	89	7.0	4.0
FTdR	94	4.8	1.1

FFUdR, 5-fluoro-1-(2′-deoxy-2′-fluoro-β-D-ribofuranosyl) uracil; FFaraU, 5-fluoro-1-(2′-deoxy-2′-fluoro-β-D-arabinofuranosyl)uracil; FTdR, 5-methyl-1-(2′-deoxy-2′-fluoro-β-D-ribofuranosyl)uracil.

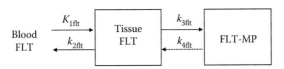

K_{1flt} represents transport

k_{3flt} represents thymidine kinase activity

k_{4flt} represents 5′-nucleotidase activity

$$\text{FLUX}_{\text{FLT}} = \frac{K_{1flt} \cdot k_{3flt}}{k_{2flt} + k_{3flt}} = \frac{K_{1flt} \cdot k_{3flt}}{K_{1flt}/V_d + k_{3flt}}$$

under conditions where k_{4flt} is negligible

FIGURE 13.6 Compartmental model for the estimation of proliferation for FLT-PET studies. In this case, a single injection is used but the blood curve is analyzed for metabolites and the input function is corrected for contamination by the metabolites. The end compartment contains all three FLT nucleotides but negligible amounts of FLT are incorporated into the DNA. The two formulas for calculation of FLUX_{FLT} are equivalent and reflect the fact that in the modeling, k_2 is reparameterized to K_1/V_d.

The additional requirement for a thymidine analog for PET imaging is that it should be a good kinase substrate (Figure 13.6). More specifically, it would be advantageous if the analog were a better substrate for cytosolic TK1 than for mitochondrial TK2. Literature

Table 13.2 Values of V_{max}/K_m for Cytosolic TK1 and Mitochondrial TK2

Nucleoside	TK1	TK2	Reference
TdR	18,910 (100)	1866 (100)	Munch-Petersen et al. (1991)
TdR	670 (100)	150 (100)	Wang and Eriksson (1996)
FUdR	4089 (22)	614 (33)	Munch-Petersen et al. (1991)
BrUdR	(80)	(100)	Eriksson et al. (1991)
FLT	1442 (7.6)	Nil	Munch-Petersen et al. (1991)
AZT	8262 (44)	160 (9)	Munch-Petersen et al. (1991)
araT	(<1)	(60)	Eriksson et al. (1991)
FMAU	31 (4.6)	38 (25.3)	Wang and Eriksson (1996)
FIAU	11 (1.7)	50 (33)	Wang and Eriksson (1996)

Values in parentheses are percentages of the ratio obtained with thymidine for the given enzyme in the individual reports listed under the References column. Eriksson et al. (1991) did not report V_{max}/K_m for each nucleoside but they reported the same ratios for thymidine as in Munch-Petersen et al. (1991).

results summarized in Table 13.2 were available for many molecules of interest and supported the advantages of 3′-fluoro-3′-deoxythymidine (FLT) for measuring TK1 activity. It is not a substrate for TP and data of Table 13.2 show that, while it is a poorer substrate than thymidine for TK1, its reactivity with TK2 is negligible. Table 13.2 also emphasizes that the arabino sugar leads to preferential reactivity with the constitutively expressed mitochondrial enzyme.

13.7 Development and Evaluation of ^{18}F Probes for Proliferation

13.7.1 Synthesis of 3′-Fluoro-3′-Deoxythymidine

Considerable support for FLT as a proliferation imaging agent has developed over the past decade. It involves a convenient radionuclide, ^{18}F, and a convenient synthesis involving a standard nucleophilic substitution with anhydrous fluoride, followed by deprotection and purification of the product by HPLC. The imaging

agent has good tumor uptake and blood clearance and this uptake reflects cellular proliferation, although it does require kinetic data analysis to distinguish transport from flux.

13.7.2 Initial Synthesis

The initial radiosynthesis of FLT was for AIDS research and was a carrier-added and low-yield

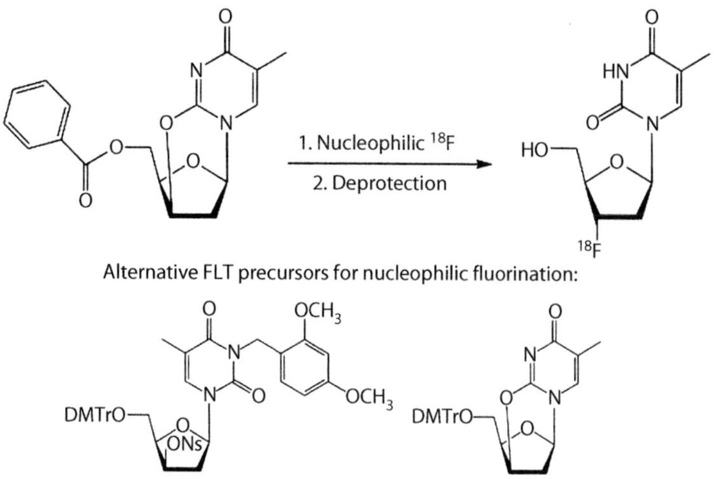

SCHEME 13.4   Synthesis of FLT.

procedure (Wilson et al. 1991). However, the first practical synthesis of high specific activity FLT for PET imaging (Scheme 13.4) was developed at the University of Washington and involved a nucleophilic displacement of a 3′-O-nosyl group on the precursor protected at two positions: a 2,4-dimethoxybenzyl group at 3-N and a O-4,4′-dimethoxytrityl group at 5′-O, followed by deprotection using ceric ammonium nitrate (Grierson and Shields 2000). The product was purified by semipreparative HPLC (C-18, 10% EtOH–water).

Shortly after this initial synthesis, several alternative and improved radiosyntheses were reported that all avoided the complications of precipitating ceric salts during the deprotection step (Machulla et al. 2000, Wodarski et al. 2000, Martin et al. 2002). This FLT synthesis could be used for labeling either 5′-O-(4,4′-dimethoxy-trityl)-2,3′-anhydrothymidine or 5′-O-benzoyl-2,3′-anhydro-thymidine as precursors. The anhydro structure serves a dual role as an

activating electrophilic center at the 3′ position and a protecting group of the N-pyrimidine site. This route used high temperature and dimethyl formamide (DMF) solvent, which proved inconvenient for HPLC purification. Additional simplifications have been introduced for routine use with PET tracer synthesizer boxes. One of the box syntheses preferred the N-tBOC-protected precursor for the best radiochemical yield and advocated purification by SepPak rather than HPLC to achieve a synthesis time of 50 min (Teng et al. 2006). A more recent report used the same precursor and hydrolysis with 1 N HCl and SepPak purification in a one-pot reaction (Tang et al. 2010). The chemical purity of these simplified and automated methods without HPLC purification of the product mixture has not been fully validated. Our laboratory has used the 5′-O-benzoyl-protected reagent with HPLC purification for all human imaging studies and the toxicity studies reported later are based on this method.

13.8   *In Vitro* and Animal Validation Studies of 3′-Fluoro-3′-deoxythymidine

Clearly, the value of FLT as an imaging agent depends on understanding the mechanism(s) by which it is trapped in cells. We first tested the hypotheses that FLT uptake reflected the enzyme activity of the proliferation-regulated cytosolic thymidine kinase, TK1 (Rasey et al. 2002). Cultured A549 human lung carcinoma cells were used while actively dividing and again when growth-arrested by keeping the cells for a week without changing the culture medium. Several measurements were compared: uptake of FLT, percentage of cells in S-phase,

and activity of cytosolic TK1. Growth-arrested A549 cells took up little FLT and had low levels of TK1 activity. When stimulated to grow actively by returning the cells to fresh medium, there was a strong correlation between increased FLT uptake and increased TK1 activity. FLT uptake better reflected the growth changes than did the uptake of [14C]-deoxyglucose. Nonproliferating A549 cells did not enter the cell cycle after they were irradiated and placed in fresh medium, nor did they show increased FLT uptake or elevated TK1 activity.

Chapter 13

Thus, FLT uptake is positively correlated with cell growth and TK1 activity and inhibition of cell cycle progression by radiation prevents both FLT uptake and increases in TK1 activity. Furthermore, these studies suggested FLT uptake could be a better measure of cell proliferation than FDG.

The similarities between FDG and FLT are remarkable in that both radiopharmaceuticals enter cells via transporters and are monophosphorylated by either hexokinase or thymidine kinase, and then are trapped because of the charge of the phosphate group. This leads to a hypothesis that there might be an FLT equivalent of the lumped constant for FDG, a scaling factor that would allow the thymidine flux for incorporation into DNA to be inferred from the FLT flux through the TK1 pathway (Krohn et al. 2005). Cell uptake studies using [^{3}H]TdR and [^{18}F]FLT showed similar values in two human and two murine lines (Grierson et al. 1998) and a more complete study involving 22 asynchronously growing tumor lines showed a correlation with $r = 0.88$ between thymidine and the analog when uptake was measured as pmol/mg protein/h (Toyohara et al. 2002). However, when the FLT/TdR uptake ratio was calculated from the published data, it was broadly distributed with a range from about 0.28 to 1.13, suggesting that a constant calibration factor would not be appropriate. This may suggest that a measure of TK1 enzyme activity does not equate with the rate of DNA synthesis. Other variables such as the relative contributions of the *de novo* and salvage pathways and the amount of nucleoside transporters might be involved as secondary factors (Paproski et al. 2008; Plotnik et al. 2010).

In vivo biodistribution studies of nucleosides in rodents can be of limited value because they have 10-fold higher levels of serum thymidine compared to humans. Biodistribution studies in dogs with PET imaging were presented in the initial report on *in vivo* imaging with [^{18}F]FLT (Shields et al. 1998a). The images were remarkable in that they are dominated by radioactivity in the spine, reflecting even higher uptake in marrow than was seen for [^{11}C]TdR. There was very little [^{18}F]FLT in the myocardium, which is the target organ for [^{11}C]TdR dosimetry. This difference reflects the fact that FLT is a poor substrate for mitochondrial TK2, which, in the case of [^{11}C]-thymidine, clearly dominates uptake in the heart. Also, there was less [^{18}F] FLT activity in the dog liver than in humans, presumably reflecting a difference in glucuronidation (Shields et al. 2002). This initial report of FLT-PET showed a conspicuous lesion for non-small-cell lung cancer, along with uptake in liver, marrow, and urinary bladder. It is not the purpose of this chapter to review the growing human experience with FLT-PET because this work is reviewed frequently (Barwick et al. 2009).

Metabolism of FLT is, as predicted, simpler than that of thymidine. In cell culture experiments, only FLT and phosphorylated FLT have been detected inside cells, with [^{18}F]FLT-MP as the dominant product (Grierson et al. 2004; Plotnik et al. 2010). The di- and triphosphate nucleotides were also produced by thymidylate kinase and nucleotide diphosphate kinase and at 1 h 14% of the intracellular pool was FLT-TP (Plotnik et al. 2010). *In vivo* studies showed that FLT was conjugated in the liver to produce the glucuronide, which was rapidly excreted into the urine, leading to intense uptake in liver and bladder as demonstrated in human FLT-PET images. *In vivo*, about two-thirds of the circulating ^{18}F after 1 h is still FLT but the level of metabolites is sufficient that it is helpful to correct the input function by the analysis of a few blood samples if compartmental analysis of the study is planned. While HPLC is the standard for this analysis, a two-step SepPak method has been reported that provides a simple metabolite assay (Shields et al. 2005). For kinetic analysis, this initial report argued that eight venous blood samples should be analyzed to provide an accurate input function. Our more recent work suggests that this number could be reduced to five without adversely affecting the FLT flux parameter derived with compartmental analysis. There is another complication related to metabolism that was not a problem with [^{11}C]TdR because the latter is rapidly incorporated into DNA. Most FLT is trapped in cells as FLT-MP and is therefore a substrate for 5′-nucleotidase, meaning that some fraction of the TK1 product can be lost over time by transport of dephosphorylated FLT out of cells (Grierson et al. 2004). The importance of this process in limiting the interpretation of FLT-PET studies has not yet been resolved.

13.9   Toxicity of 3′-Fluoro-3′-deoxythymidine

FLT was originally evaluated as a therapeutic agent because intracellular metabolism of FLT produces nucleotides that inhibit endogenous DNA polymerases and can prematurely terminate DNA chains. As part of the development of 2′,3′-dideoxynucleoside therapeutics, several analogs, including AZT and FLT, were tested for effects on proliferation and differentiation of human bone marrow cells in culture (Faraj et al. 1994). These studies showed that FLT was much more cytotoxic than AZT, which was successfully developed as an early treatment for AIDS. Therapeutic trials of FLT found evidence of hematologic and hepatic toxicity and peripheral neuropathy. The lowest therapy dose tested was 50 ng h/mL for a 12 h interval (AUC_{12}) and involved twice-daily injections for 112 days and this resulted in a mild neuropathy within 40 days of therapy in 2 of 15 patients (Flexner et al. 1994).

FDA required toxicity observations as part of the initial IND studies for [18F]FLT-PET. Two studies have been reported from the University of Washington. One study reported on the effects of [18F]FLT with a minimum specific activity of 100 Ci/mmol and a dose of 5 mCi (Turcotte et al. 2007). Blood samples from 20 patients with non-small-cell lung cancer were evaluated by a comprehensive clinical laboratory panel. Samples were taken prior to and shortly after injection of the radiopharmaceutical, and additional lab values obtained as part of clinical care were included in the analysis. In this study, patients received a maximum mass of 12.2 μg FLT, which represented an AUC_{12} exposure ranging from 0.23 to 1.34 ng h/mL for only a single injection. The lab values were carefully analyzed and showed that the mean for most values was within normal limits at all times. Albumin, RBC count, and hematocrit showed a small but significant decrease over time. Haptoglobin levels did not suggest hemolysis. An abbreviated neurological evaluation was performed and no changes were detected in signs and symptoms or complaints. There were no reports of nausea, vomiting, dizziness, or headache during the injection or the following 2.5 h.

The NCI sponsored a separate trial of safety in patients with recurrent gliomas (Spence et al. 2008). The FLT-specific activity was higher so that the maximum injected dose was 6.1 μg. This study also included a complete clinical laboratory panel and it provided much more details on neurological examinations of all patients at four time points, before and immediately after the PET study and again at 24 h and at

1 month. Vital signs and an electrocardiogram trace were recorded throughout the FLT-PET procedure and at 3 and 24 h after the infusion. The single-dose AUC_{12} value derived from these safety studies was <0.5 ng h/mL and is only 1% of the lowest single-dose therapeutic level of the HIV trial, a level that led to transient peripheral neuropathy in 2 of 15 subjects that manifests after daily dosing for about 40 days. The AUC_{12} values in arterial blood samples from the FLT-PET studies had a mean value of 0.016 ng h/mL for a two-dose study separated by about 1 month, orders of magnitude below the treatment levels. There were no symptoms in any subjects that were attributable to the study.

The potential mutagenic properties of FLT have also been tested. Ehrlich ascites cells incubated with 10 μM FLT for a range of times showed chromosome breaks and gaps but these largely resolved during a recovery time of 24 h (Wobus 1976). Exposure to 1 μM FLT resulted in levels of fragmentation that were indistinguishable from controls without FLT exposure. In these exposure studies, the fraction of FLT incorporated into DNA was low, about 10^{-6}, probably minimizing any genetic consequences from the drug.

[18F]FLT results in exposure to radiation. Biodistribution measurements from dynamic imaging studies involving 18 patients have been combined and used for dosimetry calculations (Vesselle et al. 2003). All the subjects had normal renal function, and urine was collected at the end of each imaging sequence. Regions of interest were hand-drawn around critical organs and the absorbed dose was calculated from integrated time–activity curves (MBq h/g) and S-values provided by MIRD software. From these measurements and calculations, the effective dose equivalent (EDE) is 47 mrad/mCi for men and 59 mrad/mCi for women. The critical organ is the urinary bladder wall and, depending on the time before voiding, this dose can reach as high as 650 mrad/mCi.

From these toxicity studies, the standard administration is limited to 5 mCi and <6.1 μg FLT. Our acceptance criterion for human dose is a specific activity >200 Ci/mmol. This protocol can be safely repeated three times without any harmful effects.

13.10   Quantification of 3′-Fluoro-3′-deoxythymidine

While it is not the purpose of this chapter to review clinical utility of proliferation imaging agents, conflicting results are being reported regarding the ability of FLT-PET to distinguish tumor from other pathol-

ogy. One report found that the procedure did not discriminate between reactive and metastatic lymph nodes in head and neck cancer patients when analyzed by SUV (Troost et al. 2007). In a follow-up

Chapter 13

report, the same laboratory argued that FLT showed promise for measuring early response in head and neck carcinomas (Troost et al. 2010). Imaging comparisons involving FDG-PET and FLT-PET have been particularly confusing in the literature when analysis is limited to SUV (Cobben et al. 2004; Francis et al. 2004; Chen et al. 2005; Weber 2010). The issue may be that these two imaging agents have similar size and molecular weight and partition coefficient, and so, early after administration, when biodistribution is reflecting principally delivery processes, the images can be remarkably similar. The clinical role of FLT is reviewed regularly and will not be discussed in this chapter.

Even though FLT is not metabolized to the extent of thymidine, it still requires metabolite analysis and kinetic modeling to distinguish delivery of the imaging agent from kinase flux. Because of the blood–brain barrier, we did separate model validation studies for brain tumors (Muzi et al. 2006) and somatic tumors (Muzi et al. 2005) although the model construct was the same (Figure 13.6). The transfer of FLT from blood to tissue is represented by K_{1flt}. This parameter is restricted by the blood–brain barrier and is an important reason why independent validation is required for brain and somatic tumors. The return of nonmetabolized FLT back to blood is represented by k_{2flt} and is represented in the model by K_{1flt}/V_d, where V_d represents the early distribution volume for the reversible process, $V_d = K_1/k_2$. Intracellular trapping of FLT-MP is represented by k_{3flt} and is the rate-limiting step for the retention of FLT in cells. Because nucleotides can exit cells, the model includes a reverse term, k_{4flt}. This process could be a result of nucleotidase or reversible enzyme activity or could result through nucleotide transporters but it is generally thought to be a minor process that can be ignored in implementing the model. The flux of FLT through the kinase reaction is estimated using this compartmental model. Parameters are derived from fitting the tissue and blood time curves using standard nonlinear least squares minimization algorithms; the sum of the squared difference between the PET measurements and the model solution is minimized to generate a parametric image of flux and transport (K_{1flt}) using mixture analysis (O'Sullivan 1994).

As with [^{11}C]-thymidine, it is appropriate to question whether or not modeling makes a clinically significant difference, particularly in gliomas where normal uptake of FDG is a limitation. The UCLA group has found in 15 patients that relative change in SUV between baseline and at 6 weeks predicted a favorable outcome but the change at 2 weeks was not predictive (Schiepers et al. 2010). The Cologne group preferred a compartmental analysis and showed an impressive correlation between Ki-67 staining and flux ($r = 0.79$) as well as k_3 ($r = 0.76$). The stain detects Ki-67, a nuclear protein expressed during phases G1 to M of

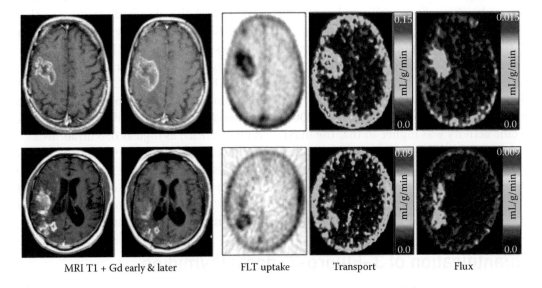

MRI T1 + Gd early & later   FLT uptake   Transport   Flux

FIGURE 13.7 MRI and FLT-PET images from two patients who were being evaluated for radionecrosis or recurrence of tumor. The patient in the top row had recurrent disease whereas the one in the bottom row did not. Parametric images show that transport (K_1) dominates the FLT uptake, but retention K_{FLT}, the proliferation parameter, is also measurable. Note from the color thermometers that the K_{FLT} scale is one-tenth that for the transport scale. The second MR images were at the time of the FLT-PET studies.

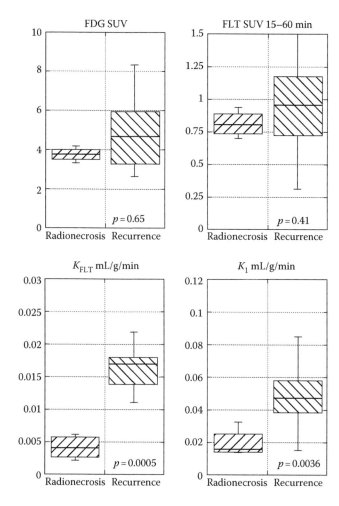

FIGURE 13.8 Box plots showing comparisons between groups of glioma patients with either radionecrosis or recurrence. The plots are for four parameters: flux or K_{FLT}, K_1, FLT SUV 15–60 min, and FDG SUV. Flux (K_{FLT}) distinguishes the two groups without overlap. SUV analysis for either FLT or FDG showed small differences in means but essentially complete overlap for the two patient groups.

the cell cycle. The transport rate constant showed a weaker correlation (Ullrich et al. 2008). We applied compartmental analysis to a group of patients where the clinical question was whether or not their previously treated brain tumors were progressing or the MR

images were suggesting progression when patients were actually responding to their therapy (Spence et al. 2009). Because these patients are often candidates for second-line drug trials, it is important to exclude those with pseudoprogression because it would give a false impression of the effectiveness of the second-line therapy if subjects were included who were already responding based on providing misleading data provided by MRI. FLT-PET and FDG-PET were compared in this study. For FLT, three levels of analysis were done: SUV_{max} for 15–60 and 60–90 min images, graphical analysis with metabolite correction over a short time interval where unidirectional transfer of FLT from blood to tissue was applicable, and quantitative analysis using the model introduced above. The time interval for graphical analysis was between 30 and 60 min because the restricted transport of FLT in brain requires a longer time for tissue pools to equilibrate with blood delivery. In this study, the FDG-PET clinical protocol was followed, with data analyzed as SUV.

The results of this study supported the value of thorough modeling analysis. Figure 13.7 shows the MRI and FLT images for two patients in this study. One patient had recurrence but the other had pseudoprogression from radionecrosis. The parametric images show that the SUV is driven by both transport and flux. The box plots in Figure 13.8 show that the SUV_{FLT} values overlapped for the patients with recurrent disease and those with pseudoprogression. The same degree of overlap was seen with the SUV_{FDG} analysis. The overlap was similar, whether it was assessed as SUV_{mean}, SUV_{max}, or SUVs based on tumor/cortex or tumor/white matter. The clear differences came from modeling the images to separately displace flux and transport. The transport parameters showed some overlap but the flux results showed a clean break between radionecrosis and true recurrence of tumor. These results showed that FLT-PET with rigorous dynamic imaging and analysis is a promising technique for separating recurrent tumor from pseudoprogression.

13.11 Alternative Thymidine Analogs to Measure Proliferation

13.11.1 ¹¹C- and ¹⁸F-Labeled 1-(2′-Deoxy-2′-Fluoro-5-Methyl-1-β-ᴅ-Arabinofuranosyl) Uracil

Because FLT is poorly incorporated into DNA and, in principle, only indirectly reflects DNA synthesis,

investigators have been developing alternative thymidine analogs that are substrates for both thymidine kinase and DNA polymerases. 1-(2′-Deoxy-2′-fluoro-5-methyl-1-β-ᴅ-arabinofuranosyl) uracil (FMAU) is not a substrate for TP but it is a substrate for thymidine kinase (Table 13.2) although it is a relatively better substrate for the constitutively expressed TK2 in

Chapter 13

mitochondria than for cytosolic TK1 (Table 13.2). One critical structural difference between the two ^{18}F-labeled nucleosides is the arabino versus the ribo configuration of the furanosyl sugar. The differential reactivity with TK2 versus TK1 probably makes it a less desirable imaging agent for measuring tumor growth. However, the fact that FMAU retains the 3′-hydroxyl makes it a substrate for DNA polymerases, so it can be incorporated into DNA, and *in vitro* studies showed that the incorporation of FMAU into DNA was proportional to DNA synthesis (Collins et al. 1999). *In vivo* studies in normal dogs showed appreciable radioactivity in the acid-precipitable DNA fraction.

FMAU has been labeled with both ^{11}C in the 5-methyl position and ^{18}F in the 2′ position. The synthesis of [^{11}C]-FMAU predated imaging with [^{18}F]FLT and paralleled the initial synthesis of [methyl-^{11}C]-thymidine (Conti et al. 1995). The 2′-fluoro was present prior to the radiosynthesis with [11]CH$_3$I, which proceeded in good yield with the dehalogenated uridine as the major by-product (Scheme 13.5a). Synthesis of the ^{18}F product proved much more challenging. The synthesis involves multiple steps after the introduction of ^{18}F and is based on the synthesis of the unlabeled molecule (Mangner et al. 2003). It starts with the sugar protected with three O-benzoyl groups and a triflate leaving group at the desired position for labeling. The ^{18}F is introduced via standard K$^+$[2.2.2]-mediated

nucleophilic radiofluorination, preferably in DMF at temperatures up to 150°C. The report emphasizes the value of heating the solvents and reagents before the precursor is introduced. This is followed by mild bromination at C1 using HBr in acetic acid under rigorously dry conditions. Traces of HBr and acetic acid are removed via an azeotrope with toluene and this protected [^{18}F]-arabinofuranose can be used without further purification in a condensation reaction with 2,4-bis-O-(trimethylsilyl)thymine. The air-sensitive reagents make this a technically challenging reaction and it takes 3 h to complete but overall decay-corrected yield is about 40% and the β/α isomer ratio is about 6.

Animal and human studies have been reported with FMAU-PET (Alauddin and Gelovani 2010). The initial evaluation involved [^{11}C]-FMAU studies in beagles implanted with brain tumors. Comparison studies were done with BrUdR detected by IHC. Trapping of [^{11}C]-FMAU detected by dynamic PET imaging correlated with tumor growth rate by BrUdR IHC, suggesting the value of this radiopharmaceutical for imaging cellular proliferation of tumors (Conti et al. 2008). There have been a few publications using [^{18}F]-FMAU in patients to evaluate various strategies for data analysis, including flux, $K_1 \cdot k_3/(k_2 + k_3)$, tumor retention ratio based on areas under the curve for tumor versus blood, and SUV parameters (Tehrani et al. 2007). This report argued that imaging was only

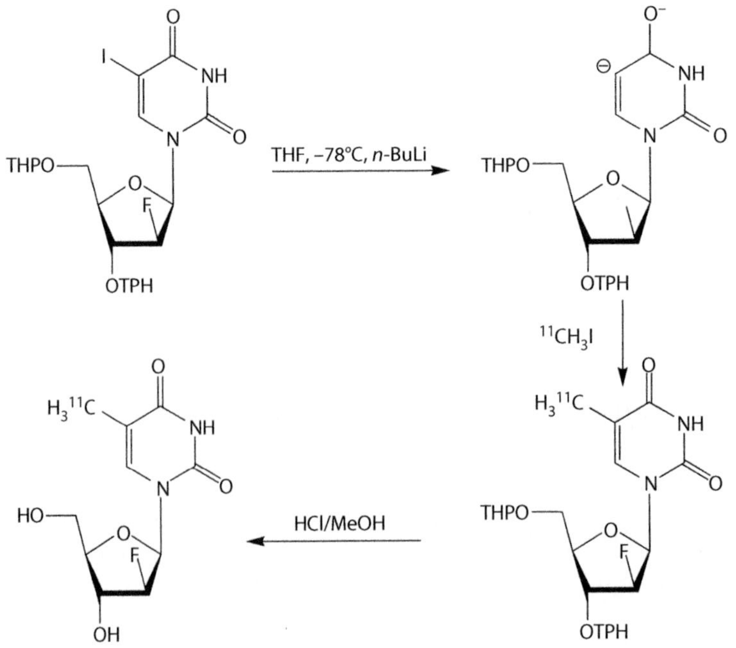

SCHEME 13.5a   Synthesis of ^{11}C-FMAU.

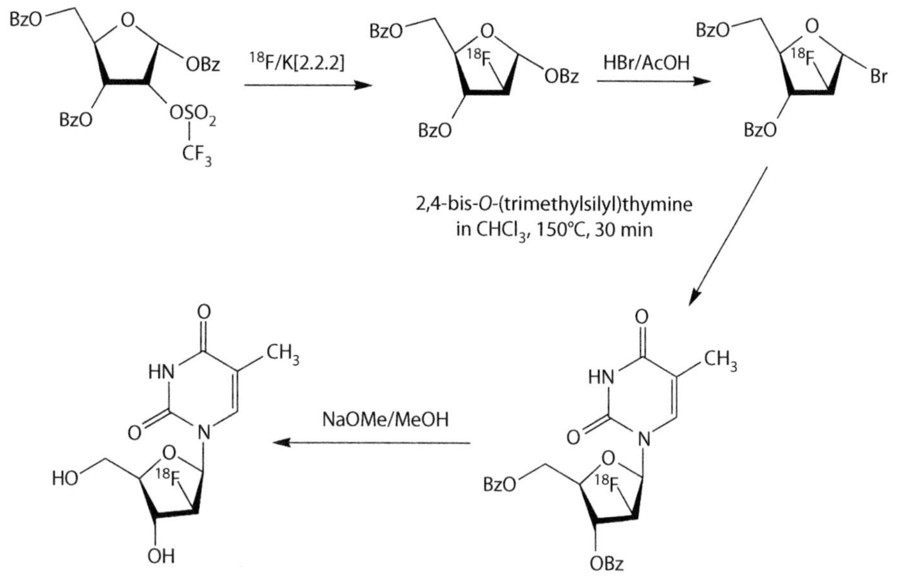

SCHEME 13.5b   Synthesis of ^{18}F-FMAU.

required up to 11 min after injection and that SUV values were desirable. Metabolite analysis was not required. Tumors in the brain, prostate, thorax, and bone have been visualized with [^{18}F]-FMAU-PET (Sun et al. 2005) (Scheme 13.5b). In fact, uptake of FMAU in the bone marrow was surprisingly low but there was a lot of normal activity in the liver and kidneys. At this time, there is very little use of FMAU-PET; however, one particularly important report compared FLT and FMAU in cancer cell lines exposed to various kinds of stress, oxidative, reductive, energy, and starvation (Tehrani et al. 2008). Nutritional stress decreased TK1 activity and FLT retention but increased FMAU. In contrast, FLT was unaffected by TK2 inhibition but FMAU retention decreased. FMAU uptake in cells correlated with TK2 activity and mitochondrial mass, as anticipated from the biochemical results in Table 13.2.

13.12   Summary and Future Direction in Proliferation Imaging

Proliferation is an important variable in assessing tumors. Imagers have shown numerous correlations between radiopharmaceutical uptake and the standard IHC assay of growth in the pathology lab, MIB-1 staining known as the Ki-67 index. But correlations are not the same as a mechanistic validation. Imaging research is plagued by numerous incidents where two factors covaried to result in a reasonable regression plot and correlation coefficient that turned out to be a coincidence. While growth can be inferred from a number of imaging agents, there is no substitute for directly imaging replication of DNA, which has been done for many decades with [^{3}H]-thymidine and can now be done with [^{11}C]-thymidine and PET imaging with compartmental analysis. The inconveniences of the [^{11}C]-thymidine PET procedure are well known and have led to attempts to substitute ^{18}F imaging agents as described in this chapter. Figure 13.9 shows the sequence of imaging agents that have been evaluated.

While considerable research has validated FLT and related nucleoside analogs, one must still worry about the extent to which they are valid in a range of pathologies and in the context of cytotoxic therapies that might disrupt regulation of TK1 production and degradation. The fact that thymidine monophosphate can come from two pathways, salvage and *de novo*, further complicates the situation. However, clever experiments described in this chapter have used this dual route to advantage by using [^{11}C]-thymidine imaging to distinguish therapy targeted at the *de novo* pathway using new versions of 5-FU, from therapy targeted at the salvage pathway. One should expect an ongoing role for [^{11}C]-thymidine PET in clinical research and further developments in alternative nucleosides directed at imaging cellular proliferation.

Chapter 13

First generation

Second generation

Third generation

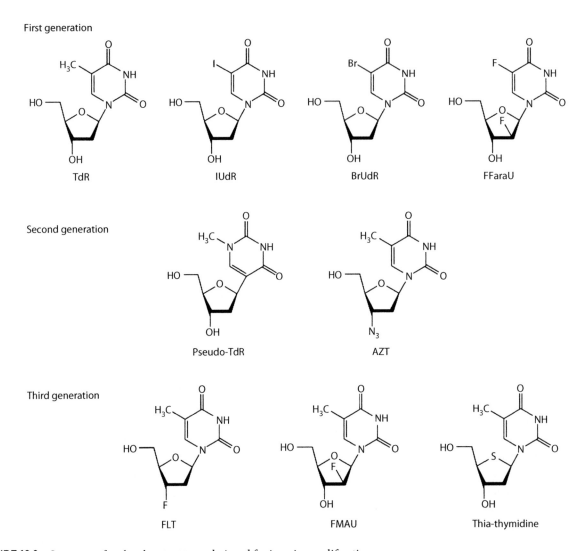

FIGURE 13.9   Sequence of molecular structures designed for imaging proliferation.

Acknowledgment

Much of the work described in this chapter has been supported by NIH through P01 CA042045, R01 CA118130.

References

Alauddin, M. M., Gelovani, J. G. 2010. Pyrimidine nucleosides in molecular PET imaging of tumor proliferation. *Curr Med Chem* 17(11):1010–29.

Barwick, T., Bencherif, B., Mountz, J. M., Avril, N. 2009. Molecular PET and PET/CT imaging of tumour cell proliferation using F-18 fluoro-L-thymidine: A comprehensive evaluation. *Nucl Med Commun* 30(12):908–17. Review.

Chen, W., Cloughesy, T., Kamdar, N. et al. 2005. Imaging proliferation in brain tumors with ^{18}F-FLT PET: Comparison with ^{18}F-FDG. *J Nucl Med* 46(6):945–52.

Christman, D., Crawford, E. J., Friedkin, M., Wolf, A. P. 1972. Detection of DNA synthesis in intact organisms with positron-emitting (methyl-11C)thymidine. *Proc Natl Acad Sci USA* 69(4):988–92.

Cobben, D. C., Elsinga, P. H., Hoekstra, H. J. et al. 2004. Is [18]F-3'-fluoro-3'-deoxy-L-thymidine useful for the staging and restaging of non-small cell lung cancer? *J Nucl Med* 45(10):1677–82.

Collins, J. M., Klecker, R. W., Katki, A. G. 1999. Suicide prodrugs activated by thymidylate synthase: Rationale for treatment and noninvasive imaging of tumors with deoxyuridine analogues. *Clin Cancer Res* 5(8):1976–81.

Conti, P. S., Hilton, J., Wong, D. F. et al. 1994. High performance liquid chromatography of carbon-11 labeled thymidine and its major catabolites for clinical PET studies. *Nucl Med Biol* 21(8):1045–51.

Conti, P. S., Alauddin, M. M., Fissekis, J. R., Schmall, B., Watanabe, K. A. 1995. Synthesis of 2'-fluoro-5-[11C]-methyl-1-beta-D-arabinofuranosyluracil ([11C]-FMAU): A potential nucleoside analog for *in vivo* study of cellular proliferation with PET. *Nucl Med Biol* 22(6):783–9.

Conti, P. S., Bading, J. R., Mouton, P. P. et al. 2008. *In vivo* measurement of cell proliferation in canine brain tumor using C-11-labeled FMAU and PET. *Nucl Med Biol* 35(1):131–41.

Crawford, E.J., Christman, D., Atkins, H., Friedkin, M., Wolf A.P. 1978. Scintigraphy with positron-emitting compounds. I. Carbon-11 labeled thymidine and thymidylate, *Int J Nucl Med Biol* 5(2–3):61–69.

Dolbeare, F., Gratzner, H., Pallavincini, M. G., Gray, J. W. 1983. Flow cytometric measurement of total DNA content and incorporated bromodeoxyuridine. *Proc Natl Acad Sci USA* 80(18):5573–77.

Eriksson, S., Kierdaszuk, B., Munch-Peterson, B., Oberg, B., Johansson, N. G. 1991. Comparison of the substrate specificities of human thymidine kinase 1 and 2 and deoxycytidine kinase toward antiviral and cytostatic nucleoside analogs. *Biochem Biophys Res Commun* 176(2):586–92.

Fabrikant, J. I. 1970. The kinetics of cellular proliferation in human tissues. Determination of duration of DNA synthesis using double labeling autoradiography. *Br J Cancer* 24(1):122–7.

Faraj, A., Fowler, D. A., Bridges, E. G., Sommadossi, J. P. 1994. Effects of 2′,3′-dideoxynucleosides on proliferation and differentiation of human pluripotent progenitors in liquid culture and their effects on mitochondrial DNA synthesis. *Antimicrob Agents Chemother* 38(5):924–30.

Flexner, C., van der Horst, C., Jacobson, M. A. et al. 1994. Relationship between plasma concentrations of 3′-deoxy-3′-fluorothymidine (alovudine) and antiretroviral activity in two concentration-controlled trials. *J Infect Dis* 170(6):1394–403.

Francis, D. L., Visvikis, D., Costa, D. C. et al. 2004. Assessment of recurrent colorectal cancer following 5-fluorouracil chemotherapy using both ¹⁸FDG and ¹⁸FLT PET. *Eur J Nucl Med Mol Imaging* 31(16):928.

Furman, P., Fyfe, J., St. Clair, M. H. et al. 1986. Phosphorylation of 3′-azido-3′-deoxythymidine and selective interaction of the 5′-triphosphate with human immunodeficiency virus reverse transcriptase. *Proc Natl Acad Sci USA* 83(21):8333–7.

Goethals, P., van Eijkeren, M., Lodewyck, W., Dams, R. 1995. Measurement of [methyl-carbon-11]thymidine and its metabolites in head and neck tumors. *J Nucl Med* 36(5):880–2.

Grierson, J. R., Shields, A. F., Zheng, M., Kozawa, S. M., Courter, J. H. 1995. Radiosynthesis of labeled beta-pseudothymidine ([C-11]- and [H-3]methyl) and its biodistribution and metabolism in normal and tumored mice. *Nucl Med Biol* 22(5):671–8.

Grierson, J. R., Vesselle, H., Hofstrand, P., Chin, L., Rasey, J. S. 1998. Comparative uptake and cell cycle measurements with [¹⁸F]FLT vs. [³H]thymidine in mammalian tumor cells. *J Nucl Med* 39(Suppl):229P–30P.

Grierson, J. R., Shields, A. F. 2000. Radiosynthesis of 3′-deoxy-3′-[(18)F]fluorothymidine: [(18)F]FLT for imaging of cellular proliferation *in vivo*. *Nucl Med Biol* 27(2):143–56.

Grierson, J. R., Schwartz, J. L., Muzi, M., Jordan, R., Krohn, K. A. 2004. Metabolism of 3′-deoxy-3′-[F-18]fluorothymidine in proliferating A549 cells: Validations for positron emission tomography. *Nucl Med Biol* 31(7):829–37.

Hughes, W. L., Bond, V. P., Brecher, G. et al. 1958. Cellular proliferation in the mouse as revealed by autoradiography with tritiated thymidine. *Proc Natl Acad Sci USA* 44(5):476–83.

Krohn, K. A., Mankoff, D. A., Muzi, M., Link, J. M., Spence, A. M. 2005. True tracers: Comparing FDG with glucose and FLT with thymidine. *Nucl Med Biol* 32(7):663–71.

Langstrom, B., Sjoberg, S., Bergson, G., Lundquist, H., Malmborg, P., Stalnocke, C.-G., Larsson, B.J. 1981. Some syntheses of 5-[¹¹C]methyl-deoxyuridine ([¹¹C]thymidine). *J Label Comp Radiopharm* 18:17–28.

Link, J. M., Grierson, J. R., Krohn, K. A. 1995. Alternatives in the synthesis of 2-[11C]-thymidine. *J Label Comp Radiopharm* 37:610–2.

Machulla, H. J., Blocher, A., Kuntzsch, M., Piert, M., Wei, R., Grierson, J. R. 2000. Simplified labeling approach for synthesizing 3′-deoxy-3′-[18F]fluorothymidine ([18F]FLT). *J Radioanal Nucl Chem* 243:843–6.

Mangner, T. J., Klecker, R. W., Anderson, L., Shields, A. F. 2003. Synthesis of 2′-deoxy-2′[¹⁸F]fluoro-beta-D-arabinofuranosyl nucleosides, [¹⁸F]FAU, [¹⁸F]FMAU, [¹⁸F]FBAU and [¹⁸F]FIAU, as potential PET agents for imaging cellular proliferation. Synthesis of [¹⁸F]labelled FAU, FMAU, FBAU, FIAU. *Nucl Med Biol* 30(3):215–24.

Mankoff, D. A., Shields, A. F., Graham, M. M. et al. 1998. Kinetic analysis of 2-[C-11]-thymidine PET imaging studies: Compartmental model and mathematical analysis. *J Nucl Med* 39(6):1043–55.

Mankoff, D. A., Shields, A. F., Link, J. M. et al. 1999. Kinetic analysis of 2-[¹¹C]-thymidine PET imaging studies: Validation studies. *J Nucl Med* 40(4):614–24.

Martiat, P., Ferrant, A., Labar, D. et al. 1988. *In vivo* measurement of carbon-11 thymidine uptake in non-Hodgkin's lymphoma using positron emission tomography. *J Nucl Med* 29:1633–37.

Martin, S. J., Eisenbarth, J. A., Wagner-Utermann, U. et al. 2002. A new precursor for the radiosynthesis of [¹⁸F]FLT. *Nucl Med Biol* 29(2):263–73.

Mendelsohn, M. L. 1962. Autoradiographic analysis of cell proliferation in spontaneous breast cancer of C3H mouse. III. The growth fraction. *J Natl Cancer Inst* 28:1015–29.

Mercer, J. R., Knaus, E. E., Wiebe, L. I. 1987. Synthesis and tumor uptake of 5-halo-1-(2′-fluoro-2′-deoxy-beta-D-ribofuranosyl) [2-14C]uracils. *J Med Chem* 30(4):670–5.

Meyer, J. S., Coplin, M. D. 1988. Thymidine labeling index, flow cytometric S-phase measurement, and DNA index in human tumors. Comparisons and correlations. *Am J Clin Pathol* 89(5):586–95.

Mitsuya, H., Weinhold, K. J., Furman, P. A. et al. 1985. 3′-Azido-3′-deoxythymidine (BW A509U): An antiviral agent that inhibits the infectivity and cytopathic effect of human T-lymphotropic virus type III/lymphadenopathy-associated virus *in vitro*. *Proc Natl Acad Sci USA* 82(20):7096–100.

Munch-Petersen, B., Cloos, L., Tyrsted, G., Eriksson, S. 1991. Diverging substrate specificity of pure human thymidine kinases 1 and 2 against antiviral dideoxynucleosides. *J Biol Chem* 266(14):9032–8.

Muzi, M., Vesselle, H., Grierson, J. R. et al. 2005. Kinetic analysis of 3′deoxy-3′-fluorothymidine PET studies: Validation studies in patients with lung cancer. *J Nucl Med* 46(2):274–82.

Muzi, M., Spence, A. M., O'Sullivan, F. et al. 2006. Kinetic analysis of 3′-deoxy-3′-¹⁸F-fluorothymidine in patients with gliomas. *J Nucl Med* 47(10):1612–21.

O'Sullivan, F. 1994. Metabolic images from dynamic positron emission tomography studies. *Stat Methods Med Res* 3(1):87–101.

Paproski, R. J., Ng, A. M., Yao, S. Y., Graham, K., Young, J. D., Cass, C. E. 2008. The role of human nucleoside transporters in uptake of 3′-deoxy-3′-fluorothymidine. *Mol Pharmacol* 74(5):1372–80.

Plotnik, D. A., Emerick, L. E., Krohn, K. A., Unadkat, J. D., Schwartz, J. L. 2010. Different modes of transport for 3H-thymidine, 3H-FLT, and 3H-FMAU in proliferating and nonproliferating human tumor cells. *J Nucl Med* 51(9):1464–71.

Quackenbush, R. C., Shields, A. F. 1988. Local re-utilization of thymidine in normal mouse tissues as measured with iododeoxyuridine. *Cell Tissue Kinet* 21(6):381–7.

Chapter 13

Rasey, J. S., Grierson, J. R., Wiens, L. W., Kolb, P. D., Schwartz, J. L. 2002. Validation of FLT uptake as a measure of thymidine kinase-1 activity in A549 carcinoma cells. *J Nucl Med* 43(9):1210–7.

Richter, F., Richter, A., Yang, K., Lipkin, M. 1992. Cell proliferation in rat colon measured with bromodeoxyuridine, proliferating cell nuclear antigen, and [3H]thymidine. *Cancer Epidemiol Biomarkers Prev* 1(7):561–6.

Schiepers, C., Dahlbom, M., Chen, W. et al. 2010. Kinetics of 3′-deoxy-3′-^{18}F-fluorothymidine during treatment monitoring of recurrent high-grade glioma. *J Nucl Med* 51(5):720–7.

Shields, A. F., Larson, S. M., Grunbaum, Z., Graham, M. M. 1984. Short-term thymidine uptake in normal and neoplastic tissues: Studies for PET. *J Nucl Med* 25(7):759–64.

Shields, A. F., Coonrod, D. V., Quackenbush, R. C., Crowley, J. J. 1987. Cellular sources of thymidine nucleotides: Studies for PET. *J Nucl Med* 28(9):1435–40.

Shields, A. F., Lim, K., Grierson, J., Link, J., Krohn, K. A. 1990. Utilization of labeled thymidine in DNA synthesis: Studies for PET. *J Nucl Med* 31(3):337–42.

Shields, A. F., Grierson, J. R., Kozawa, S. M., Zheng, M. 1996a. Development of labeled thymidine analogs for imaging tumor proliferation. *Nucl Med Biol* 23(1):17–22.

Shields, A. F., Mankoff, D., Graham, M. M. et al. 1996b. Analysis of 2-carbon-11-thymidine blood metabolites in PET imaging. *J Nucl Med* 37(2):290–6.

Shields, A. F., Grierson, J. R., Dohmen, B. M. et al. 1998a. Imaging proliferation *in vivo* with [F-18]FLT and positron emission tomography. *Nat Med* 4(11):1334–6.

Shields, A. F., Mankoff, D. A., Link, J. M. et al. 1998b. Carbon-11-thymidine and FDG to measure therapy response. *J Nucl Med* 39(10):1757–62.

Shields, A. F., Grierson, J. R., Muzik, O. et al. 2002. Kinetics of 3′-deoxy-3′[F-18]fluorothymidine uptake and retention in dogs. *Mol Imaging Biol* 4(1):83–9.

Shields, A. F., Briston, D. A., Chandupatla, S. et al. 2005. A simplified analysis of [18F]3′-deoxy-3′-fluorothymidine metabolism and retention. *Eur J Nucl Med Mol Imaging* 32(11):1269–75. Erratum in *Eur J Nucl Med Mol Imaging* 2007 34(4):613.

Slabas, A. R., MacDonald, G., Lloyd, C. W. 1980. Thymidine metabolism and the measurement of the rate of DNA synthesis in carrot suspension cultures: Evidence for degradation pathway for thymidine. *Plant Physiol* 65(6):1194–8.

Spence, A. M., Muzi, M., Link, J. M., Hoffman, J. M., Eary, J. F., Krohn, K. A. 2008. NCI-sponsored trial for the evaluation of safety and preliminary efficacy of FLT as a marker of proliferation in patients with recurrent gliomas: Safety studies. *Mol Imaging Biol* 10(5):271–80.

Spence, A. M., Muzi, M., Link, J. M. et al. 2009. NCI-sponsored trial for the evaluation of safety and preliminary efficacy of 3′-deoxy-3′-[^{18}F]fluorothymidine (FLT) as a marker of proliferation in patients with recurrent gliomas: Preliminary efficacy studies. *Mol Imaging Biol* 11(5):343–55.

Spinazzola, S., Marti, R., Nishino, I. et al. 2002. Altered thymidine metabolism due to defects of thymidine phosphorylase. *J Biol Chem* 277:4128–33.

Steel, C. J., Brady, F., Luthra, S. K. et al. 1999. An automated radiosynthesis of 2-[11C] thymidine using anhydrous [11C]urea derived from [11C]phosgene. *Appl Radiat Isot* 51(4):377–88.

Sun, H., Sloan, A., Mangner, T. J. et al. 2005. Imaging DNA synthesis *in vivo* with [^{18}F]FMAU and positron emission tomography in patients with cancer. *Eur J Nucl Med Mol Imaging* 32(1):15–22.

Sundoro-Wu, B. M., Schmall, B., Conti, P. S., Dahl, J. R., Drumm, P., Jacobsen, J. K. 1984. Selective alkylation of pyrimidyldianions: Synthesis and purification of ^{11}C labeled thymidine for tumor visualization using positron emission tomography. *Int J Appl Radiat Isot* 35(8):705–8.

Tang, G., Tang, X., Wen, F., Wang, M., Li, B. 2010. A facile and rapid automated synthesis of 3′-deoxy-3′-[(18)F]fluorothymidine. *Appl Radiat Isot* 68(9):1734–9.

Tehrani, O. S., Muzik, O., Heilbrun, L. K. et al. 2007. Tumor imaging using 1-(2′-deoxy-2′-^{18}F-fluoro-beta-D-arabinofuranosyl)thymine and PET. *J Nucl Med* 48(9):1436–41.

Tehrani, O. S., Douglas, K. A., Lawhorn-Crews, J. M., Shields, A. F. 2008. Tracking cellular stress with labeled FMAU reflects changes in mitochondrial TK2. *Eur J Nucl Med Mol Imaging* 35(8):1480–8.

Teng, B., Wang, S., Fu, Z., Dang, Y., Wu, Z., Liu, L. 2006. Semiautomatic synthesis of 3′-deoxy-3′-[18F]fluorothymidine using three precursors. *Appl Radiat Isot* 64(2):187–93.

Toyohara, J., Waki, A., Takamatsu, S., Yonekura, Y., Magata, Y., Fujibayashi, Y. 2002. Basis of FLT as a cell proliferation marker: Comparative uptake studies with [^{3}H]thymidine and [^{3}H]arabinothymidine, and cell-analysis in 22 asynchronously growing tumor cell lines. *Nucl Med Biol* 29(3):281–7.

Troost, E. G., Vogel, W. V., Merkx, M. A. et al. 2007. ^{18}F-FLT PET does not discriminate between reactive and metastatic lymph nodes in primary head and neck cancer patients. *J Nucl Med* 48(5):726–35.

Troost, E. G., Bussink, J., Hoffmann, A. L., Boerman, O. C., Oyen, W. J., Kaanders, J. H. 2010. ^{18}F-FLT PET/CT for early response monitoring and dose escalation in oropharyngeal tumors. *J Nucl Med* 51(6):866–74.

Turcotte, E., Wiens, L. W., Grierson, J. R., Peterson, L. M., Wener, M. H., Vesselle, H. 2007. Toxicology evaluation of radiotracer doses of 3′-deoxy-3′-[18F]fluorothymidine (18F-FLT) for human PET imaging: Laboratory analysis of serial blood samples and comparison to previously investigated therapeutic FLT doses. *BMC Nucl Med* 7:3.

Ullrich, R., Backes, H., Li, H. et al. 2008. Glioma proliferation as assessed by 3′-fluoro-3′-deoxy-L-thymidine positron emission tomography in patients with newly diagnosed high-grade glioma. *Clin Cancer Res* 14(7):2049–55.

Vander Borght, T., Labar, D., Pauwels, S., Lambotte, L. 1991. Production of [2-^{11}C] thymidine for quantification of cellular proliferation with PET. *Int J Rad Appl Instrum A* 42(1):103–4.

Vesselle, H., Grierson, J., Peterson, L.M., Muzi, M., Mankoff, D.A., Krohn, K.A. 2003. 18F-Fluorothymidine radiation dosimetry in human PET imaging studies. *J Nucl Med* 44:1482–88.

Wang, J., Eriksson, S. 1996. Phosphorylation of the anti-hepatitis B nucleoside analog 1-(2′-deoxy-2′-fluoro-1-beta-D-arabinofuranosyl)-5-iodouracil (FIAU) by human cytosolic and mitochondrial thymidine kinase and implications for cytotoxicity. *Antimicrob Agents Chemother* 40(6):1555–7.

Weber, W. A. 2010. Monitoring tumor response to therapy with ^{18}F-FLT PET. *J Nucl Med* 51(6):841–4.

Wells, J. M., Mankoff, D. A., Muzi, M. et al. 2002a. Kinetic analysis of 2-[^{11}C]thymidine PET imaging studies of malignant brain tumors: Compartmental model investigation and mathematical analysis. *Mol Imaging* 1(3):151–59.

Wells, J. M., Mankoff, D. A., Eary, J. F. et al. 2002b. Kinetic analysis of 2-[^{11}C]thymidine PET imaging studies of malignant brain tumors: Preliminary patient results. *Mol Imaging* 1(3):145–50.

Wells, P., Aboagye, E., Gunn, R. N. et al. 2003. 2-[11C]thymidine positron emission tomography as an indicator of thymidylate synthase inhibition in patients treated with AG337. *J Natl Cancer Inst* 95(9):675–82.

Wilson, I.K., Chatterjee, S., Wolf, W. 1991. Synthesis of 3′-fluoro-2′-deoxythymidine and studies of it 18F-labeling as a tracer for the non-invasive monitoring of the biodistribution of drugs against AIDS. *J Fluorine Chem* 55:283–289.

Wimber, D. E., Quastler, H. A. 1963. A ^{11}C and ^{3}H thymidine double labeling technique in the study of cell proliferation in transcendentia root tips. *Exp Cell Res* 30:8–22.

Wobus, A.M. 1976. Clastogenic activity of cytosine arabinoside and 3′-deoxy-3′-fluorothymidine in Ehrlich ascites tumour cells *in vitro*. *Mutat Res* 40(2):101–6.

Wodarski, C., Eisenbarth, J., Weber, K., Henze, M., Haberkorn, U., Eisenhut, M. 2000. Synthesis of 3′-deoxy-3′-[^{18}F]fluorothymidine with 2,3′-anhydro-5′-O-(4,4′-dimethoxytrityl)-thymidine. *J Label Comp Radiopharm* 43:1211–8.

Yoshii, Y., Maki, Y., Tsuboi, K., Tomono, Y., Nakagawa, K., Hoshino, T. 1986. Estimation of growth fraction with bromodeoxyuridine in human central nervous system tumors. *J Neurosurg* 65(5):659–63.

Chapter 13

14. [⁹⁹ᵐTc]TRODAT

Dopamine Transporter Imaging Agent for SPECT

Hank F. Kung and Karl Ploessl

14.1 Targeted Protein: Dopamine Transporter in the Brain

The dopamine transporter (DAT) imaging agents are useful in evaluating changes of brain function in patients with Parkinson's disease (PD), who have a selective loss of dopaminergic neurons in the basal ganglia and substantia nigra (Tatsch 2001). Developing Tc-99m-labeled DAT imaging agents for central nervous system (CNS) receptors or specific binding sites was a challenging undertaking; many have tried, but none had succeeded until 1996. Efforts in developing [⁹⁹ᵐTc]TRODAT-1 started not only out of necessity (because the supply of I-123 is limited, while Tc-99m is abundant and relatively inexpensive), but also out of curiosity in facing a seemingly overwhelming challenge. [⁹⁹ᵐTc]TRODAT-1 was developed as a useful imaging agent, which provides a simple and convenient tool for the estimation of dopamine neuronal integrity (see Figure 14.1) (Mozley et al. 2000, 2001; Acton et al. 2002; Siderowf et al. 2005).

14.1.1 Rationale for Using a Tropane Derivative

Tropane derivatives, including cocaine, are known to exhibit high binding affinity to DATs in the brain, through

which they produce various pharmacological effects. We have prepared a Tc-99m-labeled tropane as an imaging agent for the DAT (abbreviated TRODAT-1). [⁹⁹ᵐTc]TRODAT-1 was developed in 1996 (Kung et al. 1996). It was the first successful Tc-99m-labeled brain imaging agent targeting specific binding sites in the CNS. This imaging agent was designed for the early diagnosis of PD and related movement disorders. The formulation of [⁹⁹ᵐTc]TRODAT-1, developed for clinical study in the University of Pennsylvania (UPENN), was based on five different ingredients added sequentially before a final autoclaving step to sterilize the product. The labeling reaction produced two diastereoisomers with slightly different brain uptake and washout (Meegalla et al. 1997a,b, 1998). The formulation produced [⁹⁹ᵐTc]TRODAT-1 in high yields. [⁹⁹ᵐTc]TRODAT-1 used in imaging study was in an extremely small quantity, and thus it occupied a very small fraction of the DATs in the brain. The excess "cold" TRODAT-1, 200 μg for this preparation, did not show good penetration of cell membranes; therefore, it displayed no pharmacological effects. This formulation has been successfully tested in hundreds of normal and Parkinsonian subjects. The Institute of Nuclear Energy Research (INER) in Taiwan has developed a newer all-in-one-vial formulation (Wey et al. 1998). The new kit formulation, evaluated in thousands of patents, is convenient

Targeted Molecular Imaging. Edited by Michael J. Welch and William C. Eckelman © 2012 Taylor & Francis Group, LLC. ISBN: 978-1-4398-4195-2

Chapter 14

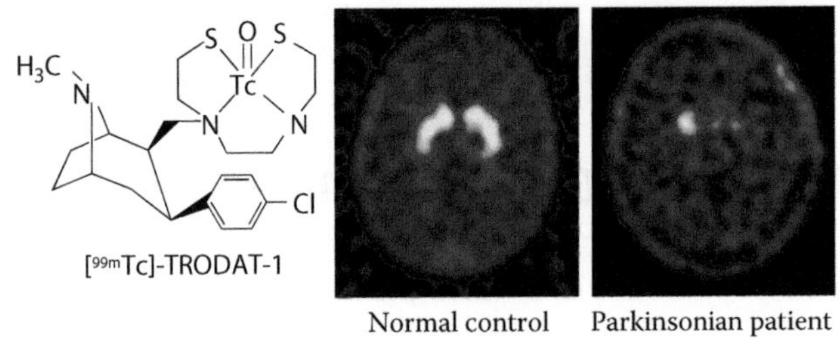

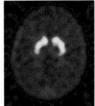
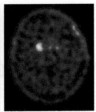

[⁹⁹ᵐTc]-TRODAT-1

Normal control   Parkinsonian patient

FIGURE 14.1 Transaxial SPECT images of human brain at 3 h post-IV injection of 20 mCi of [⁹⁹ᵐTc]TRODAT-1 for a normal and a Parkinsonian subject. In the normal subject, a high accumulation of [⁹⁹ᵐTc]TRODAT-1 was observed in caudate and putamen, where DATs are concentrated, while the Parkinsonian patient displayed a dramatic decreased uptake in this region of the brain.

and suitable for routine clinical production. There was a slight difference in the ratio of the two diastereoisomers produced by these two formulations; however, the formulations showed the same biodistribution and binding properties to the DATs. The chemical properties and pharmacological studies of the active ingredient, [⁹⁹ᵐTc] TRODAT-1, provide a clear scientific foundation for using this agent to image the DATs in the living human brain (Tzen et al. 2001; Yen et al. 2002).

14.1.2 Formulation for Making [⁹⁹ᵐTc] TRODAT-1: Statement of Quantitative Composition of Drug

Each dose contains:	
[Tc-99m] TRODAT-1	20 ± 20% mCi
TRODAT-1 ligand	180–220 µg
Stannous chloride	32 µg
Sodium glucoheptonate	320 µg
Disodium EDTA	930 µg
2.0 N hydrochloric acid	200–220 µL
Ethanol	100–200 µL
0.9% sodium chloride for injection, USP	Q.S.
Sterile buffer saline for injection, USP	Q.S.

Note: USP, U.S. Pharmacopeia; Q.S., a sufficient quantity (quantum sufficiat).

[⁹⁹ᵐTc]TRODAT-1 was prepared for the first time in April 1996 (Kung et al. 1996; Meegalla et al. 1997a). Like many new radiopharmaceuticals, the first formulation was prepared with the most rudimental or convenient method. After the initial synthesis of TRODAT-1 ligand, the labeling was successfully performed with [⁹⁹ᵐTc]pertechnetate in the presence of stannous chloride as the reducing agent. But preliminary studies on the preparation of [⁹⁹ᵐTc]TRODAT-1 showed that the radiochemical yield was not reliable. To finalize the kit formulation for human clinical trial, the procedure needed improvements and the radiolabeling procedure would have to be robust, reliable, and reproducible. Initially, the main objective was to produce [⁹⁹ᵐTc]TRODAT-1 in a consistent and reliable manner so that preclinical testing could be initiated. This was achieved by using a multicomponent kit formulation, which was not by design, but due to a limitation on the time for developing the formulation.

By preparing the product using a final autoclaving procedure, the labeling reaction and sterilization were accomplished simultaneously. It was determined earlier on that to prepare [⁹⁹ᵐTc]TRODAT-1, some type of heating would be necessary to facilitate the ⁹⁹ᵐTc complex formation. It was also recognized that to prepare a clinical dose, it was preferable to have a procedure containing a final sterilization step. By using an autoclaving procedure, both requirements were met (as outlined in Figure 14.2). Obviously, for the preparation of [⁹⁹ᵐTc]TRODAT-1, it was not necessary to use autoclaving, and heating in a boiling water bath or in a heating block at 80–100°C will likely be adequate. But autoclaving was a simple and convenient way to achieve both the heating and sterilization steps in a consistent and predictable fashion.

14.1.3 Solubility of TRODAT-1 Ligand: Adding Ethanol and Hydrochloric Acid

In 1996, the first formulation, which could provide the most reliable labeling procedure, was tested. The ligand, TRODAT-1, a thiol compound, had a tendency toward oxidation by air to form a disulfide bond. To preserve this thiol compound for efficient labeling of TcO core, it was dissolved in methanol, dispensed in glass ampoules, dried by a stream of nitrogen, and sealed under nitrogen. It was found that the ligand, TRODAT-1, in the glass ampoules, was not very soluble in water, which contributed to the inconsistency in labeling. The solubility was dramatically improved by adding ethanol and 2 N hydrochloric acid. Without the ligand being fully dissolved, a consistent labeling (radiochemical yield >90%) could not be obtained. To avoid this problem, 100–200 µL of 2 N hydrochloric acid in ethanol was first added to the vial containing the dried TRODAT-1. The dissolution was readily accomplished. The TRODAT-1 ligand was not very stable in 2 N hydrochloric acid;

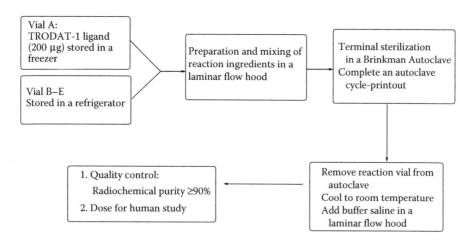

FIGURE 14.2 Flow chart on preparation of [⁹⁹ᵐTc]TRODAT-1 using an autoclaving procedure. Ingredients used in the preparation were designated as A–E in five vials.

therefore, the ligand could not be stored in the acidic solution. The ethanol/acid solution was added just prior to initiation of the labeling procedure. A buffered saline solution was added to adjust the pH to neutral, before [⁹⁹ᵐTc]TRODAT-1 was injected into human subjects.

14.1.4 Reducing Agents: Stannous Chloride

It is known that various reducing agents can be used for reducing [⁹⁹ᵐTc]pertechnetate. The reduction reaction is an essential step prior to chelating and forming any Tc-99m product. In April 1996, after evaluation of several other options, it seemed appropriate to use stannous chloride. This decision was based on the fact that the reducing agent, stannous chloride, has been successfully used in many kit formulations to make ⁹⁹ᵐTc radiopharmaceuticals. The toxicity and the effectiveness

will not be called into question. One other consideration was that the solubility of stannous chloride was sufficiently high to provide a reasonably high concentration of stannous chloride in an ethanol solution.

To stabilize the stannous chloride, we decided to use sodium glucoheptonate as a transitional chelating agent, which can form a relatively weak complex. Stannous-glucoheptonate kits have been approved for making [⁹⁹ᵐTc]glucoheptonate, a renal functional imaging agent for human study. When we investigated the potential for an impurity in the [⁹⁹ᵐTc]TRODAT-1 preparation, we found an impurity peak consistent with [⁹⁹ᵐTc]glucoheptonate (Figure 14.3). The [⁹⁹ᵐTc]glucoheptonate served as an intermediate in the labeling procedure. All of the [⁹⁹ᵐTc]pertechnetate was reduced and formed [⁹⁹ᵐTc]glucoheptonate, before it was *trans*-chelated to the more stable form, [⁹⁹ᵐTc]TRODAT-1.

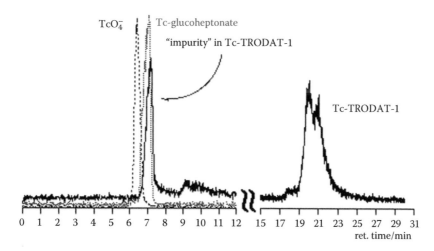

FIGURE 14.3 The HPLC profiles of [⁹⁹ᵐTc]glucoheptonate and [⁹⁹ᵐTc]TRODAT-1 obtained by a Rainin HPLC system, PRP-1 column, CH_3CN/DMGA (pH 7) 80:20, flow rate from 0.1 to 2 mL/min over 30 min. The retention time of [⁹⁹ᵐTc]glucoheptonate is consistent with the impurity peak. [⁹⁹ᵐTc]TRODAT-1 showed a double peak representing the formation of diastereoisomers.

14.1.5   The Amount of TRODAT-1 Ligand in the Formulation

Initially, the preparation of [^{99m}Tc]TRODAT-1 was always successful when using 2 mg of the ligand. However, we were mindful of the fact that TRODAT-1 is an analog of cocaine. To avoid any potential pharmacological side effect, we wanted to minimize the amount of TRODAT-1 ligand used in the kit formulation as much as possible. Later, it was found that using 200 µg of the ligand, one-tenth of the original amount, would still provide a successful labeling. There were also concerns associated with potential saturation of the binding sites (DATs) by the "unchelated" TRODAT-1 ligand in the brain, which can compete with [^{99m}Tc]TRODAT-1 in binding to the DATs at the target sites. After the formulation was in clinical trials, it was subsequently determined that the amount of the ligand could be further reduced to about 10 µg and still provide excellent labeling yields (Choi et al. 1999). However, since the clinical trial was ongoing, the formulation with lower amount of TRODAT-1 ligand was never tested in humans.

14.1.6   Single-Vial versus Multiple Component Kits

To simplify the formulation, a single-vial kit for the preparation of [^{99m}Tc]TRODAT-1 was desirable; however, attempts to combine all the ingredients together dramatically reduced the success rate of making [^{99m}Tc] TRODAT-1. When different components, TRODAT-1 ligand, stannous chloride, EDTA, and hydrochloric acid were added together, the combined solution no longer consistently produced [^{99m}Tc]TRODAT-1. Adding each component separately prior to the heating procedure, the labeling was achieved consistently. For this reason, the kit consisted of five separate vials that were combined immediately prior to the final autoclaving procedure for the preparation of [^{99m}Tc] TRODAT-1. This was not the ideal solution, but in the procedure it has been successful in making the needed doses for human trials in the United States with only a few incidental failures.

14.1.7   Chemical Identity: Diastereomers of [^{99m}Tc]TRODAT-1

In order to prepare neutral and lipophilic [Tc^{V}O]$^{3+}$ complexes based on the N$_2$S$_2$ (bis(aminoethanethiol)

(BAT)) ligands, three protons must be ionized. Our laboratory and others have demonstrated that neutral TcON$_2$S$_2$ complexes could be prepared with a predictable structure: a [Tc^{V}O]$^{3+}$ center core and a square pyramidal structure. This ligand system has been modified in a number of ways. When an N$_2$S$_2$ ligand, such as TRODAT-1, which contains two optical centers on the molecule, forms a complex with TcO^{3+} center core, the TcON$_2$S$_2$ core introduces an additional optical center, because only one of the nitrogen atoms is alkylated. The TcON$_2$S$_2$ complexes prefer a pyramidal structure with TcO at the apex, and the N$_2$S$_2$ forms the base of the pyramid. When the TcO bond points to the same direction as the alkyl–N bond, the conformation is called "Syn"; when they are pointed to opposite directions, the isomer is called "Anti" (Figure 14.4). Fortunately, in the case of TRODAT-1, only one stereo-isomer— "Syn" isomer—formed for the TcON$_2$S$_2$ core. To make the [^{99m}Tc]TRODAT-1 in a sufficient chemical quantity for characterization, we faced a different problem. Since Tc-99 is a longer-lived isotope of Tc-99m with a half-life of 250,000 years, it was not convenient to make a large chemical quantity of radioactive material. The nonradioactive rhenium (Re) derivatives were prepared as the surrogates of Tc-99m compounds. The Re has a similar chemistry to that of Tc, but has the advantage of being nonradioactive, with which a detailed chemical analysis can be performed (Figure 14.5). We prepared the [ReO]TRODAT-1 complexes in chemical scales and found two isomers (Meegalla et al. 1998). Both the isomers were purified and x-ray crystallography structure analysis was performed (Figure 14.6). The ReO complexes were then used to perform *in vitro* binding assays to measure the binding affinity of each diastereoisomer for DATs, the targeted-binding sites in the brain. It was found that both isomers bind to DAT with a similar binding affinity (inhibition constant, $K_i = 8$–15 nM) (Meegalla et al. 1998).

Biodistribution studies performed on the separated [^{99m}Tc]TRODAT-1 isomers, A and B, showed that they have distinctively different properties (see Table 14.1). Separation and biodistribution of the ^{99m}Tc-isomers showed that [^{99m}Tc]TRODAT-1A has higher initial brain uptake in rats than [^{99m}Tc]TRODAT-1B (0.5% dose/organ vs. 0.28% dose/organ, respectively). After 60 min, the ratios of striatum to cerebellum (ST/CB) were 2.72 and 3.79, for [^{99m}Tc]TRODAT-1A and B, respectively. This higher ratio of isomer B is also reflected in a slightly higher binding affinity as

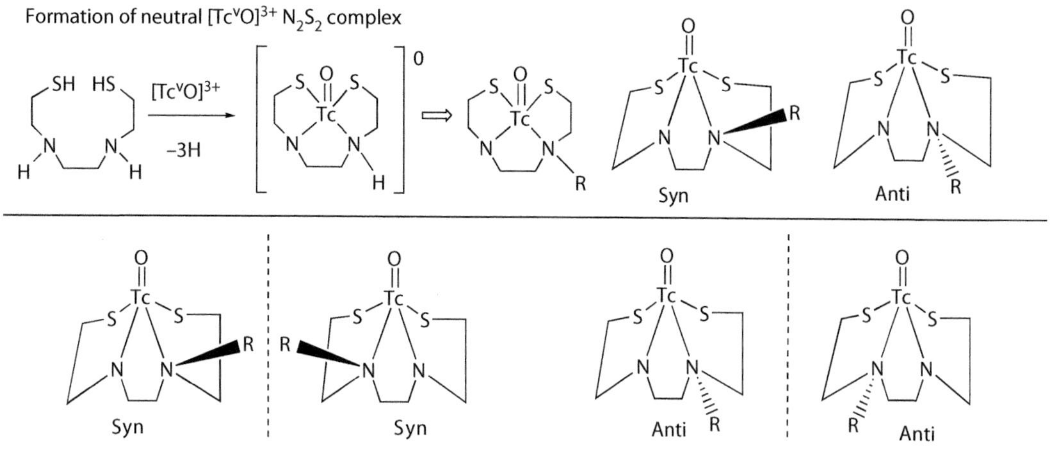

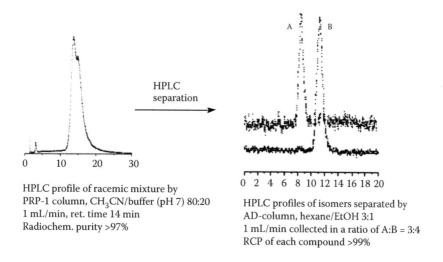

FIGURE 14.4   Formation of [Tc^VO]^{3+} complexes based on the N_2S_2 ligands. There are four possible stereoisomers. Syn and Anti plus mirror images of each. For [⁹⁹ᵐTc]TRODAT-1, only two diastereoisomers were formed.

HPLC
separation

HPLC profile of racemic mixture by
PRP-1 column, CH_3CN/buffer (pH 7) 80:20
1 mL/min, ret. time 14 min
Radiochem. purity >97%

HPLC profiles of isomers separated by
AD-column, hexane/EtOH 3:1
1 mL/min collected in a ratio of A:B = 3:4
RCP of each compound >99%

FIGURE 14.5   HPLC profiles of diastereoisomers of [⁹⁹ᵐTc]TRODAT-1: purified peak A and peak B. The positions of peak A and peak B on HPLC profiles were confirmed by the corresponding Re complexes.

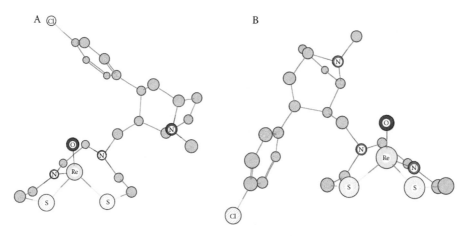

FIGURE 14.6   The *N*-alkyl substituted [Tc^VO]^{3+}N_2S_2 complexes formed diastereomers. X-ray structure analyses on the two ReO-TRODAT-1 isomers showed that both complexes have a syn configuration. Isomers A and B displayed corresponding HPLC profiles to those of peak A and peak B of [⁹⁹ᵐTc]TRODAT-1 confirming that it is likely that the Tc-99m isomers, A and B, may have similar structures. X-ray structure analyses of the ReO complexes revealed that they are both syn complexes, yet due to the chirality of the tropane moiety, they are diastereomers.

Chapter 14

Table 14.1   Brain Uptake of [^{99m}Tc]TRODAT-1 (Racemic, Peak A and Peak B) in Male Rats

Tc Complexes	Brain Uptake			Ratio (ST/CB)[b]	*In Vitro* Binding K_i (nM)[c]	
	P.C.[a]	2 min	60 min	60 min	TcOL (K_i, nM)	ReOL (K_i, nM)
Diastereomers (A + B)	227	0.43	0.12	2.66	–	14.1 ± 2.1
Peak A	305	0.50	0.21	2.72	10.2 ± 1.64	13.87 ± 1.73
Peak B	229	0.28	0.12	3.79	23.8 ± 3.83	8.42 ± 0.67

[a] P.C.: partition coefficient (*n*octanol/phosphate buffer pH 7.4); all compounds were stable in saline for >4 h.

[b] ST/CB = percentage dose/g of striatum/percentage dose/g of cerebellum.

[c] *In vitro* DAT-binding assays for ReO complexes were performed in rat striatal homogenates with [^{125}I]IPT as the ligand ($K_d = 0.3$ nM for DATs). The ratio of these two complexes appeared to be consistently reproduced (ratio [^{99m}Tc]TRODAT-1A/[^{99m}Tc]TRODAT-1B = 0.75 ± 0.09; $n = 25$). Brain uptake in rats was expressed as percent dose/organ.

determined by *in vitro*-binding studies ($K_i = 13.87$ and 8.42 nM for [ReO]TRODAT-1A and B, respectively). The lipophilicity is higher for isomer A than for B (P.C. 305 and 229, respectively). Later, in collaboration with Dr. Katzenellenbogen at the University of Illinois, Tc-99 complexes of TRODAT-1 were obtained, which showed a binding affinity of 10.2 ± 1.64 and 23.8 ± 3.83 for the A and B isomer, respectively. The twofold difference in binding affinity is not significant. These two diastereoisomers of Tc and Re complexes bind to DAT in a comparable manner. Combining the initial brain uptake and specific binding ratios at later time point, both the isomers provided a similar end result in terms of external imaging.

14.1.8   Specific Activity and Carrier Effect of TRODAT-1

The number of DATs in the brain for binding [^{99m}Tc]TRODAT-1 is limited. We needed to consider the possibility of saturating the DAT-binding sites. A conservative estimation of chemical quantity in Tc-99m pertechnetate solution produced by a commercial Tc-99m generator (Mallinckrodt Medical Inc, USP; package insert) was 0.3 μg/450 mCi. This presents that 1 mCi of Tc-99m contains 6.7×10^{-12} mol of Tc-99m plus Tc-99. If 30 mCi of [^{99m}Tc]TRODAT-1 was prepared and injected; then a total of 0.2 nmol of Tc-99/Tc-99m TRODAT-1 would be injected. The percent saturation of DAT sites by [^{99m}Tc]TRODAT-1 in the brain (basal ganglia area) could be calculated. It was estimated that DAT density in the basal ganglia (primates and humans) is about 500 nM (10 nmol of transporter in a mass of 20 g). If a maximum amount of 30 mCi was injected, then the estimated total

Tc-99m TRODAT-1 injected was 0.2 nmol. If 5% of the injected dose was taken up in the brain and totally concentrated in the basal ganglia (an overestimation), then 0.01 nmol (5% of 0.2 nmol) will be localized in the brain. It was reasonable to conclude that [^{99m}Tc]TRODAT-1 would only occupy about 0.03–0.1% of total DATs in the human brain. There is a large safety margin for using [^{99m}Tc]TRODAT-1 in human brain imaging.

We were contemplating that the TRODAT-1 ligand, which also showed potent binding to the DATs, may also compete for the same binding sites. This "extra" binding might have a serious drawback, that it may decrease the specific uptake in the basal ganglia area. We estimated the carrier effect by escalating the dose of TRODAT-1 ligand in the biodistribution study in rats. The specific uptake, as measured by target/nontarget that is, striatum/cerebellum ratio, was used to compare the control and rats coinjected with additional TRODAT-1 as the carrier. Data in Figure 14.7 and Table 14.2 suggested that the carrier effect was not significant until the carrier dose was above 500 μg per rat. The averaged body weight of a rat is about 250 g; so 500 μg per 250 g of body weight translated to human as 140 g per 70 kg of averaged human body weight. Again, there was a large safety margin, which suggests that the carrier dose of TRODAT-1 at 200 μg/dose injected into a human will not block the binding of DAT sites in the brain. It is important to recognize that the TRODAT-1 has polar functional groups and thiol and amine groups, which could impede the ability to penetrate cell membranes, including the blood–brain barrier. On the formation of [^{99m}Tc]TRODAT-1, the thiol and amine groups of the TRODAT-1 ligand are tied-in with the TcO center core to form strong coordinate

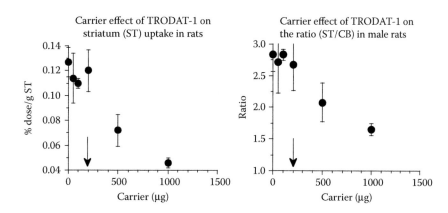

FIGURE 14.7   Carrier effect of TRODAT-1 on the specific binding of [⁹⁹ᵐTc]TRODAT-1 in the rat brain.

Table 14.2   Rat Brain Uptake and Regional Distribution of [⁹⁹ᵐTc]TRODAT-1 in the Presence of Varying Amounts of TRODAT-1 Free Ligand (Carrier Dose) (% Dose/g, Average ± SD)

Carrier Dose	0 μg (n = 6)	50 μg (n = 3)	100 μg (n = 3)	200 μg (n = 6)	500 μg (n = 3)	1000 μg (n = 3)
CB	0.046 ± 0.004	0.042 ± 0.009	0.041 ± 0.003	0.042 ± 0.004	0.034 ± 0.002	0.028 ± 0.003
ST	0.12 ± 0.012	0.11 ± 0.020	0.11 ± 0.004	0.12 ± 0.017	0.07 ± 0.013	0.046 ± 0.004
HP	0.07 ± 0.004	0.06 ± 0.010	0.06 ± 0.004	0.06 ± 0.013	0.04 ± 0.008	0.02 ± 0.005
CX	0.05 ± 0.005	0.06 ± 0.009	0.06 ± 0.010	0.06 ± 0.009	0.03 ± 0.006	0.03 ± 0.003
ST/CB	2.83 ± 0.26	2.71 ± 0.48	2.68 ± 0.10	2.83 ± 0.41	2.08 ± 0.31[a]	1.65 ± 0.01[a]

Note: CB: cerebellum; CX: cortex; HP: hippocampus; ST: striatum.

[a] The value is significantly lower than that of the control.

covalent bonds. [⁹⁹ᵐTc]TRODAT-1 is a more tightly compact molecule and as a result, it readily penetrated the intact blood–brain barrier. Inadvertently, using the blood–brain barrier, "mother nature" has provided a highly selective mechanism for filtering out TRODAT-1. The effect of this filtering mechanism was less apparent for normal cells, but the same principle should be applicable for heart and muscles cells. Potential pharmacological effects from a single injection of 200 μg of TRODAT-1 to the heart and the brain would likely be minimal.

Several lines of evidence suggest that the "carrier" dose has no effect on the imaging data in humans. First, single-photon emission computed tomography (SPECT) imaging study in baboons (15 kg body weight) using either the high-performance liquid chromatography (HPLC)-purified Tc-99m TRODAT-1 (in which TRODAT-1 ligand was removed) or the kit formulation in which 200 μg of TRODAT-1 and [⁹⁹ᵐTc]TRODAT-1 were injected, both showed the same brain uptake. Second, TRODAT-1 carrier-added experiment in male and female rats only showed competition effect when doses were higher than 200 μg/dose (Table 14.2 and

Kung et al. 1997). The potential "carrier-effect" from unlabeled TRODAT-1 appeared unlikely.

14.1.9   Clinical Application

DATs are one of the primary markers for PD in living patients. The dopamine neurons are most affected by the disease, and reflect their health (symptoms related to movement disorder). positron emission tomography (PET) and SPECT neuroimaging studies reflect the same phenomenon by visualizing decreased transporter binding in the basal ganglia, regions to which the affected nigral cells project. Neuroimaging studies in 1-methyl-4-phenyl-1,2,3,6-tetrahydropyridine (MPTP)-treated primates have consistently shown strong correlations between symptom formation and the loss of DAT-binding sites in the region of the basal ganglia. Studies of the presynaptic DATs in patients with symptomatic PD have been consistently positive. High sensitivity may reflect the observation that transporter concentrations decrease (>65%) before the PD symptoms exhibit in patients. Results from investigations with cross-sectional designs have been encouraging

Chapter 14

for the prospects of detecting therapeutically induced changes within patients. Several groups, including our own, have reported that the specific uptake values (SUVs) of both PET and SPECT tracers correlate with symptom severity and the duration of illness. This is a critical observation because it does not necessarily follow that a radiopharmaceutical that can distinguish between patients and controls will have a large-enough dynamic range to demonstrate clinically meaningful differences within groups of patients. Further evidence that many radiotracers do indeed have the necessary dynamic range comes from the consistent observation of appropriate asymmetries in patients with hemi-PD.

Timescale for changes within patients: There is some controversy regarding how long it takes patients to develop symptoms after the disease process begins, but it is clear that the rate at which PD progresses after initial presentation is variable. Some patients with hemi-Parkinson's do not develop bilateral disease for more than a decade, while most do in <3 years. The pathophysiological correlates of these variable rates are unknown, and may require longitudinal studies to detect and characterize. It is difficult to estimate precisely how long these studies would have to be conducted to detect pathophysiological correlates of benefit, such as changes in DAT concentrations. The observation that about half of all patients develop dyskinesia within 2–5 years of beginning treatment with L-DOPA suggests that 5 years may sometimes be long enough. The extended literature on the potential of neuroimaging techniques to demonstrate deteriorations in the condition has been extremely optimistic. Some investigators have demonstrated the converse by showing that some therapeutically induced changes can be observed much sooner with neuroimaging techniques.

Initial dosimetry studies (Mozley et al. 1998) showed a favorable outcome for the use of [^{99m}Tc]TRODAT-1 in human studies and allowed further evaluation of [^{99m}Tc]TRODAT-1 in humans (Mozley et al. 2000). The sample consisted of 42 patients with PD, 23 age-matched controls, and 38 healthy adults <40 years old. SPECT scans of the brain were acquired on a triple-head gamma camera 3–4 h after the intravenous injection of 740 MBq (20 mCi) [^{99m}Tc]TRODAT-1. Mean counts per pixel were measured manually in subregions of the basal ganglia and normalized to the mean background counts to give SUVs (approximately k3/k4). Patient and control groups were also compared with automated statistical parametric mapping techniques. Logistic discriminant analyses were performed to determine the optimum uptake values for differentiating patients from age-matched controls. Quantitative image analysis showed that the group mean SUVs in patients were less than the mean values in controls for all regions (all Ps < 0.000001). There was overlap in the caudate as well as in the anterior-most portion of the putamen, but not in the posterior putamen, even when the asymptomatic sides of five patients with clinically defined hemi-PD were factored in. The findings indicate that PD can be detected with [^{99m}Tc]TRODAT by simply inspecting the images for uptake in the posterior putamen. Appropriate asymmetries seem to be visible with quantification in patients with clinically defined hemi-PD, even though changes in the putamen contralateral to the clinically unaffected side in these patients appear to precede the development of symptoms.

In summary, [^{99m}Tc]TRODAT-1 is a brain imaging agent developed specifically for labeling DATs in the basal ganglia region of the brain. It may provide a simple, convenient, and efficient tool to provide diagnostic information on the status of dopamine neurons in the brain. The chemistry and pharmacology of the active ingredient, [^{99m}Tc]TRODAT-1, are reasonably understood; they provide a clear rationale and scientific foundation for using this agent for imaging the DATs in the living human brain.

References

Acton, P., S. Choi et al. 2002. Quantification of dopamine transporters in the mouse brain using ultra-high resolution single-photon emission tomography. *European Journal of Nuclear Medicine* 29(5):691–698.

Choi, S. R., M. P. Kung et al. 1999. An improved kit formulation of a dopamine transporter imaging agent: [Tc-99m]TRODAT-1. *Nuclear Medicine and Biology* 26(4):461–466.

Kung, H. F., H.-J. Kim et al. 1996. Imaging of dopamine transporters in humans with technetium-99m TRODAT-1. *European Journal of Nuclear Medicine* 23(11):1527–1530.

Kung, M. P., Stevenson, D. A., Plössl, K., Meegalla, S. K., Beckwith, A., Essman, W. D., Mu, M., Lucki, I., and Kung, H. F. 1997. [^{99m}Tc]TRODAT-1: A novel technetium-99 m complex as a DAT imaging agent. *European Journal of Nuclear Medicine* 24:372–380.

Meegalla, S. K., K. Plössl et al. 1997a. Synthesis and characterization of Tc-99m labeled tropanes as dopamine transporter imaging agents. *Journal of Medicinal Chemistry* 40:9–17.

Meegalla, S. K., K. Plössl et al. 1997b. Structure–activity relationship of [^{99m}Tc]TRODAT-1 derivatives as new dopamine transporter imaging agents. *Journal of Labelled Compounds and Radiopharmaceuticals* 40:434–436.

Meegalla, S. K., K. Plössl et al. 1998. Specificity of diastereomers of [⁹⁹ᵐTc]TRODAT-1 as dopamine transporter imaging agents. *Journal of Medicinal Chemistry* 41:428–436.

Mozley, L. H., R. C. Gur et al. 2001. Striatal dopamine transporters and cognitive functioning in healthy men and women. *American Journal of Psychiatry* 158(9):1492–1499.

Mozley, P., J. Schneider et al. 2000. Binding of [99mTc]TRODAT-1 to dopamine transporters in patients with Parkinson's disease and in healthy volunteers. *Journal of Nuclear Medicine* 41(4):584–589.

Mozley, P., J. Stubbs et al. 1998. Biodistribution and dosimetry of TRODAT-1: A technetium-99m tropane for imaging dopamine transporters. *Journal of Nuclear Medicine* 39:2069–2076.

Siderowf, A., A. Newberg et al. 2005. [99mTc]TRODAT-1 SPECT imaging correlates with odor identification in early Parkinson disease. *Neurology* 64(10):1716–1720.

Tatsch, K. 2001. Imaging of the dopaminergic system in parkinsonism with SPET. *Nuclear Medicine Communications* 22(7):819–827.

Tzen, K. Y., C. S. Lu et al. 2001. Differential diagnosis of Parkinson's disease and vascular parkinsonism by (99m)Tc-TRODAT-1. *Journal of Nuclear Medicine* 42(3):408–413.

Wey, S. P., H. Y. Tsai et al. 1998. Formulation of a lyophilized kit for preparation of 99MTC-TRODAT-1. *European Journal of Nuclear Medicine* 1164: Abstract No. PS-723.

Yen, T. C., K. Y. Tzen et al. 2002. Dopamine transporter concentration is reduced in asymptomatic Machado–Joseph disease gene carriers. *Journal of Nuclear Medicine* 43(2):153–159.

Chapter 14

15. Development of Radiotracers and Fluorescent Probes for Imaging Sigma-2 Receptors *In Vitro* and *In Vivo**

Robert H. Mach and Kenneth T. Wheeler

15.1   Introduction

Sigma receptors represent a class of proteins that were initially thought to be a subtype of the opiate receptors. However, subsequent studies revealed that sigma receptors are a distinct class of proteins that are located in the central nervous system (CNS) as well as in a variety of tissues and organs (Walker et al. 1990, Hellewell et al. 1994). Initial studies using radioligand-binding methods and biochemical analyses demonstrated that there are two subtypes of sigma receptors, termed σ_1 and $\sigma_2 \cdot \sigma_1$ receptors have a molecular weight of ~25 kDa, whereas σ_2 receptors have a molecular weight of ~21.5 kDa. The radioligand [^{3}H](+)-pentazocine and other benzomorphan analogs have a high (i.e., nM) affinity for the σ_1 receptor and a low (>1000 nM) affinity for the σ_2 receptor, whereas the guanidine analog [^{3}H]DTG binds with equal affinity to both σ_1 and σ_2 receptors. Our understanding of the functional significance of sigma receptors has been hampered by the failure to identify an endogenous ligand for these receptors. A number of studies have shown that neuroactive steroids bind with moderate affinity to σ_1 sites, and suggest that σ_1 receptors may modulate the activity of gamma aminobutyric acid (GABA) and N-methyl-D-aspartate (NMDA) receptors in the CNS (Maurice et al. 1996, 1997, Romieu et al. 2003). The nature of the endogenous ligands which interact with the σ_2 receptor is not known at this time.

The σ_1 receptor has been cloned and exhibits a 30% sequence homology with the enzyme, yeast C8-C7 sterol

Targeted Molecular Imaging. Edited by Michael J. Welch and William C. Eckelman © 2012 Taylor & Francis Group, LLC. ISBN: 978-1-4398-4195-2

* Portions of this chapter were previously published as "Development of Molecular Probes for Imaging Sigma-2 Receptors *In Vitro* and *In Vivo*" in *Central Nervous System Agents in Medicinal Chemistry,* 2009 9(3):230–45.

Chapter 15

isomerase (Hanner et al. 1996, Seth et al. 1997). However, the σ_1 receptor lacks C8-C7 isomerase activity. The σ_2 receptor has not been purified, sequenced, or cloned, and most of what is known regarding this receptor has been obtained through the use of *in vitro* receptor-binding studies. The absence of the cloned gene or the purified σ_2 receptor protein has also prevented the generation of antibodies to study the subcellular localization of this receptor using standard immunohistochemical techniques that have been employed to study the localization of the σ_1 receptor (Alonso et al. 2000). In spite of this limitation, the recent development of high-affinity σ_2 receptor radioligands and fluorescent probes has increased our understanding of the organ distribution, subcellular distribution, and intracellular trafficking of the σ_2 receptor in cancer cells.

15.2   Studies Characterizing the σ_2 Receptor as a Biomarker of the Proliferative Status of Solid Tumors

The first report demonstrating an overexpression of σ receptors in tumors cells was by Bem et al. (1991). A key observation in this study was the lower σ receptor density obtained with the σ_1 selective ligand [^{3}H]3-PPP, versus [^{3}H]DTG, a mixed σ_1/σ_2 radioligand, in biopsy samples obtained from patients with renal and colon carcinoma. Although initially interpreted as a disparity caused by the agonist ([^{3}H]3-PPP) and antagonist ([^{3}H]DTG) properties of these radioligands, the differences in binding properties of the two radioligands in tumor cells could have been caused by a higher density of σ_2 versus σ_1 receptors in the tumor samples. This hypothesis was later confirmed by Bowen and colleagues (Vilner et al. 1995), who demonstrated that there was a higher density of σ_2 versus σ_1 receptors in a wide variety of human and murine tumor cells growing under cell culture conditions. The observation that membrane homogenates from MCF-7 cells, a human breast adenocarcinoma cell line, possessed a high density of σ_2 receptors but no measurable binding of the σ_1 receptor ligand [^{3}H](+)-pentazocine suggested that the σ_2 receptor may be a potential biomarker for imaging breast tumors. However, Western blot studies have subsequently shown σ_1 receptor expression in a number of breast tumor cells lines, including MDA-MB-231, MDA-MB-361, MDA-MB-435, MCF-7, and BT20 cells (Wang et al. 2004).

These early studies clearly indicated that σ_2 receptors may serve as a biomarker for imaging tumors, but they did not investigate the relationship between the density of σ_2 receptors and the proliferative status of these tumors. The proliferative status of a solid tumor, which has been defined as the ratio of proliferating (P) cells in a solid tumor to those driven into a quiescent (Q) state by nutrient deprivation (the P:Q ratio), is an important parameter in determining how to treat a tumor with either radiation or chemotherapy (Mach and Wheeler 2007). Tumors having a high proliferative status (i.e., high P:Q ratio) generally respond better to hyperfractionated radiation therapy versus conventional radiation therapy (Fornace et al. 2001). Also, tumors having a high P:Q ratio typically respond better to cell-cycle-specific agents such as Ara-C and gemcitabine, whereas tumors having a lower proliferative status respond better to non-cell-cycle-specific agents such as cisplatin and bis-chloroethylnitrosourea (BCNU) (Mach and Wheeler 2007). Finally, the measurement of the P:Q ratio of a tumor is also expected to be useful in selecting patients that will benefit from "targeted therapies" since many of these newer cancer treatment strategies (e.g., Polo-like kinase inhibitors, Chk-1 inhibitors) target proteins that are expressed in cycling cells versus noncycling (i.e., quiescent tumor) cells (Collins and Garrett 2005, Strebhardt and Ulrich 2006).

Using the diploid mouse mammary adenocarcinoma cell line 66 (Wallen et al. 1984a,b), Wheeler and colleagues conducted a series of experiments aimed at determining if there was a difference in the density of σ_2 receptors in proliferating (66P) and quiescent (66Q) mouse adenocarcinoma cells. In their initial study, this group demonstrated that the density of σ_2 receptors in 66P cells was about 10 times greater than the density observed in 66Q cells growing under cell culture conditions (Mach et al. 1997). The density of σ_2 receptors in the 66P cells was found to be ~900,000 copies/cell versus ~90,000 receptors/cell in the 66Q cells (Figure 15.1a). Similar results were observed with the aneuploid mouse mammary adenocarcinoma line, 67 (Figure 15.1b) (Al-Nabulsi et al. 1999). However, the relatively high density of σ_2 receptors in quiescent tumor cells suggest that it still may be possible to image solid tumors having a significant Q cell fraction (i.e., low P:Q ratio) since most of the surrounding normal tissues have a very low density of σ_2 receptors (Bem et al. 1991, Vilner et al. 1995).

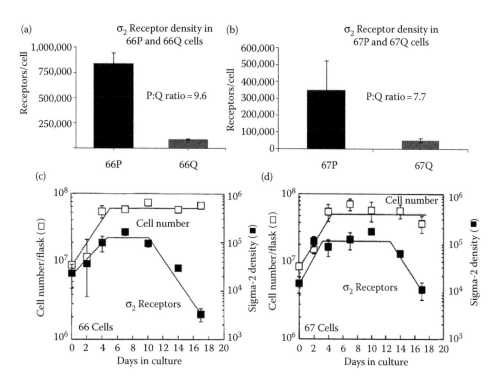

FIGURE 15.1 Differences in the density of the σ_2 receptors in 66P and 66Q cells (a) and 67P and 67Q cells (b). The upregulation kinetics of the σ_2 receptors in 66 cells (c) and 67 cells (d) during the Q to P transition and the downregulation kinetics during the P to Q transition are also shown. (Adapted with permission from Al-Nabulsi, I. et al. 1999. *Br. J. Cancer* 81:925–933.)

In a subsequent study, this group reported the upregulation and downregulation of σ_2 receptors follows the transition of 66 and 67 cells between the P and Q states (Figure 15.1c and d) (Al-Nabulsi et al. 1999). Since the downregulation of σ_2 receptors takes about 3 days to complete once the cells make the transition from P to Q, these data suggest that the σ_2 receptor is not expressed only in a single phase of the cell cycle, as are the cyclin-dependent kinases and other cell-cycle-specific proteins and is regulated in a manner similar to many membrane-bound receptors. The observation that the P:Q ratio of σ_2 receptor density in solid 66 tumors is identical to that obtained under cell culture conditions clearly demonstrated that the results obtained under cell culture conditions can be trans-lated to solid tumors xenografts of these breast cancer cell lines (Wheeler et al. 2000). These data also confirm that the σ_2 receptor is a receptor-based biomarker of cell proliferation in breast tumors. Therefore, radio-tracers having a high affinity and high selectivity for σ_2 receptors have the potential to assess the proliferative status of human breast tumors using noninvasive imaging techniques such as positron emission tomography (PET) and single-photon emission computed tomography (SPECT). It is also likely that this approach can be extended to measure the proliferative status of other human tumors, such as head and neck, mela-noma, and lung tumors, which are known to express a high density of σ_2 receptors (Vilner et al. 1995).

15.3 Development of σ_2 Selective Ligands

Although many different classes of σ receptor ligands have been reported, most of these compounds are either σ_1-selective ligands or bind with similar affinity to both σ_1 and σ_2 receptors (Walker et al. 1990). The development of ligands having a high affinity for the σ_2 versus the σ_1 receptor was not straightforward, and in some cases, the identification of σ_2 receptor-selec-tive ligands was a serendipitous discovery (Figure 15.2).

One of the first σ_2-selective ligands reported was the benzomorphan-7-one analog, CB-64D (Bowen et al. 1995). This compound was identified as part of a struc-ture–activity relationship (SAR) study aimed at opti-mizing the affinity of (–)-2-methyl-5-(3-hydroxyphenyl) morphan-7-one for the μ versus κ subtype of the opioid receptors (Bertha et al. 1994). Incorporation of an (E)-benzylidene moiety into the 8-position of the ring

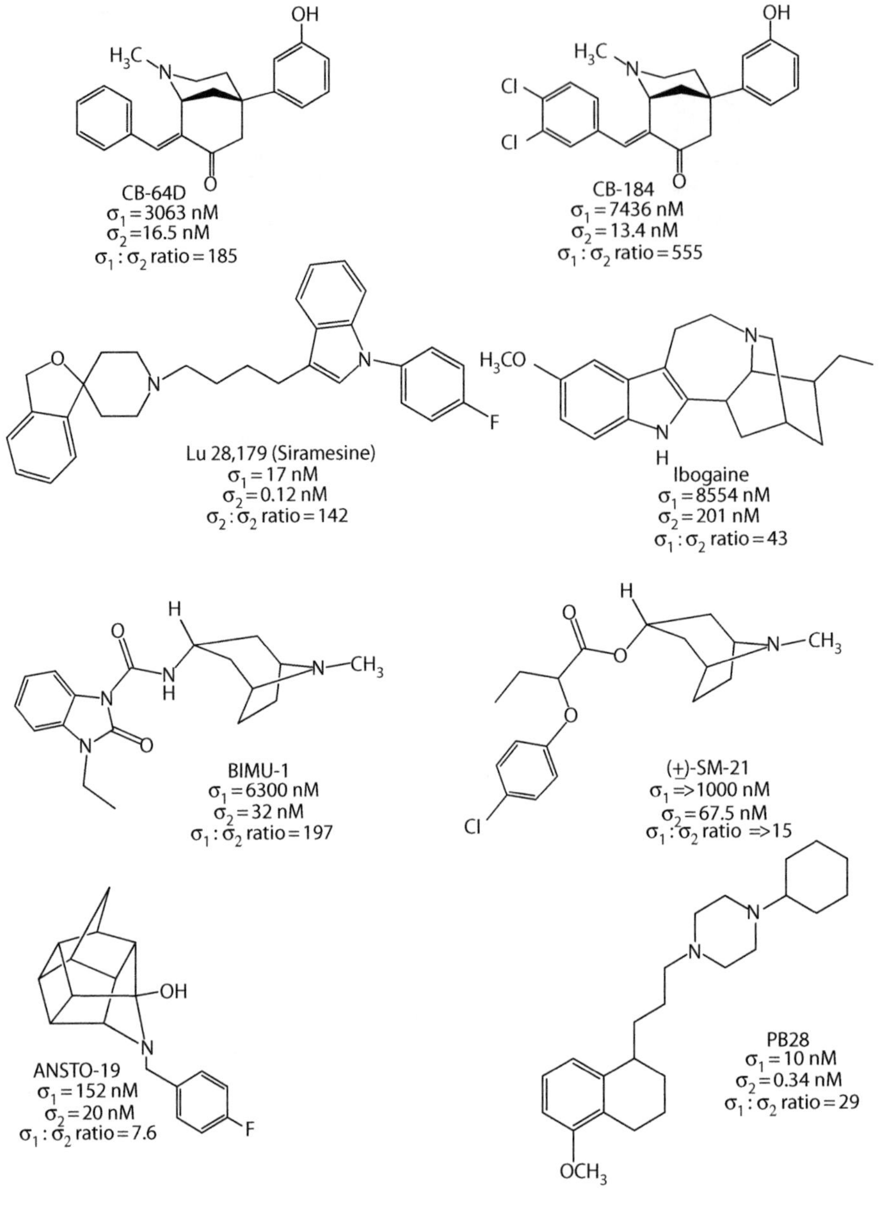

FIGURE 15.2 Structures of σ_2-selective ligands (K_i values).

system increased affinity for σ receptors. The (−)-1S,5S isomer (CB-64L) had a high affinity for σ_1 versus σ_2 receptors, whereas the (+)-1R,5R isomer (CB-64D) had a 185-fold higher selectivity for σ_2 versus σ_1 receptors (Figure 15.2). Preparation of the 3,4-dichloro analog (CB-184) resulted in an even higher affinity and selectivity for σ_2 versus σ_1 receptors (Figure 15.2). A second series of compounds having a high affinity for σ_2 receptors are the 3-(ω-aminoalkyl)-1H-indole analogs, which were originally designed to be serotonin 5-HT$_{1A}$ agonists (Perregaard et al. 1995, Moltzen et al. 1995). Additional SAR studies resulted in the synthesis of Lu 28-179 (Soby et al. 2002), which was found to have a subnanomolar affinity for σ_2 receptors and a 140-fold

selectivity for σ_2 versus σ_1 receptors (Figure 15.2). Lu 28–179 (also known as siramesine) was initially evaluated as an antidepressant and antianxiety agent, but subsequent studies have shown that this compound is cytotoxic to tumor cells by inducing a caspase-independent method of cell death (Ostenfeld et al. 2005). Similar results have been reported with CB-64D, which causes an intracellular rise in Ca^{2+} levels via the release of a thapsigargin-sensitive store in the endoplasmic reticulum (Vilner and Bowen 2000). Siramesine has been reported to cause cell death via a lysosomal leakage pathway resulting in the formation of reactive oxygen species (Ostenfeld et al. 2005). This observation suggests that σ_2 receptors are localized in the lysosomes in addition to the plasma membrane of cancer cells.

Other compounds reported to have a higher affinity for σ_2 versus σ_1 receptors are (1) the hallucinogenic natural product, ibogaine (Bowen et al. 1995, Mach et al. 1995); (2) the mixed serotonin 5-HT$_3$ antagonist/ 5-HT$_4$ agonist BIMU-1 (Bonhaus et al. 1993); (3) (±)-SM-21, an acetylcholine releaser which has been antinociceptic activity (Mach et al. 1999, Ghelardini et al. 2000); (4) the trishomocubane analog ANSTO-19 (Nguyen et al. 1996), which has a modest affinity for σ_2 receptors; and (5) the piperazine analog PB28 which has a subnanomolar affinity for σ_2 receptors, but relatively high affinity for σ_1 receptors, which results in only a moderate selectivity for σ_2 versus σ_1 receptors (σ_1:σ_2 ratio = 46, Figure 15.2) (Azzariti et al. 2006). The neurotoxicity of ibogaine has been linked to the affinity of this indole alkaloid for σ_2 receptors (Bowen et al. 1995).

A series of SAR studies using BIMU-1 as the lead compound have resulted in the identification of a series of high-affinity, high-selectivity σ_2 receptor ligands (Mach et al. 2001c, 2003, Vangveravong et al. 2006). BIMU-1 is a suitable lead compound for SAR studies since it provides a variety of regions where structural modifications can be made to optimize the σ_2 receptor affinity and reduce the affinity for serotonin 5-HT$_3$ and

5-HT$_4$ receptors (Figure 15.3). Replacement of the N-methyl of the bridgehead nitrogen with an N-benzyl group and substitution of the urea linkage of BIMU-1 with a carbamate group resulted in a dramatic increase in affinity for both σ_2 and σ_1 receptors, a loss of affinity for 5-HT$_3$ receptors, and a moderate affinity for 5-HT$_4$ receptors (Mach et al. 2001c). Expansion of the 8-azab-icyclo[3.2.1]octan-3β-yl ring system (i.e., tropane ring) to the corresponding 9-azabicyclo[3.3.1]nonan-3β-yl ring system (i.e., granatane ring) did not alter the affinity for σ_1 and σ_2 receptors relative to the tropane analogs, but eliminated affinity for the 5-HT$_4$ receptor. The most promising σ_2 receptor ligand identified from this initial SAR study was compound **1**, which had a σ_2 receptor affinity of ~3 nM and a σ_2:σ_1 selectivity of ~30 (Figure 15.3) (Mach et al. 2001c).

The granatane analog **1** was subsequently used as a secondary lead compound for the development of second-generation ligands having an improved σ_2 receptor affinity and higher σ_2:σ_1 selectivity ratio. Homologation of the N-benzyl group of **1** by one carbon atom to give the corresponding N-2-phenethyl analog, **2**, resulted in a modest improvement in σ_2 receptor affinity and increased the σ_1:σ_2 selectivity ratio to ~50 (Figure 15.3) (Mach et al. 2003). Substitution of the para-position of **1** and **2** with an amino (**3**) or dimethylamino (**4**) group resulted in further increase

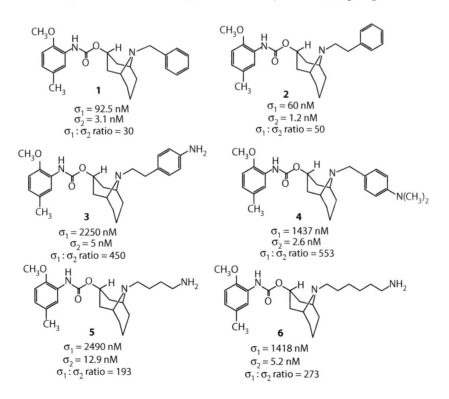

σ_1 = 92.5 nM
σ_2 = 3.1 nM
σ_1 : σ_2 ratio = 30

1

σ_1 = 60 nM
σ_2 = 1.2 nM
σ_1 : σ_2 ratio = 50

2

σ_1 = 2250 nM
σ_2 = 5 nM
σ_1 : σ_2 ratio = 450

3

σ_1 = 1437 nM
σ_2 = 2.6 nM
σ_1 : σ_2 ratio = 553

4

σ_1 = 2490 nM
σ_2 = 12.9 nM
σ_1 : σ_2 ratio = 193

5

σ_1 = 1418 nM
σ_2 = 5.2 nM
σ_1 : σ_2 ratio = 273

6

FIGURE 15.3 σ_2 Receptor ligands based on the granatane analog, **1** (K_i values).

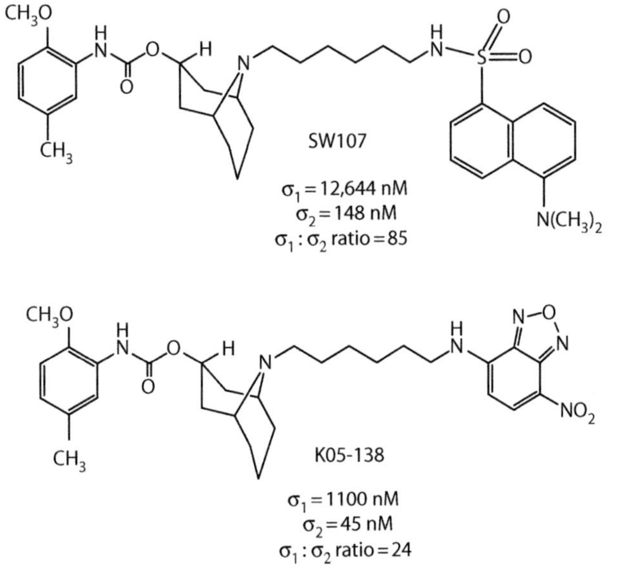

FIGURE 15.4 Fluorescent probes based on the granatane analog, **6**.

in the selectivity for σ_2 versus σ_1 receptors, primarily through a reduction in their affinity for σ_1 receptors (Mach et al. 2003). The amino group appears to be a preferred substituent for assuring a high affinity for σ_2 receptors and high σ_1:σ_2 selectivity ratio based on the

in vitro-binding properties of the aminoalkyl analogs **5** and **6** (Vangveravong et al. 2006). The high σ_2 receptor affinity of compound **6** was used in the development of the fluorescent probes SW107 and K05-138 (Figure 15.4), which have been useful in two-photon and confocal microscopy studies of σ_2 receptors in tumor cells growing under cell culture conditions (Zeng et al. 2007).

Another class of compounds having a high affinity for σ_2 receptors and excellent σ_1:σ_2 selectivity ratios are the conformationally flexible benzamide analogs (Figure 15.5). This class of compounds was initially identified by a synthesis program focused at developing radiotracers for imaging the dopamine D_3 receptor (Mach et al. 2004, Chu et al. 2005). The benzamide analog **7** was found to have a modest affinity ($K_i = 75$ nM) for σ_2 receptors and lower affinity for σ_1 receptors ($K_i \sim 800$ nM). Replacement of the N-(2,3-dichlorophenyl)piperazine ring system, which was responsible for the high D_3 affinity for this class of compounds, with a 6,7-dimethoxy-1,2,3,4-tetrahydroisoqinoline ring (i.e., compound **8**) resulted in analogs having a high affinity and excellent selectivity for σ_2 versus σ_1 receptors, and a low affinity for dopamine receptors (Figure 15.5) (Mach et al. 2004).

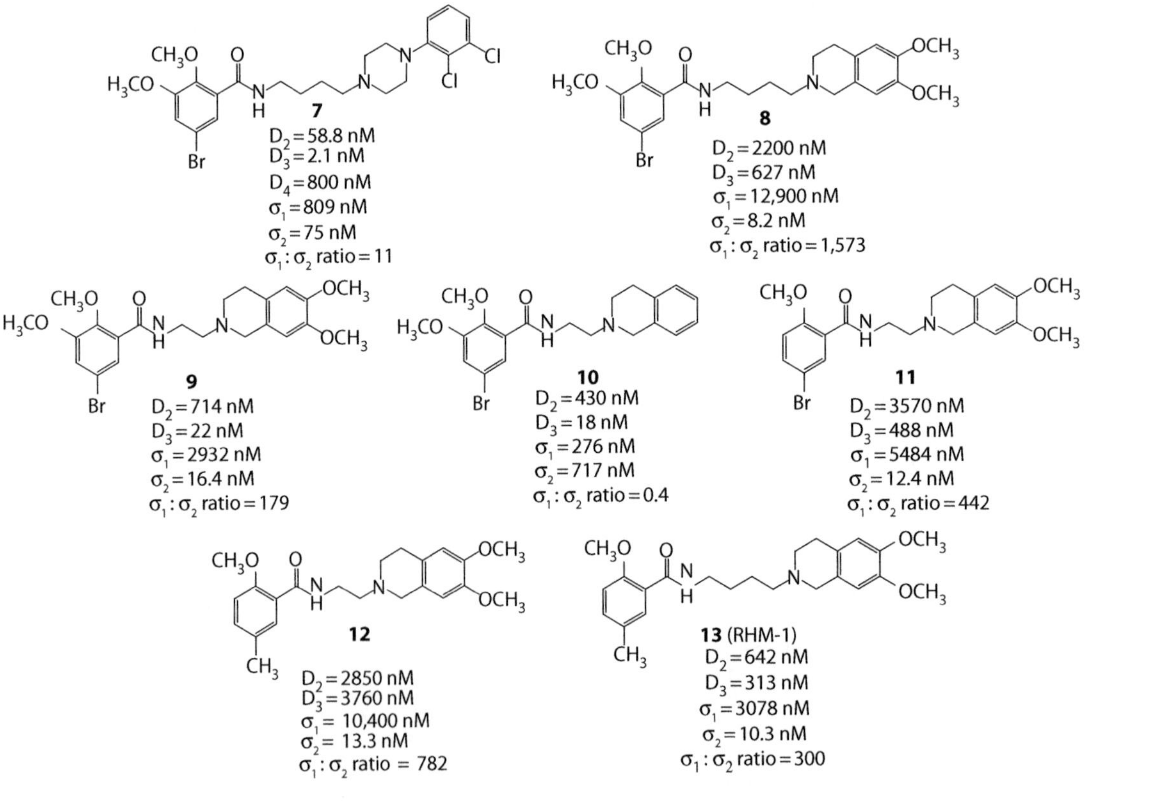

FIGURE 15.5 Structures of the conformationally flexible benzamide analogs. Receptor affinity measurements are K_i values for the respective receptors.

Additional SAR studies within this class revealed that reducing the length of the spacer group between the amide nitrogen and the nitrogen atom of the 1,2,3,4-tetrahydroisoquinoline moiety from four carbons to two carbons (i.e., compound **9**) did not change binding to the σ_2 receptor, but resulted in an unexpected increase in affinity for dopamine D_3 receptors (Mach et al. 2004). Removal of the 6,7-dimethoxy groups from **9** to give **10** resulted in a significant reduction in affinity for σ_2 receptors and no change in affinity for D_3 receptors. This observation stresses the importance of the 6,7-dimethoxy groups in the 1,2,3,4-tetrahydroisoquinoline moiety for maintaining a high affinity for σ_2 receptors (Mach et al. 2004). The 3-methoxy group in the benzamide aromatic ring

was not critical for σ_2 receptor affinity, but was important for binding to dopamine receptors since its removal resulted in a large reduction in D_3 and D_2 affinity (compare **9** versus **11**). Replacement of the 5-bromo group of **11** with a methyl group resulted in a further reduction in D_3 receptor affinity and no change in affinity for the σ_2 receptor. Extension of the two-carbon spacer of **12** to the corresponding four-carbon spacer to give compound **13** also resulted in a selective σ_2 receptor ligand. The conformationally flexible benzamide analogs with the four-carbon spacer group have proven to be an important class of σ_2-selective compounds for the preparation of radiolabeled probes to image this receptor both *in vitro* and *in vivo*.

15.4 Radioligands for *In Vitro* Binding Studies of σ_2 Receptors

The radioligand utilized most frequently in receptor-binding studies of the σ_2 receptor is [3H]di-o-tolylguanidine ([3H]DTG) (Walker et al. 1990, Hellewell et al. 1994). Although this ligand has a similar affinity for both σ_1 and σ_2 receptors, it was the difference in binding properties between [3H]DTG and radiolabeled

benzomorphan analogs such as [3H](+)-SKF 10,047 and [3H](+)-pentazocine that led to the initial description and pharmacological characterization of the σ_2 receptor (Walker et al. 1990, Hellewell et al. 1994). The photoaffinity analog [3H]azido-DTG (Figure 15.6) was also instrumental in the identification of the molecular

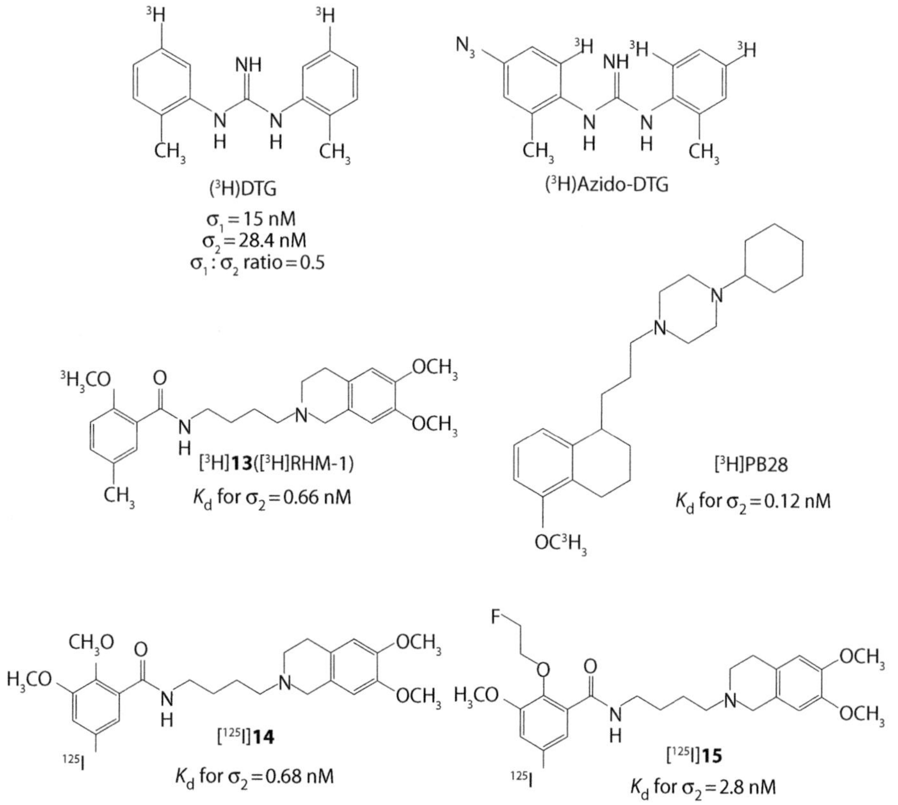

FIGURE 15.6 Radiolabeled σ_2 receptor ligands for *in vitro*-binding studies.

weight of σ_1 and σ_2 receptors (Walker et al. 1990, Hellewell et al. 1994). [^{3}H]Azido-DTG was found to label two distinct proteins in liver membranes, one with a molecular weight of ~25 kDa and a second protein with a molecular weight of ~21.5 kDa. Blocking studies using 100 nM dextrallorphan to mask σ_1-binding sites indicated that the 25 kDa protein was the σ_1 receptor, whereas the 21.5 kDa protein was the σ_2 receptor. The 25 kDa and 21.5 kDa proteins labeled by [^{3}H]azido-DTG have also been observed in other tissues, including guinea pig brain and rat phenochromocytoma (PC12) cells (Hellewell et al. 1994). The most commonly used radioligand-binding method for Scatchard and competition studies of σ receptors involve [^{3}H](+)-pentazocine to label and quantify σ_1 receptors, and [^{3}H]DTG in the presence of 100 nM of unlabeled (+)-pentazocine to measure σ_2 receptors (Hellewell et al. 1994). *In vitro* autoradiography studies measuring the relative density of σ_1 and σ_2 receptors in rat and guinea pig brain have been reported using these *in vitro*-binding methods (Walker et al. 1992, Bouchard and Quirion 1997). The use of [^{3}H]DTG in the presence of 100 nM (+)-pentazocine resulted in the identification of high σ_2 receptor density in a wide panel of murine and human tumor cells growing under cell culture conditions (Vilner et al. 1995), and the 10-fold difference in density of σ_2 receptors between P and Q of mouse mammary tumor cells (Mach et al. 1997, Al-Nabulsi et al. 1999, Wheeler et al. 2000).

Recent studies have suggested that σ_2 receptors are localized in lipid rafts of the plasma membrane (Torrence-Campbell and Bowen 1996). Lipid rafts are microdomains in the cell membrane and are believed to play a key role in both secretory and endocytotic pathways of eukaryotic cells. Lipid rafts are enriched with cholesterol, sphingolipids, and glycosylphosphatidylinositol-linked proteins that form specialized structures termed as caveolin on the incorporation of a cholesterol-binding protein. Using sucrose density centrifugation of extracts of rat liver P2 membranes, Gebreselassie and Bowen (2004) demonstrated [^{3}H] DTG-binding in protein fractions containing flotillin-2, a molecular marker of lipid rafts. Further *in vitro*-binding studies revealed that [^{3}H]DTG binding was blocked by the σ_2-selective ligand CB-64D and not by

the σ_1 receptor ligand (+)-pentazocine, confirming that σ_2 receptors are colocalized in lipid rafts. Other *in vitro*-binding studies using [^{3}H]DTG in the presence of (+)-pentazocine have shown that σ_2 receptor-binding sites are located in the mitochondria (Wang et al. 2003). This diverse distribution of σ_2 receptors between the plasma membrane and cell organelles such as the mitochondria and endoplasmic reticulum further emphasized the need for the development of fluorescent probes to study the subcellular localization of σ_2 receptors using high-resolution imaging techniques such as confocal and two-photon microscopy (Zeng et al. 2007).

Although [^{3}H]DTG is a useful ligand for characterizing the σ_2 receptor, and for screening putative σ_2-selective ligands, its rapid dissociation rate (i.e., k_{off}) is not ideal for conducting *in vitro*-binding studies. The identification of σ_2-selective ligands has resulted in the preparation of a number of tritiated compounds having more optimal σ_2 receptor-binding properties compared to [^{3}H]DTG. The first σ_2-selective ligand labeled with tritium was Lu 28-179 (siramesine). Scatchard studies in rat and human brain samples revealed that [^{3}H]Lu 28-179 has a K_d value of 1.1 nM (Søby et al. 2002). Autoradiography studies of [^{3}H]Lu 28-179 revealed a high density of σ_2 receptors in the motor cortex, hippocampus, and hind brain nuclei, which is consistent with previous autoradiography studies using [^{3}H]DTG in the presence of 1 μM dextrallorphan to mask σ_1-binding sites. Other σ_2-selective ligands that have been labeled with tritium include [^{3}H]**13**, also known as [^{3}H]RHM-1 (Xu et al. 2005), and [^{3}H]PB28 (Figure 15.6) (Colabufo et al. 2008). Of the various tritiated σ_2 receptor ligands reported to date, [^{3}H]**13** has the highest selectivity for σ_2 versus σ_1 receptors; an important property when conducting Scatchard studies of σ_2 receptors in tumors and normal tissues.

A number of sigma receptor ligands have also been radiolabeled with iodine-125 or iodine-123. However, most ^{125}I-labeled ligands developed to date have a high affinity for the σ_1 and either low or marginal affinity for the σ_2 receptor (John et al. 1996, 1999, Waterhouse et al. 1997). Two σ_2-selective ligands labeled with iodine-125 have been reported, the conformationally flexible benzamide analogs [^{125}I]**14** (Hou et al. 2006) and [^{125}I]**15** (Tu et al. 2010) (Figure 15.6).

15.5 Radioligands for *In Vivo* Imaging Studies of σ_2 Receptors

A number of structurally diverse compounds having a high affinity for σ receptors have been radiolabeled and evaluated as potential radiotracers for imaging solid

tumors with PET or SPECT. Most of the radiolabeled compounds reported to date display a higher affinity for σ_1 receptors versus σ_2 receptors, or bind with equal affinity to both σ_1 and σ_2 receptors. Since σ_1 receptors are expressed in many normal tissues (Walker et al. 1990, Hellewell et al. 1994), it is likely that a nonselective σ ligand (i.e., a ligand which binds to both σ_1 and σ_2 receptors) may have a tumor:normal tissue ratio than a σ_2-selective ligand because of the labeling of σ_1 receptors in normal tissues. This issue was addressed in a series of *in vivo* studies in tumor-bearing rodents using the radiotracer [^{18}F]N-(4′-fluorobenzyl)-4-(3-bromophenyl) acetamide, **16** (Figure 15.7) (Mach et al. 2001a).

Synthesis of the [^{18}F]**16** was accomplished by N-alkylation of the corresponding *des*-benzyl precursor with [^{18}F]4′-fluorobenzyliodide. *In vivo* studies were conducted in nude mice implanted with mouse mammary adenocarcinoma cells, line 66, a murine breast tumor cell line which expresses both σ_1 and σ_2 receptors (Mach et al. 2001a). Mice were injected with ~120 μCi of [^{18}F]**16**, and the uptake of the radiotracer in the organs of interest was measured at 1 and 2 h post-iv injection. There was a high uptake of [^{18}F]**16** in tissues known to express σ receptors (e.g., brain, lung, liver). There was also a high tumor uptake (%I.D./g = 2.5%)

and a high tumor:blood ratio. *In vivo*-blocking studies were conducted by coinjecting the σ_1-selective compound, **17** (2 mg/kg, iv), with [^{18}F]**16** to determine the effect of σ_1 receptor blockade on both the tumor uptake and the tumor:normal tissue ratio (Figure 15.7). The σ_1-blocking study reduced the tumor uptake of [^{18}F]**16** to ~1.8 %I.D./g, and lowered the tumor:blood ratio from 24.5 in the no-carrier-added study to ~15. However, blocking σ_1 receptors resulted in a dramatic increase in the tumor:lung and tumor:muscle ratios of [^{18}F]**16** relative to the no-carrier-added conditions (Figure 15.7). The improved signal:normal tissue ratios were obtained by blocking σ_1 receptors expressed in normal tissues. Similar results were also observed in microPET imaging studies of [^{18}F]**16** under no-carrier-added and σ_1-blocking conditions (Figure 15.7). These data suggest that a σ_2 selective radiotracer should yield a higher signal:normal tissue ratio for tumor imaging studies than that obtained with a radiotracer possessing a high affinity for both σ_1 and σ_2 receptors.

The high correlation between the density of σ_2 receptors and the proliferative status of solid tumors

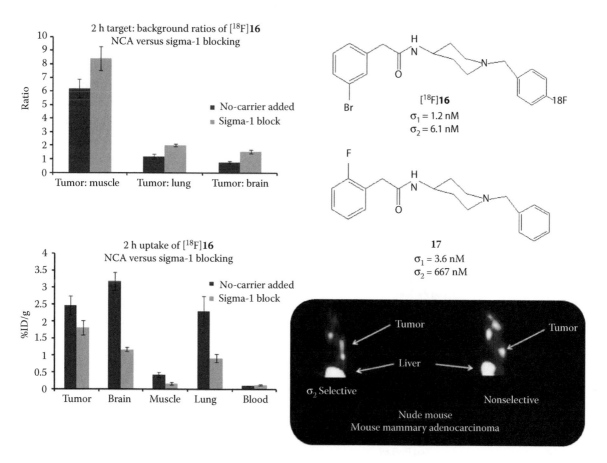

FIGURE 15.7 Tumor uptake and biodistribution studies of [^{18}F]**16** under no-carrier-added and σ_1-blocking conditions using the σ_1-selective ligand **17**. MicroPET images under no-carrier-added or σ_1-blocking conditions are also shown.

Chapter 15

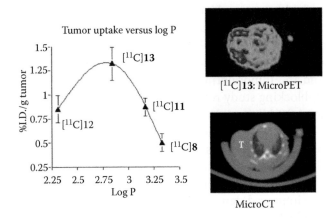

Tumor uptake versus log P
[11C]13
[11C]11
[11C]12
[11C]8
%I.D./g tumor
Log P
1.5
1.25
1
0.75
0.5
0.25
2.25 2.5 2.75 3 3.25 3.5
T
[11C]13: MicroPET
MicroCT

FIGURE 15.8 *In vivo* studies with ^{11}C-labeled conformationally flexible benzamide analogs.

indicates that σ_2-selective radiotracers are likely to be an alternative strategy for imaging cell proliferation to the thymidine-based analogs, [^{11}C]thymidine, [^{18}F]FLT, and [^{18}F]FMAU (Schwartz et al. 2003). The two classes of compounds developed to date which have a high affinity and selectivity for σ_2 versus σ_1 receptors are the 9-azabicyclo[3.3.1]nonane (granatane) analogs and the conformationally flexible benzamides. Of these, the conformationally flexible benzamide analogs have proven to be the most successful regarding the development of PET radiotracers for imaging the σ_2 receptor status of solid tumors (Tu et al. 2005). The presence of the 2-methoxy group in compounds **8**, **11**, **12**, and **13** (Figure 15.5) enabled the preparation of the corresponding ^{11}C-labeled derivatives via the *O*-alkylation of the corresponding 2-hydroxy precursors with [^{11}C]methyl iodide (Tu et al. 2005). Biodistribution and MicroPET imaging studies were conducted with [^{11}C]**8**, [^{11}C]**11**, [^{11}C]**12**, and [^{11}C]**13**, with the most promising analog proving to be [^{11}C]**13**. Although all four analogs had a high affinity for σ_2 receptors, lipophilicity proved to play a key role in the tumor uptake and tumor:normal tissue ratios of the ^{11}C-radiotracers (Figure 15.8). These data indicate that, in addition to receptor affinity, lipophilicity is an important property that must be considered in the design of receptor-based tumor imaging agents.

Although [^{11}C]**13** demonstrated promise in the imaging of breast tumors in tumor uptake and microPET imaging studies, the short half-life of carbon 11 ($t_{1/2}$ = 20.4 min) is not ideal for the development of radiotracers for use in translational PET imaging studies in cancer patients. The longer half-life of ^{18}F ($t_{1/2}$ = 109.8 min) places fewer time constraints on tracer synthesis and PET data acquisition. The longer imaging times usually result in higher tumor:normal tissue ratios for ^{18}F-labeled radiotracers relative to their ^{11}C-labeled congeners. The benzamide analogs shown in Figure 15.5 have also served as lead compounds in the development of ^{18}F-labeled σ_2-selective radiotracers. In this case, the 2-methoxy group in the benzamide ring was replaced with a 2-fluoroethoxy group (Figure 15.9) (Tu et al. 2007). The 2-fluoroethoxy- for methoxy- substitution is a frequently used strategy in

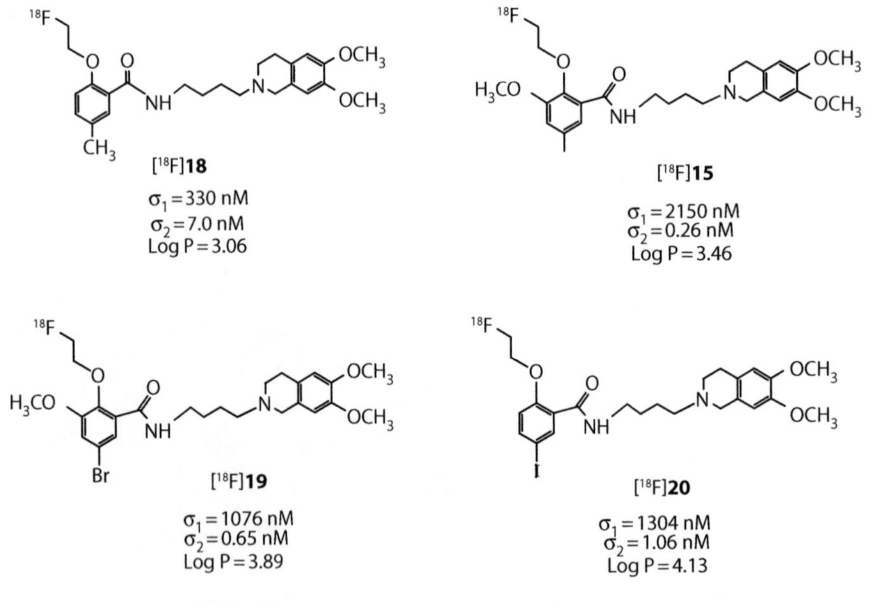

[18F]18
σ1 = 330 nM
σ2 = 7.0 nM
Log P = 3.06
[18F]15
σ1 = 2150 nM
σ2 = 0.26 nM
Log P = 3.46
[18F]19
σ1 = 1076 nM
σ2 = 0.65 nM
Log P = 3.89
[18F]20
σ1 = 1304 nM
σ2 = 1.06 nM
Log P = 4.13

FIGURE 15.9 ^{18}F-labeled σ_2 receptor imaging agents.

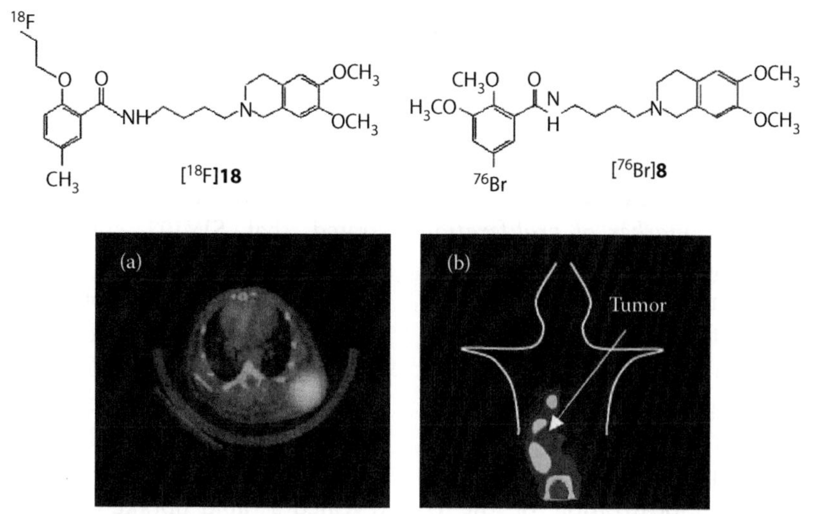

FIGURE 15.10 MicroPET/MicroCT imaging study of [^{18}F]18 (a) and microPET images of [^{76}Br]8 (b) at 2 h post-iv injection of the radio-tracer. Note the high uptake of the radiotracer in the EMT6 mammary adenocarcinoma and low uptake in the surrounding normal tissues with both radiotracers.

the development of ^{18}F-labeled radiotracers. MicroPET imaging studies indicate that [^{18}F]18, the corresponding ^{18}F-labeled congener of [^{11}C]13, is a suitable probe for imaging the σ_2 receptor status of solid tumors with PET (Figure 15.10) (Tu et al. 2007). Clinical imaging studies in cancer patients of [^{18}F]18 are currently in progress in the United States. [^{18}F]Siramesine has also been prepared via an *N*-arylation reaction of the corresponding indole precursor with [^{18}F]4-fluoroiodobenzene (Wüst and Kniess 2005). However, *in vivo* data with [^{18}F]sir-amesine have not been reported.

The presence of the bromine atom in the benzamide ring of compound **8** (Figure 15.5) also led to the prepa-ration of a Br-76-labeled radiotracer, [^{76}Br]**8** (Rowland et al. 2006). Although the limited availability and high positron energy of Br-76 (which degrades image reso-lution) limit the clinical utility of [^{76}Br]**8** as a PET radiotracer, the microPET imaging study of [^{76}Br]**8** clearly showed excellent visualization of the tumor and good tumor:normal tissue ratios when compared with [^{18}F]**18** (Figure 15.10).

A number of ^{123}I-labeled σ receptor radiotracers have been prepared and SPECT imaging studies have been conducted in breast and melanoma patients. However, these radiotracers are selective for the σ_1 receptor and have low affinity for σ_2 receptors, and will not be discussed in this chapter. Two ^{99m}Tc-labeled sigma receptor ligands have been reported in the litera-ture (Figure 15.11). [^{99m}Tc]BAT-EN6, which is an ana-log of 4-PEMP, has been shown to have a high level of binding in breast tumor membrane homogenates, but

its affinity for σ_1 and σ_2 receptors has not been reported (John et al. 1997). *In vivo* studies with [^{99m}Tc]21, which has a high affinity for σ_2 versus σ_1 receptors (Mach et al. 2001b), demonstrated a high uptake and clear visual-ization of 66 murine breast tumor xenografts in nude mice (Choi et al. 2001). These data suggest that [^{99m}Tc]21 may be a useful radiotracer for SPECT imaging studies of breast cancer patients.

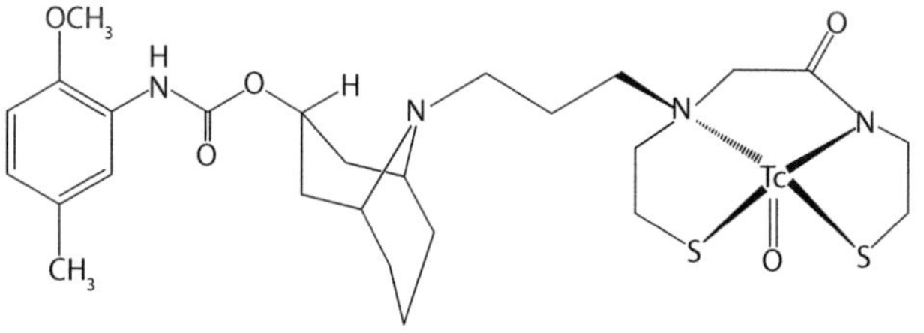

[^{99m}Tc]BAT-EN6
σ affinity = 42.7 nM

[^{99m}Tc]21
σ_1 affinity of Re complex = 1125 nM
σ_2 affinity of Re complex = 13.7 nM

FIGURE 15.11 Structures of ^{99m}Tc-labeled agents having a moderate-to-high affinity for σ_2 receptors. Affinity measurements are K_i values.

15.6 Fluorescent Probes for Confocal and Two-Photon Microscopy Studies of σ_2 Receptors in Tumor Cells

Although the studies described above indicate that the σ_2 receptor is a useful biomarker for imaging the proliferative status of solid tumors, the σ_2 receptor cannot be classified as a *molecular marker of proliferation* since its amino acid sequence is not known, and the role the σ_2 receptor plays in cell proliferation is not understood. Since the gene for the σ_2 receptor has not been cloned, and the protein has not been purified and sequenced, current knowledge of this receptor is based predominantly on receptor-binding studies with radiolabeled probes such as [³H]DTG and [³H]RHM-1. In addition, the functional localization of the σ_2 receptor has been investigated by studying the effects of σ_2 ligands on the biochemical and physiological properties of tumor cells. These properties include (1) calcium release from the endoplasmic reticulum and mitochondria (Vilner and Bowen 2000), (2) incorporation of [³H]palmitic acid to form [³H]ceramide (Crawford and Bowen 2002, Crawford et al. 2002), (3) the release of cathepsins B and L from lysosomal stores (Ostenfeld et al. 2005), and (4) the generation of reactive oxygen species (Ostenfeld et al. 2005). However, these studies only provide an indirect measure of potential sites of localization of σ_2 receptors within cancer cells. *In vitro* receptor-binding studies on subcellular fractions of brain and liver tissue have suggested that σ_2 receptors are localized on the endoplasmic reticulum, mitochondria, and plasma membrane (Hellewell et al. 1994, Wang et al. 2003).

Using the fluorescent probes shown in Figure 15.4, Zeng et al. (2007) conducted a series of confocal and two-photon microscopy studies which have provided a clearer picture of the localization of σ_2 receptors in breast tumor cells. These studies were conducted in combination with a panel of "Tracker" dyes in order to determine which organelles and subcellular compartments express σ_2 receptors. EMT-6 (mouse breast tumor) or MDA-MB-435 (metastatic human tumor) cells were incubated with 200 nM SW107 and either 50 nM MitoTracker®, 500 nM ER-Tracker®, or 75 nM LysoTracker® for 2 h at 37°C. MDA-MB-435 cells were also incubated with 200 nM SW107 and 5 µg/mL of FM®1-43FX, a membrane marker dye, for 5 min at 0°C. After the incubation period, imaging of the live cells by two-photon microscopy revealed that SW107 was distributed throughout the cytoplasm of the tumor cells, but not in the nucleus (Figure 15.12). The staining of SW107 was highly punctate, suggesting that the label was sequestered in different membrane-bound compartments. The tracker colocalization studies demonstrated that SW107 colocalizes with MitoTracker, LysoTracker, ER Tracker, and the membrane marker. This indicates that σ_2 receptors are localized in the mitochondria, lysosomes, endoplasmic reticulum, and the cytoplasmic membrane (Figure 15.12). Similar results were obtained with confocal imaging studies using the fluorescent probe, K05-138 (Figure 15.13).

Time-lapsed confocal microscopy studies also revealed a rapid uptake of K05-138 ($t_{1/2} = 16$ s), which suggested that the uptake of the fluorescent probe into tumor cells is occurred by receptor-mediated endocytosis. This was confirmed with the use of phenylarsine oxide (PAO), an endocytosis inhibitor, which reduced

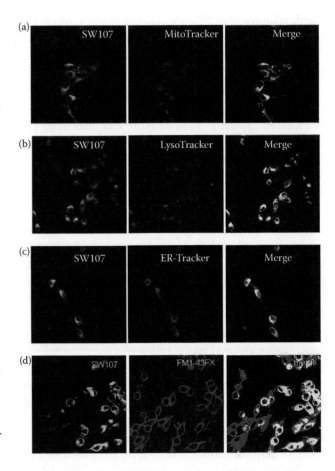

FIGURE 15.12 Two-photon microscopy studies with SW107. These "Tracker" colocalization studies reveal that σ_2 receptors are found in the mitochondria (a), lysosomes (b), endoplasmic reticulum (c), and plasma membrane (d), but not in the nucleus.

(a)

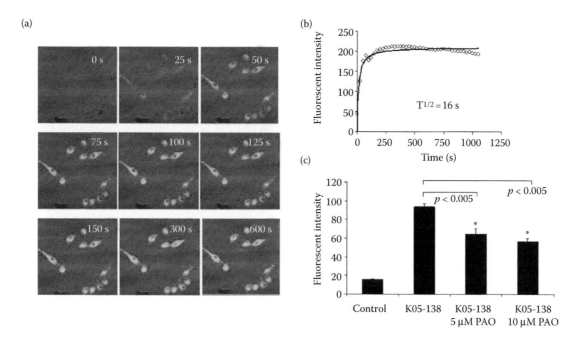

(b)

(c)

FIGURE 15.13 (a) Time-lapsed confocal microscopy studies with K05-138 showing that the σ_2 receptors are located in the cytoplasm, but not in the nucleus. (b) The rapid uptake of the fluorescent probe in MDA-MB-435 cells suggested that internalization of the probe may occur by receptor-mediated endocytosis. (c) Blocking studies with PAO, a known inhibitor of endocytosis, confirm that the internalization of the σ_2 receptor probe is, in part, by receptor-mediated endocytosis.

the uptake of K05-138 in MDA-MB-435 cells by ~40%, (Figure 15.13) (Zeng et al. 2007). These data demonstrate that ~40% of the σ_2 receptors were internalized by receptor-mediated endocytosis, while the remaining ~60% was internalized by passive diffusion across the cell membrane.

The two-photon and confocal microscopy studies conducted and described above have provided useful information for the interpretation of studies demonstrating that σ_2 receptor ligands may have a role in cancer chemotherapy. For example, cell culture studies have revealed that σ_2 receptor ligands induce apoptosis by caspase-dependent and caspase-independent pathways (Vilner and Bowen 2000, Crawford and Bowen 2002, Ostenfeld et al. 2005). These studies also suggest that subcellular organelles, such as the mitochondria and lysosomes, are involved in σ_2 ligand-induced cell death. Although the exact mechanism(s) by which σ_2 ligands induce cell death are not known, the confocal and two-photon microscopy studies provide valuable insights into the mechanisms and functions of σ_2-selective ligands and σ_2 receptors.

Mitochondria are a key organelle with respect to mediating cell death and regulating the intrinsic pathway of apoptosis. Apoptotic signals after ultraviolet irradiation or treatment with chemotherapeutic agents cause the release of cytochrome c from the mitochondria and

the subsequent activation of caspase-3 leading to apoptosis. The observation that σ_2 ligands are localized in the mitochondria and reduce mitochondrial membrane potential resulting in caspase-mediated apoptosis in SK-N-SH neuroblastoma cells suggests that σ_2 receptors play a role in the intrinsic pathway of apoptosis (Vilner and Bowen 2000).

Additional evidence suggests that lysosomal proteases, such as cathepsins, calpains, and granzymes, also contribute to σ_2-ligand-induced apoptosis (Chwieralski et al. 2006). Under physiological conditions, these proteases are found within the lysosomes but are released into the cytoplasm after exposure to cell damaging agents, thereby triggering a cascade of intracellular events leading to cell death. The σ_2-selective ligand siramesine has been reported to cause lysosomal leakage and induce cell death by a caspase-independent mechanism. Siramesine-induced cell death was also partially blocked by an inhibitor of the lysosome protease, Cathepsin B (Ostenfeld et al. 2005). The localization of fluorescent σ_2 receptor probes in the lysosomes is consistent with the observation that siramesine induces cell death partially by targeting lysosomes to cause lysosomal damage, the release of proteases, and eventually cell death (Ostenfeld et al. 2005). However, the lysosomal degradation of σ_2 receptors may be responsible for the accumulation of the fluorescent σ_2 probes in the

Chapter 15

lysosomes. Although the confocal and two-photon microscopy studies are consistent with the expression of σ_2 receptors in the lysosomal membrane, the lysosomal degradation of the fluorescent probes is an alternative explanation which cannot be ruled out at this time (Zeng et al. 2007).

As stated earlier, σ_2 receptors have also been reported to exist in lipid rafts which are mainly found in the plasma membrane (Gebreselassie and Bowen 2004). Lipid rafts play an important role in the signaling associated with a variety of cellular events including adhesion, motility, and membrane trafficking (Brown and London 1998, Simons and Toomre 2000). The observation that σ_2 fluorescent ligands are colocalized with a cytoplasmic membrane marker and undergo receptor-mediated endocytosis is consistent with their localization in lipid rafts. Thus, the fluorescent probes SW107 and K05-138 may prove to be useful tools for studying σ_2 receptors in lipid rafts using two-photon and confocal microscopy.

15.7 Conclusion

The σ_2 receptor continues to be an important target in the field of tumor biology. The high expression of this receptor in proliferating versus quiescent breast tumors indicates that the σ_2 receptor is a clinical biomarker for determining the proliferative status of solid tumors using the functional imaging techniques PET and SPECT. However, the full utility of the σ_2 receptor ligands in the imaging of cancer will not likely be achieved until (1) the gene is cloned and (2) the protein is purified and sequenced so that its functional role in normal and tumor cell biology can be understood. Until this occurs, our knowledge of this receptor will continue to rely on the use of σ_2-selective fluorescent probes and radioligands to study the expression of this protein using *in vitro* and *in vivo* models of tumors and normal tissues.

Acknowledgments

The author would also like to thank Ms Lynne A. Jones for her excellent editorial assistance. This research was supported by grants CA80452, CA81825, and CA102869 awarded by the National Cancer Institute, and grant DAMD17-01-1-0446 awarded by the Department of Defense Breast Cancer Research Program of the U.S. Army Medical Research and Materiel Command Office.

References

Al-Nabulsi, I., Mach, R.H., Wang, L.M., Wallen, C.A., Keng, P.C., Sten, K., Childers, S.R., and Wheeler, K.T. 1999. Effect of ploidy, recruitment, environmental factors, and tamoxifen treatment on the expression of sigma-2 receptors in proliferating and quiescent tumor cells. *Br. J. Cancer* 81:925–933.

Alonso, G., Phan, V., Guillemain, I., Saunier, M., Legrand, A., Anoal, M., and Maurice, T. 2000. Immunocytochemical localization of the sigma-1 receptor in the adult rat central nervous system. *Neuroscience* 97:155–170.

Azzariti, A., Colabufo, N.A., Berardi, F., Porcelli, L., Niso, M., Simone, G.M., Perrone, R., and Paradiso, A. 2006. Cyclohexylpiperazine derivative PB28, a σ_2 agonist and σ_1 antagonist receptor, inhibits cell growth, modulates P-glycoprotein, and synergizes with anthracyclines in breast cancer. *Mol. Cancer Ther.* 5:1807–1816.

Bem, W.T., Thomas, G.E., Mamone, J.Y., Homan, S.M., Levy, B.K., Johnson, F.E., and Coscia, C.J. 1991. Overexpression of sigma receptors in nonneural human tumors. *Cancer Res.* 51:6558–6562.

Bertha, C.M., Mattson, M.V., Flippen-Anderson, J.L., Rothman, R.B., Xu, H., Cha, X.Y., Becketts, K., and Rice, K.C. 1994. A marked change of receptor affinity of the 2-methyl-5-(3-hydroxyphenyl)morphans upon attachment of an (*E*)-8-benzylidene moiety: Synthesis and evaluation of a new class of σ receptor ligands. *J. Med. Chem.* 37:3163–3167.

Bonhaus, D.W., Loury, D.N., Jakeman, L.B., To, Z., DeSouza, A., Eglen, R.M., and Wong, E.H. 1993. [^{3}H]BIMU-1, a 5-hydroxytryptamine$_3$ receptor ligand in NG-108 cells, selectively labels sigma-2 binding sites in guinea pig hippocampus. *J. Pharmacol. Exp. Ther.* 267:961–970.

Bouchard, P. and Quirion, R. 1997. [^{3}H]1,3-Di-*o*-tolylguanidine and [^{3}H](+)-pentazocine binding sites in rat brain: Autoradiographic visualization of the putative sigma-1 and sigma-2 receptor subtypes. *Neuroscience* 76:467–477.

Bowen, W.D., Bertha, C.M., Vilner, B.J., and Rice, K.C. 1995. CB-64D and CB-184: Ligands with high sigma-2 receptor affinity and subtype selectivity. *Eur. J. Pharmacol.* 278:257–260.

Bowen, W.D., Vilner, B.J., Williams, W., Bertha, C.M., Kuehne, M.E., and Jacobson, A.E. 1995. Ibogaine and its congeners are σ_2 receptor-selective ligands with moderate affinity. *Eur. J. Pharmacol.* 279:R1–R3.

Brown, D.A. and London, E. 1998. Functions of lipid rafts in biological membranes. *Annu. Rev. Cell Dev. Biol.* 14:111–136.

Choi, S.-R., Yang, B., Plössl, K., Chumpradit, S., Wey, S.-P., Acton, P.D., Wheeler, K.T., Mach, R.H., and Kung, H.F. 2001. Development of a Tc-99m labeled sigma-2 receptor-specific ligand as potential breast tumor imaging agent. *Nucl. Med. Biol.* 28:657–666.

Chu, W., Tu, Z., McElveen, E., Xu, J., Taylor, M., Luedtke, R.R., and Mach, R.H. 2005. Synthesis and *in vitro* binding of *N*-phenyl piperazine analogs as potential dopamine D$_3$ receptor ligands. *Bioorg. Med. Chem.* 13:77–87.

Chwieralski, C.E., Welte, T., and Buhling, F. 2006. Cathepsin-regulated apoptosis. *Apoptosis* 11:143–149.

Colabufo, N.A., Abate, C., Contino, M., Inglese, C., Ferorelli, S., Berardi, F., and Perrone, R. 2008. Tritium radiolabeling of PB28, a potent sigma-2 receptor ligand: Pharmacokinetic and pharmacodynamic characterization. *Bioorg. Med. Chem. Lett.* 18:1484–1488.

Collins, I. and Garrett, M.D. 2005. Targeting the cell division cycle in cancer: CDK and cell cycle checkpoint kinase inhibitors. *Curr. Opin. Pharmacol.* 5:366–373.

Crawford, K.W. and Bowen, W.D. 2002. Sigma-2 receptor agonists activate a novel apoptotic pathway and potentiate antineoplastic drugs in breast tumor cell lines. *Cancer Res.* 62:313–322.

Crawford, K.W., Coop, A., and Bowen, W.D. 2002. Sigma-2 receptors regulate changes in sphingolipid levels in breast tumor cells. *Eur. J. Pharmacol.* 443:207–209.

Fornace, A.J., Fuks, Z., Weichselbaum, R.R., and Milas, L. 2001. Radiation therapy. In *The Molecular Basis of Cancer, Second Edition.* Mendelsohn, J., Howley, P.M., Israel, M.A., and Liotta, L.A. (editors), W. B. Saunders Co., Philadelphia. PA, pp. 423–466.

Gebreselassie, D. and Bowen, W.D. 2004. Sigma-2 receptors are specifically localized to lipid rafts in rat liver membranes. *Eur. J. Pharmacol.* 493:19–28.

Ghelardini, C., Galeotti, N., and Bartolini, A. 2000. Pharmacological identification of SM-21, the novel σ$_2$ antagonist. *Pharm. Biochem. Behav.* 67:659–662.

Hanner, M., Moebius, F.F., Flandorfer, A., Knaus, H.G., Striessnig, J., Kempner, E., and Glossmann, H. 1996. Purification, molecular cloning, and expression of the mammalian sigma-1 binding site. *Proc. Natl. Acad. Sci. USA* 93:8072–8077.

Hellewell, S.B., Bruce, A., Feinstein, G., Orringer, J., Williams, W., and Bowen, W.D. 1994. Rat liver and kidney contain high densities of sigma-1 and sigma-2 receptors: Characterization by ligand binding and photoaffinity labeling. *Eur. J. Pharmacol. Mol. Pharmacol. Sec.* 268:9–18.

Hou, C., Tu, Z., Mach, R.H., Kung, H.K., and Kung, M.-P. 2006. Characterization of a novel radioiodinated sigma-2 receptor ligand as a cell proliferation marker. *Nucl. Med. Biol.* 33:203–209.

John, C.S., Bowen, W.D., Fisher, S.J. Lim, B.B., Geyer, B.C., Vilner, B.J., and Wahl, R.L. 1999. Synthesis, *in vitro* pharmacologic characterization, and preclinical evaluation of N-[2-(1'-piperidinyl)ethyl]-3-[^{125}I]iodo-4-methoxybenzamide (P[^{125}I]MBA) for imaging breast cancer. *Nucl. Med. Biol.* 26:377–382.

John, C.S., Gulden, M.E., Vilner, B.J., and Bowen, W.D. 1996. Synthesis, *in vitro* validation and *in vivo* pharmacokinetics of [^{125}I]N-[2-(4-iodophenyl)ethyl]-N-methyl-2-(1-piperidinyl) ethylamine: A high affinity ligand for imaging sigma receptor positive tumors. *Nucl. Med. Biol.* 23:761–766.

John, C.S., Lim, B.B., Geyer, B.C., Vilner, B.J., and Bowen, W.D. 1997. ^{99m}Tc-Labeled sigma-binding complex: Synthesis, characterization, and specific binding to human ductal breast carcinoma (T47D) cells. *Bioconjug. Chem.* 8:304–309.

Mach, R.H. and Wheeler, K.T. 2007. Imaging the proliferative status of tumors with PET. *J. Labelled Cpd. Radiopharm.* 50:366–369.

Mach, R.H., Huang, Y., Buchheimer, N. Kuhner, R., Wu, L., Morton, T.E., Wang, L., Ehrenkaufer, R.L., Wallen, C.A., and Wheeler, K.T. 2001a. [^{18}F]N-4'-fluorobenzyl-4-(3-bromophenyl)acetamide for imaging the sigma receptor status of tumors: Comparison with [^{18}F]FDG and [^{125}I]IUDR. *Nucl. Med. Biol.* 28:451–458.

Mach, R.H., Huang, Y., Freeman, R.A., Wu, L., Vangveravong, S., and Luedtke, R.R. 2004. Conformationally-flexible benzamide analogues as dopamine D$_3$ and σ$_2$ receptor ligands. *Bioorg. Med. Chem. Lett.* 14:195–202.

Mach, R.H., Smith, C.R., Al-Nabulsi, I., Whirrett, B.R., Childers, S.R., and Wheeler, K.T. 1997. Sigma-2 receptors as potential biomarkers of proliferation in breast cancer. *Cancer Res.* 57:156–161.

Mach, R.H., Smith, C.R., and Childers, S.R. 1995. Ibogaine possesses a selective affinity for σ$_2$ receptors. *Life Sci.* 57:PL 57–62.

Mach, R.H., Vangveravong, S., Huang, Y., Yang, B., Blair, J.B., and Wu, L. 2003. Synthesis of N-substituted 9-azabicyclo[3.3.1] nonan-3α-yl phenylcarbamate analogs as sigma-2 receptor ligands. *Med. Chem. Res.* 11:380–398.

Mach, R.H., Wheeler, K.T., Blair, S., Yang, B., Day, C.S., Blair, J.B., Choi, S.R., and Kung, H.F. 2001b. Preparation of a technetium-99m SPECT agent for imaging the sigma-2 receptor status of solid tumors. *J. Labelled. Cpd. Radiopharm.* 44:899–908.

Mach, R.H., Wu, L., West, T., Whirrett, B.R., and Childers, S.R. 1999. The analgesic tropane analogue (±)-SM 21 has a high affinity for σ$_2$ receptors. *Life Sci.* 64:PL131–PL137.

Mach, R.H., Yang, B., Wu, L., Kuhner, R.J., Whirrett, B.R., and West, T. 2001c. Synthesis and sigma receptor binding affinities of 8-azabicyclo[3.2.1]octan-3α-yl and 9-azabicyclo[3.3.1]nonao-3α-yl phenylcarbamates. *Med. Chem. Res.* 10:339–355.

Maurice, T., Junien, J.-L., and Privat, A. 1997. Dehydroepiandrosterone sulfate attenuates dizocilpine-induced learning impairment in mice via sigma-1-receptors. *Behav. Brain Res.* 83:159–164.

Maurice, T., Roman, F.J., and Privat, A. 1996. Modulation by neurosteroids of the *in vivo* (+)-[^{3}H]SKF-10,047 binding to sigma-1 receptors in the mouse forebrain. *J. Neurosci. Res.* 46:734–743.

Moltzen, E.K., Perregaard, J., and Meier, E. 1995. σ Ligands with subnanomolar affinity and preference for the σ$_2$ binding site. 2. Spiro-joined benzofuran, isobenzofuran, and benzopyran piperidines. *J. Med. Chem.* 38:2009–2017.

Nguyen, V.H., Kassiou, M., Johnston, G.A., and Christie, M.J. 1996. Comparison of binding parameters of sigma 1 and sigma 2 binding sites in rat and guinea pig brain membranes: Novel subtype-selective trishomocubanes. *Eur. J. Pharmacol.* 311: 233–240.

Ostenfeld, M.S., Fehrenbacher, N., Hoyer-Hansen, M., Thomsen, C., Farkas, T., and Jaattela, M. 2005. Effective tumor cell death by sigma-2 receptor ligand siramesine involves lysosomal leakage and oxidative stress. *Cancer Res.* 65:8975–8983.

Perregaard, J., Moltzen, E.K., Meier, E., and Sanchez, C. 1995. σ Ligands with subnanomolar affinity and preference for the sigma-2 binding site.1. 3-(ω-aminoalkyl)-1H-indoles. *J. Med. Chem.* 38:1998–2008.

Romieu, P., Martin-Fardon, R., Bowen, W.D., and Maurice, T. 2003. Sigma-1 receptor-related neuroactive steroids modulate cocaine-induced reward. *J. Neurosci.* 23:3572–3576.

Rowland, D.J., Tu, Z., Xu, J., Ponde, D., Mach, R.H., and Welch, M.J. 2006. Synthesis and *in vivo* evaluation of 2 high-affinity ^{76}Br-labeled sigma2-receptor ligands. *J. Nucl. Med.* 47: 1041–1048.

Schwartz, J.L., Tamura, Y., Jordan, R., Grierson, J.R., and Krohn, K.A. 2003. Monitoring tumor cell proliferation by targeting DNA synthetic processes with thymidine and thymidine analogs. *J. Nucl. Med.* 44:2027–2032.

Seth, P., Leibach, F.H., and Ganapathy, V. 1997. Cloning and structural analysis of the cDNA and the gene encoding the murine type 1 sigma receptor. *Biochem. Biophys. Res. Commun.* 24:535–540.

Simons, K. and Toomre, D. 2000. Lipid rafts and signal transduction. *Nat. Rev. Mol. Cell Biol.* 1:31–39.

Soby, K., Mikkelsen, J.D., Meier, E., and Thomsen, C. 2002. Lu 28–179 labels a σ_2-site in rat and human brain. *Neuropharmacology* 43:95–100.

Strebhardt, K. and Ulrich, A. 2006. Targeting polo-like kinase 1 for cancer therapy. *Nat. Rev. Cancer* 6:321–330.

Torrence-Campbell, C. and Bowen, W.D. 1996. Differential solubilization of rat sigma-1 and sigma-2 receptors: Retention of sigma-2 sites in particulate fractions. *Eur. J. Pharmacol.* 304:201–210.

Tu, Z., Xu, J., Jones, L.A. Li, S., Zeng, D., Kung, M.P., Kung, H.F., and Mach, R.H. 2010. Radiosynthesis and biological evaluation of a promising sigma-2 receptor ligand radiolabeled with fluorine-18 or iodine-125 as a PET/SPECT probe for imaging breast cancer. *Appl. Radiat. Isot.* 68(12):2268–2273.

Tu, Z., Dence, C.S., Ponde, D.E. Jones, L.A., Wheeler, K.T., Welch, M.J., and Mach, R.H. 2005. Carbon-11 labeled sigma-2 receptor ligands for imaging breast cancer. *Nucl. Med. Biol.* 32:423–430.

Tu, Z., Xu, J., Jones, L.A., Li, S., Dumstorff, C., Vangveravong, S., Chen, D.L., Wheeler, K.T., Welch, M.J., and Mach, R.H. 2007. Fluorine-18 labeled benzamide analogs for imaging the sigma-2 receptor status of solid tumors with positron emission tomography. *J. Med. Chem.* 50:3194–204.

Vangveravong, S., Xu, J., Zeng, C., and Mach, R.H. 2006. Synthesis of *N*-substituted 9-azabicyclo[3.3.1]nonan-3α-yl phenylcarbamate analogs as σ_2 receptor ligands. *Bioorg. Med. Chem.* 14:6988–6897.

Vilner, B.J. and Bowen, W.D. 2000. Modulation of cellular calcium by sigma-2 receptors: Release from intracellular stores in human SK-N-SH neuroblastoma cells. *J. Pharmacol. Exp. Ther.* 292:900–911.

Vilner, B.J., John, C.S., and Bowen, W.D. 1995. Sigma-1 and sigma-2 receptors are expressed in a wide variety of human and rodent tumor cell lines. *Cancer Res.* 55:408–413.

Walker, J.M., Bowen, W.D., Goldstein, S.R., Roberts, A.H., Patrick, S.L., Hohmann, A.G., and DeCosta, B. 1992. Autoradiographic distribution of [^{3}H](+)-pentazocine and [^{3}H]1,3-di-*o*-tolylguanidine (DTG) binding sites in guinea pig brain: A comparative study. *Brain Res. 581*:33–38.

Walker, J.M., Bowen, W.D., Walker, F.O., Matsumoto, R.R., De Costa, B., and Rice, K.C. 1990. Sigma receptors: Biology and function. *Pharmacol. Rev.* 42:355–402.

Wallen, C.A., Higashikubo, R., and Dethlefsen, L.A. 1984a. Murine mammary tumour cells *in vitro*. I. The development of a quiescent state. *Cell Tissue Kinet.* 17:65–77.

Wallen, C.A., Higashikubo, R., and Dethlefsen, L.A. 1984b. Murine mammary tumour cells *in vitro*. II. Recruitment of quiescent cells. *Cell Tissue Kinet.* 17:79–89.

Wang, L.-M., Bowen, W.D., Childers, S.R., Mach, R.H., Wheeler, K.T., Williams, W., and Morton, K. 2003. Sigma-2 receptor binding activity in rat liver mitochondria. *Proc. Am. Assoc. Cancer Res.* 44:1154; (abstract).

Wang, B., Rouzier, R., Albarracin, C.T., Sahin, A., Wagner, P., Yang, Y., Smith, T.L., Meric-Bernstam, F., Marcelo Aldaz, C., and Hortobagyi G.N. 2004. Expression of sigma-1 receptor in human breast cancer. *Breast Cancer Res. Treat.* 87:205–214.

Waterhouse, R.N., Chapton, J., Izard, B., Donald A., Belbin K., O'Brien J.C., and Collier T.L. 1997. Examination of four ^{123}I-labeled piperidine-based sigma receptor ligands as potential melanoma imaging agents: Initial studies in mouse tumor models. *Nucl. Med. Biol.* 24:587–593.

Wheeler, K.T., Wang, L.M., Wallen, C.A., Childers, S.R., Cline, J. M., Keng, P.C., and Mach, R.H. 2000. Sigma-2 receptors as a biomarker of proliferation in solid tumours. *Br. J. Cancer* 82:1223–1232.

Wüst, F.R. and Kniess, T. 2005. N-arylation of indoles with [^{18}F] fluoroiodobenzene: Synthesis of ^{18}F-labelled σ_2 receptor ligands for positron emission tomography (PET). *J. Labelled. Cpd. Radiopharm.* 48:31–43.

Xu, J., Tu, Z., Jones, L.A., Wheeler, K.T., and Mach, R.H. 2005. [^{3}H] *N*-[4-(3,4-dihydro-6,7-dimethoxy-isoquinolin-2(1H)-*yl*) butyl]-2-methoxy-5-methylbenzamide: A novel sigma-2 receptor probe. *Eur. J. Pharmacol.* 525:8–17.

Zeng, C., Vangveravong, S., Xu, J., Chang, K.C., Hotchkiss, R.S., Wheeler, K.T., Shen, D., Zhuang, Z.P., Kung, H.F., and Mach, R.H. 2007. Subcellular localization of sigma-2 receptors in breast cancer cells using two-photon and confocal microscopy. *Cancer Res.* 67:6708–6716.

16. Steroid Hormone Ligands for the Estrogen Receptor

David A. Mankoff, Jeanne M. Link, and Hannah M. Linden

Targeted Molecular Imaging. Edited by Michael J. Welch and
William C. Eckelman © 2012 Taylor & Francis Group, LLC. ISBN:
978-1-4398-4195-2

Chapter 16

16.1 Introduction: Estrogen Receptor Biology

16.1.1 Estrogen Physiology Overview

Estrogens play an important role in female reproductive physiology. Although there are a number of naturally occurring compounds with estrogenic activity, human estrogens consist of the steroid compounds estriol, estradiol, and estrone. The physiologic actions of estrogens occur through the binding of estrogenic ligands to the estrogen receptor (ER, discussed in more detail in Section 16.1.2) (Katzenellenbogen 1996). Estradiol is the principal naturally occurring agonist ligand for the ER. The ER is not ubiquitously expressed in all tissues, but rather expressed selectively in the breast, uterus, ovaries, bone, and pituitary (Riggs and Hartmann 2003).

The molecular mechanism of estradiol interaction with the ER has been well studied (Katzenellenbogen 1996, Riggs and Hartmann 2003, Sledge and McGuire 1983) (Figure 16.1). Estradiol is a lipophilic molecule so it can cross cell membranes to the ER, which is located on the nuclear membrane. ER has two receptor subtypes—alpha and beta. ER alpha serves largely as an activator of downstream events related to breast and female sex organ function and is the focus of discussion in this chapter (Marino et al. 2006). The function of ER beta is less well understood and may, in some cases, inhibit ER alpha (Stein and McDonnell 2006).

The binding of estrogenic molecules to ER "activates" the receptor to interact with specific DNA

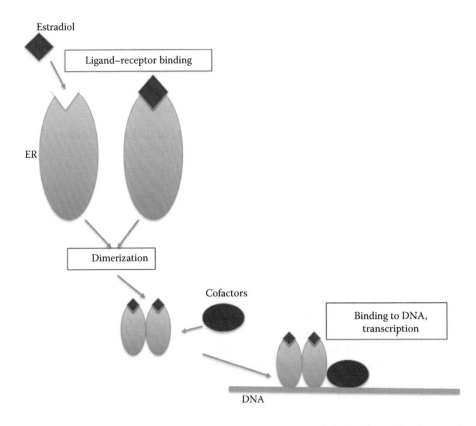

FIGURE 16.1 Diagram of estrogen binding to ER. In the classic genomic action of the ER, ligand binding to the receptor results in dimerization of the ligand–receptor complex and binding to DNA, together with cofactors. This leads to selective transcription of estrogen-related genes, or ERE.

sequences termed estrogen-response elements (ERE), and leads to selective regulation of target gene transcription (Katzenellenbogen 1996, O'Malley et al. 1968, Welboren et al. 2009). These targets include genes that are key in the proliferation, differentiation, survival, and angiogenesis of estrogen-responsive tissues, including the breast ductal epithelium (Welboren et al. 2009). ER-related genes also include ones that code for other potential therapeutic targets such as the progesterone receptor (PgR) and the insulin-like growth factor 1 receptor (IGFR-1). ER can also lead indirectly to gene activation through other transcription factors such as specificity protein 1 (Sp1), activating protein 1 (AP-1), or nuclear factor kappa β (NF-kB) (Marino et al. 2006, Welboren et al. 2009). Tissue specificity of estrogen action appears to depend in part on these cofactors as well as other coregulators that interact with the ER and the ERE to affect the pattern of gene transcription (Katzenellenbogen 1996, Welboren et al. 2009).

In the uterus, estrogens bound to ER stimulate endometrial growth and are critical in maintaining a functioning uterine-placental unit during pregnancy. Estradiol promotes new bone formation and is important in maintaining bone mineral density, especially in women (Hammond 1993, Riggs and Hartmann 2003). In the breast, estradiol promotes ductal epithelial cell proliferation and is a key component for stimulating lactation. Estrogens also have systemic effects, for example, a beneficial effect on serum lipids that has been linked to lower cardiovascular disease in premenopausal women compared with men and postmenopausal women (Bolt 1979).

Circulating levels of estrogens are variable, but tightly regulated in normal human physiology (Hammond 1993). The primary estrogenic agonist, estradiol, comes from two sources: (1) synthesis in the ovaries in premenopausal women and (2) conversion from adrenal steroids, largely through aromatization (and aromatase enzymes) (Figure 16.2) present in a variety of tissues, most notably fat, breast tissue, and breast cancers (Kuerer et al. 2001, Reed and Purohit 2001). Premenopausal levels of estradiol vary depending upon the time of day and the phase of the menstrual cycle, reaching levels as high as 500 pg/mL (1.7 nM) midcycle (Hammond 1993). During pregnancy, levels can be considerably higher. In postmenopausal women and men, levels are generally <30 pg/mL (0.1 nM). Because estradiol is lipophilic, there is nonspecific cellular uptake and estradiol is generally present in slightly higher concentration in tissues with greater fat content.

Estradiol is largely metabolized by the liver, both by oxidation to estrone and other related compounds, and

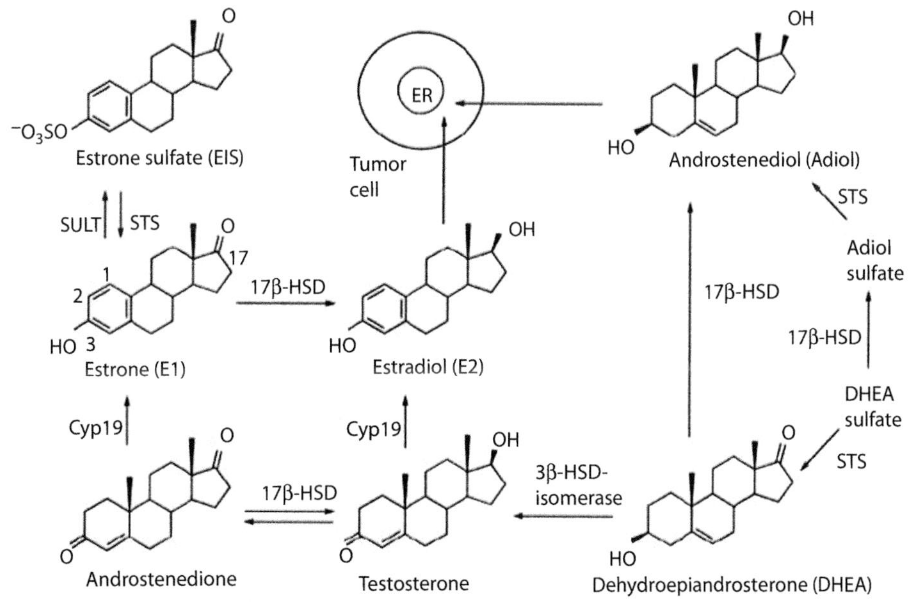

FIGURE 16.2 Estradiol metabolism. The diagram primarily focuses on metabolism leading to the generation estradiol, but also illustrates major metabolic pathways. The upper left part of the diagram shows the metabolism of estradiol to estrone and conjugation to estrone sulfate. (The reaction from estradiol to estrone is reversible, but not shown in this diagram.) The bottom part of the diagram indicates the conversion of androgens to estrogens via aromatase (Cyp19). SULT, sulfotransferase; 3-HSDisomerase, 3-hydroxysteroid dehydrogenase-C5,C4-isomerase; STS, steroid sulfatase; 17β-HSB, 17-hydroxysteroid dehydrogenase. (From Reed, M.J. et al. 2005. *Endocr Rev.* 26(2):171–202. With permission.)

by conjugation of estradiol, estrone, and other inter-mediates (Bolt 1979) that enter the enterohepatic circulation (Fink and Christensen 1981, Scharl et al. 1991, White et al. 1998) (Figure 16.2). Many tissues contain enzymes that allow rapid interconversion between estrone and estradiol. The conjugated estrogens are excreted into the bile and can be reabsorbed and also deconjugated by both intestinal enzymes and intestinal flora, resulting in enterohepatic circulation (Bolt 1979, Hammond 1993, Scharl et al. 1991). The oral administration of conjugated estrogen, with subsequent deconjugation and absorption, has been the main method for hormone supplementation in post-menopausal women. This dosing strategy provides relatively efficient use of estrogens while allowing regulation and variation in circulating plasma levels.

Circulating estradiol is largely protein bound with high affinity but low capacity to sex hormone-binding globulin (SHBG or SBP), and with low affinity but high capacity to albumin (Pan et al. 1985, Petra 1991). Most circulating estradiol is bound to SHBG, with the remainder bound to albumin (Pan et al. 1985). Binding to both SHBG and ER appears to be important for normal estrogen physiology, and also appears to be necessary for ER imaging agents (Jonson et al. 1999, Petra 1991, Tewson et al. 1999) (discussed in more detail in Section 16.2.2). One of the physiologic roles of SHBG appears to be regulation of estrogen metabolism (Plymate et al. 1990). Binding to SHBG, extraction by the liver, and reabsorption of conjugated estrogens in the small intestine all play important roles in regulating estradiol levels in normal physiology (Pan et al. 1985, Plymate et al. 1990).

16.1.2   The Target Protein, the Estrogen Receptor, and Estrogen–ER Interaction

The ER translates estrogenic signals into physiologic action. The ER is a protein receptor typically found in or close to the cell nucleus; however, some data suggest that it can also be present in the cytoplasm and associated with the cell membrane (Osborne et al. 2005a). The ER is part of the family of nuclear hormone receptors, which also includes the androgen receptor, PgR, glucocorticoid receptor, thyroid hormone receptor, and receptors for vitamin D and retinoic acid (Welboren et al. 2009).

ER alpha is a 64 kD protein that, in its unbound state, is complexed to HSP90 protein (Santen et al. 1990). Cloning of the ER increased our understanding of its structure and function (Greene et al. 1986). It has

a ligand-binding domain responsible for estrogen binding, a hinge region responsible for binding to HSP90 and for dimerization, and a DNA-binding domain responsible for the effect of activated ER on transcription (Marino et al. 2006). In the absence of estrogens, the ER exists as a single unit complex to HSP90. Estrogen binding to ER alpha results in dissociation from HSP90 and conformational changes leading to dimerization of ligand-bound receptors (Welboren et al. 2009). The binding of estradiol to ER alpha occurs in the ligand-binding domain, which is approximately at residues 200–300 from the C-terminus (Marino et al. 2006).

In the classic action of ER (Figure 16.1), estrogen binding leads to receptor phosphorylation and dimerization with another estrogen–ER complex, which in turn leads to the binding of the ligand–receptor dimer to specific regions of the ERE genes. Coactivators such as AIB1 interact with ER to facilitate conformational changes in the chromosomes that allow increased transcription of certain target genes (Anzick et al. 1997, Katzenellenbogen 1996, Osborne et al. 2001, Schiff et al. 2005). Activated ER can also lead to decreased transcription of some genes through core-pressors that can limit DNA transcription, often in antagonism to the coactivators. When they bind to the ER, ligands other than estradiol may engender different physiologic responses mediated, in part, by their effect on the coactivators and corepressors (Osborne et al. 2001). For example, the selective estrogen receptor modulator (SERM) drugs exhibit varying degrees of either ER agonist or antagonist behavior in different tissues, thought to be the result of differential actions of coregulators and corepressors in different tissues (Riggs and Hartmann 2003).

ER may also work through nonclassical mechanisms such as protein–protein interactions, leading to activation through transcription factors and to direct activation of transcription (Osborne et al. 2005b, Schiff et al. 2005) (Figure 16.3). Nongenomic effects of ER activation are proposed, such as activation of phosphatidylinositol 3 kinase (PI3k), and Akt. These nongenomic effects are thought to be mediated by ER localized to the plasma membrane or cytosol, with actions occurring more rapidly in response to activation than is likely by transcriptional activity (Linden and Mankoff 2010, Osborne et al. 2005b). These non-classical actions of ER may help explain correlation of estrogens with other growth factor receptor systems (Marino et al. 2006, Osborne et al. 2005b), which may mediate resistance to endocrine treatments of cancer,

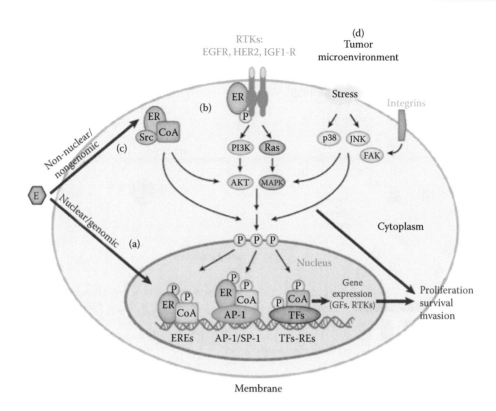

FIGURE 16.3 ER crosstalk. An illustration of nongenomic versus genomic actions of the ER. Nongenomic ER may be associated with the cellular membrane, rather than the nucleus in the genomic mode of action. Nongenomically mediated actions of ER likely occur, directly or indirectly, through interactions with other growth factor pathways, especially the EGFR/HER2 and IGFR pathways for breast cancer. (From Osborne, C.K. and Schiff, R. 2011. *Annu Rev Med.* 62: 233–247. With permission.)

in some instances. The nongenomic effects of ER on PI3k and Akt may moderate rapid increases in glycolysis of breast tumor cells in response to estradiol (Linden and Mankoff 2010), which may account for rapid increases in fluorodeoxyglucose (FDG) uptake in images after estradiol infusion in patients (Dehdashti et al. 2009).

The concentration of ER in tissue varies widely. Many tissues have low or nil ER expression. The highest levels are often found in the uterus, reaching 30 nM in the immature rat uterus and 3 nM in the midproliferative human uterus (Katzenellenbogen 1981, 1992, 1995). Expression in breast tumor also varies widely, ranging from <0.3 to 3 nM. The binding of estradiol to ER alpha is relatively high-affinity, estimated to have a K_d of 0.3 nM at 0°C, and likely lower at higher temperatures (Katzenellenbogen 1981, 1992, 1995). Dissociation of estradiol from ER alpha is slow and for imaging, the relatively rapid process of ligand binding and dimerization results in kinetic curves more consistent with trapping of the ligand in the tissue, as opposed to more classic membrane receptor competitive binding studies that measure reversible binding (Mankoff et al. 1997).

16.1.3 *In Vitro* Assays for the Estrogen Receptor

Measuring the level of ER expression is clinically important, in particular for breast cancer (Figure 16.4). Classically, this was done using competitive receptor-binding methods where structurally similar molecules compete for ER binding (McCarty et al. 1980). In this technique, fresh or fresh-frozen tissue is homogenized. Samples of the tissue homogenate are incubated with a fixed amount of ligand, typically ^{3}H-estradiol and varying concentrations of nonradioactive estradiol until the reaction is effectively at equilibrium, assuming that the amount of receptor is the limiting reagent. Nonbound (free) ^{3}H-estradiol and estradiol are separated using a variety of methods, classically using dextran-coated charcoal and the ^{3}H is counted. The amount of bound ^{3}H-estradiol in the competition assay is plotted as bound ligand on the ordinate and nonradioactive estradiol on the abscissa. Nonspecific binding is also accounted for. The quantity of ER, normalized to grams of tissue protein, is measured from the inflection point in the binding curve, yielding the quantity of ER in the preparation that can be

Chapter 16

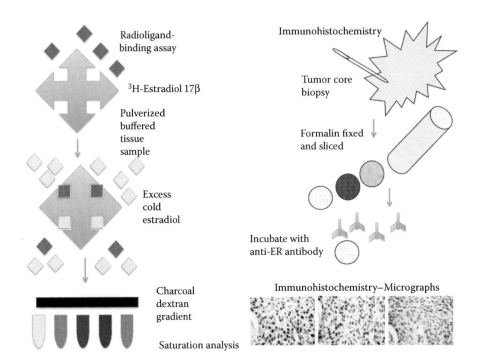

FIGURE 16.4 Illustration of *in vitro* assay methods for ER expression. On the left, the radioligand-binding approach is illustrated, where tissue is incubated with ³H-estradiol in the presence of varying levels of cold estradiol, and the quantity of ER in the sample is taken from the fractional-binding curves using a scatchard-like analysis. For immunohistochemistry, tissue is exposed to an antibody directed to the ER, which is stained using a second antibody directed against the anti-ER antibody. This results in brown nuclear staining, indicative of ER, in the background of counterstaining (blue) to show the tissue cellular structure.

determined from classic ligand-binding analysis or mathematical algorithms (Katzenellenbogen 1992, McCarty et al. 1980). This method has the advantage that it yields a direct measure of the moles of ER per unit tissue, providing a quantitative and reproducible measure of ER expression. Radioligand assays have been shown to be highly predictive of endocrine responsiveness, in proportion to the level of ER expression (Campbell et al. 1981). However, radioligand methods have the disadvantage that they require fresh or fresh-frozen tissue, making tissue sampling and transport more challenging, and they can suffer from interference from circulating estrogens (McCarty et al. 1980, Parl and Posey 1988). Furthermore, since tissue samples often contain a mix of tumor and normal tissue, and normal breast tissue also expresses ER, it may be difficult to separate normal tissue and tumor estradiol binding (Parl and Posey 1988).

In the current clinical practice for the assay of ER expression, radioligand-binding methods have been almost entirely replaced by immune-based assays, mostly by immunohistochemistry (IHC) (Allred et al. 1990, 2009, McCarty et al. 1980). In this method, tissue is exposed to an antibody preparation directed against ER epitopes. This is followed by a secondary reaction,

typically with another antibody, directed against the ER-directed antibody to produce specific tissue staining indicative of ER expression. IHC has the advantage that it works on fixed (archival) tissues, can directly visualize ER expression on tumor cells (vs. normal breast cells), and is not affected by circulating estrogens (Allred et al. 1990, 2009, McCarty et al. 1980). A number of studies have shown good correspondence between IHC-based methods and radioligand-binding assays (Allred et al. 1990, McCarty et al. 1985). IHC has the disadvantage that it is less quantitative and less quantitatively reproducible than radioligand binding, although some scoring systems (Allred et al. 1990) and image analysis methods (Lehr et al. 1997, McCabe et al. 2005) can improve reproducibility and provide semi-quantitative measures of the level of ER expression in the order of a 1–8 scale (Allred et al. 1990), where 8 corresponds to ~300 fmol/mg of protein.

16.1.4   Endocrine Therapy for Breast Cancer

The association of breast cancer with estrogen levels has led to several approaches to reduce tumor estrogen (Osborne et al. 2000, Santen et al. 1990). The first generation of these endocrine therapies uses molecules that essentially irreversibly bind to the ER and antagonize the action of estrogens (antagonists) (Osborne and Schiff 2000). The most widely used of these compounds is tamoxifen and fulvestrant.

A second approach has been the use of aromatase inhibitors (AIs) (Goss and Wu 2007). Aromatase is an enzyme that is found in several tissues, including fat and normal breast tissue, and converts steroids to estrogens, for example, testosterone to estrogen and androstenedione to estrone (Reed and Purohit 2001). AIs are molecules that block aromatase from making the conversion. There are two kinds of AIs, irreversible steroidal inhibitors, for example, exemestane, and reversible nonsteroidal inhibitors, for example, anastrozole and letrozole. Because AIs do not block estrogen production by the ovary, they are best suited for postmenopausal patients, where ovarian estrogen production has ceased (Goss and Wu 2007). Recent studies have shown the AIs to be superior to tamoxifen in postmenopausal patients, and they have become first-line treatment for postmenopausal patients with ER-expressing cancers (Buzdar et al. 2001, Mouridsen et al. 2001).

Hormonal therapies have changed the paradigm for breast cancer therapy but still have limitations. While the absence of ER expression reliably predicts that endocrine therapy will fail, for patients with ER-expressing tumors, only 50–75% of newly diagnosed patients will respond to endocrine therapy, and even fewer (~25%) of previously treated patients will respond (Osborne and Schiff 2011). These are some of the challenges where positron emission tomography (PET) ER imaging can help.

16.1.5   What Questions Should ER Imaging Address?

The primary clinical role of ER imaging to date has been in the application to ER-expressing cancers. Estrogens are established growth factors for endometrial and many breast cancers. Over 70% of breast cancers express ER, and estradiol and other estrogens provide a key stimulus for tumor growth and as target for endocrine-based therapy (Cigler and Goss 2007, Dunnwald et al. 2007, Li et al. 2003, Osborne et al. 1980, Pujol et al. 1994, Sledge and McGuire 1983). Imaging can serve several roles in ER-expressing cancers. It can provide a fairly specific means for tumor identification. Since relatively few normal tissues express ER, the identification of sites of aberrant ER expression in a patient with a known ER+

(ER-expressing) tumor can provide a basis for specific disease localization by PET or single-photon emission computed tomography (SPECT) imaging. However, tumor detection by this approach has limitations in that, especially in breast cancer, the expression of ER in the tumor may be heterogeneous, and ER imaging will miss tumor sites that have lost ER expression.

The most important role for ER imaging relates to ER as a therapeutic target. This is best established for breast cancer, where the presence or lack of ER expression in breast cancers is the primary determinant for selecting, and more importantly not selecting, endocrine therapy (Allred et al. 2009). Less than 5% of ER-negative tumors will respond to endocrine therapy, and therefore endocrine therapy is not considered for tumors that lack ER expression (Allred et al. 2009, McCarty et al. 1984). For this reason, assay for ER expression in tumor biopsy material is part of standard clinical care and is key in therapeutic decision making. ER imaging can play an especially important role in advanced or persistent tumors where ER expression may be spatially or temporally heterogeneous. The ability to measure regional ER expression in the entire burden of disease and to track changes over time offers potential for guiding endocrine therapy in a way not currently possible through biopsy-based assay methods.

In addition, the ability to assess regional estrogen–ER binding noninvasively in serial studies offers the capability of measuring the pharmacodynamics of endocrine therapy. One example is the ability to measure the extent of ER blockade by imaging before and after therapy with ER antagonists such as tamoxifen and fulvestrant (Linden et al. 2005, Mortimer et al. 2001). ER imaging can play a key role in clinical dose-finding for such drugs. A recent example where this approach could be helpful is for the pure antagonist fulvestrant. Recent studies have suggested that the initial recommended dosing was too low for optimal efficacy, supported by later studies showing improved efficacy at higher doses (Di Leo et al. 2010, Pritchard et al. 2010).

16.1.6   Particular Challenges of ER Imaging

The ligand–receptor interaction is a bimolecular chemical reaction. The concentration of the receptor in tumors is typically quite low; the level of ER expression in breast cancer is in the range of 3–100 fmol/mg protein (Campbell et al. 1981). Furthermore, as noted, receptor-specific ligands bind to the receptor with high affinity, subnanomolar, and with low ligand–receptor dissociation rates (k_{off}) (Krohn 2001). The combination of low receptor concentration and high ligand–receptor affinity leads to low overall capacity for ligand–receptor-binding. The high-affinity, low-capacity ligand–receptor-binding reaction presents a challenge for imaging in that the number of molecules that can contribute to the specific receptor image is small. This is distinct from imaging based upon an enzymatic reaction, such as glucose phosphorylation, where a single target enzyme molecule can lead to the metabolism and trapping of many probe molecules. Nonspecific binding of ligand to plasma proteins and nontarget tissues can also limit imaging agent delivery and contribute to nontarget image background. These issues make receptor imaging in general and ER imaging in particular, quite challenging. It is important to administer only very low molar amounts of ER-imaging probes. Even small molar quantities of the imaging agent may saturate the receptor, limiting the ability to visualize receptor expression and increasing the background of nonspecific binding (Katzenellenbogen et al. 1997), and could conceivably lead to physiologic consequences. As such, ER has been largely confined to radionuclide imaging (PET and SPECT), where it is possible to generate images with nanomolar to picomolar concentrations of the imaging probe.

It is important to note that the criteria for a good ER imaging agent are different than those for ER-targeting drugs. While selectivity for the drug requires an effect on the tumor in the absence of appreciable toxicity from non-target-tissue drug action, the requirement for high target uptake and low image background in imaging places a burden for radiopharmaceutical selectivity and background clearance than can be more stringent than for therapeutic drugs. For example, tamoxifen is a fairly selective and effective drug, even though its affinity for the ER is considerably less than estradiol (Osborne et al. 2000). Tamoxifen has been a highly effective therapeutic agent; however, its biochemical characteristics make it a challenging probe for ER imaging (Yang et al. 1994). The low affinity of tamoxifen for the receptor relative to estradiol suggests that labeled tamoxifen would not be a good radiopharmaceutical for imaging the ER. Its utility would be likely limited to the ability to measure the pharmacokinetic and transport of SERMs.

16.2 Imaging the ER: Developmental Work in ER Imaging Radiopharmaceuticals and Chemical Identity of the Probe

16.2.1 Early Work in the Development of Probes for ER Imaging

Considerable efforts have gone into the development of radiopharmaceuticals for ER imaging, as reviewed in Katzenellenbogen (1995). Early efforts to develop a labeling approach focused on steroids labeled with halogens, largely iodine and bromine isotopes (Figure 16.5). Katzenellenbogen and colleagues generated a series of compounds labeled with [77]Br, including close steroid analogs of estradiol and nonsteroid compounds such as hexestrols. These efforts demonstrated the feasibility of imaging ER expression *in vivo* in animals and some compounds, for example, [77]Br-16α-17β-estradiol showed interesting early results in patients (Katzenellenbogen et al. 1981, McElvany et al. 1982). There were parallel efforts to develop ^{123}I-labeled compounds that included 16α-[123]I-17β-estradiol that showed promising results in animals (Pavlik et al. 1990) and [123]I-*cis*-11β-methoxy-17α-iodovinylestradiol (Z-MIVE) (Bennink et al. 2001), and some promising results in humans (Rijks et al. 1998, Zielinski et al. 1989). The Sherbrooke research group evaluated multiple structure–activity studies of halogen substitution on estradiol (Ali et al. 1991, 1993). Radioiodinated 16 alpha estradiols combined with fluorine substituted at the 2 and 4 positions and found that the 4-fluorosubstitution retained affinity for ER and the 2-F substitution did not. In general, 17 alpha radioiodoestradiols had lower ER affinity than the 16 alpha estradiols.

16.2.2 ^{18}F-Labeled Compounds

Despite considerable work with other halides for ER imaging (Katzenellenbogen et al. 1975, Katzenellenbogen 1981a, 1995), the emergence of PET, especially using ^{18}F, provided strong impetus for ^{18}F-labeled compounds for PET ER imaging (Katzenellenbogen et al. 1997). Fluoride is a small halogen that can be substituted in several positions of the estrogen while preserving binding affinity to both ER and SHBG (Kiesewetter et al. 1984, VanBrocklin et al.

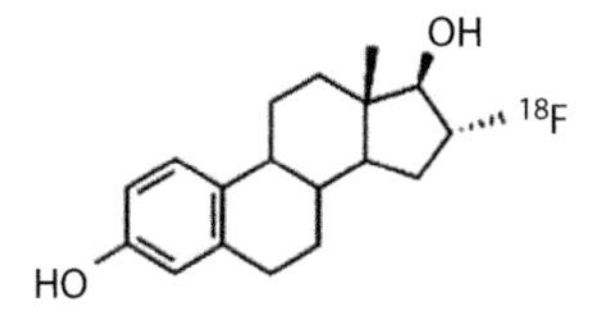

16α-[18F]-Fluoro-17β-estradiol

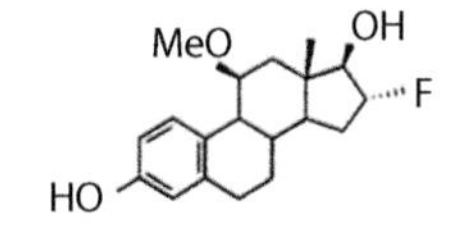

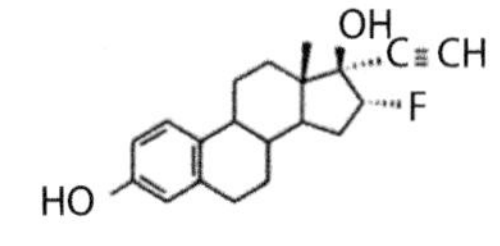

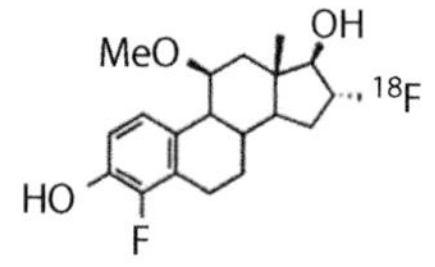

11β-Methoxy-16α-
[18F]-fluoro-17β-estradiol

16α-[18F]-Fluoro-17α-
ethynyl-17β-estradiol

4-Fluoro-11β-methoxy-
16α-[18F]-fluoro-17β-estradiol

FIGURE 16.5   The structure of FES and some of the alternative-fluorinated estradiols suggested for proposed for imaging ER.

1992). Furthermore, ^{18}F has a sufficiently long half-life (110 min) to permit multistep synthesis of ligands (Kiesewetter et al. 1984, Lim et al. 1996) and uptake by target tissue and elimination by nontarget tissue during imaging (Katzenellenbogen et al. 1997, Mankoff et al. 1997), and the use of PET permits quantitative imaging of regional receptor binding. Elegant studies from the laboratories of Katzenellenbogen and Welch (Katzenellenbogen 1992, Kiesewetter et al. 1984, Pomper et al. 1990a) explored a variety of label positions and substitutions on the estradiol molecule. The classic study of Kieswetter (Kiesewetter et al. 1984) tested binding to ER for a range of compounds, including estradiol and hexestrols labeled at various positions. This analysis identified a number of compounds with relative binding to ER that was close to, and in some case higher than, estradiol. Of these compounds, 16α-[18F]-fluoro-17β-estradiol (FES) had the combination of selective uptake in immature rat uteri, the highest uterus to background ratio, and the ability to be blocked by an excess of cold estradiol—all properties making it attractive as an ER-imaging agent (Figure 16.5).

Further studies sought to identify compounds with more attractive properties for imaging, including higher binding to ER, and possibly slower metabolism. Pomper tested the effect of different substitutions on binding and identified some compounds with higher ER binding than FES, especially 11β-methoxy or 17 alpha-ethynyl compounds (Pomper et al. 1990b). In many cases, these had higher ER binding by *in vitro* assays, but not always by *in vivo* studies. One compound, ^{18}F-labeled moxestrol ([18F] betaFMOX), was developed with the goal of decreased metabolism and increased ER binding for ER, and preclinical studies

in vitro and in rats demonstrated better ER-binding *in vitro* and increased uterine uptake in immature rats for FMOX compared to FES (Jonson and Welch 1998, Jonson et al. 1999). However, this compound performed poorly in human studies. The hypothesis to explain these findings is that poor binding of FMOX to the steroid transport protein, SHBG, likely limited its utility in humans. Rats lack SHBG (Petra 1991), and therefore, FMOX was an effective imaging agent for rats; however, in humans, it was performed poorly compared to FES, which has modest SHBG binding (Pomper et al. 1990a). In this case, a change to the imaging molecule that promoted increased ER binding unfavorably altered the binding to SHBG, resulting in a compound with poorer performance. This example illustrates the demanding nature of radiopharmaceutical design for tumor receptor imaging and the need for validation at each step of development from lab bench to bedside and the limitation of rat studies in this context.

Recently, after considerable preclinical evaluation, human studies of 4-fluoro-11β-methoxy-16α-[18F]-fluoroestradiol (4F-M[18F]FES) have been reported (Beauregard et al. 2009, Benard et al. 2008). The rationale for studying this radiopharmaceutical in humans is that in rodent studies compared with FES, it showed equal to slightly better uptake in tumors, greater selective uptake in uterus, and lower background (Benard et al. 2008). However, the binding of 4F-M[18F]FES to SHBG is poor. In humans, comparison of normal dosimetry suggests that the 4F-M[18F]FES has much larger clearance through the gallbladder with greater hepatobiliary excretion but lower muscle uptake than FES (Beauregard et al. 2009). Tumor comparison data remain to be reported and the future utility of this molecule for human cancer remains to be tested.

16.2.3 16α-[¹⁸F]-Fluoro-17β-Estradiol

Although a variety of ER-imaging agents have been tested, the most successful ER-imaging radiopharmaceutical to date is FES (Cummins 1993, Katzenellenbogen et al. 1997) (Figure 16.5). FES has binding characteristics similar to estradiol for both the ER and the transport protein, SHBG (Kiesewetter et al. 1984, Tewson et al. 1999). Its binding affinity for ER is about 80% that of estradiol, and its *in vitro*-binding affinity for SHBG is ~10% that of estradiol (Kiesewetter et al. 1984, Pomper et al. 1990a), although *in vivo* data, presented later in this chapter, suggest that FES and estradiol may have similar SHBG binding. Preclinical and clinical studies for FES are reviewed in subsequent sections.

The original synthesis published by Landvatter and Kieswetter (Kiesewetter et al. 1984, Landvatter et al. 1983) used a precursor with triflate leaving groups at the 4 and 17 positions and produced a 43% radiochemical yield and specific activity of 166 Ci/mmol in a total synthesis time of 75–90 min. Although this synthesis was used for many of the early animal and human studies of FES, it was somewhat difficult to reproduce and used a fairly unstable precursor. Lim and Tewson (Lim et al. 1996) developed a synthesis using a relatively stable 3-O-methoxymethyl-16,17-O-sulfuryl-16-epiestriol precursor. Nucleophilic substitution fluorination is followed by acid hydrolysis to remove the remaining sulfate and protecting groups. This reported method yielded robust, reproducible synthesis, but was time consuming and required two preparative high-performance liquid chromatography (HPLC) steps. Romer and colleagues refined the Lim synthesis to develop a "one-pot" synthesis requiring only a single preparative HPLC, which was well suited for more automated synthesis (Romer et al. 2001). This approach has been further developed by Link (Peterson et al. 2008) adapting commercial synthesis units, and is the basis for a current US NCI IND (US IND 79,005).

16.3 Characterizing the Probe: Measuring Specific Activity

16.3.1 Why Is Specific Activity Important for ER Imaging?

The physiologic actions of estrogen/ER occur at low concentrations of both the agonist ligand and the receptor. Circulating estradiol levels in plasma in premenopausal women vary over the course of the menstrual cycle, reaching levels of 1.7 nM midcycle. Postmenopausal plasma levels are typically 0.1 nM or less in postmenopausal women. The concentration of the receptor is typically quite low; the level of ER expression in breast cancer is in the range of 3–100 fmol/mg protein (Campbell et al. 1981). This is distinct from other commonly used PET agents, for example, FDG, where physiologic levels of the traced substance (glucose for FDG) are at micromolar or higher levels and the target enzyme is abundant. The low levels of normal estrogens and ER impose a fairly strict upper limit on the mass of an ER-imaging agent for two reasons (Katzenellenbogen 1995): (1) The concentration of circulating estrogen associated with the use of a steroid tracer as an imaging agent should be below physiologic steroid levels to avoid physiologic effects by the imaging agent. (2) High mass of the radiopharmaceutical may occupy a considerable fraction of ER, falsely lowering the apparent binding of the imaging agent to ER. This reduces the sensitivity for visualizing ER-expressing tissue and leads to underestimates of ER expression. Based upon these considerations and expected levels of ER expression in tissues, Katzenellenbogen and colleagues (Katzenellenbogen 1995) estimated that radiopharmaceutical specific activities of >1000 mCi/μmol are needed, leading to a limit for injections of 20 nmol or less for typical 5 mCi of radiopharmaceutical for PET imaging. Recent studies indicating the effect of mass on uptake are described in more detail later in this section. Consideration of the injected mass leads to the need to assay radiopharmaceutical mass, and thus specific activity, for each batch of radiopharmaceutical for ER imaging.

16.3.2 Competitive-Binding Assays to Determine Specific Activity

The original and most highly validated assays for measuring radiopharmaceutical effective mass are based upon classic competitive-binding methods as described in Section 16.1.3. This approach has been widely used to test imaging compounds developed for ER-imaging (Katzenellenbogen 1995) and continues to be the gold standard for assays of specific activity for ER-imaging agents (Katzenellenbogen 1995). However, the time, labor, and complexity of the assay make it impractical for "on-line," rapid measures of specific activity needed for patient imaging with short-lived radiopharmaceuticals.

16.3.3   Measurement of Specific Activity Using HPLC–Mass Spectrometry

Specific activity is the amount of radioactivity per mass of radiotracer. Due to contamination with atoms of the same element as the radioelement used, for most PET radiopharmaceuticals, there are 100 or more nonradioactive atoms of the pharmaceutical for each radioactive atom. The typical method to determine specific activity of PET radiopharmaceuticals is through the use of HPLC using UV absorbance detection. The specificity depends upon the HPLC to achieve separation from impurities and gain specificity of the signal. Most groups report "effective" specific activity. This is an acknowledgment that radioactivity can be measured accurately and precisely using a dose calibrator for a known amount of radioactivity but UV absorbance may include signal from not only the radiopharmaceutical but also from other impurities in the product that have UV absorbance, including the "pure" reagents used for formulating a product dose. In many cases, this is sufficient because the administration of up to an mg of unlabeled material does not affect the biodistribution of the radiopharmaceutical. However, for many receptor–ligand studies, as stated previously, the specific signal of receptor binding will decrease if the receptor is saturated with too much ligand circulating through the body.

UV absorbance has good but not high sensitivity for estrogens at the <1 mcg/mL level. At the low end of the regression curve, the ability to estimate the concentration of a sample becomes quite poor. Another challenge is that UV absorbance detection has no specificity for different steroid side-products from the radiosynthesis, and so even with a product signal well within the precise region of a regression curve, UV-absorbing impurities or side-products will be included as part of the mass injected. Mass spectrometry (MS) has high sensitivity and is both accurate and precise at the level of <0.01 mcg/mL FES dose. MS is also specific and will measure only the mass of FES and does not include any miscellaneous impurities or side-products. However, there may be steroid by-products of the synthesis that would compete with FES for binding to SHBG or ER. Thus, MS, while more accurate, may not provide a good measure of potential efficacy of the radiopharmaceutical as more "biologic" assays such as the competitive-binding assay (Katzenellenbogen 1995). For this reason, when the FES synthesis, purification, and quality assurance methods are developed, it is essential that the final product as assayed by the quality assurance chemical purity and specific activity test be tested against the biological activity for binding to the ER.

We therefore compared HPLC-MS estimates of FES mass to measures of estradiol equivalent mass by a competitive estrogen receptor-binding assay. For these assays, samples of FES made by standard radiopharmaceutical production methods, as described in Section 16.2.3, were used. For the binding assay, purified estrogen receptor α (PanVera, now Invitrogen) was incubated with ^{3}H-estradiol (Perkin Elmer) and either estrone at different concentrations or the different FES products as a competitor. After incubation, hydroxylapatite (HAP) was added and the bound ER separated with the HAP by centrifugation. The resuspended washed pellet was counted for tritium. Negative and positive controls were also run in a modification of the method of PanVera. An example of data from a set of competitive-binding experiments is shown in Figure 16.6a. Nonlinear regression was used to determine 50% of maximal competition (IC_{50}) (Figure 16.6b). These figures show that the assayed FES competed for the ERα at about 50% of the level of estradiol. This is consistent with previously reported values (see Section 16.4.1) (Yoo et al. 2005).

16.3.4   How Much Mass Can Be Tolerated in Human ER Imaging with FES?

Early studies with FES demonstrated, as expected, that an excess of estradiol blocks FES uptake into ER-rich tissues such as immature rat uterus (Kiesewetter et al. 1984, Mathias et al. 1987, Pomper et al. 1990a). However, measurements were needed to establish how much injected mass could be tolerated without limiting FES uptake in ER-expressing tumors in patients, especially in tumors, where expression levels are typically lower than in normal uterus. Through a series of calculations based upon specific and nonspecific binding, and estimated concentration of ER in normal uterus and ER-expressed breast cancer, Katzenellenbogen estimated that specific activities <1000 Ci/mmol would lead to reduced FES uptake (Katzenellenbogen 1992). Specific activity was also studied in animal

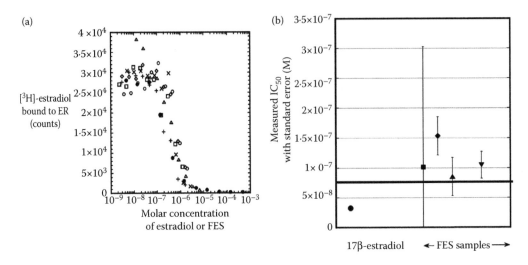

FIGURE 16.6 For receptor-binding studies, it is important that the specific activity, radioactivity per mass of the radioligand, be known accurately. Some analytical techniques may not accurately assess the mass of the compound of interest due to additional compounds being included in the analysis. (a) Raw data from competition assays for ERα for estradiol (solid circles) and several FES product batches (open symbols) plotted as the molar concentration of the standard estradiol or the molarity of the FES as measured by HPLC-MS. (b) A single validation for the accuracy of the analytical technique by assessing measured mass of FES with varying amounts of standard added versus the biological effective mass. If the measured mass is incorrect, a competitive-binding assay will show that the effective mass activity is different from that measured by HPLC/UV or mass spectrometry. When IC$_{50}$ (50% inhibition) values for a set of FES runs and an estradiol standard are plotted (mean and 1 SD) against the measured mass by HPLC-MS, the IC50 is greater than for estradiol (b). This is because FES has a lower affinity ~0.54 (Yoo et al. 2005) for binding to the estrogen receptor (alpha) than does estradiol. From the literature, the pure FES would be expected to be at the line on the graph. None of these samples are statistically different from that expected for a pure sample, and the analytical technique for measuring FES mass is valid. However, one sample has high variability and should be repeated if possible and the second value is high which is the type of outcome that would be seen if there was another chemical in addition to FES or and estrogenic compound in the product and further tests using dilutions, including various amounts of cold compound can determine this.

models, with similar conclusions (Benard et al. 2008). For a typical 5 mCi injection, a 1000 Ci/mmol limit implies that a maximum injected mass of 5 nmol leads to a peak concentration in a typical 56 kg female of 1 nM (Mankoff et al. 1997). This, by analogy to calculations and animal studies, led to the concern that FES imaging might not work in premenopausal females, where plasma levels of estradiol can reach close to 2 nM.

In humans, it is difficult to assess the impact of specific activity. The most direct method of studying the effect of injected FES mass would require paired studies and without the coinjection of "cold" FES. This would require studies on two difference days, to allow clearance of the ^{18}F label, and microgram quantities of cold FES, both of which present considerable practical challenges. Some studies have, however, provided an indication of the effect of FES mass and circulating estrogens on FES uptake. For example, studies failed to show that FES uptake is lower in premenopausal versus postmenopausal patients. A comparison of FES uptake with *in vitro* assay that include both pre- and postmenopausal females showed good correlation across both groups (Peterson et al. 2008). This study indirectly suggested that 1000 Ci/mmol might not necessarily be a strict lower limit for FES-specific activity. We have recently tested the association of FES uptake and tracer-specific activity and injected mass by statistical methods in a large number of patients ($N \sim 300$) with ER-expressing tumors (Peterson et al. 2010). This analysis showed no effect of injected mass for injections corresponding to specific activities of >1000 Ci/mmol, as predicted by Katzenellenbogen. However, below 500–1000 Ci/mmol, the effect of lower-specific activity (more injected mass) was small, only an estimated average 5–10% decline per factor of 2 lower specific activity. Results suggested that at a specific activity of <500–1000 Ci/mmol, injected mass of >0.2 nmol/kg could lead to lower FES uptake, but that the effect of a larger injected mass is likely to be small. Overall, these results suggest that the specific activity of FES should be assayed prior to use in humans, but that the effect of low-specific activity on FES uptake in humans in less than might have been predicted on the basis of animal studies.

16.4   *In Vivo* Specificity and Biochemistry of the Probe: Preclinical and Human Studies Characterization and Validation Studies

16.4.1   Cell- and Receptor-Binding Studies with FES

Competitive-binding assays using cytosolic preparations of rat or calf uterus have been used to measure the binding affinity of FES relative to estradiol. This yielded relative-binding affinity (RBA), expressed as (K_d for estradiol)/(K_d for FES) × 100 equal to 80 (80% binding affinity of estradiol) (Katzenellenbogen 1992, Kiesewetter et al. 1984). To measure the affinity of FES to SHBG relative to estradiol, competitive-binding assays were also performed using third trimester human plasma as a source of SHBG (Katzenellenbogen 1992, Kiesewetter et al. 1984). SHBG has considerable species variation, making it necessary to obtain a human source of SHBG for a valid test. This showed an FES RBA for SHBG that was 9.5% of the RBA for estradiol; however, this result was challenged somewhat by the human studies outlined in Section 16.4.4.

16.4.2   *In Vivo* Biodistribution in Animals

Early studies of FES uptake and biodistribution were performed using immature female rats. Uterine binding in immature female rats, where concentration of ER can be as high as 30 nM, serves as a valid and readily available measure of FES uptake into ER-rich tumors. These studies showed that FES had prolonged retention in rat uteri with high target–background ratios (uterus-to-blood 39 ± 16 at 1 h after injection (Kiesewetter et al. 1984). Injection of an excess of estradiol blocked FES uptake into rat uteri, confirming the specificity of binding and providing an estimate of the relatively modest level of nonspecific binding (Katzenellenbogen 1992, Kiesewetter et al. 1984, Pomper et al. 1990a). Studies extracting FES bound to uterus after injection confirmed that the radioactive species bound to uterine ER was FES and not a labeled metabolite (Mathias et al. 1987). Biodistribution studies in the rat also showed some nonspecific binding in lipid-rich tissues, high uptake in liver, and excretion in the urine (Mathias et al. 1987). It should be noted, however, that the rat model can be misleading. Rats lack SHBG and, instead, alpha fetal protein (AFP) serves as the major protein carrier of estradiol in rat plasma. Thus, the kinetics and biodistribution of estrogen ligands may be different in rat than in human. This is believed to be an important factor for some compounds that looked promising in rat models, but did not perform well in humans, including the FMOX studies described in Section 16.2.2 (Jonson and Welch 1998, Jonson et al. 1999).

16.4.3   Metabolism in Animal Models

The studies of Mathias (Mathias et al. 1987) examined the metabolism of FES in rats containing carcinogen-induced tumors and examined the nature of labeled species. These studies showed that within 30 min after injection, <20% of circulating radioactivity was in the form of intact FES. Extraction of radioactivity from tissue showed that labeled metabolites of FES predominated in the blood and in the muscle at 1 h after injection, but that FES was the dominant-labeled species in the uterus. However, there was some decline in uterine uptake at later times, suggesting either release of FES or transient nonspecific binding of metabolites. Radioactivity taken from the blood of rats at 1 h after injection was almost all in the form of FES metabolites. Radioactivity obtained from the 1 h blood samples was re-injected into other rats and was not specifically taken up in the uterus, suggesting that the radioactive metabolites do not bind to ER.

Scharl (Scharl et al. 1991) examined the metabolism and biodistribution of ^{125}I-16α-iodo-17β-estradiol, expected to be similar to that of FES, in a swine model. These studies found water-soluble metabolites labeled with ^{125}I in the blood and the bile, identified as sulfate or glucuronide conjugates of the parent compound. When radioactivity recovered from the bile was reinjected and followed in the intestines, the radioactive species had been nearly completely extracted from intestinal contents by the time they reached distal colon, suggesting very efficient enterohepatic circulation.

16.4.4   Biodistribution, Kinetics, and Metabolism in Humans

Mankoff and Tewson studied the clearance and metabolism of FES in breast cancer patients (Mankoff et al. 1997). As in animals, FES was rapidly metabolized, and by 20 min, <20% of circulating radioactivity was in the form of intact FES (Figure 16.7). There was rapid initial blood clearance of FES, with a plateau in decay-corrected radioactivity concentration after 30 min, largely due to a nearly constant level of labeled metabolites.

Chapter 16

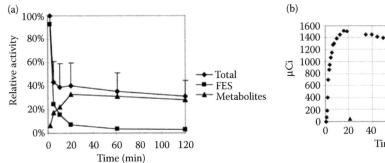

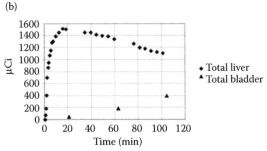

FIGURE 16.7 FES metabolism in humans. (a) A plot of percent of circulating activity present as FES versus labeled metabolites shows rapid metabolism of FES. (b) A plot of total liver and bladder radioactivity over time in a single patient, obtained from PET images, shows approximately equal release of FES metabolites from the liver and clearance by the kidneys, results in a relatively constant level of total blood radioactivity after initial clearance. (From Mankoff, D.A., Tewson, T.J., and Eary, J.F. 1997. *Nucl Med Biol.* 24(4):341–348. With permission.)

There was rapid initial uptake in liver and then slower clearance. Imaging showed that the rate of clearance from the liver, likely indicative of the release of labeled metabolites, was approximately the same as the rate of clearance into the urine (also labeled metabolites, see below), leading to an approximately constant level of labeled metabolites in the blood. Uptake in ER-expressing tumors also reached a relatively constant level by ~30 min after injection. These results suggested that imaging times as early as 30 min postinjection were feasible and that imaging at later times after injection were not likely to lead to significantly increased target-to-background ratios in humans.

Analysis of radioactive species in blood and urine in humans by HPLC largely showed two-labeled species in the blood and urine of patients injected with FES, thought to be most likely sulfate and glucuronide conjugates (Mankoff et al. 1997). Treatment of urine and blood by acid hydrolysis eliminated the peak thought to be sulfate, and treatment of urine with glucuronidase eliminated the other peak, confirming it to be a glucuronide. Interestingly, almost all of the label recovered after glucuronidase appeared to be FES itself, rather than an oxidation product, which might have been expected by analogy to estradiol metabolism (Bolt 1979). While the difference could reflect differences in the metabolism of FES and estradiol, it may simply relate to the timescale of the FES metabolism studies (<2 h) versus published estradiol metabolism studies (hours to days).

16.4.5 Biodistribution and Radiation Dosimetry in Patients

Human biodistribution studies showed, as expected, high initial uptake in liver and excretion into the bile with high uptake in the gallbladder and small intestine (Mankoff et al. 2001). As in the swine studies, radioactivity was not seen in the distal colon in patients, confirming efficient enterohepatic circulation. Uptake was seen in the normal uterus and some nonspecific uptake could be seen in lipid-rich tissues, but only at low levels. Radiation dosimetry estimates show that organ doses with FES PET are comparable with those associated with other commonly performed nuclear studies and potential radiation risks are well within acceptable limits. The effective dose equivalent is 0.022 mSv/Bq (80 mrem/mCi) and the organ that received the highest dose was the liver at 0.13 mSv/Bq (470 mrad/mCi) (Mankoff et al. 2001). The recommended injection is 6 mCi (222 MBq) or less.

16.4.6 Protein Binding

Tewson (Tewson et al. 1999) studied FES protein binding in breast cancer patients. This study obtained blood samples and used concentrated acetic acid to specifically precipitate SHBG (Petra 1991). This was compared with binding in the presence of an excess of dihydrotestosterone (DHT) to identify nonspecific protein binding. This study showed that an average of 45% of FES in circulating plasma is bound to SHBG with a range of ~30–55%, and with the remainder nonspecifically and more weakly bound to albumin (Tewson et al. 1999). The labeled metabolites had relatively little SHBG binding. The transfer of FES from SHBG to albumin-binding appeared rapid. There was some concern that FES could locally saturate SHBG at the time of FES injection, and it was noted on images that FES could bind to the venous vessel wall at the time of infusion, but only in the peripheral vein, stopping once the tracer reached the central circulation and the higher quantity of SHBG. With this consideration in mind, the recommended infusion time for FES was at least 2 min.

In the studies of FES protein binding in patients, the natural variation in SHBG concentration led to variations in the fractional binding of FES to SHBG in circulating plasma; an expected curvilinear relationship of fractional binding to total blood SHBG concentration was observed. Fitting this curve to a stoichiometric model of SHBG versus albumin binding, using the measured albumin and SHBG concentrations, and assuming a K_d value for FES binding to albumin similar to the published value for estradiol and albumin, it was possible to estimate a K_d for FES–SHBG binding. This approach yielded a relative affinity for FES to SHBG much closer to the published value for estradiol than the 9.5% found *in vitro* competitive-binding assay (Kiesewetter et al. 1984, Pomper et al. 1990a). This suggests that the binding of FES to SHBG *in vivo* in patients may be quite similar to estradiol, making FES an overall excellent mimic of estradiol binding and kinetics in humans.

16.4.7   Toxicity Studies

There have been extensive toxicity studies of estradiol (reviewed in (FES US IND 79,005). These studies showed toxicities only at estradiol concentrations that were 2–3 orders of magnitude higher than those encountered in FES imaging in patients, and only after weeks to months of exposure. For FES, a 6 mCi injection typically leads to a 1 µCi/mL peak concentration and an activity concentration of 0.05 µCi/mL by 1 h after injection (Mankoff et al. 1997). For a typical specific activity of 1000 Ci/mmol, this corresponds to peak blood FES levels of 290 pg/mL (1 pmol/mL) and 15 pg/mL (50 fmol/mL) at 60 min. Toxicity studies for estradiol revealed toxicity only after chronic exposures of 2 mg/mL or greater. Genotoxicity and mutagenicity were seen only after chronic exposure to concentrations of 25 µg/mL or greater (Wheeler et al. 1986). Therefore, toxicity or genotoxicity from FES used for PET imaging is considered highly unlikely.

Additional toxicity studies were undertaken as part of the submission of an IND for FES sponsored by the US National Cancer Institute (NCI) Cancer Imaging Program (CIP) (IND 79,005, Jacobs; IND 101202, Link). A limited study of the toxicity of FES was carried out as a joint undertaking involving the

PET radiochemistry program at UW, the Clinical Monitoring Research Program of the Cancer Imaging Program at NCI, and RTI International, a contract-testing laboratory. From this study, the no-observed-effect-level for intravenous administration of fluoroestradiol to rats for 14 consecutive days was >51 µg/kg/day. (This should be compared to a maximal single dose of 5 µg for a typical 56 kg woman for an FES PET study.) The studies included a 14 day intravenous repeat-dose toxicology study in Sprague–Dawley rats, nominally 6–8 weeks old at initiation of dosing. Four groups were compared: vehicle only, fluoroestradiol at 13 and 51 µg/kg and, when appropriate, cyclophosphamide at 30 mg/kg as a positive control. There were no statistical differences in toxicity assessed at pathology between the vehicle control and the two FES dose levels. This included studies of bone marrow, showing no significant effects on blood-forming elements compared to vehicle control. In addition to the rat studies, the mutagenic potential of FES was tested by measuring its ability to induce reverse mutations at selected loci of several strains of bacteria, the Ames test; where fluoroestradiol was negative in the Bacterial Reverse Mutation Assay. Fluoroestradiol was also tested in the L5178Y/TK$^{+/-}$ Mouse Lymphoma Mutagenesis Assay in the absence and presence of Aroclor-induced rat liver S9 and fluoroestradiol was also negative for mutagenic effects by this tests. The potential for effect on cardiac rhythm was also tested using cell lines transfected with the cardiac potassium channel, human ether-á-go-go-related gene (hERG). FES showed mild inhibitory effects at 8 ng/mL but this is 10 times the maximal level expected in human studies. In summary, studies undertaken as part of an NCI-sponsored IND for FES showed no significant toxicity or mutagenicity and only very mild effects on cardiac potassium channels at concentrations significantly higher than those encountered in human imaging. These studies confirm the safety of FES for PT imaging.

The safety of FES has also been supported by experience in published human imaging studies (Table 16.1). To date, results of FES PET imaging for over 500 patients have been reported in published studies or regulatory documentation, with no reported adverse events (IND 79,005, Jacobs; IND 101202, Link). A review of clinical records for over 250 patient studies at our center showed no reported toxicities or changes in liver function, renal function, or blood counts as a result of FES PET.

Chapter 16

Table 16.1   Published Human FES Studies

Clinical Condition	n^a	MBq Injected	Specific Activity	μmoles Injected	Reference
Primary endometrial cancer	19	185 MBq	2700–5400 mCi/μmol	0.001–0.002 μmol	Tsujikawa et al. (2011)
Advanced primary or metastatic breast cancer	51	222 MBq	Not reported	Not reported	Dehdashti et al. (2009)
Suspicious endometrial lesions	31	185 MBq	2700–5400 mCi/μmol	0.001–0.002 μmol	Tsujikawa et al. (2009)
Primary of metastatic breast cancer	17	222 MBq	≥1000 mCi/μmol	≤0.006 μmol	Peterson et al. (2008)
Suspicious endometrial lesions	38	185 MBq	2700–5400 mCi/μmol	0.001–0.002 μmol	Tsujikawa et al. (2008)
Healthy volunteers (uterine imaging)	16	185 MBq	≥297 mCi/μmol	≤0.017 μmol	Tsuchida et al. (2007)
Metastatic breast cancer	20	300–400 MBq	48–350 mCi/μmol	0.01–0.10 μmol	Kumar et al. (2007)
Metastatic breast cancer	47	222 MBq	≥1000 mCi/μmol	≤0.006 μmol	Linden et al. (2006)
Primary and metastatic breast cancer	49	56–296 MBq	1000–2000 mCi/μmol	0.0015–0.004 μmol	Mankoff et al. (2001)
Metastatic breast cancer	40	222 MBq	Not reported	0.006 μmol	Mortimer et al. (2001)
Primary and metastatic breast cancer	18	≤222 MBq	Not reported	≤0.006 μmol	Tewson et al. (1999)
Metastatic breast cancer	11	222 MBq	Not reported	0.006 μmol	Dehdashti et al. (1999)
Primary or metastatic breast cancer	15	≤222 MBq	Not reported	≤0.006 μmol	Mankoff et al. (1997)
Meningioma	6	148–296 MBq	4.3–11.1 Ci/μmol	Not reported	Moresco et al. (1997)
Primary or metastatic breast cancer	43	222 MBq	Not reported	0.006 μmol	Mortimer et al. (1996)
Primary or metastatic breast cancer	53	222 MBq	Not Reported	0.006 μmol	Dehdashti et al. (1995)
Metastatic breast cancer	16	222 MBq	Not reported	≤0.006 μmol	McGuire et al. (1991)
Primary breast cancer	13	92.5–222 MBq	Not reported	≤0.006 μmol	Mintun et al. (1988)

Note: The table includes studies published through the end of 2010, excluding case reports.

ᵃ Patients are reported in more than one publication in many cases.

16.5   Clinical Results: Patients Studies with the Probe

16.5.1   Validation and Approaches to Quantification

FES uptake has been validated as a measure of ER expression in breast tumors (see Figure 16.8). Mintun et al. (1988) showed a correlation between FES uptake in the primary tumor measured on PET images and the tumor ER concentration measured *in vitro* by radioligand binding after excision in 13 patients with primary breast masses. In this study, FES uptake was quantified from static images taken ~90 min after injection, using the %ID/g measure. This measure is similar to an standardized uptake value (SUV) (see below), but without compensation for variable patient weight.

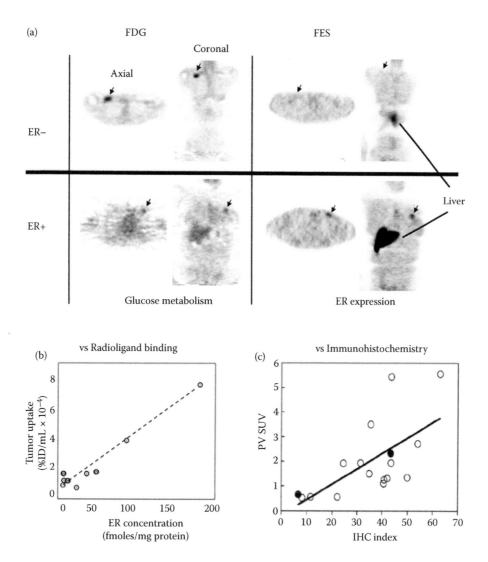

FIGURE 16.8 FES PET as an indicator of ER expression. (a) Two patients with ER-negative recurrent breast cancer (top) and primary ER+ breast cancer (bottom) in both axial and coronal views. Dark areas indicate increased uptake. The top patient had recurrent supraclavicular nodal disease from an ER+ primary tumor, with FDG uptake, but no FES uptake (small arrows). The recurrent tumor was ER-negative by IHC of biopsy of the lesion. The bottom patient had a new primary tumor with modest FDG uptake but high FES uptake (small arrows) and was highly ER+ by IHC. Normal liver uptake is seen for both patients. A comparison of FES uptake with *in vitro* assay by radioligand binding (Adapted from Mintun, M.A. et al. 1988. *Radiology.* 169(1):45–48.) (b) and *in vitro* assay by quantitative IHC (Adapted from Peterson, L.M. et al. 2008. *J Nucl Med.* 49(3):367–374.) (c) demonstrates close correspondence between FES uptake and *in vitro* measures of ER expression.

Our laboratory also performed a comparison of FES uptake with *in vitro* assay, but using different measures both for FES uptake and for *in vitro* assay (Peterson et al. 2008). For this study, based upon the analysis of tumor uptake and blood clearance in an earlier paper (Mankoff et al. 1996, 1997), FES images were obtained using dynamic imaging over a single-imaging field from injection to 60 min. Uptake curves for tumor, normal tissue, and blood (left ventricle) were obtained from region-of-interest analysis. Several measures of uptake were tested. The SUV for the 30–60 min summed data from the tumor time–activity curve was calculated using the standard formula:

$$SUV = \frac{\bar{A}}{ID/wt} \tag{16.1}$$

$\bar{A}$ is defined as the average tissue uptake (μCi/g) from 30–60 min after injection, ID is the injected dose (mCi), and wt is the patient weight (kg). We also tested the additional measure of FES uptake, flux$_{tot}$, to account for variable FES blood clearance seen in prior studies (Mankoff et al. 1997) as defined by

$$Flux_{tot} = \frac{\bar{A}}{\int C_{b_{FES}} dt} \tag{16.2}$$

Chapter 16

where C_b is the blood clearance curve (total blood radioactivity concentration) integrated over time. The flux measure has units of mL/min/g and is similar to SUV, but uses the area under the blood clearance curve as an indicator of tracer availability to the tumor instead of (ID/wt). We also calculated a flux measurement where the blood clearance curve was corrected for labeled metabolites, as previously described (Mankoff et al. 1997), and was defined as

$$\text{Flux}_{\text{corr}} = \frac{\overline{A}}{\int C_{b_{\text{FES}}} dt} \tag{16.3}$$

where C_b is the blood clearance curve corrected for FES metabolites over time.

In addition, rather than radioligand binding, these studies used the more common method of IHC to measure ER expression in *in vitro* assay of biopsy material. For this application, a semiquantitative index of IHC staining, previously validated against the radioligand-binding assay (Lehr et al. 1997), was used.

In comparing imaging measures of FES uptake with *in vitro* assay for IHC, imaging results would not be expected to correlate perfectly with biopsy results since, unlike the comparison of FES uptake with *in vitro* radioligand binding in the Mintun paper (Mintun et al. 1988), the imaging and *in vitro* assays differ (*in vivo* radioligand binding versus *in vitro* immune assay). The correlation between FES uptake measures and the IHC index at the level expected from prior studies comparing *in vitro* IHC to radioligand binding (McCarty et al. 1985, Parl and Posey 1988). There was not a significantly better correlation between the FES flux measures and IHC compared to SUV, suggesting that the simpler SUV may be an adequate measure for most studies, and that metabolite analysis may not be needed for routine FES imaging. In a comparison between FES uptake and standard IHC scoring (ER negative vs. ER positive [ER– vs. ER+]), an FES SUV value of 1.1 appeared to be at the boundary between ER– and ER+ by standard IHC scoring.

These two studies established FES PET as a quantitative measure of regional ER expression; however, further studies from a variety of metastatic tissues and clinical settings would lend support to these earlier findings.

16.5.2   Early Studies of Patient Imaging

In the earliest reported study of FES PET in patients, FES uptake was seen at sites of primary carcinoma, and in axillary nodes and one distant metastatic site (Mintun et al. 1988). The investigators then extended the use of this radiopharmaceutical for imaging of metastatic breast cancer. Sixteen patients with metastatic disease underwent FES PET imaging with increased uptake seen on 53 of 57 metastatic lesions, resulting in a 93% sensitivity and only 2 apparent false-positives (McGuire et al. 1991). Imaging results were reported as percentage uptake of injected dose per mL, ratio of lesion to soft tissue, and as the ratio of lesion to uninvolved bone. The same group found similar results in a later study of FES imaging in 21 patients with metastatic breast cancer with 88% overall agreement between *in vitro* ER assays and FES PET (Dehdashti et al. 1995). In addition to subjective analysis, FES uptake was reported as an SUV. Using an SUV > 1 to identify ER-expressing disease, the sensitivity of FES imaging was 76% with no false-positives in 21 metastatic breast cancer patients (Mortimer et al. 1996).

One of the chief advantages of ER imaging over tissue sampling to determine ER expression is the ability to evaluate the heterogeneity of ER expression. Mortimer et al. (1996) found that 4 of 17 (24%) patients with metastatic breast cancer had discordance in FES uptake between sites in individuals. Mankoff et al. (2002) found the absence of FES uptake in one or more metastatic sites in 10% of patients who had primary ER-positive tumors. In this same preliminary study, the quantitative site-to-site variability in FES uptake in individuals was high (COV of ~30%). Thirteen percent of patients (6 of 47) with ER-positive primaries had one or more sites of FES-negative disease in a subsequent study by the same group (Linden et al. 2006). The rate of complete loss of ER expression at metastatic sites from ER+ tumors by FES was comparable with, but slightly lower than, values obtained from tissue samples reported in the literature, typically more in the range of 25% (Kuukasjarvi et al. 1996, Spataro et al. 1992), suggesting that sampling error may contribute to apparent heterogeneity in tissue-based assay studies.

16.5.3   Studies Testing FES PET as a Predictor of Tumor Responsiveness to Hormonal Therapy

In clinical practice, the primary use of *in vitro* ER assay is a predictive assay for endocrine therapy. Although FES has not been prospectively tested as a predictive assay in clinical trials, comparison of FES uptake versus response with endocrine therapy in some groups of patients has indicated the likely value of FES PET as a predictive assay. Mortimer et al. (2001) showed that the level of FES uptake predicted response to tamoxifen, demonstrating the potential utility of FES PET for predicting response in the locally advanced and metastatic setting. Forty women with biopsy-proven ER-positive breast cancer had FES PET before and 7–10 days after initiation of tamoxifen therapy and tumor FES PET was assessed with the SUV method. Both percentage decrease in FES (responders = 55 ± 14%, nonresponders = 19% ± 17%) and absolute change in tumor SUV (responders = 2.5 decrease ± 1.8; nonresponders = 0.5 decrease ± 0.6 SUV units) predicted response to tamoxifen. The level of FES uptake pretherapy also predicted response to tamoxifen. The positive and negative predictive values for baseline FES uptake using an arbitrary cut-off for SUV of 2.0 were 79% and 88%, respectively (Mortimer et al. 2001). No patient with an SUV less than ~1.5 responded. (See the discussion below for more information on this cut-off value.)

Linden et al. (2006) showed that initial FES uptake measurements in patients with ER-positive tumors was correlated with subsequent response to 6 months of hormonal therapy (Figure 16.9). Forty-seven heavily pretreated patients with ER-positive, metastatic breast cancer were given predominantly salvage AI therapy. Objective response was seen in 11/47 (23%) patients. FES PET was assessed qualitatively and quantitatively using SUV and Flux calculations. Although no patient without FES uptake at known tumor sites responded, qualitative FES PET results did not significantly predict response to hormonal therapy. However, quantitative results were predictive of response in that no patient with an SUV of <1.5 responded to treatment. In this study, 0/15 patients with initial SUV < 1.5 responded to hormonal therapy, compared with 11/32 (34%) patients with initial SUV > 1.5 ($P < .01$). Similar results were seen using FES flux to measure the uptake ($P < .005$). Interestingly, no patient whose tumor overexpressed HER2 had an objective response, including patients with SUV > 1.5. In the subset of patients without HER2 overexpression, 11/24 (46%) of patients with SUV > 1.5 responded to hormonal therapy. Hypothetically, the use of FES PET to select patients could have increased the response rate from 23% to 34% overall, and from 29% to 46% in the subset of patients lacking HER2 overexpression. The timing of FES imaging may be a confounder in this study since patients underwent FES imaging while on AI therapy but preliminary data from the same group show that

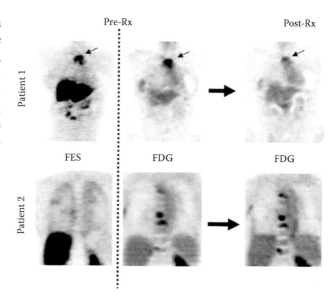

FIGURE 16.9 Images illustrating the correlation between FES uptake and subsequent response to hormonal therapy. Coronal images of FES uptake (left column) and FDG (middle column) uptake pretherapy, along with FDG uptake posthormonal therapy (right column), are shown for two patients (top and bottom row). The patient in the top row is a 44-year-old woman who was previously treated with adjuvant tamoxifen and had a sternal recurrence of breast cancer 4 years after primary tumor treatment. Her lesion had high pretherapy FES uptake in the lesion (arrow; image also shows liver and bowel uptake, both normal findings). FDG images taken before and after 6 weeks of letrozole treatment show a significant decline in FDG uptake, with subsequent excellent clinical response. The patient in the bottom row is a 69-year-old woman with newly diagnosed metastatic breast cancer that had not previously been treated. Her primary tumor was ER+ by immunohistochemistry and showed FES uptake (not shown). However, her pretherapy FES uptake showed no uptake at bone metastases documented by multiple imaging modalities, including FDG PET. The patient received multiple hormonal treatments with no response of the bone metastases, indicated by the post-therapy FDG PET, despite response by the primary tumor. The patient ultimately had progression of bony metastases and succumbed to her disease.

serial FES measurements change <20% in patients early after the start of AI therapy (Linden et al. 2005).

Interestingly, in the Linden and Mortimer studies (Linden et al. 2006, Mortimer et al. 2001), no patient with an average tumor SUV under 1.5 responded; however, the Peterson study (Peterson et al. 2008) showed that an SUV of 1.1 matches the value commonly accepted for an ER+ tumor by IHC. This discrepancy is not unexpected, as prior studies have that the chance of response increases with higher tumor ER expression (Allred et al. 1990, Campbell et al. 1981). This demonstrates the benefit of quantitative imaging of ER expression by PET, where the level of FES uptake, and not simply the presence or absence of uptake, adds predictive value.

Chapter 16

16.5.4 PET to Measure Changes in Response to ER-Directed Therapy

Serial PET ER imaging can be used to measure the pharmacodynamic affect of ER-directed endocrine therapies (Figure 16.10). Similar to studies of AR receptor blockade, McGuire demonstrated tamoxifen blockade of the ER in serial FES PET scans in early patient studies (McGuire et al. 1991). Mortimer (Mortimer et al. 2001) later showed a lower level of blockade occurring as early as 1 week after starting tamoxifen. Linden et al. (2005) analyzed serial FES PET in patients with metastatic disease undergoing treatment with tamoxifen ($n = 2$), AI ($n = 14$), or fulvestrant ($n = 5$). Patients were imaged a median of 29 days after starting treatment. The decline in FES SUV was greater for antagonists (tamoxifen and fulvestrant) than for AIs, which lower the agonist concentration but do not block the receptor. Interestingly, posttreatment qualitative FES scans showed complete blockage with tamoxifen but incomplete blockage with fulvestrant in four of the five

patients, suggesting that standard doses of fulvestrant might not be fully effective in ER blockade. This early finding has been borne out by subsequent clinical studies of fulvestrant at higher doses, which showed improved efficacy (Di Leo et al. 2010).

An alternative approach for predicting endocrine sensitivity has been to assess tissue response after short-term exposure to ER agonists or antagonists and test whether it is predictive of subsequent clinical response and disease-free survival. This was originally performed by serial tumor biopsy and assay for proliferation by the Ki-67 index and showed that a decline in proliferation predicted better clinical outcome (Dowsett et al. 2005, 2007). In a series of studies at Washington University in St. Louis, Dehdashti and colleagues (Dehdashti et al. 2009, Ellis et al. 2009, Mortimer et al. 2001) showed that an early rise in FDG uptake following an ER agonist (early tamoxifen "flare" or estradiol administration) predicted response to endocrine therapy, in some studies more accurately than FES uptake (Dehdashti et al. 2009, Mortimer et al. 2001). Recently,

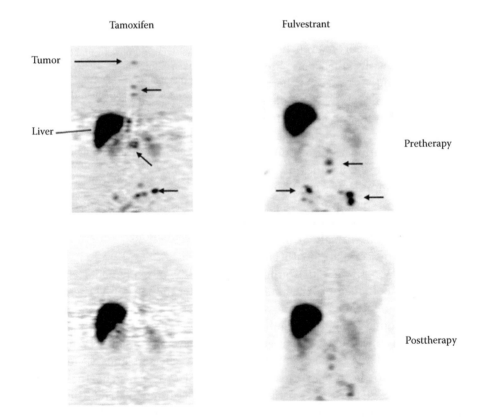

FIGURE 16.10 Blockade of the ER indicated by serial FES PET. FES PET images of two patients are shown before (top) and after (bottom) endocrine therapy. The pretherapy images indicate FES uptake in multiple bony lesions (arrow), indicative of ER+ metastases. Normal liver uptake is also seen. The patient on the left received tamoxifen, and a posttherapy scan shows an absence of FES uptake at all sites of disease, despite evidence of disease by other imaging, including FDG PET (not shown) indicating complete ER blockade by tamoxifen. The patient on the right received fulvestrant, but had persistent FES uptake posttherapy, indicating incomplete ER blockade by fulvestrant.

Linden and colleagues have shown that an early decline in FDG uptake following the withdrawal of estrogens after AI administration was also predictive of good response (Linden et al. 2009).

16.5.5 FES PET for Imaging Other Organs and Tumors

Some preliminary studies have evaluated FES PET imaging in settings other than breast cancer. Moresco studied FES uptake in normal brain tissue and meningiomas (Moresco et al. 1995, 1997) based upon some observations of ER expression in meningiomas and studies showing that tamoxifen could be effective in limiting growth. The study used a measure of FES uptake similar to the flux measure defined above. While FES uptake in normal brain tissue was too low to quantify estradiol binding reliably by PET, significant FES uptake was seen in some meningiomas.

Selective FES uptake by uterine endometrium has been shown in human imaging, with cyclic changes mirroring the menstrual cycle (Tsuchida et al. 2007). FES uptake in endometrial cancer has also been reported (Tsujikawa et al. 2009, Yoshida et al. 2007). Another study showed that the ratio of FES/FDG uptake in endometrial lesions was predictive of the pathologic features of uterine lesions (Tsujikawa et al. 2009). Lesions with less FES binding, indicative of lower ER expression, relative to FDG uptake had more aggressive, less differentiated features by histopathology.

16.5.6 Clinical Imaging Summary

Overall, early studies of PET ER imaging show its promise as a tool for directing breast cancer treatment. The promising early results of the Mortimer and Linden studies (Linden et al. 2006, Mortimer et al. 2001) pointing toward the potential utility of FES PET as a predictive assay need to be confirmed in larger trials involving more institutions and objective determination of the appropriate FES SUV cut-off for predicting response to endocrine therapy. Ongoing studies, including prospective clinical trials, should define its potential use in clinical trials and clinical practice for breast cancer and possibly other ER-expressing tumors.

16.6   Summary and Future Directions in ER Imaging

Thus far, preclinical and clinical studies have established FES PET as a useful method for quantitative imaging of ER expression. Studies have confirmed that FES is a safe drug with acceptable radiation dosimetry and that its uptake correlates with *in vitro* assay of ER expression that is widely used for clinical breast cancer treatment decisions. Early studies support FES PET as a predictive assay for endocrine therapy response and as a tool for measuring the effect of endocrine therapy on estradiol binding in breast cancer.

Although other compounds for ER imaging have been tested, and continue to be developed (see Section 16.2), none has exceeded the performance of FES in patients. FES has a structure close to estradiol and it has binding affinity for both ER and SHBG that is close to the native compound. Studies suggest that both these characteristics are important for a human ER imaging agent, and given the extremely constrained biochemistry for estrogen in human physiology, successful ER-imaging agents will need to closely mimic estradiol properties, as does FES. In this context, it is not clear whether there is much margin to improve on FES.

The data supporting FES thus far have come from relatively small single-center trials. While these studies strongly support the utility of FES for endocrine therapy clinical trials and clinical practice, larger multicenter trials are needed for further validation and regulatory approval. The creation of an NCI IND (IND79, 2005) is an important step in this direction, and ongoing trials will hopefully lead to more widespread availability and use of the potentially valuable imaging tool. FES is a clear example of success in rationale design and development of a PET-imaging agent.

References

Ali, H., Rousseau, J., and van Lier, J.E. 1993. 7 alpha-Methyl- and 11 beta-ethoxy-substitution of [^{125}I]-16 alpha-iodoestradiol: Effect on estrogen receptor mediated target tissue uptake. *J Med Chem.* 36:264–271.

Ali, H., Rousseau, J., Ghaffari, M.A., and van Lier, J.E. 1991. Synthesis, receptor binding, and tissue distribution of 7 alpha- and 11 beta-substituted (17 alpha,20E)- and (17 alpha,20Z)-21-[125I]iodo-19-norpregna-1,3,5(10),20-tetraene-3,17-diols. *J Med Chem.* 34:854–860.

Allred, D.C., Bustamante, M.A., Daniel, C.O., Gaskill, H.V., and Cruz, A.B., Jr. 1990. Immunocytochemical analysis of estrogen receptors in human breast carcinomas. Evaluation of 130 cases and review of the literature regarding concordance with biochemical assay and clinical relevance. *Arch Surg.* 125(1):107–113.

Allred, D.C., Carlson, R.W., Berry, D.A. et al. 2009. NCCN task force report: Estrogen receptor and progesterone receptor testing in breast cancer by immunohistochemistry. *J Natl Compr Canc Netw.* 7 Suppl 6:S1–S21; quiz S22–S23.

Anzick, S.L., Kononen, J., Walker, R.L. et al. 1997. AIB1, a steroid receptor coactivator amplified in breast and ovarian cancer. *Science.* 277(5328):965–968.

Beauregard, J.M., Croteau, E., Ahmed, N., van Lier, J.E., and Benard, F. 2009. Assessment of human biodistribution and dosimetry of 4-fluoro-11beta-methoxy-16alpha-^{18}F-fluoroestradiol using serial whole-body PET/CT. *J Nucl Med.* 50(1):100–107.

Benard, F., Ahmed, N., Beauregard, J.M. et al. 2008. [^{18}F]Fluorinated estradiol derivatives for oestrogen receptor imaging: Impact of substituents, formulation and specific activity on the biodistribution in breast tumour-bearing mice. *Eur J Nucl Med Mol Imaging.* 35(8):1473–1479.

Bennink, R.J., Rijks, L.J., van Tienhoven, G., Noorduyn, L.A., Janssen, A.G., and Sloof, G.W. 2001. Estrogen receptor status in primary breast cancer: Iodine 123-labeled cis-11beta-methoxy-17alpha-iodovinyl estradiol scintigraphy. *Radiology.* 220(3):774–779.

Bolt, H.M. 1979. Metabolism of estrogens—Natural and synthetic. *Pharmacol Ther.* 4(1):155–181.

Buzdar, A., Douma, J., Davidson, N. et al. 2001. Phase III, multicenter, double-blind, randomized study of letrozole, an aromatase inhibitor, for advanced breast cancer versus megestrol acetate. *J Clin Oncol.* 19(14):3357–3366.

Campbell, F.C., Elston, C.W., Blamey, R.W. et al. 1981. Quantitative oestradiol receptor values in primary breast cancer and response of metastases to endocrine therapy. *Lancet.* 1:1317–1319.

Cigler, T. and Goss, P.E. 2007. Breast cancer adjuvant endocrine therapy. *Cancer J.* 13(3):148–155.

Cummins, C.H. 1993. Radiolabeled steroidal estrogens in cancer research. *Steroids.* 58(6):245–259.

Dehdashti, F., Flanagan, F.L., Mortimer, J.E., Katzenellenbogen, J.A., Welch, M.J., and Siegel, B.A. 1999. Positron emission tomographic assessment of "metabolic flare" to predict response of metastatic breast cancer to antiestrogen therapy. *Eur J Nucl Med.* 26(1):51–56.

Dehdashti, F., Mortimer, J.E., Siegel, B.A. et al. 1995. Positron tomographic assessment of estrogen receptors in breast cancer: Comparison with FDG-PET and *in vitro* receptor assays. *J Nucl Med.* 36(10):1766–1774.

Dehdashti, F., Mortimer, J.E., Trinkaus, K. et al. 2009. PET-based estradiol challenge as a predictive biomarker of response to endocrine therapy in women with estrogen-receptor-positive breast cancer. *Breast Cancer Res Treat.* 113(3):509–517.

Di Leo, A., Jerusalem, G., Petruzelka, L. et al. 2010. Results of the CONFIRM phase III trial comparing fulvestrant 250 mg with fulvestrant 500 mg in postmenopausal women with estrogen receptor-positive advanced breast cancer. *J Clin Oncol.* 28(30):4594–4600.

Dowsett, M., Smith, I.E., Ebbs, S.R. et al. 2005. Short-term changes in Ki-67 during neoadjuvant treatment of primary breast cancer with anastrozole or tamoxifen alone or combined correlate with recurrence-free survival. *Clin Cancer Res.* 11(2 Pt 2):951s–958s.

Dowsett, M., Smith, I.E., Ebbs, S.R. et al. 2007. Prognostic value of Ki67 expression after short-term presurgical endocrine therapy for primary breast cancer. *J Natl Cancer Inst.* 99(2):167–170.

Dunnwald, L.K., Rossing, M.A., and Li, C.I. 2007. Hormone receptor status, tumor characteristics, and prognosis: A prospective cohort of breast cancer patients. *Breast Cancer Res.* 9(1):R6.

Ellis, M.J., Gao, F., Dehdashti, F. et al. 2009. Lower-dose vs high-dose oral estradiol therapy of hormone receptor-positive, aromatase inhibitor-resistant advanced breast cancer: A phase 2 randomized study. *JAMA.* 302(7):774–780.

Fink, B.J. and Christensen, M.S. 1981. Bioavailability of oestradiol and oestriol administered orally to oophorectomized women. *Maturitas.* 3(3–4):289–294.

Goss, P. and Wu, M. 2007. Application of aromatase inhibitors in endocrine responsive breast cancers. *Breast.* 16(Suppl2):S114–9.

Greene, G.L., Gilna, P., Waterfield, M., Baker, A., Hort, Y., and Shine, J. 1986. Sequence and expression of human estrogen receptor complementary DNA. *Science.* 231(4742):1150–1154.

Hammond, G. 1993. Steroid hormone action. In: Parker M, ed. *Extracellular Steroid-Binding Proteins.* New York: Oxford University Press.

Jonson, S.D. and Welch, M.J. 1998. PET imaging of breast cancer with fluorine-18 radiolabeled estrogens and progestins. *Q J Nucl Med.* 42(1):8–17.

Jonson, S.D., Bonasera, T.A., Dehdashti, F., Cristel, M.E., Katzenellenbogen, J.A., and Welch, M.J. 1999. Comparative breast tumor imaging and comparative *in vitro* metabolism of 16alpha-[^{18}F]fluoroestradiol-17beta and 16beta-[18F]fluoromoxestrol in isolated hepatocytes. *Nucl Med Biol.* 26(1):123–130.

Katzenellenbogen, B.S. 1996. Estrogen receptors: Bioactivities and interactions with cell signaling pathways. *Biol Reprod.* 54(2):287–293.

Katzenellenbogen, J. 1992. The pharmacology of steroid radiopharmaceuticals: Specific and non-specific binding and uptake selectivity. In: Nunn A, ed. *Radiopharmaceuticals: Chemistry and Pharmacology.* New York: Marcel Dekker; pp. 297–331.

Katzenellenbogen, J.A., Hsiung, H.M., Carlson, K.E., McGuire, W.L., Kraay, R.J., and Katzenellenbogen, B.S. 1975. Iodohexestrols. II. Characterization of the binding and estrogenic activity of iodinated hexestrol derivatives, *in vitro* and *in vivo*. *Biochemistry.* 14(8):1742–1750.

Katzenellenbogen, J.A. 1981. The development of gamma-emitting hormone analogs as imaging agents for receptor-positive tumors. *Prog Clin Biol Res.* 75B:313–327.

Katzenellenbogen, J.A. 1995. Designing steroid receptor-based radiotracers to image breast and prostate tumors. *J Nucl Med.* 36(6 Suppl):8S–13S.

Katzenellenbogen, J.A., Senderoff, S.G., McElvany, K.D., O'Brien, H.A., Jr., and Welch, M.J. 1981. 16 alpha-[^{77}Br]bromoestradiol-17 beta: A high specific-activity, gamma-emitting tracer with uptake in rat uterus and uterus and induced mammary tumors. *J Nucl Med.* 22(1):42–47.

Katzenellenbogen, J.A., Welch, M.J., and Dehdashti, F. 1997. The development of estrogen and progestin radiopharmaceuticals for imaging breast cancer. *Anticancer Res.* 17(3B):1573–1576.

Kiesewetter, D.O., Kilbourn, M.R., Landvatter, S.W., Heiman, D.F., Katzenellenbogen, J.A., and Welch, M.J. 1984. Preparation of four fluorine- 18-labeled estrogens and their selective uptakes in target tissues of immature rats. *J Nucl Med.* November 25(11):1212–1221.

Krohn, K.A. 2001. The physical chemistry of ligand-receptor binding identifies some limitations to the analysis of receptor images. *Nucl Med Biol.* 28(5):477–483.

Kuerer, H.M., Buzdar, A.U., and Singletary, S.E. 2001. Biologic basis and evolving role of aromatase inhibitors in the management of invasive carcinoma of the breast. *J Surg Oncol.* 77(2):139–147.

Kumar, P., Mercer, J., Doerkson, C., Tonkin, K., and McEwan, A.J. 2007. Clinical production, stability studies and PET imaging with 16-alpha-[18F]fluoroestradiol ([18F]FES) in ER positive breast cancer patients. *J Pharm Sci.* 10(2): 256s–265s.

Kuukasjarvi, T., Kononen, J., Helin, H., Holli, K., and Isola, J. 1996. Loss of estrogen receptor in recurrent breast cancer is associated with poor response to endocrine therapy. *J Clin Oncol.* 14(9):2584–2589.

Landvatter, S.W., Kiesewetter, D.O., Kilbourn, M.R., Katzenellenbogen, J.A., and Welch, M.J. 1983. (2R*, 3S*)-1-[18F]fluoro-2,3-bis(4-hydroxyphenyl)pentane [(18F)fluoronorhexestrol), a positron-emitting estrogen that shows highly-selective, receptor-mediated uptake by target tissues in vivo. *Life Sci.* 33(19):1933–1938.

Lehr, H.A., Mankoff, D.A., Corwin, D., Santeusanio, G., and Gown, A.M. 1997. Application of photoshop-based image analysis to quantification of hormone receptor expression in breast cancer. *J Histochem Cytochem.* 45(11):1559–1565.

Li, C.I., Daling, J.R., and Malone, K.E. 2003. Incidence of invasive breast cancer by hormone receptor status from 1992 to 1998. *J Clin Oncol.* 21(1):28–34.

Lim, J.L., Zheng, L., Berridge, M.S., and Tewson, T.J. 1996. The use of 3-methoxymethyl-16 beta, 17 beta-epiestriol-O-cyclic sulfone as the precursor in the synthesis of F-18 16 alpha-fluoroestradiol. *Nucl Med Biol.* 23(7):911–915.

Linden, H.M. and Mankoff, D.A. 2010. Breast cancer and hormonal stimulation: Is glycolysis the first sign of response? *J Nucl Med.* 51(11):1663–1664.

Linden, H., Kurland, B., Livingston, R. et al. 2009. Early Assessment of Response to Aromatase Inhibitor (AI) Therapy. Paper presented at American Society of Clinical Oncology 2009 Annual Meeting; June, 2009; Orlando, FL.

Linden, H.M., Link, J.M., Stekhova, S. et al. 2005. Serial 18F-fluoroestradiol positron emission tomography (FES PET) measures estrogen receptor binding during endocrine therapy. *Breast Cancer Res Treat.* 94S1:S237.

Linden, H.M., Stekhova, S.A., Link, J.M. et al. 2006. Quantitative fluoroestradiol positron emission tomography imaging predicts response to endocrine treatment in breast cancer. *J Clin Oncol.* 24(18):2793–2799.

Mankoff, D.A., Peterson, L.M., Petra, P.H. et al. 2002. Factors affecting the level and heterogeneity of uptake [F-18] fluoroestradiol [FES] in patients with estrogen receptor positive breast cancer. *J Nucl Med.* 43:286–287.

Mankoff, D.A., Peterson, L.M., Tewson, T.J. et al. 2001. [18F]fluoroestradiol radiation dosimetry in human PET studies. *J Nucl Med.* 42(4):679–684.

Mankoff, D.A., Shields, A.F., Graham, M.M., Link, J.M., and Krohn, K.A. 1996. A graphical analysis method to estimate blood-to-tissue transfer constants for tracers with labeled metabolites. *J Nucl Med.* 37(12):2049–2057.

Mankoff, D.A., Tewson, T.J., and Eary, J.F. 1997. Analysis of blood clearance and labeled metabolites for the estrogen receptor tracer [F-18]-16 alpha-fluoroestradiol (FES). *Nucl Med Biol.* 24(4):341–348.

Marino, M., Galluzzo, P., and Ascenzi, P. 2006. Estrogen signaling multiple pathways to impact gene transcription. *Curr Genomics.* 7(8):497–508.

Mathias, C.J., Welch, M.J., Katzenellenbogen, J.A. et al. 1987. Characterization of the uptake of 16 alpha-([18F]fluoro)-17 beta-estradiol in DMBA-induced mammary tumors. *Int J Rad Appl Instrum.* 14:15–25.

McCabe, A., Dolled-Filhart, M., Camp, R.L., and Rimm, D.L. 2005. Automated quantitative analysis (AQUA) of *in situ* protein expression, antibody concentration, and prognosis. *J Natl Cancer Inst.* 97(24):1808–1815.

McCarty, K.S., Jr., Hiatt, K.B., Budwit, D.A. et al. 1984. Clinical response to hormone therapy correlated with estrogen receptor analyses. Biochemical v histochemical methods. *Arch Pathol Lab Med.* 108(1):24–26.

McCarty, K.S., Jr., Miller, L.S., Cox, E.B., Konrath, J., and McCarty, K.S., Sr. 1985. Estrogen receptor analyses. Correlation of biochemical and immunohistochemical methods using monoclonal antireceptor antibodies. *Arch Pathol Lab Med.* 109(8):716–721.

McCarty, K.S., Jr., Woodard, B.H., Nichols, D.E., Wilkinson, W., and McCarty, K.S., Sr. 1980. Comparison of biochemical and histochemical techniques for estrogen receptor analyses in mammary carcinoma. *Cancer.* 46(12 Suppl):2842–2845.

McElvany, K.D., Carlson, K.E., Welch, M.J., Senderoff, S.G., and Katzenellenbogen, J.A. 1982. *In vivo* comparison of 16 alpha[77Br]bromoestradiol-17 beta and 16 alpha-[125I]iodoestradiol-17 beta. *J Nucl Med.* 23(5):420–424.

McGuire, A.H., Dehdashti, F., Siegel, B.A. et al. 1991. Positron tomographic assessment of 16 alpha-[18F] fluoro-17 beta-estradiol uptake in metastatic breast carcinoma. *J Nucl Med.* 32(8):1526–1531.

Mintun, M.A., Welch, M.J., Siegel, B.A. et al. 1988. Breast cancer: PET imaging of estrogen receptors. *Radiology.* 169(1):45–48.

Moresco, R.M., Casati, R., Lucignani, G. et al. 1995. Systemic and cerebral kinetics of 16 alpha [18F]fluoro-17 beta-estradiol: A ligand for the *in vivo* assessment of estrogen receptor binding parameters. *J Cereb Blood Flow Metab.* 15(2):301–311.

Moresco, R.M., Scheithauer, B.W., Lucignani, G. et al. 1997. Oestrogen receptors in meningiomas: A correlative PET and immunohistochemical study. *Nucl Med Commun.* 18(7):606–615.

Mortimer, J.E., Dehdashti, F., Siegel, B.A., Katzenellenbogen, J.A., Fracasso, P., and Welch, M.J. 1996. Positron emission tomography with 2-[18F]fluoro-2-deoxy-D-glucose and 16alpha-[18F] fluoro-17beta-estradiol in breast cancer: Correlation with estrogen receptor status and response to systemic therapy. *Clin Cancer Res.* 2(6):933–939.

Mortimer, J.E., Dehdashti, F., Siegel, B.A., Trinkaus, K., Katzenellenbogen, J.A., and Welch, M.J. 2001. Metabolic flare: Indicator of hormone responsiveness in advanced breast cancer. *J Clin Oncol.* 19(11):2797–2803.

Mouridsen, H., Gershanovich, M., Sun, Y. et al. 2001. Superior efficacy of letrozole versus tamoxifen as first-line therapy for postmenopausal women with advanced breast cancer: Results of a phase III study of the International Letrozole Breast Cancer Group. *J Clin Oncol.* 19(10):2596–2606.

O'Malley, B.W., McGuire, W.L., and Middleton, P.A. 1968. Altered gene expression during differentiation: Population changes in hybridizable RNA after stimulation of the chick oviduct with oestrogen. *Nature.* 218(5148):1249–1251.

Chapter 16

Osborne, C.K. and Schiff, R. 2011. Mechanisms of endocrine resistance in breast cancer. *Annu Rev Med.* 62:233–247.

Osborne, C.K., Schiff, R., Arpino, G., Lee, A.S., and Hilsenbeck, V.G. 2005a. Endocrine responsiveness: Understanding how progesterone receptor can be used to select endocrine therapy. *Breast.* 14(6):458–465.

Osborne, C.K., Schiff, R., Fuqua, S.A., and Shou, J. 2001. Estrogen receptor: Current understanding of its activation and modulation. *Clin Cancer Res.* 7(12 Suppl):4338s–4342s; discussion 4411s–4412s.

Osborne, C.K., Shou, J., Massarweh, S., and Schiff, R. 2005b. Crosstalk between estrogen receptor and growth factor receptor pathways as a cause for endocrine therapy resistance in breast cancer. *Clin Cancer Res.* 11(2 Pt 2):865s–870s.

Osborne, C.K., Yochmowitz, M.G., Knight, W.A., 3rd, and McGuire, W.L. 1980. The value of estrogen and progesterone receptors in the treatment of breast cancer. *Cancer.* 46(12 Suppl):2884–2888.

Osborne, C.K., Zhao, H., and Fuqua, S.A. 2000. Selective estrogen receptor modulators: Structure, function, and clinical use. *J Clin Oncol.* 18(17):3172–3186.

Pan, C.C., Woolever, C.A., and Bhavnani, B.R. 1985. Transport of equine estrogens: Binding of conjugated and unconjugated equine estrogens with human serum proteins. *J Clin Endocrinol Metab.* 61(3):499–507.

Parl, F.F. and Posey, Y.F. 1988. Discrepancies of the biochemical and immunohistochemical estrogen receptor assays in breast cancer. *Hum Pathol.* 19(8):960–966.

Pavlik, E.J., Nelson, K., Gallion, H.H. et al. 1990. Characterization of high specific activity [16 alpha-123I]Iodo-17 beta-estradiol as an estrogen receptor-specific radioligand capable of imaging estrogen receptor-positive tumors. *Cancer Res.* 50(24):7799–7805.

Peterson, L.M., Mankoff, D.A., Lawton, T. et al. 2008. Quantitative imaging of estrogen receptor expression in breast cancer with PET and ^{18}F-fluoroestradiol. *J Nucl Med.* 49(3):367–374.

Peterson, L.M., Kurland, B.F., Link, J.M. et al. 2010. Factors influencing the uptake of [F-18]fluorestradiol (FES) in patients with estrogen receptor positive (ER+) breast cancer. *J Nucl Med.* 51. (suppl 2):178.

Petra, P. 1991. The plasma sex steroid binding protein (SBP or SBHG). A critical review of recent developments on the structure, molecular biology and function. *J Steroid Biochem Mol Biol.* 40:735–753.

Plymate, S.R., Namkung, P.C., Metej, L.A., and Petra, P.H. 1990. Direct effect of plasma sex hormone binding globulin (SHBG) on the metabolic clearance rate of 17 beta-estradiol in the primate. *J Steroid Biochem.* 36(4):311–317.

Pomper, M.G., Pinney, K.G., Carlson, K.E. et al. 1990a. Target tissue uptake selectivity of three fluorine-substituted progestins: Potential imaging agents for receptor-positive breast tumors. *Int J Rad Appl Instrum B.* 17(3):309–319.

Pomper, M.G., VanBrocklin, H., Thieme, A.M. et al. 1990b. 11 beta-methoxy-, 11 beta-ethyl- and 17 alpha-ethynyl-substituted 16 alpha-fluoroestradiols: Receptor-based imaging agents with enhanced uptake efficiency and selectivity. *J Med Chem.* 33(12):3143–3155.

Pritchard, K.I., Rolski, J., Papai, Z. et al. 2010. Results of a phase II study comparing three dosing regimens of fulvestrant in postmenopausal women with advanced breast cancer (FINDER2). *Breast Cancer Res Treat.* 123(2):453–461.

Pujol, P., Hilsenbeck, S.G., Chamness, G.C., and Elledge, R.M. 1994. Rising levels of estrogen receptor in breast cancer over 2 decades. *Cancer.* 74(5):1601–1606.

Reed, M.J. and Purohit, A. 2001. Aromatase regulation and breast cancer. *Clin Endocrinol (Oxf).* 54(5):563–571.

Reed, M.J., Purohit, A., Woo, L.W., Newman, S.P., and Potter, B.V. 2005. Steroid sulfatase: Molecular biology, regulation, and inhibition. *Endocr Rev.* 26(2):171–202.

Riggs, B.L. and Hartmann, L.C. 2003. Selective estrogen-receptor modulators—Mechanisms of action and application to clinical practice. *N Engl J Med.* 348(7):618–629.

Rijks, L.J., Busemann, S.E., Stabin, M.G., de Bruin, K., Janssen, A.G., and van Royen, E.A. 1998. Biodistribution and dosimetry of iodine-123-labelled Z-MIVE: An oestrogen receptor radioligand for breast cancer imaging. *Eur J Nucl Med.* 25(1):40–47.

Romer, J., Fuchtner, F., Steinbach, J., and Kasch, H. 2001. Automated synthesis of 16alpha-[18F]fluoroestradiol-3,17beta-disulphamate. *Appl Radiat Isot.* 55(5):631–639.

Santen, R.J., Manni, A., Harvey, H., and Redmond, C. 1990. Endocrine treatment of breast cancer in women. *Endocr Rev.* 11(2):221–265.

Scharl, A., Beckman, M.W., Artwohl, J.E., Kullander, S., and Holt, J.A. 1991. Rapid liver metabolism, urinary and biliary excretion, and enterohepatic circulation of 16 alpha-radioiodio-17beta-estradiol. *Int J Radiat Oncol Biol Phys.* 21:1235–1240.

Schiff, R., Massarweh, S.A., Shou, J. et al. 2005. Advanced concepts in estrogen receptor biology and breast cancer endocrine resistance: Implicated role of growth factor signaling and estrogen receptor coregulators. *Cancer Chemother Pharmacol.* 56 Suppl 1:10–20.

Sledge, G.J. and McGuire, W. 1983. Steroid hormone receptors in human breast cancer. *Adv Cancer Res.* 38:61–75.

Spataro, V., Price, K., Goldhirsch, A. et al. 1992. Sequential estrogen receptor determinations from primary breast cancer and at relapse: Prognostic and therapeutic relevance. The International Breast Cancer Study Group (formerly Ludwig Group). *Ann Oncol.* 3(9):733–740.

Stein, R.A. and McDonnell, D.P. 2006. Estrogen-related receptor alpha as a therapeutic target in cancer. *Endocr Relat Cancer.* 13 Suppl 1:S25–S32.

Tewson, T.J., Mankoff, D.A., Peterson, L.M., Woo, I., and Petra, P. 1999. Interactions of 16alpha-[18F]-fluoroestradiol (FES) with sex steroid binding protein (SBP). *Nucl Med Biol.* 26(8):905–913.

Tsuchida, T., Okazawa, H., Mori, T. et al. 2007. *In vivo* imaging of estrogen receptor concentration in the endometrium and myometrium using FES PET—Influence of menstrual cycle and endogenous estrogen level. *Nucl Med Biol.* 34(2):205–210.

Tsujikawa, T., Yoshida, Y., Kiyono, Y. et al. 2011. Functional oestrogen receptor alpha imaging in endometrial carcinoma using 16alpha-[(18)F]fluoro-17beta-oestradiol PET. *Eur J Nucl Med Mol Imaging.* 38(1):37–45.

Tsujikawa, T., Yoshida, Y., Kudo, T. et al. 2009. Functional images reflect aggressiveness of endometrial carcinoma: Estrogen receptor expression combined with ^{18}F-FDG PET. *J Nucl Med.* 50(10):1598–1604.

Tsujikawa, T., Yoshida, Y., Mori, T. et al. 2008. Uterine tumors: Pathophysiologic imaging with 16alpha-[18F]fluoro-17beta-estradiol and 18F fluorodeoxyglucose PET—Initial experience. *Radiology.* 248(2):599–605.

VanBrocklin, H.F., Pomper, M.G., Carlson, K.E., Welch, M.J., and Katzenellenbogen, J.A. 1992. Preparation and evaluation of 17-ethynyl-substituted 16 alpha-[18F]fluoroestradiols: Selective receptor-based PET imaging agents. *Int J Rad Appl Instrum B.* 19(3):363–374.

Welboren, W.J., Sweep, F.C., Span, P.N., and Stunnenberg, H.G. 2009. Genomic actions of estrogen receptor alpha: What are the targets and how are they regulated? *Endocr Relat Cancer.* 16(4):1073–1089.

Wheeler, W.J., Cherry, L.M., Downs, T., and Hsu, T.C. 1986. Mitotic inhibition and aneuploidy induction by naturally occurring and synthetic estrogens in Chinese hamster cells *in vitro*. *Mutat Res.* 171(1):31–41.

White, C.M., Ferraro-Borgida, M.J., Fossati, A.T. et al. 1998. The pharmacokinetics of intravenous estradiol—A preliminary study. *Pharmacotherapy.* 18(6):1343–1346.

Yang, D.J., Li, C., Kuang, L.R. et al. 1994. Imaging, biodistribution and therapy potential of halogenated tamoxifen analogues. *Life Sci.* 55(1):53–67.

Yoo, J., Dence, C.S., Sharp, T.L., Katzenellenbogen, J.A., and Welch, M.J. 2005. Synthesis of an estrogen receptor â-selective radioligand: 5-[¹⁸F]Fluoro-(2*R**,3*S**)-2,3-bis(4-hydroxyphenyl) pentanenitrile and comparison of *in vivo* distribution with 16-alpha-[¹⁸F]Fluoro-17-beta-estradiol. *J Med Chem.* 48:6366–6378.

Yoshida, Y., Kurokawa, T., Sawamura, Y. et al. 2007. The positron emission tomography with F18 17beta-estradiol has the potential to benefit diagnosis and treatment of endometrial cancer. *Gynecol Oncol.* 104(3):764–766.

Zielinski, J.E., Larner, J.M., Hoffer, P.B., and Hochberg, R.B. 1989. The synthesis of 11 beta-methoxy-[16 alpha-¹²³I] iodoestradiol and its interaction with the estrogen receptor *in vivo* and *in vitro*. *J Nucl Med.* 30(2):209–215.

Chapter 16

17. Targeting Norepinephrine Transporters in Cardiac Sympathetic Nerve Terminals

David M. Raffel

17.1 Target Protein

The norepinephrine transporter (NET) is the primary molecular target of the cardiac sympathetic nerve imaging agents 2-(3-iodanylbenzyl)guanidine (*meta*-iodobenzylguanidine, MIBG; radiolabeled with ^{131}I or ^{123}I) and (−)-(1*R*,2*S*)-1-(*meta*-hydroxyphenyl)-2-([11C]methylamino)-1-propanol ([11C]*meta*-hydroxyephedrine, HED). Localized in the cell membranes of terminal sympathetic nerve axons in the heart, NET is a protein made up of 617 amino acids, with a molecular mass of ~69,000, and known to have 12 transmembrane domains (Pacholczyk et al. 1991). In the normal heart, more than 90% of norepinephrine molecules released by nerve stimulation are recovered by NET transport back into presynaptic nerve terminals (Schömig et al. 1989). MIBG and HED are "substrates" for NET, and thus they are actively transported by NET into the axoplasm of sympathetic nerve terminal axons (Figure 17.1). Because MIBG and HED are structural analogs of the endogenous neurotransmitter norepinephrine, they also interact with a secondary molecular target, the vesicular monoamine transporter (second isoform; VMAT2). These are expressed in high concentrations in the membranes of norepinephrine storage vesicles. VMAT2 transport is the mechanism responsible for repackaging norepinephrine molecules recovered by NET into storage vesicles for subsequent reuse by the neuron. While NET is the primary molecular target of MIBG and HED, VMAT2 transport of these tracers into storage vesicles is an important process that increases their neuronal distribution volumes and slows their efflux from neuronal compartments.

NET was chosen as a molecular target because in the heart the transporter is exclusively found on presynaptic sympathetic nerve terminals, and it is expressed in high densities on the terminal nerve axons. Several high-affinity NET inhibitors have been radiolabeled with tritium and successfully used for *in vitro* assays of NET density, including [3H]desipramine (Raisman et al. 1982) and [3H]mazindol (Raffel and Chen 2004). However, the consistently high lipophilicities of

Targeted Molecular Imaging. Edited by Michael J. Welch and William C. Eckelman © 2012 Taylor & Francis Group, LLC. ISBN: 978-1-4398-4195-2

these compounds (log P > 5) causes radiolabeled analogs of NET inhibitors to suffer from very high nonspecific cardiac uptake *in vivo*, preventing their successful application as neuronal imaging agents

(Van Dort et al. 1997). Because of this, the most successful imaging agents targeting cardiac presynaptic nerve populations have been NET substrates rather than NET inhibitors.

17.2 Chemical Identity and Quantitation

The chemical structures of MIBG and HED are shown in Figure 17.2. MIBG was not specifically designed and conceived as a cardiac nerve imaging agent. This fortuitous secondary application of the tracer can be considered a stroke of luck. The invention of MIBG was in fact the successful culmination of a longstanding research project at the University of Michigan aimed at developing an imaging agent that would selectively localize in the medullary tissue of the adrenal gland (Sisson 2000). The main clinical goals of this research were to use nuclear scintigraphy to visualize adrenomedullary

hyperplasias and associated elusive neoplasms, such as pheochromocytoma and neuroblastoma.

The chromaffin granules in adrenal medulla tissue are a major site of the synthesis, storage, metabolism, and secretion of the endogenous catecholamines epinephrine and norepinephrine. At the time MIBG was developed, it was well known that the "monoamine uptake system" present in the adrenal medulla would avidly take up and store many structural analogs of these catecholamines. For example, early studies at Michigan directed at developing adrenomedullary imaging agents showed that intravenous injection of [^{14}C]-labeled epinephrine, dopamine, norepinephrine, and other related compounds all accumulated to varying degrees in the canine adrenal medulla (Morales et al. 1967). Among these, [^{14}C]dopamine was found to have the highest adrenal medulla uptake, achieving tissue-to-blood ratios >1000:1 at 24 h. These studies demonstrated the feasibility of using radiolabeled substrates of the chromaffin granule "monoamine uptake system" as an effective mechanism for localizing high concentrations of radiotracer in adrenal medulla. After several years of further research in this area at Michigan, radioiodinated MIBG was developed by Donald M. Wieland, PhD, as a clinically viable adrenomedullary imaging agent (Wieland et al. 1979). As part of MIBG's early characterization as an adrenal imaging agent, its avid uptake and retention in canine heart tissue was noted as being "consistent with the rich sympathetic innervation of the heart" (Wieland et al. 1980). Thus, MIBG's potential as a cardiac sympathetic nerve imaging agent was recognized early during its development, and this secondary imaging application of MIBG was methodically studied by Wieland and coworkers (Raffel and Wieland 2010). Ultimately, MIBG's application as a cardiac nerve marker has proven to be important one, allowing noninvasive scintigraphic assessments of the damage to cardiac sympathetic neurons in a wide range of cardiac diseases (Higuchi and Schwaiger 2006).

Interest at the University of Michigan in developing an MIBG-like radiotracer for positron emission tomography (PET) studies ultimately led to the invention of

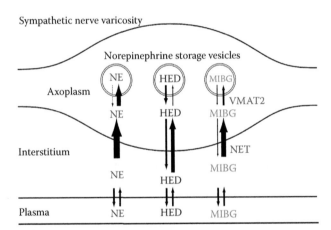

FIGURE 17.1 Schematic illustration of the uptake of norepinephrine (NE), [^{11}C]*meta*-hydroxyephedrine (HED), and [^{123}I]*meta*-iodobenzylguanidine (MIBG) into cardiac sympathetic nerve varicosities. After extraction from plasma, HED and MIBG are transported into sympathetic nerve terminal axons by the norepinephrine transporter (NET). Once in the neuronal axoplasm, they are subsequently transported into norepinephrine storage vesicles by the second isoform of the vesicular monoamine transporter (VMAT2). Arrow thicknesses are drawn in approximate proportion to the relative rate of each process.

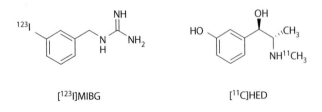

FIGURE 17.2 Chemical structures of MIBG and HED.

HED (Rosenspire et al. 1990). Don Wieland noted from literature reports that the sympathomimetic amine metaraminol was a "false neurotransmitter" that was avidly taken up into sympathetic neurons and vesicles, but resistant to metabolism by monoamine oxidase (MAO) in the neuronal axoplasm (Leitz and Stefano 1971). The first metaraminol analog tested was 6-[^{18}F] fluorometaraminol (6-FMR), which localized in sympathetic neurons in the rat heart with very high selectivity (Mislankar et al. 1988). PET studies in dog hearts with variable amounts of regional denervation induced by epicardial application of the neurotoxin phenol demonstrated that regional 6-FMR uptake was highly correlated with local tissue concentrations of norepinephrine (Wieland et al. 1990). However, these promising imaging studies also revealed that the low-specific activity of 6-FMR (<370 GBq/mmol = 10 Ci/mmol), which was

synthesized using an electrophilic substitution reaction, meant that enough unlabeled 6-FMR was injected to cause a significant increase in blood pressure (pressor effect). Radiosynthetic methods for preparing high-specific-activity 6-FMR were not readily available in 1990, and so this roadblock precluded the use of 6-FMR in human subjects. Instead, Wieland chose to prepare a high-specific-activity carbon-11-labeled analog of metaraminol: N-[^{11}C]methyl-metaraminol, or, equivalently, [^{11}C]meta-hydroxyephedrine (HED). Like 6-FMR, HED was found to have very high selectivity for cardiac sympathetic neurons in rats (Rosenspire et al. 1990). HED has been adopted by several PET research centers and has provided groundbreaking clinical information on changes in regional sympathetic innervation in numerous disease states (Lautamäki et al. 2007).

17.3 Specific Activity

Methods for directly measuring the regional cardiac NET density in human subjects are not currently available, but in vitro measurements of heart tissue samples suggest that NET densities in human left ventricular tissue are high (Böhm et al. 1995). There is also evidence that cardiac NET densities may be different in various mammalian species (Raffel and Chen 2004), and this needs to be kept in mind when interpreting results from different animal models. In a study of the effects of hypoxia on adrenergic neurotransmission in the hearts of Wistar rats, Mardon et al. (1998) measured presynaptic NET densities and postsynaptic β-adrenergic receptor (βAR) densities using in vitro-binding assays. In a membrane preparation made from purified left ventricle (LV) homogenates, control values for NET density were 523 ± 48 fmol/mg (using [^{3}H]mazindol as the radioligand) and for βAR density were 33 ± 2.9 fmol/mg (using [^{3}H]CGP 12177). These data suggest that NET density is approximately 16-fold higher than βAR density in the hearts of Wistar rats. In normal human hearts, cardiac βAR densities have been estimated to be ~6.6 pmol/g tissue, based on tracer kinetic modeling studies with the PET radioligand [^{11}C]CGP-12177 (Merlet et al. 1993). If the relative proportion of presynaptic NET/postsynaptic βAR in humans is comparable to that in the rat, this would suggest that NET densities in human heart is of the order of (16)(6.6) = 106 pmol/g. Assuming a cardiac tissue density of 1.05 g/mL, this would correspond to a tissue NET concentration of ~100 nM.

Radioiodinated MIBG was originally synthesized using an isotopic exchange reaction that yielded specific

activities <185 MBq/mg, or ~63 GBq/mmol (<5 mCi/mg, or ~1.6 Ci/mmol) (Mangner et al. 1982). Mainly driven by a desire to improve MIBG uptake into adrenergic tumors, several groups have successfully developed methods of preparing no-carrier-added (n.c.a.) MIBG at very high-specific activities, in excess of 74 TBq/mmol (2000 Ci/mmol) (DeGrado et al. 1998, Donovan and Valliant 2008, Mairs et al. 1995, Vaidyanathan and Zalutsky 1993).

The transport kinetics of NET substrates like MIBG and HED are well described by Michaelis–Menten kinetics, with a half-saturation concentration (K_m) and maximum transport velocity (V_{max}) specific for each substrate (Schömig et al. 1988). The "initial velocity of transport" of the substrate $V_{init} = S_o V_{max}/(K_m + S_o)$, with S_o representing the concentration of the substrate outside neurons in the synaptic cleft. Similarly, VMAT2 transport of the tracers from neuronal axoplasm into storage vesicles is also characterized by a set of Michaelis–Menten transport constants (K_m^*, V_{max}^*) that are unique to each substrate.

Theoretically, one would want the specific activity of a neuronal tracer to be high enough so that (a) tracer concentrations in the extracellular space (C_{ecs}) would be much lower than the K_m value for NET transport (i.e., $C_{ecs} \ll K_m$), and (b) tracer concentrations in the neuronal axoplasm (C_{axo}) would be much lower than the K_m^* of VMAT2 transport (i.e., $C_{axo} \ll K_m^*$). If these conditions are fulfilled, "effective transport rate constants" for NET transport and VMAT2 transport would be given by the ratios $k_3 = V_{max}/K_m$ and $k_5 = V_{max}^*/K_m^*$, respectively

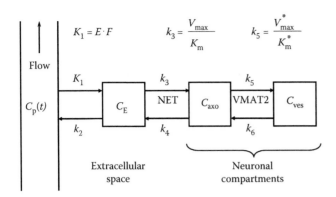

FIGURE 17.3 Comprehensive compartmental model of the neuronal uptake and retention of MIBG and HED by cardiac sympathetic neurons. The extraction of tracer in plasma (C_p) into extracellular space is characterized by the rate constant K_1 (mL/min/g) $= E \cdot F$, where E is the unidirectional extraction fraction and F is myocardial blood flow (mL/min/g). For high-specific-activity tracers, the rate constant for NET transport into sympathetic neurons k_3 (min^{-1}) $= (V_{max}/K_m)$, where K_m and V_{max} are the Michaelis–Menten transport constants for NET transport of the tracer. Similarly, the rate constant for uptake into storage vesicles k_5 (min^{-1}) $= (V_{max}^*/K_m^*)$, where K_m^* and V_{max}^* are the Michaelis–Menten transport constants for VMAT2 transport. Diffusion from each compartment is also possible (k_2, k_4, k_6, all with units min^{-1}).

(Figure 17.3). Conversely, if specific activities are low, it is possible that either NET or VMAT2 transport could be saturated, altering the neuronal uptake and retention of the tracer.

The Michaelis–Menten transport constants for NET and VMAT2 transport in cardiac sympathetic neurons are difficult to measure *in vivo*, and are likely to be highly species dependent due to variation in cardiac NET and VMAT2 densities. Using high-specific-activity [131I]MIBG with variable amounts of cold MIBG added, DeGrado and coworkers measured the K_m and V_{max} values for NET transport of MIBG in an isolated working rat heart preparation (DeGrado et al. 1998). Expressed in terms of the concentration of [131I]MIBG in heart perfusate, a $K_m = 52$ nM was estimated, with a corresponding $V_{max} = 0.23$ nmol/min/g wet.

Using this measured K_m value to derive a maximum desirable arterial plasma concentration for human imaging studies is challenging. However, using an alternative approach, DeGrado et al. observed that cardiac MIBG retention levels following a 1-h-washout study in the isolated rat heart were not significantly different when initial MIBG loading doses were <0.5 nmol/g. Extrapolating these tissue values into typical MIBG imaging parameters in human subjects, it was estimated that specific activities above 185 MBq/mg, or ~63 GBq/mmol (5 mCi/mg, or ~1.6 Ci/mmol) would not be likely to result in MIBG loading doses in cardiac tissue above 0.5 nmol/g. Supporting this estimate, a direct comparison of [123I]MIBG at a specific activity of 229 MBq/mg = 74 GBq/mmol (6.2 mCi/mg = 2.0 Ci/mmol) and n.c.a. [123I]MIBG (with specific activity 100,000 times higher) demonstrated that while n.c.a. [123I]MIBG had modestly higher heart uptake, both formulations provided comparable image quality and similar heart-to-liver and heart-to-blood ratios (Farahati et al. 1997).

HED is prepared by [11C]-methylation of (–)-metaraminol using either [11C]methyl iodide (Rosenspire et al. 1990) or [11C]methyl triflate (Jewett 1992). Initially at our institution, specific activities of 18.5–55.5 TBq/mmol (500–1500 Ci/mmol) were typically achieved, with >98% chiral purity. Currently, using a General Electric Medical Systems PETtrace cyclotron and FX C-Pro-automated radiosynthesis module, we routinely prepare HED at specific activities >92.5 TBq/mmol (2500 Ci/mmol).

The NET transport parameters (K_m, V_{max}) for HED are likely to be of a magnitude similar to those of MIBG. The ratio of V_{max}/K_m, which is measured as a neuronal uptake rate constant (K_{up}) in the isolated rat heart, was found to be 3.6 mL/min/g wet for MIBG versus 2.7 mL/min/g wet for HED (DeGrado et al. 1993, 1995). In light of the much higher-specific activity of HED relative to MIBG, issues related to the mass associated with injected doses for imaging should not be a major concern for HED.

17.4 Specificity

MIBG and HED are each highly selective for presynaptic sympathetic nerve terminals in the heart. When interpreting data from MIBG studies performed *in vivo*, it is important to bear in mind that a second "extraneuronal" monoamine uptake system, referred to as "uptake-2," exists in the myocytes of some nonhuman mammalian species. Studies in the isolated rat heart demonstrated that MIBG is a good substrate for

uptake-2 (DeGrado et al. 1995). Cardiac uptake-2 activity is important in the rat (Iversen 1965), cat (Graefe 1981), and dog (Eisenhofer et al. 1992), but is absent or poorly developed in the rabbit (Graefe et al. 1978). Imaging studies with a radioiodinated bretylium analog, [131I]-*o*-iodobenzyltrimethylammonium iodide, which proved to be a selective uptake-2 substrate, provided direct evidence that uptake-2 activity is present in pig hearts, but is absent in the hearts of monkeys and humans (Carr et al. 1979). MIBG does not localize in the denervated hearts of cardiac transplant patients, consistent with uptake-2 being negligible in human heart (Dae et al. 1992, Glowniak et al. 1989). HED, on the other hand, is a very poor substrate for uptake-2 (DeGrado et al. 1993). This is not completely surprising as the radiosynthetic precursor of HED, (–)-metaraminol, is known to have very low potency for inhibiting uptake-2 (Burgen and Iversen 1965).

17.4.1 *Meta*-Iodobenzylguanidine

Early studies of the specificity of cardiac MIBG uptake for sympathetic neurons in the rat heart compared the effects of cardiac denervation produced by administration of the neurotoxin 6-hydroxydopamine (6-OHDA) and pharmacological blockade of NET transport by the NET-selective inhibitor desipramine (DMI). 6-OHDA (100 mg/kg, i.p.), which is known to produce almost complete denervation of the rat heart, was administered 5 days prior to intravenous injection of [^{125}I]MIBG or [^{3}H]norepinephrine (Sisson et al. 1987). Cardiac levels of the tracers were determined 2 h after injection, and 6-OHDA reduced [^{3}H]norepinephrine levels to 12% of control levels, while [^{125}I]MIBG uptake was reduced to 31% of controls. In comparison, NET inhibition by injection of DMI (10 mg/kg, i.p.) 2 h prior to tracer injection resulted in reductions in cardiac uptake to 6% of controls for [^{3}H]norepinephrine and 50% of controls for [^{125}I]MIBG. Thus, while 6-OHDA and DMI block were both highly effective at reducing the neuronal uptake of [^{3}H]norepinephrine, extraneuronal uptake into myocytes, mediated by uptake-2, led to higher residual cardiac uptake of [^{125}I]MIBG.

Pretreatment of dogs with the selective VMAT2 inhibitor reserpine (1 mg/kg, i.m.) led to a 33% reduction of the cardiac retention of [^{125}I]MIBG 2 h after tracer injection (Wieland et al. 1981). In rats, pretreatment with reserpine (4 mg/kg, i.p.) caused a 50% reduction in cardiac uptake of [^{131}I]MIBG 4 h after injection (Nakajo et al. 1986). Again, extraneuronal uptake of MIBG in these animal species complicates interpretation of these studies, but it can be inferred from the data that VMAT2 transport of MIBG into norepinephrine storage vesicles significantly increases the neuronal retention of the tracer.

More recently, studies of MIBG retention in the rabbit heart, which lacks uptake-2 activity, were performed (Nomura et al. 2006). In rabbits, chemical sympathectomy of the heart using 6-OHDA caused a large reduction of [^{123}I]MIBG retention, down to just 10% of control levels. Pretreatment of rabbits with reserpine (2 mg/kg, i.p.) 2 h before imaging only caused a modest reduction of [^{123}I]MIBG retention at 40 min after injection (18% less than control levels), suggesting a relatively slow leakage rate of MIBG from neurons in the absence of vesicular storage. The very large reduction in [^{123}I]MIBG retention observed in the 6-OHDA-treated hearts demonstrates MIBG's high selectivity for sympathetic nerve terminal axons.

As touched on previously, some of the most compelling evidence for the neuronal selectivity of MIBG in human heart comes from studies in patients that have received heart transplants. In the hearts of patients studied within 6 months of receiving their transplant, none retained any significant accumulation of MIBG (Dae et al. 1992, Glowniak et al. 1989). Thus, in human heart, cardiac MIBG retention is exclusively due to NET transport into presynaptic sympathetic neurons.

MIBG is chemically stable inside sympathetic neurons. Thin-layer chromatographic analysis of tissue extracts of rat hearts excised 30 min after injection of [^{123}I]MIBG demonstrated that 100% of the radioactivity was associated with intact parent MIBG (Pissarek et al. 2002). This indicates that MIBG is resistant to metabolism by neuronal enzymes such as mitochondria-bound MAO and dopamine-β-hydroxylase inside storage vesicles.

MIBG has been shown to impair mitochondrial respiration by inhibiting complex I of the respiratory chain (Loesberg et al. 1990). Studies using beef heart submitochondrial particles demonstrated that the maximal effect of MIBG on mitochondrial respiration was comparable to that of rotenone (Cornelissen et al. 1997). Electron spectroscopic imaging studies of the localization of MIBG uptake in cells from the human neuroblastoma cell line NB1-G determined that a large fraction of MIBG uptake was within mitochondria, in addition to uptake in small vesicular structures and the nuclear membrane (Gaze et al. 1991). Considering these reports, it is conceivable that interaction of MIBG with mitochondria in the axoplasm of cardiac sympathetic neurons could be an important mechanism for prolonging the neuronal retention of MIBG, independent of MIBG's vesicular uptake. To our knowledge, this hypothesis has not yet been tested.

In terms of any potential binding of MIBG to cardiac postsynaptic βARs, it has been shown that MIBG has no activity at these receptors, as evidenced by its

Chapter 17

inability to inhibit binding of [^{125}I]iodocyanopindolol to βAR populations in rat heart and spleen tissue sections (Pissarek et al. 2002).

17.4.2 Hydroxyephedrine

The high selectivity of HED for cardiac sympathetic nerve varicosities has been consistently demonstrated in several animal models and in human studies. Initial studies focused on biodistribution and pharmacologic-blocking studies in rats. For example, pretreatment of rats with the potent NET inhibitor DMI (10 mg/kg, i.p.) prior to tracer injection reduced HED retention at $t = 30$ min in rat LV to only 8% of control values, confirming high selectivity for sympathetic neurons (Rosenspire et al. 1990). The importance of the vesicular uptake of HED on its neuronal retention was demonstrated by pretreating rats with the VMAT2 inhibitor reserpine (1 mg/kg) 3 h before tracer administration, which caused an 82% reduction in HED retention at $t = 30$ min (Schwaiger et al. 1990). HED is metabolically stable inside sympathetic neurons, with HED representing >99% of radioactivity in rat hearts 60 min after tracer injection (Law et al. 1997).

Another early study in rats tested the ability of HED to track cardiac reinnervation. Chemical sympathectomy of the heart was effected in a group of rats ($n = 25$) using 6-OHDA (100 mg/kg, i.p.). HED retention in the LV at $t = 30$ min was measured in a sham-injected control group ($n = 6$), and compared with the same measurement in five subgroups of the 6-OHDA-treated rats, performed at different time points following the 6-OHDA-induced denervation (Figure 17.4).

Twenty-four hours after 6-OHDA injection (Day 1), cardiac retention of HED was only 11% of control levels, again demonstrating the high selectivity of HED for cardiac sympathetic neurons. Cardiac HED retention was then seen to steadily increase over the following weeks, until at Day 63 after denervation, HED retention was back to 72% of the control levels. This study confirmed that HED was capable of tracking the relatively rapid reinnervation that occurs in rat hearts following chemical sympathectomy (Nori et al. 1995).

Further preclinical studies of HED included extensive PET-imaging studies in dogs. Control studies demonstrated that HED was rapidly taken up by cardiac sympathetic neurons, with essentially no subsequent clearance from the heart. When cardiac NET transporters were blocked by pretreatment with DMI (5 mg/kg, i.v.) cardiac HED retention was greatly reduced (Figure 17.5), again demonstrating the high selectivity of HED for sympathetic nerve terminals. Important insights into the dynamics of the neuronal retention of HED were obtained in pharmacological "chase" studies. In a dog in which the normal rapid cardiac uptake and long retention of HED was allowed to develop, at $t = 30$ min, DMI (5 mg/kg, i.v.) was administered to effect a rapid blockade of NET transporters. This DMI "chase" led to a rapid efflux of HED from the heart (Figure 17.5), indicating that HED is continually leaking from sympathetic nerve terminals,

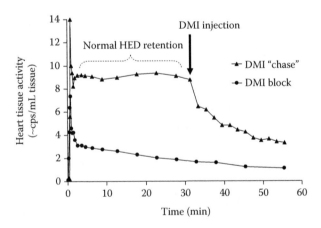

FIGURE 17.5 Kinetic studies of the effect of pharmacologically blocking NET transport in cardiac sympathetic neurons with desipramine (DMI) in the dog heart. The top curve shows a normal HED study initially, with DMI (5 mg/kg) administered as an intravenous bolus at $t = 30$ min (DMI "chase"). Blocking reuptake at NET causes a rapid efflux of HED from the heart, indicating that HED molecules normally undergo a continuous leak-and-reuptake cycle at NET. DMI block before administration of HED greatly reduces the cardiac uptake and retention of HED, showing its high selectivity for cardiac sympathetic nerve varicosities.

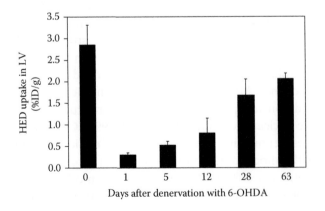

FIGURE 17.4 Data showing the ability of HED to track reinnervation of rat hearts following chemical denervation of the heart by systemic injection of the neurotoxin 6-hydroxydopamine (6-OHDA). HED retention in the LV is expressed as the percentage of injected dose per gram (%ID/g).

only to be taken back up into the neurons by NET transport. Thus, the normal long retention time of HED in control studies masks a continuous leak–reuptake cycle of the HED molecules. Similar studies in dogs using cocaine, another potent NET inhibitor, showed that cocaine "chase" was just as effective as DMI in blocking neuronal reuptake of HED, leading to a rapid efflux of HED from the heart (Melon et al. 1994).

The ability of DMI to "chase" HED from cardiac sympathetic neurons was also seen in detailed studies of HED retention mechanisms in the isolated rat heart (DeGrado et al. 1993). In this experimental model, because the Krebs–Henseleit heart perfusion buffer cannot carry as much oxygen as blood, the heart autoregulates its coronary flow rate to values more than 10 times physiological levels, enhancing the clearance rate of HED over that seen in intact animals. In control studies, HED efflux from the perfused rat heart occurs with a clearance half-time of ~63 min. If DMI is added to the perfusate during the washout study, HED clearance is greatly accelerated, to a half-time of only 3.6 min. In light of the rapid turnover of the extracellular space in this model, the measured 3.6 min half-time during DMI chase largely reflects the rate of HED leakage from sympathetic neurons. DMI-blocking studies in the isolated rat heart also confirmed HED's high selectivity for sympathetic neurons, as well as the low affinity of HED for extraneuronal uptake (DeGrado et al. 1993). Also, pretreatment of rats with the VMAT2 inhibitor reserpine (1 mg/kg, i.p.) 3 h before heart isolation leads to a reduction in the neuronal uptake of HED, and greatly enhances its clearance from the isolated rat heart (Raffel and Wieland 2001). This shows that the leak–reuptake cycling of HED at the neuronal NET transporter is not sufficiently efficient by itself to maintain the long neuronal residence times observed *in vivo* in intact animals. HED uptake into storage vesicles (from which HED leaks at a significant rate as well) is an important process that increases the neuronal distribution volume of HED and also serves to prolong the neuronal retention of the tracer.

The selectivity of HED for cardiac sympathetic nerve terminals has also been shown in human subjects. In an early study, cardiac HED retention measured in recent heart transplant patients was only 23% of the levels seen in healthy controls (Schwaiger et al. 1990). The very low retention of HED in recently transplanted hearts has been consistently documented in studies that have used HED to track the relatively slow and incomplete reinnervation that occurs in these patients (Bengel et al. 1999, Überfuhr et al. 2000). A study of the effect of DMI (<1 mg/kg, p.o.) in a small group of normal volunteers caused a reduction of cardiac HED retention to <34% of control values (Raffel et al. 1996). More recently, PET studies with HED in patients with Parkinsonian syndromes showed complete cardiac denervation in several patients with Parkinson's disease, with residual HED retention levels <33% of controls (Raffel et al. 2006b). Together, these clinical data indicate that HED is highly selective for sympathetic nerve terminals in the human heart.

17.5   Lipophilicity

Radiotracers for the heart do not face the normal blood–brain barrier concerns that impose a "log P window" on brain imaging agents (Dischino et al. 1983). Usually, hydrophilic compounds with low (or negative) log P values are advantageous for heart imaging agents, as this tends to greatly reduce nonspecific binding to myocardial cells. As mentioned earlier, the very high log P values of most potent NET inhibitors like desipramine and mazindol (log P > 5) cause radiolabeled analogs of these compounds to fail as cardiac-imaging agents due to their very high levels of nonspecific uptake in the heart. MIBG is an interesting molecule in that the iodobenzyl end of the molecule is very lipophilic, while the high pK_a of the guanidine moiety in the side chain makes that part of the molecular highly polar and hydrophilic. A calculated log P (CLOGP) value of 2.3 has been reported for MIBG (Advanced Chemistry Development ACD/Labs software V11.02). Cardiac uptake of MIBG in recent heart transplant recipients, even within a few minutes after tracer injection, has been reported to be very low, demonstrating that the tracer has very little nonspecific uptake in human heart (Dae et al. 1992). This finding is also consistent with the view that "extraneuronal uptake" (uptake-2) is either absent or poorly developed in human heart. A log P of 0.31 has been estimated for HED (Raffel and Wieland 1999), which makes it more lipophilic than the endogenous catecholamines norepinephrine (log P = −1.74) and epinephrine (log P = −1.34). HED has consistently demonstrated very low nonspecific cardiac uptake in animal and human studies. However, HED has sufficiently high lipophilicity that it readily diffuses across the membranes of the storage vesicles and sympathetic neurons, as described above. Based on kinetic studies in

Chapter 17

the isolated rat heart, MIBG does not appear to diffuse across cell membranes as rapidly as HED, probably due to the high polarity of the guanidine group (DeGrado et al. 1995, Raffel et al. 1998).

17.6 Kinetic Studies

Transport rates of NET substrates like MIBG and HED are described by the Michaelis–Menten equation, $V_{init} = S_o V_{max}/(K_m + S_o)$, with S_o representing the tracer concentration outside neurons in the synaptic cleft. However, if "tracer" concentrations of a radiolabeled substrate are used for imaging studies, then $S_o \ll K_m$ at all times, simplifying the transport rate equation to $V_{init} = (V_{max}/K_m)S_o$. This equation shows that it is the ratio of a substrate's Michaelis–Menten transport parameters V_{max} and K_m that determines its *in vivo* neuronal uptake rate. Thus, a "neuronal uptake rate constant" $k_{uptake} = V_{max}/K_m$ can be defined for each NET substrate. Furthermore, the maximum velocity of transport, V_{max}, is directly proportional to NET density. Obviously, in diseases in which sympathetic denervation occurs, regional NET density will decrease. This will cause V_{max} to decline in direct proportion to reductions in NET density. Hence the value of $k_{uptake} = V_{max}/K_m$ will also decline in direct proportion to the degree of NET density deficits. It can thus be argued that the most sensitive measure of cardiac NET density that could be obtained from imaging studies with radiolabeled NET substrates is an accurate estimate of k_{uptake} from the kinetic data.

The fact is, MIBG and HED are both excellent NET substrates, meaning that they possess very high values of the tracer-level transport constant $k_{uptake} = V_{max}/K_m$. Their rapid neuronal uptake rates are evidenced by the very rapid loading of cardiac neurons seen in clinical imaging studies with these tracers. While this is good for providing high-quality scintigraphic images of the heart, it also means that their neuronal uptake is rate limited not by NET transport, but rather by the delivery of the tracer from plasma to interstitium. In other words, MIBG and HED are "flow-limited tracers." One drawback of using flow-limited tracers is that the application of tracer kinetic modeling methods fails, because it is not possible to accurately estimate the rate constant $k_{uptake} = V_{max}/K_m$ from the measured tracer kinetics.

Since standard kinetic modeling methods with MIBG and HED have proven unsuccessful, it has been necessary to use semiquantitative measures of tracer retention to assess cardiac sympathetic nerve populations. For planar scintigraphy, a commonly employed measure of cardiac MIBG retention is the heart-to-mediastinum ratio (H/M), which is obtained by dividing the mean counts/pixel in a region of interest (ROI) drawn around the entire heart (H) to those in a reference ROI placed on the upper mediastinum (M) in a planar chest image acquired 4 h after MIBG injection (Merlet et al. 1992). Another approach, often calculated in addition to H/M, is to compare MIBG uptake in an early image (15 min) and a delayed image (4 h) and calculate a "washout rate": WR (%) = {[$(H - M)_e - (H - M)_d$]/$(H - M)_e$} × 100%.

For analysis of cardiac HED retention, a more detailed regional analysis of tracer retention is performed, and the results statistically compared with a healthy control database. Currently, at our institution, PET images are acquired in a 40 min dynamic sequence (frame rates 12 × 10 s, 2 × 30 s, 2 × 60 s, 2 × 150 s, 2 × 300 s, and 2 × 600 s). After image reconstruction, the LV is divided into many subregions or "sectors" (480 total), and the tissue concentration of HED in the final image frame is extracted for each LV sector. Also, a small region of interest is placed in the LV blood chamber at the base of the heart to determine the time–activity curve of radioactivity in blood. A "retention index" (RI), with units of mL blood/min/mL tissue, is calculated for each LV sector by normalizing the final HED tissue concentration in the region to the integral of the HED time–activity curve in blood:

$$\text{RI (mL blood/min/mL tissue)} = \frac{\text{regional tissue radioactivity in the final image frame}}{\int_{t=0}^{t=40} \text{blood radioactivity in the LV chamber}}.$$

The calculated RI values are saved in a file and can be displayed and visually inspected in the conventional "polar map" format used throughout nuclear cardiology. For statistical assessments, the RI values for a particular subject are compared to a database of RI values for healthy controls using "z-score analysis." For this analysis, a z-score is calculated for each sector in a patient's polar map as $z_i = (\mu_i - q_i)/\sigma_i$, where q_i is the patient's RI value for the ith sector of the polar map, μ_i is the healthy control population mean RI for that sector, and σ_i is the corresponding across-subject standard deviation of the healthy control population for that sector. Left ventricular sectors with *z*-scores more than 2.5 (i.e., their RI values are more than 2.5 standard deviations below the normal mean) are considered to have "abnormal" HED retention. The fraction of all sectors in each patient's polar map that are abnormal is calculated as a measure of the "extent" of abnormal HED retention in the LV, and this used as a measure of LV denervation. Some investigators have also reported the average *z*-score value for those sectors that test as "abnormal," and used this as a measure of the "severity" of the denervation, as a higher *z*-score value would imply a larger HED retention deficit (i.e., a lower HED retention level).

17.7 Sensitivity

The sensitivities of the retention of MIBG and HED to changes in cardiac NET density have been studied in a few animal models, but extrapolation of the results of these studies to clinical cardiac neuroimaging in humans is not straightforward. Studies in the rat support a linear relationship between the cardiac retention of HED or MIBG and NET density, while data from clinical studies with HED and MIBG tend to support a nonlinear relationship in human subjects.

Focusing on HED for the moment, we had previously demonstrated that NET density in the rat heart could be decreased in a dose-dependent manner by treatment with increasing doses of the sympathetic neurotoxin 6-OHDA (Raffel and Chen 2004). Using 6-OHDA doses ranging from 3 to 100 mg/kg i.p., we found an EC_{50} for 6-OHDA reduction of cardiac NET density of 12.4 ± 0.5 mg/kg, with a Hill slope of -2.8. We used this 6-OHDA rat model of cardiac denervation to map the dependence of HED retention on NET density by correlating *in vivo* retention of HED at $t = 30$ min after HED injection to *in vitro* measurements of cardiac NET density in the same hearts (Raffel et al. 2006a). A strong linear correlation ($r^2 = 0.95$) between cardiac HED retention and NET density was observed for data from all rats, over the range of control animals down to almost complete denervation at the highest 6-OHDA doses. These data demonstrated that cardiac HED retention was a very good surrogate measure of NET density in the rat heart.

Comparable studies measuring HED retention versus NET density have not been performed in other animal species, but one early study of HED retention in varying levels of denervation in the dog heart may provide some insights. HED retention in canine heart was determined as a function of varying degrees of denervation induced by "painting" the surface of the heart with varying concentrations of the sympathetic neurotoxin phenol, which penetrates a few mm into the epicardial wall and causes sympathetic nerve destruction. Animals ($n = 2$) were killed immediately following an HED imaging study and their hearts sectioned

(20 sections each, $N = 40$ total). Tissue norepinephrine concentrations were determined for each tissue section as a surrogate measure of nerve density and the data regionally correlated to corresponding PET-derived measures of HED retention (Figure 17.6). The data are consistent with HED behaving as a flow-limited tracer in the dog heart, giving rise to a nonlinear relationship between HED retention and nerve density. Figure 17.6 illustrates the problem with using retention measures of a flow-limited nerve tracer: nerve losses need to become significant before the observed tracer retention begins to decline. For a flow-limited tracer, modest levels of nerve losses that occur early in the course of a disease that causes progressive cardiac denervation would be missed using semiquantitative measures of tracer retention.

There is indirect evidence of a nonlinear relationship between HED retention and NET density in human heart. Ungerer et al. (1998) assessed cardiac sympathetic innervation with HED and perfusion with [^{13}N]ammonia in eight patients with severe heart failure who were awaiting heart transplantation. The

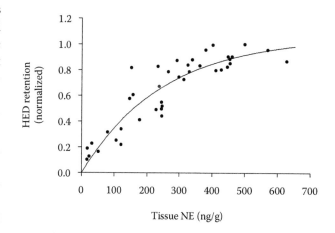

FIGURE 17.6 Correlation of regional HED retention with tissue norepinephrine (NE) concentrations in dog hearts with varying degrees of regional denervation induced by the application of different concentrations of the neurotoxin phenol to the epicardial wall.

PET studies were done <3 months before the patients underwent transplantation. After surgery, tissue samples were taken from nine regions of the explanted hearts and assayed for NET density, which were then correlated to HED "RI" values in the corresponding regions of the heart. The data showed a linear correlation between the HED retention and NET density. However, this linear relationship may have represented the gently sloping "plateau" region of a nonlinear relationship between HED retention and NET density. In going from the highest measured regional NET density values (~283 fmol/mg protein) to the lowest (~89 fmol/mg protein), a 69% decline in NET density, HED retention values decreased only 20%. This suggests a lack of sensitivity of HED retention as a measure of NET density in patients with severe heart failure. It is also consistent with HED exhibiting retention characteristics of a flow-limited tracer in human heart, since large reductions in NET density resulted in only mild reductions in the observed HED retention.

Without further data, we are left to speculate on why HED has a linear dependence on NET density in rat heart but a nonlinear relationship in the hearts of higher mammalian species like the dog or human. The main determinant of whether or not a particular sympathetic nerve tracer will be a flow-dependent tracer is the relative magnitudes of its rate constants k_2 and k_3 as shown in Figure 17.3. If the NET transport rate k_3 is much faster than the clearance rate back into plasma ($k_3 \gg k_2$), then the fraction of tracer molecules delivered into extracellular space that are subsequently transported into neurons approaches 100%, since this fraction is equal to the ratio $k_3/(k_2 + k_3)$. If $k_3 \gg k_2$, then $k_3/(k_2 + k_3) \cong 1$. Rapid NET transport causes the neuronal uptake of the tracer to be "rate limited" by delivery from plasma to interstitium (K_1 in Figure 17.3) rather than by NET transport (k_3). Since $K_1 = E \cdot F$, the product of the dimensionless unidirectional extraction fraction (E) and myocardial blood flow (F; mL blood/min/g tissue), such a tracer is called a "delivery-limited" or "flow-limited" tracer. In this case, in areas of the heart with moderate declines in regional nerve density (with corresponding declines in NET density), the surviving neurons would have sufficient capacity to transport most of the tracer molecules delivered from plasma to interstitium, and tracer retention measures would not significantly decline. Only once nerve losses became more severe would retention measures begin to drop off significantly. This is what leads to the nonlinear relationship

between retention and NET density for flow-limited tracers. The rather modest declines in HED retention versus NET density reported in heart failure patients (Ungerer et al. 1998) would be consistent with HED behaving as a flow-limited tracer. If, on the other hand, the NET transport rate k_3 was less than clearance rate back into plasma ($k_3 < k_2$), then the neuronal uptake of the tracer would be rate limited by NET transport, and the relationship between tracer retention and NET density would tend to be linear.

Based on the above arguments, it appears that for HED in the rat heart, the magnitudes of the rate constants k_2 and k_3 satisfy the relationship $k_3 < k_2$, leading to a linear relationship between HED and NET density in this species. On the other hand, in dogs and humans, the rate constants for HED have the relationship $k_3 \gg k_2$, leading to a nonlinear relationship between cardiac HED retention and NET density.

Two factors may be contributing to the differences in the relative magnitudes of the rate constants k_2 and k_3 in different species. First, resting myocardial perfusion in the rat heart is much higher than in the dog and human heart. Resting myocardial perfusion values of 3.5–3.7 mL/min/g are typical in the rat heart (Croteau et al. 2004, Waller et al. 2000), while reported values in dog heart ranged 0.85–1.0 mL/min/g (Deussen et al. 1996, Wei et al. 2001) and those of human heart ranged 0.70–0.97 mL/min/g (Hutchins and Schwaiger 1993). Since the rate constant k_2 is a flow-related parameter that increases with myocardial blood flow, it would be expected that k_2 values for HED in rat heart would be considerably higher than those in dog or human heart. Second, there is indirect evidence that NET densities in dog and human heart may be higher than those in the rat heart. In terms of the relative cardiac NET densities in different species, note that comparing *in vitro-* binding assay measurements between different laboratories is frequently unreliable. Each lab uses different methods for tissue homogenization, preparation of purified membranes using differential centrifugation, and assay filtration techniques, which lead to different absolute quantitative measurements. However, comparing assay results within the same lab under identical assay methods may be more reliable in terms of estimating relative concentrations between species. For example, in our lab we used common binding assay methods to measure cardiac NET densities (B_{max}) in Sprague–Dawley rats (461 ± 58 fmol/mg protein; $n = 6$), New Zealand white rabbits (250 ± 49 fmol/mg protein; $n = 3$), and mongrel dogs (881 ± 197 fmol/mg protein; $n = 5$)

(Raffel and Chen 2004). Thus, it appears that NET density follows the rank order: dog > rat > rabbit. Böhm et al. reported cardiac NET densities of 1102 ± 37 fmol/mg protein in LV samples from human donors who were brain dead following traumatic injury (Böhm et al. 1995). Using an identical tissue preparation of rat heart tissue, Böhm's laboratory measured NET densities of only 71 ± 6 fmol/mg protein in Sprague–Dawley rat hearts, suggesting that NET density in human heart could be as high as 15-fold higher than rat heart (Böhm et al. 1998). Together with our data, these findings suggest that NET density follows the rank order: human > dog > rat > rabbit. Since the NET transport parameter V_{max} is directly proportional to NET density (B_{max}), a higher NET density in human heart implies that V_{max} values in human heart may be much higher than in rat heart. Since $k_3 = V_{max}/K_m$, the neuronal uptake rates of sympathetic nerve tracers like HED and MIBG could be much higher in human heart than in rat heart. If in humans the k_3 value for NET transport is 15-fold higher and k_2 is threefold lower than in rats, this would mean that the ratio k_3/k_2 would be 45-fold higher in humans than in rats. Hence, the combined species differences in resting myocardial perfusion and cardiac NET densities on the particular values of rates constants k_2 and k_3 could explain why experiments tend to show a linear relationship between cardiac retention HED and NET density in rat hearts, whereas in human hearts, interpretation of the data suggests a nonlinear relationship.

These studies demonstrate the limitations of some animal models in predicting the kinetic behavior of cardiac sympathetic nerve tracers in humans. In addition to the above considerations, the high level of extraneuronal uptake (uptake-2) in species like the rat further complicates the interpretation of experimental results.

Switching focus to MIBG, studies in rats have shown that cardiac MIBG retention is reduced in proportion to losses of NET density that occurs as a result of aging (Kiyono et al. 2002) and diabetes (Kiyono et al. 2001). This would again be consistent with a linear relationship between MIBG retention and NET

density in rats. One interesting clinical study supports the hypothesis that MIBG is a flow-limited tracer with a nonlinear dependence on NET density in human hearts. Estorch et al. (2004) examined six patients with movement disorders who were expected to exhibit MIBG retention deficits, based on their clinical history. In baseline MIBG scans, no regional deficits in MIBG uptake were observed. A second MIBG scan was performed in each patient within a 1 week period, following an oral administration of 25 mg of amitriptyline, a tricyclic antidepressant which acts as a NET inhibitor. In these follow-up MIBG scans, regional MIBG retention deficits were revealed in all six patients. This suggests that in the baseline MIBG scans, even though regional nerve losses were present, they could not be detected because the remaining nerve populations were capable of retaining MIBG at normal levels. By inducing a partial pharmacological blockade of NET transporters with amitriptyline, these regional nerve deficits could be detected, since neuronal uptake rates of MIBG were significantly reduced throughout the heart, to the point where MIBG retention would tend to track more linearly with the levels of free NET.

Clearly, more studies of this kind are needed to better delineate the sensitivity of HED and MIBG retention to changes in nerve density. A better understanding of the sensitivity of MIBG and HED retention measures to changes in NET density suffers from the lack of a "gold standard" measurement of regional NET density in living humans. To this end, our laboratory is currently working on the development of new sympathetic nerve tracers with more optimal kinetic properties, which would allow accurate quantitative measures of regional nerve density using tracer kinetic modeling (Raffel et al. 2007). In particular, our goal is to design a tracer with a slower NET transport rate so that it is not a "flow-limited" tracer. If a successful tracer is developed, it will be of great interest to determine if it is capable of detecting regional losses of sympathetic nerves earlier in the progression of denervation than is currently possible with HED or MIBG. Such a tracer would also be a powerful tool in tracking changes in cardiac nerve density in response to therapies designed to halt or reverse cardiac denervation.

17.8 Expected Impact on Clinical Care

Cardiac neuroimaging studies with MIBG and HED have provided important insight into a critical component of cardiac function, complementing clinical

assessments of cardiac perfusion and metabolism. Deleterious effects of altered cardiac autonomic function have been implicated as underlying factors contributing

Chapter 17

to the high morbidity and mortality associated with conditions such as sudden cardiac death (Schwartz 1998), congestive heart failure (Packer 1992), diabetic autonomic neuropathy (DAN) (Ewing 1996), myocardial ischemia (Armour 1998), and cardiac arrhythmias (Zipes 1995). With more than 300,000 deaths/year in the United States attributed to sudden cardiac death alone (Virmani et al. 2001), increasing our understanding of the role of autonomic dysfunction in these diseases is an important clinical goal. Available methods for assessing sympathetic function have traditionally been limited to global measures such as hemodynamic responses to cold pressor stress or plasma norepinephrine levels. Newer techniques, such as heart rate variability measurements and [^{3}H]norepinephrine spillover techniques, have provided more cardiac-specific measures of sympathetic function (Kingwell et al. 1994). However, none of these methods provide detailed regional information about cardiac sympathetic nerve function.

MIBG has been used by many research centers around the world to investigate changes in cardiac sympathetic innervation in many disease states. The information that has been gained from these studies is remarkable, laying a foundation for more detailed investigations in the future (Flotats and Carrió 2004). For example, in patients with ischemic or idiopathic cardiomyopathy, cardiac MIBG retention expressed as the heart-to-mediastinum ratio (*H/M*) was found to be the most accurate prognostic indicator of 24 month survival (Merlet et al. 1999). Another study demonstrated that MIBG retention measures, quantified as *H/M* or as the % washout rate, are predictive of congestive heart failure episodes and sudden cardiac death in patients with ischemic heart disease, dilated cardiomyopathy, and hypertensive heart disease (Nagamatsu et al. 2007). Cardiac MIBG retention (*H/M*) was found to be an independent predictor of lethal cardiac events in a group of heart patients with surgically implanted cardioverter defibrillators, suggesting a role for MIBG imaging in risk stratifying patients for this surgery (Nagahara et al. 2008).

Since the first clinical studies with HED were reported around 20 years ago (Schwaiger et al. 1990), several PET centers have used the tracer to characterize regional changes in cardiac sympathetic nerve populations (Lautamäki et al. 2007). Valuable insights into the effects of diseases on cardiac innervation have been garnered from these studies, and support a role for cardiac neuroimaging in the management of cardiac patients (Henneman et al. 2008).

One of the noteworthy clinical findings from HED studies includes the characterization of the evolution of cardiac denervation in patients with DAN. Patients with DAN can experience selective sympathetic denervation of the heart, causing abnormal vasoconstriction and impaired myocardial perfusion and contractility. If parasympathetic denervation also occurs, patients tend to have a fixed heart rate and inadequate capacity to increase heart rate with physiological demands. Reduced cardiovascular autonomic function in diabetics with DAN is associated with increased risk of silent ischemia and mortality. Patients with severe dysautonomia are at risk of sudden cardiac death secondary to cardiac dysrhythmia (Pop-Busui 2010). Figure 17.7 shows orthogonal slice PET images of cardiac HED scans and corresponding polar maps of HED RI values in a normal subject. Also shown are representative HED data from a diabetic patient with DAN, showing profound denervation of the apex, inferior wall, and lateral wall. This pattern of left ventricular denervation is remarkably consistent in the hearts of DAN patients, with denervation beginning in the apex, moving up the inferior wall and eventually the lateral wall. The denervation never appears to spread to encompass the entire LV. An "island" of preserved innervation remains in proximal segments of the septal and anterior wall (Stevens et al. 1998). Figure 17.8 shows the progression of cardiac denervation in a female diabetic patient with poor glycemic control who was scanned with HED at baseline (age 21 years) and then again 3.5 years later (age 24 years). This young diabetic patient with DAN died from sudden cardiac death

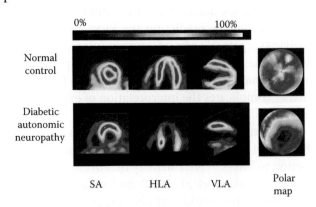

FIGURE 17.7 Representative PET images and polar maps of HED retention in a normal control subject and a patient with DAN. In normal heart, HED uptake is uniform throughout the LV. In the DAN patient, note the profound denervation of the apex, inferior wall, and lateral wall, with preserved innervation in the proximal septal and anterior walls. SA: short axis; HLA: horizontal long axis; VLA: vertical long axis. Maximum uptake of 100% is equivalent to an RI of 0.90 mL blood/min/mL tissue.

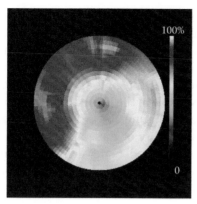

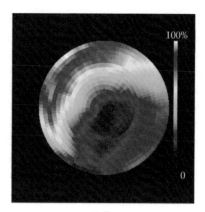

Baseline

Follow-up
3.5 years later

FIGURE 17.8 Polar maps of HED retention in a diabetic patient with DAN (female, age 21 at baseline) for two HED scans acquired 3.5 years apart. At baseline, moderate denervation of the apex and up along the entire inferior wall were observed. After 3.5 years of poor glycemic control in this patient, extensive denervation was observed, including the apex, inferior wall, and lateral wall. Maximum uptake of 100% is equivalent to an RI of 0.90 mL blood/min/mL tissue.

<1 year after the follow-up HED scan. These data illustrate the ability of HED to track the progression of denervation in these patients, providing unique clinical information on changes in the regional distribution of functional cardiac sympathetic neurons in patients with DAN.

Another classic example of the kind of insights that HED imaging has provided is the detailed mapping of regional nerve damage caused by myocardial infarction (Allman et al. 1993). HED and [¹³N]ammonia were used to assess innervation and perfusion in patients with a first acute myocardial infarction at baseline (~1 week after infarction) and again at follow-up 5–11 months later. HED imaging demonstrated that

the area of damage to nerves in the infarct zone exceeded the extent of tissue necrosis, as evidenced by deficits in resting perfusion abnormalities (Figure 17.9). The follow-up studies showed no changes in the HED and perfusion abnormalities, indicating that the neuronal damage persisted in the infarct or neighboring peri-infarct zones. This suggests that cardiac sympathetic nerve populations are even more sensitive to the ischemic insults of infarction than myocytes.

To date, nuclear cardiac neuroimaging has remained largely as a research tool. However, a few commercial vendors are currently taking major steps toward bringing this class of radiopharmaceuticals to the market. For example, GE Healthcare has previously marketed

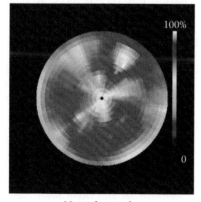
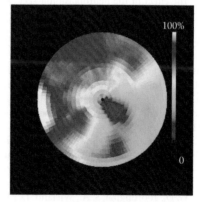

Normal control

Myocardial infarction

FIGURE 17.9 Cardiac denervation induced by myocardial infarction. HED studies showed that the area of damage to cardiac sympathetic nerves was larger than the damage to infarcted heart tissue. Follow-up studies 5–11 months later showed no changes in the HED retention abnormalities, demonstrating that the nerve damage persisted many months after the infarct event. Maximum uptake of 100% is equivalent to an RI of 0.90 mL blood/min/mL tissue.

Chapter 17

[^{123}I]-MIBG for imaging of adrenergic tumors under the trade name AdreView™. GE Healthcare has funded a multicenter clinical trial assessing MIBG imaging as prognostic tool in heart failure patients. Results of their 3 year prospective clinical trial (the "ADMIRE-HF" trial) have recently been reported (Jacobson et al. 2010). The *H/M* ratio for cardiac MIBG uptake, used as a measure of the severity of sympathetic neuronal dysfunction, was found to be significantly associated with the risk of cardiac death in heart failure patients. MIBG measures were complementary with other clinical measures such as left ventricular ejection fraction and B-type natriuretic peptide levels, improving the prediction of adverse outcomes. Thus, it appears that cardiac neuronal imaging agents may soon be commercially available in the United States for clinical assessments of cardiac sympathetic nerve dysfunction. The wider availability of these agents should generate additional data that will better define the clinical value of cardiac neuroimaging in patient management.

References

Allman, K. C., Wieland, D. M., Muzik, O., Degrado, T. R., Wolfe, E. R., and Schwaiger, M. 1993. Carbon-11 hydroxyephedrine with positron emission tomography for serial assessment of cardiac adrenergic neuronal function after acute myocardial infarction in humans. *J. Am. Coll. Cardiol.* 22: 368–75.

Armour, J. A. 1998. Myocardial ischaemia and the cardiac nervous system. *Cardiovasc. Res.* 41: 41–54.

Bengel, F. M., Überfuhr, P., Ziegler, S. I., Nekolla, S., Reichart, B., and Schwaiger, M. 1999. Serial assessment of sympathetic reinnervation after orthotopic heart transplantation. *Circulation* 99: 1866–71.

Böhm, M., Castellano, M., Flesch, M. et al. 1998. Chamber-specific alterations of norepinephrine uptake sites in cardiac hypertrophy. *Hypertension* 32: 831–37.

Böhm, M., La Rosée, K., Schwinger, R. H. G., and Erdmann, E. 1995. Evidence for reduction of norepinephrine uptake sites in the failing human heart. *J. Am. Coll. Cardiol.* 25: 145–53.

Burgen, A. S. V. and Iversen, L. L. 1965. The inhibition of noradrenaline uptake by sympathomimetic amines in the rat isolated heart. *Br. J. Pharmacol.* 25: 34–49.

Carr, E. A., Carroll, M., Counsell, R. E., and Tyson, J. W. 1979. Studies of uptake of the bretylium analogue, iodobenzyltrimethylammonum iodide, by non-primate, monkey and human heart. *Br. J. Clin. Pharmacol.* 8: 425–32.

Cornelissen, J., Van Kuilenburg, A. B. P., Voûte, P. A., and Van Gennip, A. H. 1997. The effect of the neuroblastoma-seeking agent *meta*-iodobenzylguanidine (MIBG) on NADH-driven superoxide formation and NADH-driven lipid peroxidation in beef heart submitochondrial particles. *Eur. J. Cancer* 33: 421–24.

Croteau, E., Bénard, F., Bentourkia, M., Rousseau, J., Paquette, M., and Lecomte, R. 2004. Quantitative myocardial perfusion and coronary reserve in rats with ^{13}N-ammonia and small animal PET: Impact of anesthesia and pharmacologic stress agents. *J. Nucl. Med.* 45: 1924–30.

Dae, M. W., De Marco, T., Botvinick, E. H. et al. 1992. Scintigraphic assessment of MIBG uptake in globally denervated human and canine hearts—Implications for clinical studies. *J. Nucl. Med.* 33: 1444–50.

DeGrado, T. R., Hutchins, G. D., Toorongian, S. A., Wieland, D. M., and Schwaiger, M. 1993. Myocardial kinetics of carbon-11-*meta*-hydroxyephedrine: Retention mechanisms and effects of norepinephrine. *J. Nucl. Med.* 34: 1287–93.

DeGrado, T. R., Zalutsky, M. R., Coleman, R. E., and Vaidyanathan, G. 1998. Effects of specific activity on *meta*-[^{131}I]iodobenzylguanidine kinetics in isolated rat heart. *Nucl. Med. Biol.* 25: 59–64.

DeGrado, T. R., Zalutsky, M. R., and Vaidyanathan, G. 1995. Uptake mechanisms of *meta*-[^{123}I]iodobenzylguanidine in isolated rat heart. *Nucl. Med. Biol.* 22: 1–12.

Deussen, A., Flesche, C. W., Lauer, T., Sonntag, M., and Schrader, J. 1996. Spatial heterogeneity of blood flow in the dog heart. II. Temporal stability in response to adrenergic stimulation. *Pflügers Arch. Eur. J. Physiol.* 432: 451–61.

Dischino, D. D., Welch, M. J., Kilbourn, M. R., and Raichle, M. E. 1983. Relationship between lipophilicity and brain extraction of C-11-labeled radiopharmaceuticals. *J. Nucl. Med.* 24: 1030–38.

Donovan, A. C. and Valliant, J. F. 2008. A convenient solution-phase method for the preparation of *meta*-iodobenzylguanidine in high effective specific activity. *Nucl. Med. Biol.* 35: 741–46.

Eisenhofer, G., Smolich, J. J., and Esler, M. D. 1992. Disposition of endogenous adrenaline compared to noradrenaline released by cardiac sympathetic nerves in the anaesthetized dog. *Naunyn-Schmiedeberg's Arch. Pharmacol.* 345: 160–71.

Estorch, M., Carrió, I., Mena, E. et al. 2004. Challenging the neuronal MIBG uptake by pharmacological intervention: Effect of a single dose of oral amitriptyline on regional cardiac MIBG uptake. *Eur. J. Nucl. Med. Mol. Imaging* 31: 1575–80.

Ewing, D. J. 1996. Diabetic autonomic neuropathy and the heart. *Diabetes Res. Clin. Pract.* 30(Suppl): 31–36.

Farahati, J., Bier, D., Scheubeck, M. et al. 1997. Effect of specific activity on cardiac uptake of iodine-123-MIBG. *J. Nucl. Med.* 38: 447–51.

Flotats, A. and Carrió, I. 2004. Cardiac neurotransmission SPECT imaging. *J. Nucl. Cardiol.* 11: 587–602.

Gaze, M. N., Huxham, I. M., Mairs, R. J., and Barrett, A. 1991. Intracellular localization of *meta*iodobenzylguanidine in human neuroblastoma cells by electron spectroscopic imaging. *Int. J. Cancer* 47: 875–80.

Glowniak, J., Turner, F., Gray, L., Palac, R., Lagunas-Solar, M., and Woodward, W. 1989. Iodine-123 *meta*iodobenzylguanidine imaging of the heart in idiopathic congestive cardiomyopathy and cardiac transplants. *J. Nucl. Med.* 30: 1182–91.

Graefe, K.-H. 1981. The disposition of ^{3}H-(–)-noradrenaline in the perfused cat and rabbit heart. *Naunyn-Schmiedeberg's Arch. Pharmacol.* 318: 71–82.

Graefe, K.-H., Bönisch, H., and Keller, B. 1978. Saturation kinetics of the adrenergic neurone uptake system in the perfused rabbit heart: A new method for determination of initial rates of amine uptake. *Naunyn-Schmiedeberg's Arch. Pharmacol.* 302: 263–73.

Henneman, M. M., Bengel, F. M., van der Wall, E. E., Knuuti, J., and Bax, J. J. 2008. Cardiac neuronal imaging: Application in the evaluation of cardiac disease. *J. Nucl. Cardiol.* 15: 442–55.

Higuchi, T. and Schwaiger, M. 2006. Imaging cardiac neuronal function and dysfunction. *Curr. Cardiol. Rep.* 8: 131–38.

Hutchins, G. D. and Schwaiger, M. 1993. In *Nuclear Cardiology: State of the Art and Future Directions* (Eds, Zaret, B. L. and Beller, G. A.) Mosby, St. Louis, MO, pp. 305–13.

Iversen, L. L. 1965. The uptake of catechol amines at high perfusion concentrations in the rat isolated heart: A novel catechol amine uptake process. *Br. J. Pharmacol.* 25: 18–33.

Jacobson, A. F., Senior, R., Cerqueira, M. D. et al. 2010. Myocardial iodine-123 *meta*-iodobenzylguanidine imaging and cardiac events in heart failure: Results of the prospective ADMIRE-HF (AdreView myocardial imaging for risk evaluation in heart failure) study. *J. Am. Coll. Cardiol.* 55: 2212–21.

Jewett, D. M. 1992. A simple synthesis of [^{11}C]-methyl triflate. *Appl. Radiat. Isot.* 43: 1383–85.

Kingwell, B. A., Thompson, J. M., Kaye, D. M., McPherson, G. A., Jennings, G. L., and Esler, M. D. 1994. Heart rate spectral analysis, cardiac norepinephrine spillover, and muscle sympathetic nerve activity during human sympathetic nervous activation and failure. *Circulation* 90: 234–40.

Kiyono, Y., Iida, Y., Kawashima, H., Tamaki, N., Nishimura, H., and Saji, H. 2001. Regional alterations of myocardial norepinephrine transporter density in streptozotocin-induced diabetic rats: Implications for heterogeneous cardiac accumulation of MIBG in diabetes. *Eur. J. Nucl. Med.* 38: 894–99.

Kiyono, Y., Kanegawa, N., Kawashima, H. et al. 2002. Age-related changes of myocardial norepinephrine transporter density in rats: Implications for differential cardiac accumulation of MIBG in aging. *Nucl. Med. Biol.* 29: 679–84.

Lautamäki, R., Tipre, D., and Bengel, F. M. 2007. Cardiac sympathetic neuronal imaging using PET. *Eur. J. Nucl. Med. Mol. Imaging* 34: S74–S85.

Law, M. P., Osman, S., Davenport, R. J., Cunningham, V. J., Pike, V. W., and Camici, P. G. 1997. Biodistribution and metabolism of [N-methyl-^{11}C]-*m*-hydroxyephedrine in the rat. *Nucl. Med. Biol.* 24: 417–24.

Leitz, F. H. and Stefano, F. J. E. 1971. The effect of tyramine, amphetamine and metaraminol on the metabolic disposition of ^{3}H-norepinephrine released from the adrenergic neuron. *J. Pharmacol. Exp. Ther.* 178: 464–73.

Loesberg, C., Van Rouij, H., Nooijen, W. J., Meijer, A. J., and Smets, L. A. 1990. Impaired mitochondrial respiration and stimulated glycolysis by *m*-iodobenzylguanidine (MIBG). *Int. J. Cancer* 46: 276–81.

Mairs, R. J., Russel, J., Cunningham, S. et al. 1995. Enhanced tumour uptake and *in vitro* radiotoxicity of no-carrier-added [^{131}I] *meta*-iodobenzylguanidine: Implications for the targeted radiotherapy of neuroblastoma. *Eur. J. Cancer Res.* 31A: 576–81.

Mangner, T. J., Wu, J., and Wieland, D. M. 1982. Solid-phase exchange radioiodination of aryl iodides. Facilitation by ammonium sulfate. *J. Org. Chem.* 47: 1484–88.

Mardon, K., Merlet, P., Syrota, A., and Maziere, B. 1998. Effects of altitude hypoxia on myocardial beta-adrenergic pathway in rats. *J. Appl. Physiol.* 85: 890–97.

Melon, P. G., Nguyen, N., DeGrado, T. R., Mangner, T. J., Wieland, D. M., and Schwaiger, M. 1994. Imaging of cardiac neuronal function after cocaine exposure using carbon-11 hydroxyephedrine and positron emission tomography. *J. Am. Coll. Cardiol.* 23: 1693–99.

Merlet, P., Benvenuti, C., Moyse, D. et al. 1999. Prognostic value of MIBG imaging in idiopathic dilated cardiomyopathy. *J. Nucl. Med.* 40: 917–23.

Merlet, P., Delforge, J., Syrota, A. et al. 1993. Positron emission tomography with ^{11}C CGP-12177 to assess β-adrenergic receptor concentration in idiopathic dilated cardiomyopathy. *Circulation* 87: 1169–78.

Merlet, P., Valette, H., Dubois-Randé, J.-L. et al. 1992. Prognostic value of cardiac *meta*iodobenzylguanidine imaging in patients with heart failure. *J. Nucl. Med.* 33: 471–77.

Mislankar, S. G., Gildersleeve, D. L., Wieland, D. M., Massin, C. C., Mulholland, G. K., and Toorongian, S. A. 1988. 6-[^{18}F] Fluorometaraminol: A radiotracer for *in vivo* mapping of adrenergic nerves of the heart. *J. Med. Chem.* 31: 362–66.

Morales, J. O., Beierwaltes, W. H., Counsell, R. E., and Meier, D. H. 1967. The concentration of radioactivity from labeled epinephrine and its precursors in the dog adrenal medulla. *J. Nucl. Med.* 8: 800–09.

Nagahara, D., Nakata, T., Hashimoto, A. et al. 2008. Predicting the need for an implantable cardioverter defibrillator using cardiac *meta*iodobenzylguanidine activity together with plasma natriuretic peptide concentration or left ventricular function. *J. Nucl. Med.* 49: 225–33.

Nagamatsu, H., Momose, M., Kobayashi, H., Kusakabe, K., and Kasanuki, H. 2007. Prognostic value of ^{123}I-*meta*iodobenzylguanidine in patients with various heart diseases. *Ann. Nucl. Med.* 21: 513–20.

Nakajo, M., Shimabukuro, K., Yoshimura, H. et al. 1986. Iodine-131 *meta*iodobenzylguanidine intra- and extravesicular accumulation in the rat heart. *J. Nucl. Med.* 27: 84–89.

Nomura, Y., Matsunari, I., Takamatsu, H. et al. 2006. Quantitation of cardiac sympathetic innervation in rabbits using ^{11}C-hydroxyephedrine PET: Relation to ^{123}I-MIBG uptake. *Eur. J. Nucl. Med. Mol. Imaging* 33: 871–78.

Nori, S. L., Gaudino, M., Alessandrini, F., Bronzetti, E., and Santarelli, P. 1995. Immunohistochemical evidence of sympathetic denervation and reinnervation after necrotic injury in rat myocardium. *Cell Mol. Biol.* 41: 799–807.

Pacholczyk, T., Blakely, R. D., and Amara, S. G. 1991. Expression cloning of a cocaine- and antidepressant-sensitive human noradrenaline transporter. *Nature* 350: 350–54.

Packer, M. 1992. The neurohormonal hypothesis: A theory to explain the mechanism of disease progression in heart failure. *J. Am. Coll. Cardiol.* 20: 248–54.

Pissarek, M., Ermert, J., Oesterreich, G., Ing, D., Bier, D., and Coenen, H. H. 2002. Relative uptake, metabolism, and β-receptor binding of (1R,2S)-4-^{18}F-fluorometaraminol and ^{123}I-MIBG in normotensive and spontaneously hypertensive rats. *J. Nucl. Med.* 43: 366–73.

Pop-Busui, R. 2010. Cardiac autonomic neuropathy in diabetes: A clinical perspective. *Diabetes Care* 33: 434–41.

Raffel, D., Loc'h, C., Mardon, K., Mazière, B., and Syrota, A. 1998. Kinetics of the norepinephrine analog [Br-76]-*meta*-bromobenzylguanidine in isolated working rat heart. *Nucl. Med. Biol.* 25: 1–16.

Raffel, D. M. and Chen, W. 2004. Binding of [^{3}H]mazindol to cardiac norepinephrine transporters: Kinetic and equilibrium studies. *Naunyn-Schmiedeberg's Arch. Pharmacol.* 370: 9–16.

Raffel, D. M., Chen, W., Sherman, P. S., Gildersleeve, D. L., and Jung, Y. W. 2006a. Dependence of cardiac ^{11}C-*meta*-hydroxyephedrine retention on norepinephrine transporter density. *J. Nucl. Med.* 47: 1490–96.

Chapter 17

Raffel, D. M., Corbett, J. R., del Rosario, R. B. et al. 1996. Clinical evaluation of carbon-11-phenylephrine: MAO sensitive marker of cardiac sympathetic neurons. *J. Nucl. Med.* 37: 1923–31.

Raffel, D. M., Jung, Y. W., Gildersleeve, D. L. et al. 2007. Radiolabeled phenethylguanidines: Novel imaging agents for cardiac sympathetic neurons and adrenergic tumors. *J. Med. Chem.* 50: 2078–88.

Raffel, D. M., Koeppe, R. A., Little, R. et al. 2006b. PET Measurement of cardiac and nigrostriatal denervation in parkinsonian syndromes. *J. Nucl. Med.* 47: 1769–77.

Raffel, D. M. and Wieland, D. M. 1999. Influence of vesicular storage and monoamine oxidase activity on [^{11}C]phenylephrine kinetics: Studies in isolated rat heart. *J. Nucl. Med.* 40: 323–30.

Raffel, D. M. and Wieland, D. M. 2001. Assessment of cardiac sympathetic nerve integrity with positron emission tomography. *Nucl. Med. Biol.* 28: 541–59.

Raffel, D. M. and Wieland, D. M. 2010. Development of mIBG as a cardiac innervation imaging agent. *JACC Cardiovasc. Imaging* 3: 111–16.

Raisman, R., Sette, M., Pimoule, C., Briley, M., and Langer, S. Z. 1982. High-affinity [^{3}H]desipramine binding in the peripheral and central nervous system: A specific site associated with the neuronal uptake of noradrenaline. *Eur. J. Pharmacol.* 78: 345–51.

Rosenspire, K. C., Haka, M. S., Van Dort, M. E. et al. 1990. Synthesis and preliminary evaluation of carbon-11-*meta*-hydroxyephedrine: A false transmitter agent for heart neuronal imaging. *J. Nucl. Med.* 31: 1328–34.

Schömig, E., Fischer, P., Schönfeld, C.-L., and Trendelenburg, U. 1989. The extent of neuronal re-uptake of ^{3}H-noradrenaline in isolated vasa deferentia and atria of the rat. *Naunyn-Schmiedeberg's Arch. Pharmacol.* 340: 520–08.

Schömig, E., Korber, M., and Bönisch, H. 1988. Kinetic evidence for a common binding site for substrates and inhibitors of the neuronal noradrenaline carrier. *Naunyn-Schmiedeberg's Arch. Pharmacol.* 337: 626–32.

Schwaiger, M., Kalff, V., Rosenspire, K. et al. 1990. Noninvasive evaluation of sympathetic nervous system in human heart by positron emission tomography. *Circulation* 82: 457–64.

Schwartz, P. J. 1998. The autonomic nervous system and sudden death. *Eur. Heart J.* 19(Suppl. F): F72–F80.

Sisson, J. C. 2000. Radiopharmaceuticals for nuclear endocrinology at the University of Michigan. *Cancer Biother. Radiopharm.* 15: 305–18.

Sisson, J. C., Wieland, D. M., Sherman, P., Mangner, T. J., Tobes, M. C., and Jacques, S., Jr. 1987. *Meta*iodobenzylguanidine as an index of the adrenergic nervous system integrity and function. *J. Nucl. Med.* 28: 1620–24.

Stevens, M. J., Raffel, D. M., Allman, K. C. et al. 1998. Cardiac sympathetic dysinnervation in diabetes: Implications for enhanced cardiovascular risk. *Circulation* 98: 961–68.

Überfuhr, P., Ziegler, S., Schwaiblmair, M., Reichart, B., and Schwaiger, M. 2000. Incomplete sympathetic reinnervation of the orthotopically transplanted human heart: Observation up to 13 years after heart transplantation. *Eur. J. Cardiothorac. Surg.* 17: 161–68.

Ungerer, M., Hartmann, F., Karoglan, M. et al. 1998. Regional *in vivo* and *in vitro* characterization of autonomic innervation in cardiomyopathic human heart. *Circulation* 97: 174–80.

Vaidyanathan, G. and Zalutsky, M. R. 1993. No-carrier-added synthesis of *meta*-[^{131}I]iodobenzylguanidine. *Appl. Radiat. Isot.* 44: 621–28.

Van Dort, M. E., Kim, J. H., Tluczek, L., and Wieland, D. M. 1997. Synthesis of carbon-11 labeled desipramine and its metabolite 2-hydroxydesipramine: Potential radiotracers for PET studies of the norepinephrine transporter. *Nucl. Med. Biol.* 24: 707–11.

Virmani, R., Burke, A. P., and Farb, A. 2001. Sudden cardiac death. *Cardiovasc. Pathol.* 10: 275–82.

Waller, C., Kahler, E., Hiller, K.-H. et al. 2000. Myocardial perfusion and intracapillary blood volume in rats at rest and with coronary dilation: MR imaging *in vivo* with use of spin-labeling technique. *Radiology* 215: 189–97.

Wei, K., Le, E., Bin, J.-P., Coggins, M., Jayawera, A. R., and Kaul, S. 2001. Mechanism of reversible ^{99m}Tc-sestamibi perfusion defects during pharmacologically induced vasodilation. *Am. J. Physiol. Heart Circ. Physiol.* 280: H1896–H904.

Wieland, D. M., Brown, L. E., Rogers, W. L. et al. 1981. Myocardial imaging with a radioiodinated norepinephrine storage analog. *J. Nucl. Med.* 22: 22–31.

Wieland, D. M., Rosenspire, K. C., Hutchins, G. D. et al. 1990. Neuronal mapping of the heart with 6-[^{18}F]fluorometaraminol. *J. Med. Chem.* 33: 956–64.

Wieland, D. M., Swanson, D. P., Brown, L. E., and Beierwaltes, W. H. 1979. Imaging the adrenal medulla with an I-131-labeled anti-adrenergic agent. *J. Nucl. Med.* 20: 155–58.

Wieland, D. M., Wu, J., Brown, L. E., Mangner, T. J., Swanson, D. P., and Beierwaltes, W. H. 1980. Radiolabeled adrenergic neuron-blocking agents: Adrenomedullary imaging with [^{131}I]iodobenzylguanidine. *J. Nucl. Med.* 21: 349–53.

Zipes, D. P. 1995. In *Cardiac Electrophysiology: From Cell to Bedside* (Eds, Zipes, D. P. and Jalife, J.) W.B. Saunders, Philadelphia, PA, pp. 441–53.

18. PET Quantification of Myocardial Metabolism

Challenges and Opportunities

Arun K. Thukkani and Robert J. Gropler

18.1 Introduction

Normal myocardial function is predicated on the ability of the myocardium to alter its substrate utilization to meet changing metabolic demands and conditions. This loss in flexibility of substrate consumption can lead to an overreliance on one substrate that initiates a cascade of events that results in cardiac dysfunction. Although metabolic perturbations have been described in various important diseases, including diabetic and dilated cardiomyopathy, the exact mechanisms and signaling pathways activated in these pathologies as well as key factors that may beneficially modulate these disturbances have not yet been elucidated (see Figure 18.1). This is particularly important given that the currently available therapeutics conventionally used to treat various aspects of cardiovascular disease primarily have effects on the neurohormonal, chronotropic, or inotropic state of the heart. Although some of these medications have now been shown to alter some aspects of cardiac metabolism beneficially and this may partially explain their utility, the modulation of myocardial metabolic perturbations contributing to cardiac dysfunction may represent an as-of-yet-unrecognized therapeutic approach further complementing these more traditional treatments (Andersson et al. 1993, Bristow et al. 1996, Colucci et al. 1996, Waagstein et al. 1993). Moreover, a more complete characterization of such changes may add important insights into disease prognosis beyond the currently used structural or functional measures of cardiac function.

The study of myocardial metabolism has been advanced by the advent of positron emission tomography (PET). PET-imaging techniques allow for a very sensitive,

Targeted Molecular Imaging. Edited by Michael J. Welch and William C. Eckelman © 2012 Taylor & Francis Group, LLC. ISBN: 978-1-4398-4195-2

Chapter 18

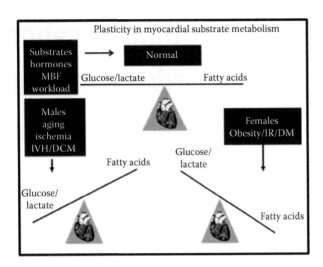

FIGURE 18.1 Summary of the determinants of myocardial substrate metabolism and its perturbations associated with disease. DCM, dilated cardiomyopathy; IR, insulin resistance; DM, diabetes mellitus; IR, ischemia–reperfusion; LVH, left ventricular hypertrophy. (From Herrero, P. and Gropler, R. J. 2005. *Journal of Nuclear Cardiology* 12: 345–58. With permission.)

noninvasive assessment of cardiac metabolic processes *in vivo* in both normal and disease states. Through the rational design of novel PET tracers and their rigorous validation in newly developed animal models of human disease, PET imaging in conjunction with next-generation radiotracers will likely fundamentally advance our understanding of these maladaptive processes and accelerate the development of new therapies.

The first portion of this chapter summarizes the fundamentals of myocardial metabolism, its important regulatory mechanisms, its determinants, as well as describes the metabolic perturbations associated with cardiovascular disease. The second portion details the relative advantages of PET imaging to assess myocardial metabolism, describes currently available PET radiotracers designed to evaluate particular aspects of cardiac metabolism, and summarizes the concepts associated with general radiotracer design and preclinical validation studies.

18.2   Overview of Myocardial Metabolism

While a detailed discussion on myocardial bioenergetics is beyond the focus of this chapter, a review of normal myocardial substrate metabolism, its major regulatory mechanisms influencing substrate utilization, and, importantly, perturbations of these normal processes now thought to contribute to clinically important myocardial disease states, is fundamentally essential for the rational development of PET tracers to assess the cardiac metabolism.

In normal states, almost all myocardial ATP production (~98%) is generated by mitochondrial oxidative phosphorylation with the remainder derived from glycolytic pathways feeding the Krebs cycle (Opie 1991). Reducing equivalents provided by NADH and FADH$_2$, the products of the Krebs cycle enter different portions of the mitochondrial electron transport chain. Specifically, while electrons derived from NADH enter mitochondrial complex 1, those from FADH$_2$ are instead transferred to complex 2 resulting in lower ATP generation per substrate molecule (~3 ATP/NADH vs. ~2 ATP/FADH$_2$, respectively). Both NADH and FADH$_2$ are produced by the carefully regulated, stepwise catabolism of different myocardial nutrients, most notably fatty acids, which in normal heart usually accounts for 60–90% of mitochondrial ATP generation while the balance is provided by the metabolism of carbohydrates (glucose and lactate) to

pyruvate (Stanley et al. 1997, Stanley and Chandler 2002). While these various substrates ultimately generate acetyl-CoA subsequently entering the Krebs cycle, various intermediary biochemical reactions and intracellular transport pathways, catalyzed by important enzymes and transport proteins controlled by complex allosteric mechanisms, represent the key regulatory steps governing the balance between fatty acid and pyruvate metabolism—thus providing fine, dynamic, and oftentimes uniquely reciprocal control of substrate flux through myocardial fatty acid β-oxidation and glycolytic pathways as dictated by ever-changing physiological conditions.

18.2.1   Myocardial Fatty Acid Metabolism

Myocardial uptake of nonesterified fatty acids is primarily determined by its concentration in the circulation, which can vary widely, as well as the expression of a specific sarcolemmal fatty acid transporters such as CD36, the plasma membrane-associated fatty acid-binding protein (FABP$_{pm}$) and FATPs (van der Vusse et al. 2000). Upon uptake, these insoluble, nonesterified free fatty acids are immediately bound by cytosolic fatty acid-binding proteins and through the action of fatty acyl-CoA synthase, form fatty acyl-CoA. While soluble and able to diffuse across the outer mitochondrial membrane, these species are either converted by carnitine palmitoyltransferase I (CPT-1) into long-chain fatty acylcarnitine species or diverted toward augmenting myocardial triglyceride pool via esterification pathways (Coleman et al. 2000). The long-chain fatty acylcarnitine species are transported across the inner mitochondrial membrane by carnitine acylcarnitine translocase (CAT) where long-chain acyl-CoAs are then regenerated (see Figure 18.2). Fatty acid catabolism takes place then in the mitochondrial matrix through repeated oxidative cleavage of two-carbon acetyl-CoA through four reactions catalyzed by separate enzymes with varied isoforms specific for short-, medium-, and long-chain fatty intermediates ultimately resulting in acetyl-CoA entering the Krebs cycle and producing NADH, $FADH_2$, and CO_2 (Eaton 2002).

While uptake of free fatty acids is certainly a determinant of myocardial fat utilization (Arthur 1968), with the latter also being driven by important transcriptional control mechanisms involving peroxisome proliferator activated-receptor alpha (PPARα) and PPARβ/δ, the interaction of CPT-1 with malonyl-CoA represents an example demonstrating the reciprocal control of fatty acid and glucose metabolism (Finck et al. 2002). Malonyl-CoA, produced by carboxylation of acetyl-CoA by acetyl-CoA carboxylase (ACC) and degraded back to acetyl-CoA by malonyl-CoA decarboxylase (MCD), is a potent inhibitor of CPT-1 (Kerner and Bieber 1990, Lopaschuk et al. 1994). While CPT-1 exists in two isoforms with CPT-1α and CPT-1β predominant in the liver and heart, respectively, the CPT-1β isoform is ~30-fold more sensitive to the inhibitory effect of malonyl-CoA. This suggests that levels of malonyl-CoA may play a key role in the regulation of these pathways: as acetyl-CoA derived from carbohydrate metabolism accumulates and is converted to malonyl-CoA, rates of fatty acid oxidation are decreased (see Figure 18.3) (McGarry et al. 1983).

Studies have demonstrated that during modest ischemia, cardiac metabolism shifts more toward anaerobic glycolysis (derived from cardiac glycogen stores and accumulating lactate) rather than fatty acid oxidation, though the latter is still predominant,

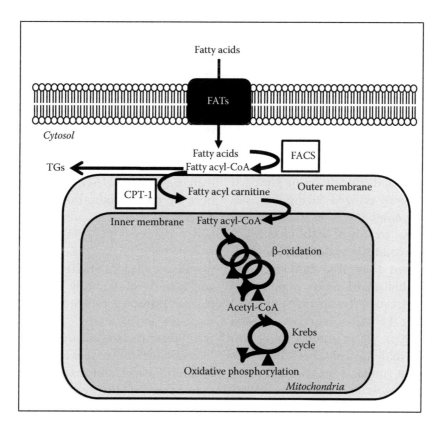

FIGURE 18.2 Overview of myocardial fatty acid metabolism. Uptake of circulating fatty acids are facilitated by various fatty acid transporters, and, through subsequent cytosolic and mitochondrial reactions, undergo β-oxidation to generate acetyl-CoA shunted into the Krebs cycle. FATs, fatty acid transporters; FACS, fatty acyl-CoA synthase; CPT-1, carnitine palmitoyltransferase 1; TG, triglycerides.

Chapter 18

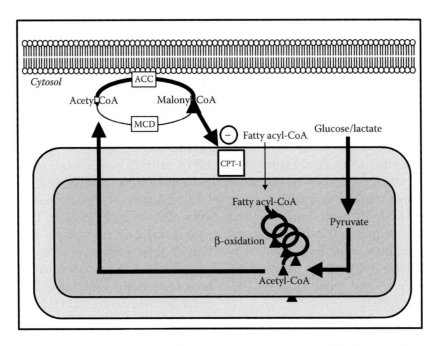

FIGURE 18.3 Summary of regulation of fatty acid metabolism by malonyl-CoA. Increased levels of acetyl-CoA are converted to malonyl-CoA by ACC. If MCD activity is relatively decreased, increasing levels of malonyl-CoA will inhibit CPT-1, leading to decreased rats of fatty acid oxidation. ACC, acetyl-CoA carboxylase; MCD, malonyl-CoA decarboxylase; CPT-1, carnitine palmitoyltransferase 1.

perhaps in part due to relatively conserved malonyl-CoA levels in ischemic myocardium (Opie 1991). However, upon reperfusion, malonyl-CoA levels decrease by ~98% secondary to a concomitant reduction in ACC activity without a parallel decline in MCD activity. Together, this leads to reduced inhibition of CPT-1 and increased fatty acid metabolism through which its accumulating downstream products (i.e., NADH and acetyl-CoA) downregulate carbohydrate metabolic flux: any available precursors (i.e., accumulated lactate and any remaining glycogen stores) are thus inefficiently utilized during the postischemic reperfusion period (Dyck et al. 2004). In normal rat myocardium, treatment with novel MCD inhibitors increased myocardial malonyl-CoA levels resulting in a decreased fatty acid oxidation and increased glucose metabolism. Moreover in rat hearts rendered ischemic, this treatment resulted in improved cardiac function during the postischemic reperfusion period (Dyck et al. 2004). Other studies using etomoxir and oxfenicine, inhibitors of CPT-1, have also been shown to mediate similar beneficial effects in various animal models as well (Higgins et al. 1980, Wahr et al. 1994). In sum then, these studies would suggest modulating myocardial metabolism during ischemia and reperfusion toward utilizing more glucose and lactate, myocardial substrates both more oxygen-efficient compared to fatty acids (Opie 1991),

may represent a novel therapeutic strategy to help optimize cardiac function following ischemia.

18.2.2 Myocardial Carbohydrate Metabolism

The healthy heart consumes the carbohydrates glucose and lactate, albeit at significantly lower rates versus fatty acids, converting them via glycolysis to pyruvate, which readily enter the Krebs cycle. The NADH and pyruvate produced by these pathways are then converted to NAD^+ and CO_2 under aerobic conditions by mitochondrial oxidative phosphorylation; if anaerobic conditions prevail, NAD^+ and lactate are both produced (Opie 1991). Myocardial glucose uptake is dependent on arterial glucose content, the transmembrane glucose gradient, as well as the expression and activity of a family of sarcolemmal membrane glucose transport proteins, GLUT-1 and GLUT-4. In response to increased circulating glucose, insulin secretion stimulates the translocation of GLUT-4 from intracellular storage vesicles to the sarcolemmal membrane; these mechanisms also seem to be active during ischemia as well (Russell et al. 1998, Young et al. 1997). Upon entry, absorbed glucose is readily phosphorylated by hexokinase to form glucose-6-phosphate. Intracellular glycogen stores are also a source of glucose-6-phosphate with the glycogenolysis pathway activated in response to low myocardial ATP content, increased intracellular inorganic phosphate, and adrenergic stimulation, all of which are generally present during ischemia or intense exercise (Goldfarb et al. 1986, Stanley et al. 1997). Given a constant supply of glucose-6-phosphate generated by glucose uptake and glycogenolysis, the initial mechanism controlling flux through the glycolytic pathway is the highly regulated activity of phosphofructokinase-1 (PFK-1) which catalyzes the first irreversible reaction by generating fructose 1,6-biphosphate from fructose-6-phosphate. PFK-1 is activated by increased myocardial content of ATP precursors, including ADP, AMP, and inorganic phosphate through allosteric mechanisms (Stanley and Chandler 2002). Thus, this feed-forward mechanism allows for ready pathway activation in response to dynamic energetic demands. Feedback mechanisms are also operational as PFK-1 activity is readily inhibited by markers of high myocardial energetic states including accumulating levels of ATP as well as citrate, the first product of the Krebs cycle (Stanley and Chandler 2002). The inhibition of PFK-1 by accumulating citrate derived from fatty acids, first proposed by Randle et al. in 1963 as part of a "glucose-fatty acid cycle" describing fuel flux and selection by tissues, again emphasizes a reciprocal relationship in the regulation of substrate metabolism (see Figure 18.4) (Hue and Taegtmeyer 2009).

The next major regulatory step in the glycolytic pathway involves glyceraldehyde-3-phosphate dehydrogenase (GAPDH) which catalyzes both the conversion of glyceraldehyde-3-phosphate to 1,3-diphosphoglycerate and NAD^+ to NADH. The enzyme is subject to both

feed-forward and feedback mechanisms incumbent on dynamic concentrations of substrates and products, respectively. Indeed, the accumulation of cytosolic NADH or 1,3-diphosphoglycerate or both inhibit GAPDH activity while increasing levels of NAD^+ have the converse effect. In the presence of lactate which inhibits NAD^+ regeneration from NADH, flux through GADPH is reduced so much so that during severe myocardial ischemia, GAPDH becomes the rate-controlling step along the glycolytic pathway (Stanley et al. 2005).

Subsequent reactions result in the generation of pyruvate. The activity of pyruvate dehydrogenase (PDH), a key mitochondrial enzymatic complex catalyzing the irreversible decarboxylation of pyruvate to acetyl-CoA, is under complex control through covalent modification. Indeed, PDH catalytic activity is regulated by its phosphorylation state which is incumbent on the relative activities of a family of pyruvate dehydrogenase kinases (PDK) in balance with that of a family of pyruvate dehydrogenase phosphatases (PDP). The mechanism of regulation of PDH is dependent on the expression of different PDK and PDP isoforms, which vary across different types of tissue (see Figure 18.5). Moreover, these various tissue-specific kinase and phosphatase isoforms act upon one of three different sites of the PDH complex to render it inactive when phosphorylated and active when modification is reversed. The balance between kinase and phosphatase activity is also tightly controlled. PDP has been shown to be activated by increased intracellular calcium and magnesium levels as might occur during high adrenergic states (McCormack and Denton 1989). On the other

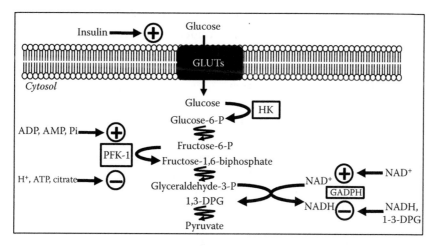

FIGURE 18.4 The regulation of glucose metabolism. After uptake by the glucose transporters GLUT-1 and GLUT-4, glucose undergoes an irreversible phosphorylation by HK. PFK-1 and GADPH both catalyze important glycolytic reactions and are heavily regulated steps in this process. HK, hexokinase; PFK-1, phosphofructokinase-1; GADPH, glyceraldehyde-3-phosphate dehydrogenase; DPG, diphosphoglycerate.

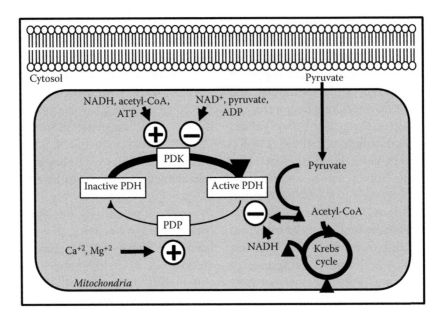

FIGURE 18.5 The complex regulation of PDH. The activity of PDH, catalyzing the conversion of pyruvate to acetyl-CoA, is dependent on its phosphorylation state as determined by the relative activity of PDK and PDP. PDH, pyruvate dehydrogenase; PDK, pyruvate dehydrogenase kinase; PDP, pyruvate dehydrogenase phosphatase.

hand, pyruvate, the substrate of PDH, as well as ADP both inhibit PDK thus allowing increased flux into the Krebs cycle as PDH is activated. While declining intracellular levels of downstream products, specifically acetyl-CoA and NADH, also exert this effect, conversely, high levels of these products signifying a high myocardial energetic state activate PDK-mediated inhibition of PDH, ultimately reducing pyruvate levels and therefore, carbohydrate utilization (Kerbey et al. 1976, Randle 1986). The predominant isoform of PDK present in the heart is PDK4 and its expression is induced by starvation and diabetes (Bowker-Kinley et al. 1998, Wu et al. 2001). Again exemplifying the reciprocal regulation controlling myocardial substrate utilization, these conditions, characteristically associated with high circulating levels of free fatty acids and increased myocardial fat metabolism, lead to greater PDK4 activity thereby reducing pyruvate oxidation derived from carbohydrate precursors as PDH is inhibited. In contrast, declining plasma free fatty acid content or inhibition of CPT-1 increases PDH activity and helping to shift toward increased carbohydrate metabolism (Higgins et al. 1980, Lopaschuk et al. 1994, Randle 1986). While the details of this complex regulatory mechanism are beyond the scope of this chapter, several detailed reviews are available (Patel and Korotchkina 2001, Sugden and Holness 2006).

Normally, under resting conditions, lactate uptake supplies ~50% of the myocardial pyruvate oxidized *in vivo* with the balance supplied by glucose (Stanley 1991). Incumbent on arterial lactate content, myocardial lactate uptake occurs through monocarboxylate transporter-1 (MCT-1), the predominant sarcolemmal lactate transporter expressed in the heart; absorbed lactate is ultimately consumed through sequential reactions producing pyruvate as an intermediate product but culminating in the generation of CO_2 and ATP (Garcia et al. 1994). Upon uptake, lactate is readily oxidized by lactate dehydrogenase to generate pyruvate and thus the lactate and glucose metabolic pathways converge prior to pyruvate oxidation by PDH. By inference then, the same regulatory mechanisms governing pyruvate utilization at the level of PDH likely govern lactate metabolism under normal conditions.

18.2.3 Reciprocal Regulation of Myocardial Substrate Utilization; Role of AMPK

The balance between myocardial fatty acid and carbohydrate metabolism is tightly regulated and, as discussed previously, oftentimes reciprocal: dynamic substrate and product levels working to increase metabolic flux through one pathway oftentimes result in a corresponding decrease through the other. Over the last decade, the role of AMP-activated protein kinase (AMPK) in reciprocal regulation myocardial substrate utilization has been investigated in detail. AMPK has been found to regulate myocardial fuel selection and energy generation through pleiotropic mechanisms particularly active under stressful conditions. For example during myocardial ischemia, rapidly depleting ATP stores result in the accumulation of ADP, AMP, creatine, and inorganic phosphate. AMPK is readily stimulated by the binding of AMP to its γ regulatory subunit and this activation leads both to the mobilization of GLUT-4 from intracellular storage vesicles to the sarcolemma as well as inhibits their removal by endocytosis (Kim et al. 2009, Yang and Holman 2006, Young et al. 1997). Tremendous increases in the rates of glycolysis during ischemia have also been partially attributed to AMPK-mediated stimulation of phosphofructose kinase-2 (PFK-2) generating fructose 2,6-biphosphate, an allosteric activator of the key glycolytic enzyme PFK-1 (Marsin et al. 2000). Earlier on during the postischemic reperfusion period, AMPK activation continues to enhance myocardial glucose uptake and utilization yet persistent AMPK activity leads to the inactivation of ACC; subsequent decreases in malonyl-CoA levels during reperfusion relieves CPT-1 inhibition, allowing increased rates of mitochondrial fatty acid oxidation (Russell et al. 2004, Stanley et al. 1997).

18.2.4 Metabolic Perturbations in the Failing and Ischemic Heart: Preclinical and Clinical Studies

While not yet fully elucidated, it is clear that altered substrate utilization is present in animal models of severe ventricular dysfunction and heart failure. In studies in the canine rapid pacing model of heart failure, the rates of fatty acid and glucose metabolism are generally near normal levels during the early and middle stages of the development of heart failure (Recchia et al. 1998). However, continued progression to more severe ventricular dysfunction leading to end-stage heart failure demonstrated a relative increase in glucose oxidation with a measured transmyocardial respiratory quotient (an index of substrate selection and utilization) corroborating this shift; the activity of key enzymes of fatty acid metabolism, most notably CPT-1, was also reduced (Osorio et al. 2002). In studies using a canine microembolization model simulating early, hemodynamically compensated stages of heart failure, no significant changes in carbohydrate or fatty acid metabolism were detected, the latter reflected by relatively preserved levels of CPT-1 activity, again suggesting that metabolic alterations accompany only the later, decompensated stages of contractile dysfunction (Chandler et al. 2004, Panchal

et al. 1998). Interestingly, upregulation of the proteins involved in the glycolytic pathway is yet to be reported and the elucidation of such is confounded by studies in various animal models demonstrating a paradoxical decrease in the expression or activity of key proteins, such as GLUT-4 and PDH, despite an increase in glucose uptake and oxidation (Osorio et al. 2002, Rosenblatt-Velin et al. 2001). Of note, this shift toward glucose utilization in the failing heart is partially believed to result from the renewed expression of metabolic genes typically active in the fetal state (Razeghi et al. 2001).

The evidence documenting these alterations in heart failure patients is limited and oftentimes conflicting. In a small study of patients with moderate-to-severe heart failure, PET imaging using ^{11}C-glucose and ^{11}C-palmitate demonstrated increased glucose utilization with a reciprocal decrease in fatty acid utilization (Dávila-Román et al. 2002). However, others have reported entirely contrasting findings (Paolisso et al. 1994). Additionally, widely varied levels of myocardial glucose uptake as assessed by various techniques have been described as well (Paolisso et al. 1994, Tadamura et al. 1998, Uehara et al. 1998). These inconsistencies together suggest that confounding factors such as patient comorbidities (i.e., insulin resistance) and the etiology of cardiac dysfunction (i.e., ischemic vs. non-ischemic) likely mediate as of yet not clearly elucidated effects on metabolism in clinical heart failure. Despite this, several lines of indirect evidence suggest that modulation of metabolism in heart failure may have beneficial effects. Supraphysiological arterial concentrations of pyruvate have been shown to acutely improve ventricular function in patients with moderate-to-severe heart failure (Hermann et al. 1999). Similarly, treatment with dichloroacetate to activate PDH and increase pyruvate oxidation had similar effects in another cohort with comparable severity of heart failure (Bersin et al. 1994). Studies of chronic therapy with beta-adrenergic receptor antagonists, a class of therapeutics with proven mortality benefits in heart failure patients, have demonstrated significant decreases in myocardial fatty acid oxidation with corresponding increases in carbohydrate consumption (Andersson et al. 1991, Eichhorn et al. 1990, 1994, Wallhaus et al. 2001). While far from being clear, the development of novel PET tracers to both help more precisely delineate these metabolic alterations related to heart failure and identify potential targets amenable to therapeutic modulation remains a focus of active investigation.

As discussed previously, ischemic myocardium demonstrates increased rates of pyruvate oxidation

Chapter 18

though fatty acid metabolism still remains predominant. The available circulating free fatty acids work to inhibit myocardial consumption of glucose, lactate, and pyruvate through various mechanisms. This effect is detrimental during ischemia when the rate of glycolysis is significantly increased; the suppression of pyruvate oxidation leads to increased myocardial lactate production and accumulation, decreasing cellular pH levels and further contributing to ensuing cellular injury and dysfunction. This maladaptive process is hypothesized to represent a remnant of the evolution of the metabolic stress response: prior selection pressures in mammalian precursor species resulted in fatty acids becoming the preferred fuel during stress given the fact that fatty acids are more energetically efficient (more ATP per molecule generated) yet less oxygen-efficient, compared to carbohydrates (Stanley and Chandler 2002, Yael et al. 2010). However, as primordial metabolic stress responses evolved in mammalian species not largely afflicted by cardiopulmonary disease, ischemia or hypoxemia likely had little or no role in shaping this response. Consequently, this persistent dependence on fatty acid substrates during stress represents a long-standing yet inferior response to the ischemia that is now commonly seen: by predominantly relying on fatty acids at the expense of the more oxygen-efficient abundant carbohydrate substrates available, myocardial energy generation in the presence of an interrupted oxygen supply remains wasteful and suboptimal (Wolff et al. 2002). This concept has led to preclinical and clinical studies focusing on modulating this maladaptive response as an adjunct to conventionally used therapies.

Several preclinical studies using a variety of inhibitors of fatty acid oxidation in various animal models of ischemia and reperfusion have reported decreased levels of fatty acid oxidation accompanied by improved cardiac function and these encouraging results have led to clinical investigations. Ranolazine is a partial inhibitor of fatty oxidation mediating a number of effects thought to beneficially modulate cardiac metabolism, including activating glycolytic pathways, limiting fatty acid oxidation, and reducing lactate release while preserving ATP levels in the myocardium during both ischemia and reperfusion (Clarke et al. 1996, Gralinski et al. 1994, McCormack et al. 1996). Subsequent clinical studies in patients with ischemic heart disease have demonstrated significant effects in decreasing the time to onset of chest pain symptoms and have helped lead to its widespread use as an adjunctive antianginal therapy (Chaitman et al. 2004, Cocco et al. 1992, Pepine and Wolff 1999, Wolff et al. 2002). Moreover, the administration of the fatty acid oxidation inhibitor, trimetizidine, to patients with dilated cardiomyopathy results in a significant improvement in left ventricular ejection fraction. However, the improvement in left ventricular function is paralleled by only a mild decrease in myocardial fatty acid oxidation (Tuunanen et al. 2008). Thus, it appears that the improvement in left ventricular function reflects more than a shift in metabolism and is likely influenced by other factors such as improved whole-body insulin resistance and synergestic effects with β-blockade. Ongoing preclinical and clinical studies will likely identify further therapeutic approaches to beneficially modulate perturbations in cardiac metabolism as related to clinically important cardiovascular disease.

18.3   PET Imaging of Myocardial Metabolism

PET is a powerful imaging modality and when applied to biologic systems can provide quantitative measures of biochemical parameters. PET imaging, with its inherent advantages, is particularly well suited to noninvasively assess myocardial metabolism. The radionuclides commonly utilized in PET imaging, including isotopes of oxygen (^{15}O), nitrogen (^{13}N), carbon (^{11}C), and fluorine (^{18}F), are positron emitters that can be fairly easily incorporated into important biochemical substrates that participate in vital metabolic pathways without fundamentally altering their inherent chemical or biochemical properties. Furthermore, the signal emitted by positrons is uniquely quantifiable, allowing for the exquisitely sensitive measurement of tissue radioactive concentration on a picomolar scale with sufficient temporal yet albeit relatively low spatial resolution. In conjunction with a detailed analysis of the kinetics of *in vivo* radiotracer uptake and elimination, sophisticated tracer mathematical modeling techniques fitting the observed *in vivo* behavior of the radiotracer can be applied to generate good-fit models allowing for quantification of specific metabolic processes. While the topic of tracer kinetic modeling is beyond the focus of this chapter, the reader is directed to several available detailed reviews (Ichise et al. 2001, Shoghi 2009). Even with these relative strengths, PET has several important limitations to consider as well.

These include the use of ionizing radiation, the relatively high cost of PET-imaging systems thus limiting their widespread availability, the need for an on-site cyclotron to generate the required radionuclides, as well as sufficient capabilities in radiochemical methods to incorporate these radionuclides into PET radiotracers. Additionally, sophisticated correction techniques accounting for scatter, attenuation, and partial-volume effects, partly due to its low relative spatial resolution, are required to offset the sometimes significant quantitative bias prevalent with this imaging modality (Soret et al. 2007). Despite these limitations, the number of imaging centers dedicated to specialized PET techniques continues to grow (Garcia et al. 2007).

18.3.1 Choice of PET Radiotracer to Assess Myocardial Metabolism

A diverse variety of PET radiotracer analogs and substrates are currently available to assess different aspects of myocardial metabolism with the choice of radiotracer predicated on various important factors, including the biochemical parameter of interest as well as the metabolic fate of the radiotracer. The components of myocardial metabolism primarily assessed by PET imaging include myocardial oxygen consumption, carbohydrate and fatty acid metabolism, and myocardial blood flow.

Myocardial oxygen consumption (MV_{O_2}) can be measured by PET imaging techniques utilizing ^{15}O-oxygen since oxygen is the final electron receptor in all pathways of aerobic myocardial metabolism. However, (MV_{O_2}) can only be calculated in conjunction with measurements of arterial oxygen content and myocardial blood flow, the latter of which can only be determined with additional tracer-imaging studies using ^{15}O-H_2O, ^{13}N-ammonia, or ^{82}rubidium. This, along with its complex kinetic modeling, represents the major drawbacks of this technique (Laine et al. 1999, Lida et al. 1996).

Alternatively, ^{11}C-acetate can be used to measure (MV_{O_2}). Acetate, a two-carbon fatty acid, is normally rapidly converted to acetyl-CoA and ultimately metabolized through the Krebs cycle. Since the activity of the Krebs cycle is tightly coupled to the rate of mitochondrial oxidative phosphorylation, myocardial consumption of ^{11}C-acetate provides an approximation of overall oxidative metabolism or (MV_{O_2}). Again, the kinetic modeling when using this radiotracer is

quite complex, must account for the generation of $^{11}CO_2$, and its accuracy varies depending on cardiac output (Brown et al. 1987, 1988, Buck et al. 1991, Sun et al. 1998).

Myocardial carbohydrate metabolism has been largely studied by utilizing ^{18}F-fluorodeoxyglucose (^{18}F-FDG) and ^{11}C-glucose, respectively. Both radiotracers possess strengths and limitations in their ability to infer important parameters regarding myocardial glucose metabolism mainly determined by their respective identities as analogs or substrates. Additionally, very preliminary studies utilizing the radiotracer L-3-^{11}C-lactate have been reported and may represent another approach to more comprehensively assess myocardial carbohydrate metabolism (Herrero et al. 2007).

Fatty acid metabolism has been measured by both radiolabeled substrates and analogs. ^{11}C-palmitate has been previously used to measure myocardial triglyceride metabolism in a canine model with its chief advantage being that it accurately reflects *in vivo* levels of lipid oxidation. However, its widespread adoption has been largely limited by its suboptimal image quality and specificity as well as the need for highly specialized radiochemistry synthetic techniques and fairly sophisticated kinetic modeling (Bergmann et al. 1996). Aside from this, various radiolabeled fatty acid analogs have been developed for similar purposes. Most of these analogs have been designed to reflect myocardial β-oxidation with 14-(R,S)-^{18}F-fluoro-6-thiaheptadecanoic acid (FTHA) being the first. Using a variety of animal models, myocardial FTHA retention was shown to correlate well with changing levels of arterial substrate content, blood flow, and work load; however, its continued development and validation has been hampered by its relative insensitivity in reflecting decreased rates of β-oxidation in the presence of hypoxia (DeGrado et al. 1991, Schulz et al. 1996, Taylor et al. 2001). Newer analogs including 16-^{18}F-fluoro-4-thia-palmitate (FTP) and *trans*-9(RS)-^{18}F-fluoro-3,4(RS,RS) methyleneheptadecanoic acid (FCPHA) have been designed and their use is predicated on correction by a lumped constant similar to ^{18}F-FDG to account for differences in the kinetics between these radiotracers and unlabeled palmitic acid. The validation of these newer tracers is still ongoing.

18.3.2 Validation of Novel PET Radiotracers in Small Animals

The development, translation, and application of novel PET radiotracers to assess cardiac bioenergetic flux in

Chapter 18

humans is predicated on a fundamental understanding of myocardial metabolism and its perturbations as well as a candidate radiotracer whose function, specificity, pharmacokinetics, and safety profile have all been thoroughly validated in animal models of incrementally increasing complexity.

Initial studies to elucidate myocardial metabolism have relied upon *ex vivo* isolated perfusion working heart models typically in rodents. By systematically altering loading conditions, oxygen tension, and the content of substrates or hormones in the perfusate, the use of ^{3}H and ^{14}C tracers have been applied to elucidate the metabolic fate of both glucose and fatty acid substrates (How et al. 2005). While the radioactivity of the perfusate relative to the effluent roughly reflects the rate of myocardial substrate uptake, tissue homogenates derived from the various treatment conditions allow for a more exact specification of the fate of accumulated radioactivity through autoradiography of the variously fractionated biochemical intermediates and products; pertinent tissue sections or euthanized whole animals can be examined in a similar fashion. During the initial stages of validation of a newly developed radiotracer using these methods, correlations between observed tissue radiotracer content and the expression of its targets are accomplished by *in vitro* analysis of tissue homogenates for target enzyme or ligand expression as assayed by Western blotting, quantitative PCR, or immunohistochemical techniques, respectively. Additional treatment conditions in the presence of specific inhibitors with particular attention to the modulation of accumulated radiotracer content and target enzyme expression in a correspondingly similar degree and direction are crucial.

Limited by these methods, a comprehensive understanding of *in vivo* tracer behavior over a specified time period requires the compilation of static images from numerous, individual animals sacrificed at multiple time points. Greatly increasing both the number of animals required and total study cost, the conclusions extrapolated from such a tedious, labor-intensive strategy may not only inaccurately represent the continuous molecular process under investigation, but is also more subject to the inherent variability derived from the sheer multitude of separate animals being analyzed. However, on the other hand, small-animal PET imaging has circumvented many of these limitations and has fundamentally advanced the field of preclinical radiotracer validation. Endowed with unmatched sensitivity at near-picomolar levels of detection, small animal PET allows for the longitudinal, nondestructive assessment of dynamic *in vivo* tracer behavior in a single animal; the modulation of radiotracer behavior in response to therapeutics or inhibitors can also be monitored. Moreover, small-animal PET offers a more cost- and time-effective means of preclinical radiotracer validation by significantly reducing the aggregate number of animals required, the total dose of radiotracer administered, as well as enhancing study accuracy and reliability.

The use of various types of transgenic mouse models has proven useful to this endeavor. Small-animal PET studies of knockout mice lacking the target of interest compared to wild type can be useful in elaborating the amount of specific binding of a radiotracer *in vivo*, especially when specific inhibitors or sufficient amounts of unlabeled competitor are both not available. The use of knockout mice, however, may be limited depending on the target ligand or process of interest. Pertaining to developing novel radiotracers for metabolic imaging purposes, a relatively few number of null strains are available, especially considering that engineered absence of a key factor participating in an important metabolic pathway may lead to untoward effects, including developmental abnormalities or even embryonic lethality. However, certain genetic knockout mice of various acyl-CoA dehydrogenase isoforms, PPARα, and GLUT-4 have been generated, exhibiting particular metabolic abnormalities and cardiac phenotypes (Neubauer 2007). Taken together, knockout mice relevant to myocardial metabolism or any other molecular or biochemical process under investigation are important tools that can aid in initial radiotracer evaluation.

Continued innovation in molecular and cell biology techniques have also resulted in transgenic mice harboring tissue-specific overexpression of factors active in pathways of interest. As an example, cardiac-restricted overexpression of PPARα in transgenic mice (MHC-PPAR) demonstrates increased rates of myocardial fatty acid oxidation while reciprocally repressing glucose utilization. These metabolic alterations coupled with the later development of ventricular hypertrophy and dysfunction closely resemble the collective biochemical, structural, and functional changes seen in diabetic cardiomyopathy; accompanying imaging studies using ^{11}C-palmitate and ^{18}F-FDG have corroborated this shift in substrate utilization (see Figure 18.6) (Finck et al. 2002).

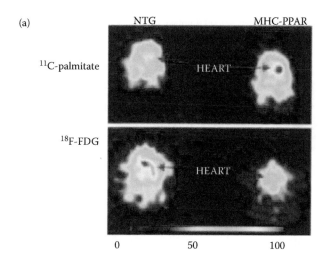

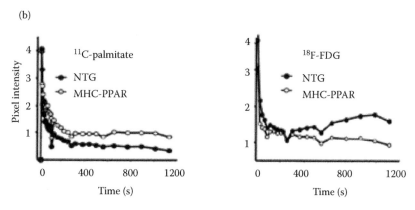

FIGURE 18.6 MHC-PPAR mice demonstrate altered myocardial metabolism. (a) Myocardial PET images using ^{11}C-palmitate and ^{18}F-FDG of normal and MHC-PPAR show relatively increased myocardial fatty acid uptake while the levels of glucose uptake are reciprocally decreased. (b) Time–activity curves of the cardiac fields for both radiotracers corroborate this finding. (From Finck, B. N. et al. 2002. *The Journal of Clinical Investigation* 109: 121–30. With permission.)

While useful for the initial proof of concept evaluation of novel radiotracers, transgenic strains often do not demonstrate all the important hallmarks of human diseases. For example, considering that the MHC-PPAR mouse strain manifests only the cardiac changes attributed to diabetes and does not spontaneously develop hyperglycemia, insulin resistance, or obesity, studies extending the evaluation of candidate radiotracers to more biologically relevant animal models of human disease are a necessary subsequent step. As such, the Zucker diabetic fatty rat (ZDF), by developing the major systemic markers of the type 2 diabetic phenotype, has been used for such a purpose. As compared to lean littermate control rats, studies on the ZDF rats demonstrated higher plasma glucose, insulin, and free fatty acid levels. Small-animal PET studies demonstrated that the myocardial metabolic phenotype could be imaged in the rodent heart. That is an increase in

myocardial fatty acid uptake and oxidation due to primarily increased delivery of fatty acids to the heart, but relatively preserved levels of glucose utilization reflecting the counterbalancing effects of reduced myocardial glucose transport processes and hyperglycemia (see Figure 18.7) (Welch et al. 2006). In this way, these types of studies in both diseased rodent models and healthy matched controls allow for direct comparative analysis of tracer behavior *in vivo* and can be correlated to the expression level of targets or relevant factors in tissue. Fundamental conclusions regarding the utility of the candidate radiotracer to measure the parameter or target of interest can then be made in a context that more closely resembles and encompasses the important aspects of the disease under investigation. Thus, small-animal PET studies of a novel radiotracer using animal models that better recapitulate biologically relevant human disease offer an early and

Chapter 18

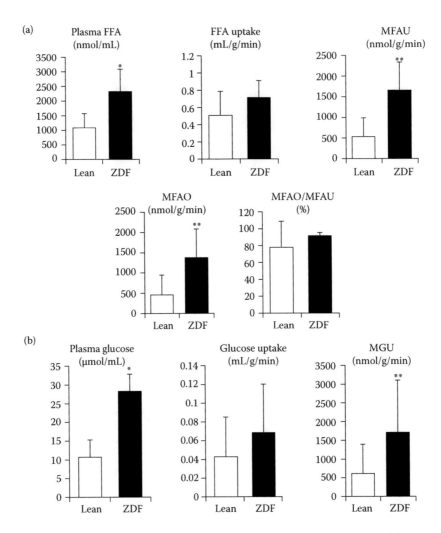

FIGURE 18.7 (a) Fatty acid metabolism measurements obtained in ZDF and lean rats by compartmental modeling of ¹¹C-palmitate PET data. MFAUP, myocardial fatty acid uptake; MFAU, myocardial fatty acid utilization; MFAO, myocardial fatty acid oxidation; MFAO/MFAU, myocardial FFA that was oxidized. *$P < 0.001$; **$P < 0.01$ (■, ZDF; □, Lean). (b) Glucose metabolism measurements obtained in ZDF and lean rats by compartmental modeling of 1-¹¹C-glucose PET data. MGUP, myocardial glucose uptake; MGU, myocardial glucose utilization. *$P < 0.0001$; **$P = 0.06$. (From Welch, M. J. et al. 2006. *Journal of Nuclear Medicine* 47: 689–97. With permission.)

important gauge of both imaging utility and its likelihood of translation and application into higher-order species.

A particularly relevant example of this validation strategy demonstrates these fundamental concepts. Recently, it has been shown that myocardial glucose uptake derived by ¹⁸F-FDG PET imaging was significantly decreased in ZDF rats compared to lean controls and this correlated well to the myocardial expression of GLUT-4 (Shoghi et al. 2008). Rosiglitazone, a widely used insulin-sensitizing medication that activates the PPAR system, beneficially modulated these metabolic perturbations by increasing glucose uptake while reciprocally decreasing myocardial fatty acid oxidation as measured by ¹¹C-palmitate; the former correlated to higher levels of GLUT-4 expression while the latter was associated with the diminished expression of two key mediators of myocardial fatty acid metabolism, CPT-1 and medium-chain acyl-CoA dehydrogenase (MCAD) (Shoghi et al. 2009). These studies not only demonstrate the utility of PET imaging to detect alterations in myocardial metabolism or its modulation with therapeutics but through correlation with adjunctive tissue expression analysis of targets or related factors, insights regarding the mechanisms responsible for pathologic changes associated with various human diseases can begin to be elucidated.

To achieve the full potential of small-animal PET, particularly in mice, several important physical and biological challenges must be overcome. Particularly pertinent to small-animal imaging is the impact of limited spatial resolution resulting in partial volume and spillover effects due to the small size of the mouse heart. Accurate quantifiable PET imaging can also be limited by insufficient temporal resolution in the face of the very rapid heart rate present in mouse heart. Finally, the biological or metabolic processes of interest occur much more rapidly than in large animal or humans providing further challenges to accurate kinetic modeling. To meet these challenges, numerous approaches are being devised which are discussed in Chapter 4.

18.3.3   Translation of Novel Radiotracers into Higher-Order Species

The advent of small-animal PET has advanced the field of preclinical evaluation of novel radiotracers by allowing preliminary studies in rodent models of relevant human disease. Despite the limitations inherent of small-animal PET, it allows for the assessment of candidate tracer pharmacokinetics, safety, and utility at an early and less costly stage. If the results of these small-animal studies are largely favorable, the validation of the radiotracer in higher-order, more complex, large animal models is often an intermediate step toward its potential application in humans.

The use of large-animal models is appealing for several reasons especially related to PET imaging. The relative size and structure of the organ of interest (i.e., heart, brain, etc.) in large versus small animals much more closely approximates that seen in humans. Moreover, compared to rodents, important parameters in large animals relevant to imaging such as heart rate or myocardial oxygen consumption, as well as target enzymes or structural proteins, are much more closely related to humans (Ginis et al. 2004, Haghighi et al. 2003). Many relevant large-animal models recapitulating several important human cardiovascular pathologies including dilated and ischemic cardiomyopathy, myocardial infarction, and pressure or volume overload have been described and are readily available for study though transgenic models are not (Dixon and Spinale 2009). Lastly, specialized methods associated with studying myocardial metabolism related to developing novel radiotracers are only feasible in large ani-

mals. In conjunction with arterial substrate content, catheterization of the coronary sinus in large animals as well as humans allows for the most direct measurement of myocardial substrate uptake as reflected by the calculated arteriovenous substrate difference (AVD) and, from a technical standpoint, cannot be reliably performed in rodents (Bing et al. 1953). Additionally, hyperinsulinemic–euglycemic clamp protocols to precisely alter arterial glucose levels can easily be performed in larger animals. Its use in rodents has been described, it is technically more challenging, and is subject to marked variation based on the exact methods utilized (Ayala et al. 2006). These considerations regarding the translation of tracers into large animals are well exemplified in a recent study (Herrero et al. 2006). Using the hyperinsulinemic–euglycemic clamp technique in dogs, a novel compartmental model validating the use of ^{11}C-glucose to measure myocardial glucose metabolism was developed. These studies were heavily dependent on the accurate and comprehensive assessment of the metabolic fate of ^{11}C-glucose as determined through both AVD measures of ^{11}C-glucose and ^{11}C-lactate myocardial uptake as well as myocardial tissue ^{11}C-glycogen content over a wide range of physiological conditions.

The translation of novel radiotracers for clinical application ultimately requires pilot studies in patients with the disease of interest. Results describing the tracer accumulation in a patient cohort can then be compared to the preliminary results derived from preclinical animal studies. As an example, ZDF rats have demonstrated relatively increased rates of myocardial fatty acid oxidation as assessed by PET imaging (see Figure 18.7) (Welch et al. 2006). Using ^{11}C-glucose and ^{11}C-palmitate radiotracers, recent PET studies of patients with type I diabetes demonstrated comparable results (Herrero et al. 2006). So, the general agreement of these clinical findings with those from animal models helps validate the expanded investigational use of these particular radiotracers in other contexts of myocardial metabolic disease. For radiotracer development in general, such corroborative studies in conjunction with a rigorous assessment of the candidate tracer's safety and pharmacokinetic profiles in both animals and humans are crucially important requirements needed before later phases of clinical investigation can be pursued.

18.4   Conclusions

While prior myocardial metabolic PET-imaging studies have greatly enhanced the understanding of disease

mechanisms, continued investigative focus is required to more fully elucidate the total contribution of altered

Chapter 18

metabolism in clinically relevant cardiovascular disease. Indeed, the concept of modulating cardiac metabolism in disease remains an active area of investigation and it is likely that additional pathways or factors may be identified that may be suitable targets for therapeutic intervention. Attaining greater insights into these mechanisms has been fundamentally advanced by the emergence of small-animal PET imaging, a powerful and sensitive modality allowing noninvasive assessment of *in vivo* radiotracer behavior.

Both the development of new transgenic animal models capitulating important aspects of human disease and continued innovation in small-animal PET technology and image reconstruction will accelerate this process and help circumvent some of the limitations commonly encountered. As such, the development and validation of novel PET radiotracers to specifically interrogate other aspects of myocardial metabolism is needed and continues to be a central focus of contemporary PET-imaging research.

References

Andersson, B., Blomström-Lundqvist, C., Hedner, T., and Waagstein, F. 1991. Exercise hemodynamics and myocardial metabolism during long-term beta-adrenergic blockade in severe heart failure. *Journal of the American College of Cardiology* 18: 1059–66.

Andersson, B., Lomsky, M., and Waagstein, F. 1993. The link between acute haemodynamic adrenergic beta-blockade and long-term effects in patients with heart failure. *European Heart Journal* 14: 1375–85.

Arthur, A. S. 1968. The transport and utilization of free fatty acid. *Annals of the New York Academy of Sciences* 149: 768–83.

Ayala, J. E., Bracy, D. P., McGuinness, O. P., and Wasserman, D. H. 2006. Considerations in the design of hyperinsulinemic-euglycemic clamps in the conscious mouse. *Diabetes* 55: 390–97.

Bergmann, S. R., Weinheimer, C. J., Markham, J., and Herrero, P. 1996. Quantitation of myocardial fatty acid metabolism using PET. *Journal of Nuclear Medicine* 37: 1723–30.

Bersin, R. M., Wolfe, C., Kwasman, M. et al. 1994. Improved hemodynamic function and mechanical efficiency in congestive heart failure with sodium dichloroacetate. *Journal of the American College of Cardiology* 23: 1617–24.

Bing, R. J., Siegel, A., Vitale, A. et al. 1953. Metabolic studies on the human heart in vivo: Studies on carbohydrate metabolism of the human heart. *The American Journal of Medicine* 15: 284–96.

Bowker-Kinley, M. M., Davis, W. I., Wu, P., Harris, R. A., and Popov, K. M. 1998. Evidence for existence of tissue-specific regulation of the mammalian pyruvate dehydrogenase complex. *Biochemistry Journal* 329: 191–96.

Bristow, M. R., Gilbert, E. M., Abraham, W. T. et al. 1996. Carvedilol produces dose-related improvements in left ventricular function and survival in subjects with chronic heart failure. *Circulation* 94: 2807–16.

Brown, M., Marshall, D. R., Sobel, B. E., and Bergmann, S. R. 1987. Delineation of myocardial oxygen utilization with carbon-11-labeled acetate. *Circulation* 76: 687–96.

Brown, M. A., Myears, D. W., and Bergmann, S. R. 1988. Noninvasive assessment of canine myocardial oxidative metabolism with carbon-11 acetate and positron emission tomography. *Journal of the American College of Cardiology* 12: 1054–63.

Buck, A., Wolpers, H. G., Hutchins, G. D. et al. 1991. Effect of carbon-11-acetate recirculation on estimates of myocardial oxygen consumption by PET. *Journal of Nuclear Medicine* 32: 1950–57.

Chaitman, B. R., Skettino, S. L., Parker, J. O. et al. 2004. Anti-ischemic effects and long-term survival during ranolazine monotherapy in patients with chronic severe angina. *Journal of the American College of Cardiology* 43: 1375–82.

Chandler, M. P., Kerner, J., Huang, H. et al. 2004. Moderate severity heart failure does not involve a downregulation of myocardial fatty acid oxidation. *American Journal of Physiology Heart and Circulatory Physiology* 287: H1538–43.

Clarke, B., Wyatt, K. M., and McCormack, J. G. 1996. Ranolazine increases active pyruvate dehydrogenase in perfused normoxic rat hearts: Evidence for an indirect mechanism. *Journal of Molecular and Cellular Cardiology* 28: 341–50.

Cocco, G., Rousseau, M. F., Bouvy, T. et al. 1992. Effects of a new metabolic modulator, ranolazine, on exercise tolerance in angina pectoris patients treated with beta-blocker or diltiazem. *Journal of Cardiovascular Pharmacology* 20: 131–38.

Coleman, R. A., Lewin, T. M., and Muoio, D. M. 2000. Physiological and nutritional regulation of enzymes of triacylglycerol synthesis. *Annual Review of Nutrition* 20: 77–103.

Colucci, W. S., Packer, M., Bristow, M. R. et al. 1996. Carvedilol inhibits clinical progression in patients with mild symptoms of heart failure. *Circulation* 94: 2800–06.

Dávila-Román, V. í. G., Vedala, G., Herrero, P. et al. 2002. Altered myocardial fatty acid and glucose metabolism in idiopathic dilated cardiomyopathy. *Journal of the American College of Cardiology* 40: 271–77.

DeGrado, T. R., Coenen, H. H., and Stocklin, G. 1991. 14(R,S)-[18F]Fluoro-6-thia-heptadecanoic acid (FTHA): Evaluation in mouse of a new probe of myocardial utilization of long chain fatty acids. *Journal of Nuclear Medicine* 32: 1888–96.

Dixon, J. A. and Spinale, F. G. 2009. Large animal models of heart failure: A critical link in the translation of basic science to clinical practice. *Circulation Heart Failure* 2: 262–71.

Dyck, J. R. B., Cheng, J.-F., Stanley, W. C. et al. 2004. Malonyl coenzyme A decarboxylase inhibition protects the ischemic heart by inhibiting fatty acid oxidation and stimulating glucose oxidation. *Circulation Research* 94: e78–84.

Eaton, S. 2002. Control of mitochondrial [beta]-oxidation flux. *Progress in Lipid Research* 41: 197–239.

Eichhorn, E. J., Bedotto, J. B., Malloy, C. R. et al. 1990. Effect of beta-adrenergic blockade on myocardial function and energetics in congestive heart failure. Improvements in hemodynamic, contractile, and diastolic performance with bucindolol. *Circulation* 82: 473–83.

Eichhorn, E. J., Heesch, C. M., Barnett, J. H. et al. 1994. Effect of metoprolol on myocardial function and energetics in patients with nonischemic dilated cardiomyopathy: A randomized, double-blind, placebo-controlled study. *Journal of the American College of Cardiology* 24: 1310–20.

Finck, B. N., Lehman, J. J., Leone, T. C. et al. 2002. The cardiac phenotype induced by PPAR-alpha overexpression mimics that caused by diabetes mellitus. *The Journal of Clinical Investigation* 109: 121–30.

Garcia, C. K., Goldstein, J. L., Pathak, R. K., Anderson, R. G. W, and Brown, M. S. 1994. Molecular characterization of a membrane transporter for lactate, pyruvate, and other monocarboxylates: Implications for the Cori cycle. *Cell* 76: 865–73.

Garcia, E. V., Faber, T. L., Cooke, C. D. et al. 2007. The increasing role of quantification in clinical nuclear cardiology: The Emory approach. *Journal of Nuclear Cardiology* 14: 420–32.

Ginis, I., Luo, Y., Miura, T. et al. 2004. Differences between human and mouse embryonic stem cells. *Developmental Biology* 269: 360–80.

Goldfarb, A. H., Bruno, J. F., and Buckenmeyer, P. J. 1986. Intensity and duration effects of exercise on heart cAMP, phosphorylase, and glycogen. *Journal of Applied Physiology* 60: 1268–73.

Gralinski, M. R., Black, S. C., Kilgore, K. S. et al. 1994. Cardioprotective effects of ranolazine (RS-43285) in the isolated perfused rabbit heart. *Cardiovascular Research* 28: 1231–37.

Haghighi, K., Kolokathis, F., Pater, L. et al. 2003. Human phospholamban null results in lethal dilated cardiomyopathy revealing a critical difference between mouse and human. *The Journal of Clinical Investigation* 111: 869–76.

Hermann, H.-P., Pieske, B., Schwarzmöller, E. et al. 1999. Haemodynamic effects of intracoronary pyruvate in patients with congestive heart failure: An open study. *Lancet* 353: 1321–23.

Herrero, P, and Gropler, R. J. 2005. Imaging of myocardial metabolism. *Journal of Nuclear Cardiology* 12: 345–58.

Herrero, P., Peterson, L. R., McGill, J. B. et al. 2006. Increased myocardial fatty acid metabolism in patients with type 1 diabetes mellitus. *Journal of the American College of Cardiology* 47: 598–604.

Herrero, P., Dence, C. S., Coggan, A. R. et al. 2007. L-3-11C-Lactate as a PET tracer of myocardial lactate metabolism: A feasibility study. *Journal of Nuclear Medicine* 48: 2046–55.

Higgins, A. J., Morville, M., Burges, R. A. et al. 1980. Oxfenicine diverts rat muscle metabolism from fatty acid to carbohydrate oxidation and protects the ischaemic rat heart. *Life Sciences* 27: 963–70.

How, O.-J., Aasum, E., Kunnathu, S. et al. 2005. Influence of substrate supply on cardiac efficiency, as measured by pressure-volume analysis in *ex vivo* mouse hearts. *American Journal of Physiology Heart and Circulatory Physiology* 288: H2979–85.

Hue, L, and Taegtmeyer, H. 2009. The Randle cycle revisited: A new head for an old hat. *American Journal of Physiology Endocrinology and Metabolism* 297: E578–91.

Ichise, M., Meyers, J. H., and Yonekura, Y. 2001. An introduction to PET and SPECT neuroreceptor quantification models. *Journal of Nuclear Medicine* 42: 755–63.

Kerbey, A. L., Randle, P. J., Cooper, R. H. et al. 1976. Regulation of pyruvate dehydrogenase in rat heart. mechanism of regulation of proportions of dephosphorylated and phosphorylated enzyme by oxidation of fatty acids and ketone bodies and of effects of diabetes: Role of coenzyme A, acetyl-coenzyme A and reduced and oxidized nicotinamide-adenine dinucleotide. *Biochemistry Journal* 154: 327–48.

Kerner, J, and Bieber, L. 1990. Isolation of a malonyl-CoA-sensitive CPT-/beta-oxidation enzyme complex from heart mitochondria. *Biochemistry* 29: 4326–34.

Kim, A. S., Miller, E. J., and Young, L. H. 2009. AMP-activated protein kinase: A core signalling pathway in the heart. *Acta Physiologica* 196: 37–53.

Laine, H., Katoh, C., Luotolahti, M. et al. 1999. Myocardial oxygen consumption is unchanged but efficiency is reduced in patients with essential hypertension and left ventricular hypertrophy. *Circulation* 100: 2425–30.

Lida, H., Rhodes, C. G., Araujo, L. I. et al. 1996. Noninvasive quantification of regional myocardial metabolic rate for oxygen by use of 15O₂ inhalation and positron emission tomography: Theory, error analysis, and application in humans. *Circulation* 94: 792–807.

Lopaschuk, G. D., Belke, D. D., Gamble, J., Itoi, T., and Schonekess, B. O. 1994. Regulation of fatty acid oxidation in the mamalian heart in health and disease. *Biochemistry Biophysics Acta* 1213: 263–76.

Marsin, A. S., Bertrand, L., Rider, M. H. et al. 2000. Phosphorylation and activation of heart PFK-2 by AMPK has a role in the stimulation of glycolysis during ischaemia. *Current Biology* 10: 1247–55.

McCormack, J. G, and Denton, R. M. 1989. Influence of calcium ions on mammalian intramitochondrial dehydrogenases. *Methods in Enzymology* 174: 95–118.

McCormack, J. G., Barr, R. L., Wolff, A. A., and Lopaschuk, G. D. 1996. Ranolazine stimulates glucose oxidation in normoxic, ischemic, and reperfused ischemic rat hearts. *Circulation* 93: 135–42.

McGarry, J. D., Mills, S. E., Long, C. S., and Foster, D. W. 1983. Observations on the affinity for carnitine, and malonyl-CoA sensitivity, of carnitine palmitoyltransferase I in animal and human tissues. *Biochemistry Journal* 214: 21–28.

Neubauer, S. 2007. The failing heart—An engine out of fuel. *New England Journal of Medicine* 356: 1140–51.

Opie, L. H. 1991. *The Heart: Physiology and Metabolism.* New York: Raven Press.

Osorio, J. C., Stanley, W. C., Linke, A. et al. 2002. Impaired myocardial fatty acid oxidation and reduced protein expression of retinoid X receptor-{alpha} in pacing-induced heart failure. *Circulation* 106: 606–12.

Panchal, A. R., Stanley, W. C., Kerner, J., and Sabbah, H. N. 1998. Beta-receptor blockade decreases carnitine palmitoyl transferase I activity in dogs with heart failure. *Journal of Cardiac Failure* 4: 121–26.

Paolisso, G., Gambardella, A., Galzerano, D. et al. 1994. Total-body and myocardial substrate oxidation in congestive heart failure. *Metabolism* 43: 174–79.

Patel, M. S, and Korotchkina, L. G. 2001. Regulation of mammalian pyruvate dehydrogenase complex by phosphorylation: Complexity of multiple phosphorylation sites and kinases. *Experimental and Molecular Medicine* 33: 191–97.

Pepine, C. J, and Wolff, A. A. 1999. A controlled trial with a novel anti-ischemic agent, ranolazine, in chronic stable angina pectoris that is responsive to conventional antianginal agents. *The American Journal of Cardiology* 84: 46–50.

Randle, P. J. 1986. Fuel selection in animals. *Biochemical Society Transactions* 14: 799–806.

Razeghi, P., Young, M. E., Alcorn, J. L. et al. 2001. Metabolic gene expression in fetal and failing human heart. *Circulation* 104: 2923–31.

Chapter 18

Recchia, F. A., McConnell, P. I., Bernstein, R. D. et al. 1998. Reduced nitric oxide production and altered myocardial metabolism during the decompensation of pacing-induced heart failure in the conscious dog. *Circulation Research* 83: 969–79.

Rosenblatt-Velin, N., Montessuit, C., Papageorgiou, I., Terrand, J., and Lerch, R. 2001. Postinfarction heart failure in rats is associated with upregulation of GLUT-1 and downregulation of genes of fatty acid metabolism. *Cardiovascular Research* 52: 407–16.

Russell, R. R., Yin, R., Caplan, M. J. et al. 1998. Additive effects of hyperinsulinemia and ischemia on myocardial GLUT1 and GLUT4 translocation in vivo. *Circulation* 98: 2180–86.

Russell, R. R., Li, J., Coven, D. L. et al. 2004. AMP-activated protein kinase mediates ischemic glucose uptake and prevents postischemic cardiac dysfunction, apoptosis, and injury. *The Journal of Clinical Investigation* 114: 495–503.

Schulz, G., von Dahl, J., Kaiser, H. J. et al. 1996. Imaging of beta-oxidation by static PET with 14(*R,S*)-[18F]-fluoro-6-thiaheptadecanoic acid (FTHA) in patients with advanced coronary heart disease: A comparison with 18FDG-PET and 99Tcm-MIBI SPET. *Nuclear Medicine Communications* 17: 1057–64.

Shoghi, K. I., Gropler, R. J., Sharp, T. L. et al. 2008. Time course of alterations in myocardial glucose utilization in the Zucker diabetic fatty rat with correlation to gene expression of glucose transporters: A small-animal PET investigation. *Journal of Nuclear Medicine* 49: 1320–27.

Shoghi, K. I. 2009. Quantitative small animal PET. *Quarterly Journal of Nuclear Medicine and Molecular Imaging* 53: 365–73.

Shoghi, K. I., Finck, B. N., Schechtman, K. B. et al. 2009. *In vivo* metabolic phenotyping of myocardial substrate metabolism in rodents: Differential efficacy of metformin and rosiglitazone monotherapy. *Circulation Cardiovascular Imaging* 2: 373–81.

Soret, M., Bacharach, S. L., and Buvat, I. 2007. Partial-volume effect in PET tumor imaging. *Journal of Nuclear Medicine* 48: 932–45.

Stanley, W. C. 1991. Myocardial lactate metabolism during exercise. *Medicine and Science in Sports and Exercise* 23: 920–24.

Stanley, W. C., Lopaschuk, G. D., and McCormack, J. G. 1997. Regulation of energy substrate metabolism in the diabetic heart. *Cardiovascular Research* 34: 25–33.

Stanley, W. C, and Chandler, M. P. 2002. Energy metabolism in the normal and failing heart: Potential for therapeutic interventions. *Heart Failure Reviews* 7: 115–30.

Stanley, W. C., Recchia, F. A., and Lopaschuk, G. D. 2005. Myocardial substrate metabolism in the normal and failing heart. *Physiological Reviews* 85: 1093–129.

Sugden, M. C, and Holness, M. J. 2006. Mechanisms underlying regulation of the expression and activities of the mammalian pyruvate dehydrogenase kinases. *Archives of Physiology and Biochemistry* 112: 139–49.

Sun, K. T., Yeatman, L. A., Buxton, D. B. et al. 1998. Simultaneous measurement of myocardial oxygen consumption and blood flow using [1-carbon-11]acetate. *Journal of Nuclear Medicine* 39: 272–80.

Tadamura, E., Kudoh, T., Hattori, N. et al. 1998. Impairment of BMIPP uptake precedes abnormalities in oxygen and glucose metabolism in hypertrophic cardiomyopathy. *Journal of Nuclear Medicine* 39: 390–96.

Taylor, M., Wallhaus, T. R., DeGrado, T. R. et al. 2001. An evaluation of myocardial fatty acid and glucose uptake using PET with [18F]fluoro-6-thia-heptadecanoic acid and [18F]FDG in patients with congestive heart failure. *Journal of Nuclear Medicine* 42: 55–62.

Tuunanen, H., Engblom, E., Naum, A. et al. 2008. Trimetazidine, a metabolic modulator, has cardiac and extracardiac benefits in idiopathic dilated cardiomyopathy. *Circulation* 118: 1250–58.

Uehara, T., Ishida, Y., Hayashida, K. et al. 1998. Myocardial glucose metabolism in patients with hypertrophic cardiomyopathy: Assessment by F-18-FDG PET study. *Annals of Nuclear Medicine* 12: 95–103.

van der Vusse, G. J., van Bilsen, M., and Glatz, J. F. C. 2000. Cardiac fatty acid uptake and transport in health and disease. *Cardiovascular Research* 45: 279–93.

Waagstein, F., Bristow, M. R., Swedberg, K. et al. 1993. Beneficial effects of metoprolol in idiopathic dilated cardiomyopathy. Metoprolol in dilated cardiomyopathy (MDC) trial study group. *Lancet* 342: 1441–46.

Wahr, J. A., Childs, K. F., and Bolling, S. F. 1994. Dichloroacetate enhances myocardial functional and metabolic recovery following global ischemia. *Journal of Cardiothoracic and Vascular Anesthesia* 8: 192–97.

Wallhaus, T. R., Taylor, M., DeGrado, T. R. et al. 2001. Myocardial free fatty acid and glucose use after carvedilol treatment in patients with congestive heart failure. *Circulation* 103: 2441–46.

Welch, M. J., Lewis, J. S., Kim, J. et al. 2006. Assessment of myocardial metabolism in diabetic rats using small-animal PET: A feasibility study. *Journal of Nuclear Medicine* 47: 689–97.

Wolff, A. A., Rotmensch, H. H., Stanley, W. C., and Ferrari, R. 2002. Metabolic approaches to the treatment of ischemic heart disease: The clinicians' perspective. *Heart Failure Reviews* 7: 187–203.

Wu, P., Peters, J. M., and Harris, R. A. 2001. Adaptive increase in pyruvate dehydrogenase kinase 4 during starvation is mediated by peroxisome proliferator-activated receptor [alpha]. *Biochemical and Biophysical Research Communications* 287: 391–96.

Yael, Y., Magdalena, J., Nuss, H. B. et al. 2010. Matching ATP supply and demand in mammalian heart. *Annals of the New York Academy of Sciences* 1188: 133–42.

Yang, J, and Holman, G. D. 2006. Long-term metformin treatment stimulates cardiomyocyte glucose transport through an AMP-activated protein kinase-dependent reduction in GLUT4 endocytosis. *Endocrinology* 147: 2728–36.

Young, L. H., Renfu, Y., Russell, R. R. et al. 1997. Low-flow ischemia leads to translocation of canine heart GLUT-4 and GLUT-1 glucose transporters to the sarcolemma *in vivo*. *Circulation* 95: 415–22.

19. Development of Receptor-Avid Peptide Radiotracers for Targeting Human Cancers

Wynn A. Volkert, Thomas P. Quinn, C. Jeffrey Smith, and Timothy J. Hoffman

19.1 Introduction

The development of receptor-specific radiolabeled peptides for diagnosis and treatment of cancers in human patients are being pursued by an increasing number of investigators across the globe (Ginj and Maecke 2004; Weiner and Thakur 2005; Li et al. 2008; Mankoff et al. 2008; Fleischman et al. 2009; Garcia-Garayoa et al. 2009). Numerous approaches for designing new radiolabeled peptide constructs that specifically target receptors that are uniquely expressed or overexpressed by human cancer cells are being employed. For example, technological advances in molecular biology, combinatorial chemistry, interrogation of the human genome, and peptide biochemistry are being used as a molecular basis to identify and characterize receptor expression and develop cognate receptor-avid radiolabeled peptide conjugates (Reubi 2003; Beneditt et al. 2004; Buchsbaum

et al. 2004; Landon et al. 2004). The design of effective cancer-specific molecular imaging radiotracers requires optimizing the balance between specific *in vivo* targeting of tumors and clearance of radioactivity from relevant nontarget tissues. To this end, it is critical that the radiolabeling strategies associated with the receptor-targeting vectors will normally alter or perhaps enhance their capability for high-specific binding to the receptor with high affinity (Knight 2003; Liu 2004; Smith et al. 2005; Wester and Kessler 2005). Several issues must be addressed when designing site-specific imaging peptide-based radiopharmaceuticals. These include problems related to selective delivery of the radioactive drug to tumors, maximization of residualization of radionuclide in tumors, optimization clearance of the radioactivity from the blood and normal tissues, minimization of catabolism or metabolism in the bloodstream, and maximizing the specific activity of the radiolabeled peptide (Meares et al. 1988; Eckelman et al. 2008; Mankoff et al. 2008; Garcia-Garayoa et al. 2009; Vanderheyden 2009). This latter issue is particularly critical when the

Targeted Molecular Imaging. Edited by Michael J. Welch and William C. Eckelman © 2012 Taylor & Francis Group, LLC. ISBN: 978-1-4398-4195-2

number of cognate receptors expressed by the cancer cells is low since the presence of significant levels of unlabeled or cold peptide conjugates that target the receptor with similar affinities will competitively inhibit *in vivo*-specific binding of the radiolabeled analog to the cancer cells (Eckelman et al. 2008; Mankoff et al. 2008). Since multiple parameters must be considered during the design and development of cancer-specific radiopharmaceuticals as diagnostic imaging agents suitable for human applications, simply randomly linking a radionuclide to a peptide-based vector is not a rational approach.

Following the identification of receptor-avid peptides that hold promise for further development as molecular imaging radiopharmaceuticals, the chemistry involved in conjugation at a particular site on the peptide, and the ease and robustness of the labeling method are essential components of the radioactive drug design process (Anderson et al. 1999; Maina et al. 2002; Wild et al. 2003; Boswell et al. 2004; Liu 2004; Prasanphanich et al. 2007). Low-molecular-weight receptor-specific peptides will be sensitive to side-chain additions or changes in amino acid sequences or changes in the physicochemical properties of regions involved with receptor specificity and affinity. Clearly, radiometal chelates must be appended at a site on the peptide that is compatible with the specificity and retention in the targeted tumors. At a minimum, the radiometal chelate must not sterically interfere with binding of the peptide vector component with its cognate receptor. These considerations necessitate that the selections of appropriate radionuclides, radiometal chelate, conjugation site, and labeling chemistry are critical elements in designing effective peptide conjugate receptor-specific radiopharmaceuticals (Beneditt et al. 2004; Ginj and Maecke 2004; Weiner and Thakur 2005; Mankoff et al. 2008). Low-molecular-weight peptides and their corresponding radiolabeled conjugates are usually rapidly cleared from the bloodstream via renal or hepatobiliary excretion, or both, depending on their structure and physiochemical properties (Lister-James et al. 1997; Wester and Kessler 2005; Giron 2009). The hydrophobic characteristics of the conjugate make important contributions to their pharmacokinetic and tissue distribution. As lipophilicity increases, it generally results in increased elimination by the hepatobiliary route, while increasing hydrophilicity favors preferential removal by the kidneys (Fritzberg et al. 1994; Akazawa et al. 2001; Beneditt et al. 2004). It is possible to increase tumor uptake by increasing lipophilicity to produce higher protein binding and slower

blood clearance. This strategy generally leads to greater hepatobiliary clearance that results in higher levels of liver and abdominal radioactivity which interferes with single-photon emission computed tomography (SPECT) or positron emission tomography (PET) imaging studies (Lister-James et al. 1997; Knight 2003).

Low-molecular-weight radiolabeled peptide conjugates offer several advantages compared with larger radiolabeled tumor-targeting constructs. In contrast to larger molecular entities (e.g., Monoclonal antibodies (MAbs) or nanostructures), small peptides are not immunogenic and can be readily and inexpensively synthesized using automated solid-state technology to produce well-defined and high-purity products (Reubi 2003; Weiner and Thakur 2005). The peptide conjugates are usually efficiently radiolabeled and high-specific-activity radiotracers can be prepared via high-performance liquid chromatography (HPLC) separation of the radiolabeled peptide conjugate from the unlabeled or "cold" peptide conjugate (Smith et al. 2005; Eckelman et al. 2008). The low-molecular-weight molecular entities can be designed to clear efficiently from the blood via the renal/urinary pathway and produce minimal hepatobiliary clearance while exhibiting effective penetration into tumors (Knight 2003; Beneditt et al. 2004; Eberle et al. 2004; Giblin et al. 2004). Disadvantages associated with low-molecular-weight peptides include their potential susceptibility to proteolysis in the blood and limitations on the magnitude of tumor uptake that is related to their relatively rapid clearance from the blood (Reubi 2003; Weiner and Thakur 2005). Many of endogenous receptor-avid peptides have plasma half-lives that prohibit their use in directly developing *in vivo* diagnostic radiopharmaceuticals (Buchegger et al. 2003; Reubi 2003; Whetstone 2004). Stabilized analogs can be formulated to provide resistance to *in vivo* catabolism by substitution of D-amino acids or pseudo-amino acids for L-amino acids in the peptide sequence (Sawyer et al. 1980; Reubi et al. 2002; Ginj and Maecke 2004; Garcia-Garayoa et al. 2006, 2009) and synthesis of cyclized analogs (Gali et al. 2002, 2006; Reubi 2003) that maintain the three-dimensional (3D) structure of the receptor-binding motif of the peptide while inhibiting proteolysis. The three radiolabeled peptides exemplified in this chapter provide examples for these approaches and their effectiveness in producing *in vivo*-targeted cancer-specific imaging radiotracers.

Developing an effective strategy for linking radiometals to the receptor-avid peptide to produce cancer-specific radiolabeled bioconjugates is a critical component for initiating translation of the approach for eventual clinical application. The radiometallated chelate must be positioned at a site on the peptide vector that does not significantly hinder receptor-binding affinity or specificity. A linker, spacer group, or tether is usually needed to accomplish this requirement. In addition, the physicochemical properties of the tether will play an essential role in tuning the *in vivo* pharmacokinetics of the radiolabeled bioconjugates (Knight 2003; Smith et al. 2003; Beneditt et al. 2004; Eberle et al. 2004). Since most peptide conjugates have relatively low molecular weights, an optimum balance between the properties of the radiometal chelate, the peptide-targeting vector, and the spacer group must be achieved to produce a molecular imaging radiotracer that exhibits the appropriate tumor uptake and retention and corresponding pharmacokinetics to maximize *in vivo*-imaging properties (e.g., high-specific uptake values). Studies to develop radiolabeled conjugates of radiometallated bioconjugates of three types of peptide-binding moieties are described in this chapter. The specific structures and physicochemical characteristics of various analog play a pivotal role in determining the *in vivo* specificity and magnitude of the deposition (both early and long term) of the radiotracer in the tumor and its route and route of clearance from the blood and nontarget tissues and organs.

Numerous peptide receptors that are uniquely expressed or overexpressed on human cancer cells have been studied as a molecular basis and promising targets for designing cancer-specific molecular imaging radiopharmaceuticals (Reubi 2003; Landon et al. 2004; Weiner and Thakur 2005). The pioneering development of radiolabeled peptide vectors that specifically bind to somatostatin receptors demonstrated the capability of using radiometal chelate-conjugated peptides for imaging a variety of human cancers (including carcinoid, neuroblastoma pheochromocytoma, medullary thyroid cancer, and others) (Krenning et al. 1993; Reubiet et al. 1996; deJong et al. 1998; Virgolini et al. 2000). Research efforts with cyclical somatostatin analogs resulted in the development of an effective molecular imaging agent that targets cancerous tumors that overexpress somatostatin receptors resulted in its translation to a Food and Drug Administration (FDA)-approved radiopharmaceutical (i.e., ^{111}In-DTPA-octreotide—Octreoscan®) that is widely used in clinical imaging of neuroendocrine tumors (Krenning et al. 1995; Kwekkboom et al. 2005). There exists extensive literature describing the early and subsequent development and utilization of this radiopharmaceutical and numerous other

radiolabeled octreotide analogs for imaging somatostatin receptor expressing tumors in animals and humans (Heppler et al. 1999; Maina et al. 2002; Virgolini et al. 2002; Reubi 2003).

The design and development of radiometallated peptide conjugates as potential cancer-specific targeting radiopharmaceuticals that selectively bind with three cognate receptor types (that is, melanocortin-1 (MC-1), bombesin 1–4 (BBN1–4), and guanylin/guanylate cyclase-C (GC-C) receptors) are presented in this chapter. All the three regulatory receptor systems are classified as G protein-coupled receptors (Drewett and Garbers 1994; Miao et al. 2003; Reubi 2003). One characteristic of G protein-coupled receptors that is relevant to the development of receptor-avid diagnostic agents is that they are capable of efficiently transporting the radiotracers intracellularly via a receptor-mediated endocytosis mechanism (Harden et al. 1998; Duvernay et al. 2005). This process usually occurs following a receptor–agonist interaction on the exterior surface of the cancer cell membrane that initiates the associated signaling cascades. It is important to recognize that even though the agonist can behave as autocrines, the high-specific-activity radiolabeled peptide conjugates are administered at tracer levels (Eckelman et al. 2008) for molecular imaging and produce no measurable autocrine effects. A benefit of internalization is that it facilities trapping of the radiotracer at the tumor site for extended periods of time. Even though internalized, they may exhibit delayed excretion from the cell or can be metabolized or catabolized by intracellular enzymes to produce metabolic products that are residualized inside of the cell or excreted from the cell (deJong et al. 1998; Reubi et al. 2002; Reubi 2003). Radiolabeled peptides that bind to the G protein-coupled receptors as antagonists are generally not transported intracellularly and are susceptible to displacement or dissociation from the cell surface receptor resulting in shorter tumor retention time. Radiolabeled receptor-avid antagonists have, however, been shown to hold important potential as cancer-specific molecular imaging radiopharmaceuticals (Ginij et al. 2006; Abd-Elgaliel et al. 2008; Cescato et al. 2008).

Receptor-avid peptide conjugates used as diagnostic imaging agents can play an important role in the era of personalized medicine (Eckelman et al. 2008). A major advantage offered by radiolabeled peptides that target cell surface receptors expressed on human cancer cells is that the analogous molecular constructs can also be labeled with therapeutic radionuclides for use to develop corresponding targeted radiotherapeutic

(TRT) radiopharmaceuticals (Anderson and Welch 1999; Smith et al. 2003; Krenning et al. 2004). The radiolabeled peptide conjugates described in this chapter offer this potential, and research is being performed to develop effective TRT radiotracers, based on the *in vivo* targeting and pharmacokinetic properties of the receptor-avid peptides developed as molecular imaging probes (Giblin et al. 2006b; Johnson et al. 2006; Miao et al. 2008). The *in vitro* and *in vivo* studies outlined in the following discussions to validate specificity of *in vivo* radiotracer targeting of specific receptors expressed by cancer cells with the relevant target affinity and *in vivo* pharmacokinetic properties in normal and tumor-bearing animal models are critical steps in the translation of these radiotracers for use in human clinical trials.

19.2   Metal-Cyclized Melanotropin Peptides of Melanoma Imaging

The development of metal-cyclized melanotropin peptides began as a structure–activity design program to develop receptor-avid peptides that could be directly radiolabeled at specific locations to produce radiochemical stable and biologically active imaging probes. The target receptor was the melanocortin-1 receptor (MC1-R), a 7-helix transmembrane G-protein-linked receptor overexpressed on the surface of melanoma tumor cells (Siegrist et al. 1989; Abdel-Malek 2001; Lin and Fisher 2007). The natural ligand for the MC1-R is the 13-amino acid (Ac–Ser^1–Tyr^2–Ser^3–Met^4–Glu^5–His^6–Phe^7–Arg^8–Trp^9–Gly^{10}–Lys^{11}–Pro^{12}–Val^{13}–NH$_2$) α-melanocyte-stimulating hormone (α-MSH), proteolytically processed from proopiomelanocortin (Sawyer et al. 1990; Cone et al. 1996). By the early 1990s, the literature was rich in α-MSH structure–activity studies, which provided a strong foundation for the design of radiolabeled α-MSH analogs (Hruby et al. 1993). The core MC1-R-binding amino acids (His^6–Phe^7–Arg^8–Trp^9) had been defined and numerous α-MSH analogs with improved receptor affinity and prolonged stability had been reported, including the superpotent NDP analog Ac–[Nle^4,D-Phe^7]–α-MSH that displayed picomolar receptor affinity (Sawyer et al. 1980). Attempts to target the MC1-R with radiolabeled NDP peptides were disappointing due to poor biodistribution properties and low tumor retention (Garg et al. 1996; Vaidyanathan and Zalutsky 1997; Chen et al. 1999). However, one clinical case study was reported with a bidentate NDP analog ^{111}In-(bisMSH-DTPA) (Wraight et al. 1992). ^{111}In-(bisMSH-DTPA) imaged 89% of the melanoma tumors in patients but suffered from high nonspecific liver uptake and slow kidney clearance. An analysis of MC1-R expression on 10 different melanoma cell lines revealed that there were ~1000–8000 receptors (1.7×10^{-21}–1.3×10^{-20} M) per cell (Miao et al. 2003). It is estimated that the MC1-R receptor concentration on a tumor with a 1 mL volume would be 10^{-11}–10^{-12} M. The average molecular weight of the MC1-R-targeting peptide chelator conjugates described is 2000, and then 1–10 ng of peptide would saturate the receptor-binding sites. The relatively low cellular concentration of MC1-Rs meant that radiolabeled ligands would have to have high-specific activity and prolonged tumor localization to be effective imaging or therapeutic agents. Despite the challenges associated with low receptor concentrations, the combination of a robust melanotropin structure–activity database coupled with reports on a few preclinical and clinical melanoma imaging probes provided a rich environment for the design and characterization of a new class of MC1-R-targeting agents.

The focus of the new melanotropin design was to incorporate the metal (radiometal) center into the 3D structure of molecule to form a compact chemically stable MC1-R-avid molecule peptide. Two families of chemically constrained melanotropin analogs, which were cyclized by disulfide and lactam bond formation, exhibited low nanomolar to picomolar affinities for the MC1-R and prolonged bioactivity (Cody et al. 1985; Al-Obeidi et al. 1989). The disulfide cyclized structure Ac–[Cys^4,D-Phe^7,Cys^{10}]–α-MSH was selected as the template for molecular-modeling studies. Modeling studies revealed that simply reducing the disulfide bond and incorporating a ^{99m}TcO (V), ReO (V), or ^{188}ReO (V) core via S$_2$N$_2$ coordination did not result in a structure that was predicted to be energetically stable. The results indicated that several equally stable Re-MSH structures could be formed which would result in variable-length receptor-binding loops. The second-generation metal-cyclized molecule, Re-CCMSH, contained two N-terminal geminal cysteine residues that were designed to provide three of the four donor atoms in an S$_3$N coordination geometry with the fourth provided by the Cys^{10} thiol. Structurally, the Re-CCMSH design was predicted to be stable and entropically it was likely that cyclization would be favored if only one donor atom interaction with the metal complex was required to close the circle.

The Re-cyclized versions of the two melanotropin analogs were synthesized and characterized by mass spectrometry and multidimensional nuclear magnetic resonance (NMR) (Giblin et al. 1998). Re-MSH was cyclized by the two cysteine thiols (Cys[4,10]) and two amides from Trp[9] and Cys[10], yielding an 18-member receptor-binding loop compared to a 24-member receptor-binding loop present in the original disulfide-bonded analog. Considerable strain was noted in the molecule during NMR structural studies, including rearrangements within the molecule. The second-generation Re-CCMSH molecule was cyclized by the three cysteine thiols (Cys[3,4,10]) and the Cys[4] amide nitrogen coordination of the ReO core, resulting in a 23-member receptor-binding loop (Figure 19.1). Re-CCMSH was stable for a month in solution. MC1-R cell-binding assays with [^{125}I-Tyr[3]]-NDP and B16/F1 murine melanoma cells were performed to assess the receptor avidity. Competitive inhibition (K_i) values for Re-MSH and Re-CCMSH were 6.6×10^{-8} and 2.9×10^{-9}, respectively. The ^{99m}Tc-MSH and ^{99m}Tc-CCMSH molecules were prepared by a glucoheptonate transchelation reaction and purified via C-18 reverse-phase high-performance liquid chromatography (Giblin et al. 1998). ^{99m}Tc-MSH was less stable in phosphate buffer and more susceptible to cysteine challenge than ^{99m}Tc-CCMSH. Results from the initial *in vitro* studies revealed a strong correlation between structural and radiochemical stability and receptor affinity leading to the selection of ^{99m}Tc-CCMSH for animal biodistribution and imaging studies.

Additional *in vitro* cell binding, internalization, and efflux studies demonstrated that ^{99m}Tc-CCMSH was rapidly bound and internalized by B16/F1 melanoma cells. Efflux rates for internalized ^{99m}Tc-CCMSH were dramatically less than ^{99m}Tc-labeled linear NDP analogs

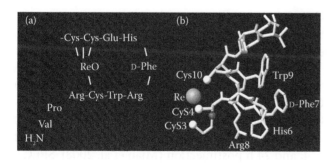

FIGURE 19.1 The structure of Re(Arg[11])CCMSH. (a) A schematic of the rhenium (Re) cyclized α-MSH analog illustrating the metal coordination by three cysteine residues. (b) A solution structure model of Re(Arg[11])CCMSH derived from two-dimensional NMR studies. The highly conserved MC1-R-binding residues His[6], D-Phe[7], Arg[8], and Trp[9] are labeled as well as the Re coordinating cysteine residues with the three thiols (white spheres) and one backbone amide nitrogen (dark sphere).

(Giblin et al. 1998; Chen et al. 1999). Biodistribution studies of ^{99m}Tc-CCMSH (1.21×10^{10} MBq/g) in B16/F1 melanoma-bearing mice demonstrated rapid tumor uptake with prolonged retention. Tumor uptake at 1, 4, and 24 h was 11.64 ± 1.54, 9.51 ± 1.97, and $2.59 \pm 0.83\%$ID/g, respectively. Disappearance of radioactivity from normal organs and tissue was rapid with the exception of the kidneys. Uptake of radioactivity in the kidneys was 19.71 ± 1.18, 14.60 ± 1.88, and $1.22 \pm 0.15\%$ID/g at 1, 4, and 24 h postinjection. Greater than 80% of the injected dose was in the urine 4 h postinjection. Coinjection of 2 μg of NDP reduced tumor uptake by >80% but did not affect normal organ uptake demonstrating that radioactivity in the tumor was receptor mediated, while radioactivity in the kidneys was not receptor related. To broaden the scope of radionuclides that could be targeted to the MC1-R, the metal chelator 1,4,7,10-tetraazacyclododecane-1,4,7,10-tetraacetic acid (DOTA) was appended to the amino terminus of the Re-cyclized peptide (Chen et al. 2001). The DOTA-conjugated peptides were radiolabeled with numerous radionuclides, including [111]In (Chen et al. 2001; Cheng et al. 2002) [64]Cu (McQuade et al. 2005; Wei et al. 2007a), [86]Y (McQuade et al. 2005), [203]Pb (Miao et al. 2008), and [68]Ga (Wei et al. 2007b; Cantorias et al. 2009) for imaging and [177]Lu (Miao et al. 2006, 2007), [90]Y (Miao et al. 2006), [212]Bi, and [212]Pb (Miao et al. 2005) for therapy.

Tumor uptake and retention of ^{111}In-DOTA-Re-CCMSH (8.12×10^8 MBq/g) were similar to ^{99m}Tc-CCMSH (Chen et al. 2001). Disappearance of radioactivity from normal organs and tissues was rapid with the exception of the kidneys, which exhibited radioactivity retention similar to that of ^{99m}Tc-CCMSH. While the tumor uptake of ^{99m}Tc-CCMSH and ^{111}In-DOTA-Re-CCMSH was impressive, the kidney retention was troubling. Two strategies were employed to reduce kidney retention. First, lysine coinjection was examined to reduce renal reabsorption of the radiolabeled peptide. Coinjection of 20 mg of lysine reduced kidney retention by >50% without reducing tumor uptake. The second strategy involved replacing Lys[11] with various amino acids to determine if altering the peptide structure outside the conserved receptor-binding region could reduce renal retention. Replacement of Lys[11] with neutral, hydrophobic, or negatively charged amino acids reduced kidney retention but also significantly reduced tumor uptake (Chen et al. 1999 Chen 2000). Substitution of Lys[11] with Arg also reduced renal retention and improved tumor uptake (Cheng et al. 2002). For example, 4 h postinjection, tumor and kidney uptake values for ^{99m}Tc-CCMSH versus [99m]Tc-(Arg[11])CCMSH were 9.51 ± 1.97 and 14.60 ± 1.88%ID/g versus 11.16 ± 1.77 and 5.53 ± 1.17%ID/g, respectively (Miao et al. 2007). A similar trend was observed in the tumor and kidney uptake values for ^{111}In-DOTA-Re-CCMSH versus ^{111}In-DOTA-Re(Arg[11])CCMSH 4 h post-injection which were 9.49 ± 0.90 and 9.27 ± 2.65%ID/g versus 17.41 ± 5.63 and 7.73 ± 1.13%ID/g, respectively (Cheng et al. 2002). The Lys[11]/Arg[11] substitution improved tumor-to-kidney as well as tumor-to-other normal tissue ratios resulting in superior SPECT imaging in melanoma mouse models with [99m]Tc-(Arg[11])CCMSH (Figure 19.2) and ^{111}In-DOTA-Re(Arg[11])CCMSH (Figure 19.3).

The melanoma imaging analog [203]Pb-DOTA-Re(Arg[11])CCMSH was developed as a matched pair imaging agent for the α-particle-emitting melanoma therapeutic [212]Pb-DOTA-Re(Arg[11])CCMSH (Miao et al. 2008). Lead-203 ($t_{1/2} = 51.8$ h) has a 279 keV gamma suitable for SPECT imaging (Miao et al. 2008). In conversations with the FDA, it was made clear that a matched pair imaging agent was necessary to support clinical translation of the α-emitter therapeutic [212]Pb-DOTA-Re(Arg[11])CCMSH. In a preclinical therapy study with [212]Pb-DOTA-Re(Arg[11])CCMSH, 20% of the melanoma-bearing mice in the 3.7 MBq treatment group and 45% of the mice in the 7.4 MBq group survived the entire 120 day therapy study (Miao et al. 2005). Postmortem histopathological examination of the tumor site and other major organs showed no sign of primary or metastatic melanoma or melanoma-associated S100 antigen, allowing the mice to be classified as complete remissions or cures. [203]Pb-DOTA-Re(Arg[11])

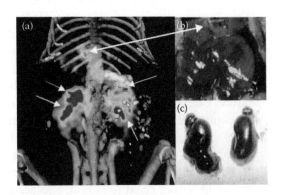

FIGURE 19.2 Scintigraphic/CT imaging of B16 melanoma tumors with [99m]Tc-(Arg[11])CCMSH. (a) Scintigraphic/CT image of a B16 melanoma-bearing mouse 2 h postinjection of [99m]Tc-(Arg[11]) CCMSH. Tumor deposits were identified in the liver, adrenals, and kidneys. Visual conformation of the tumor deposits visualized with [99m]Tc-(Arg[11])CCMSH upon postmortem necropsy in the (b) liver and (c) adrenals and kidneys.

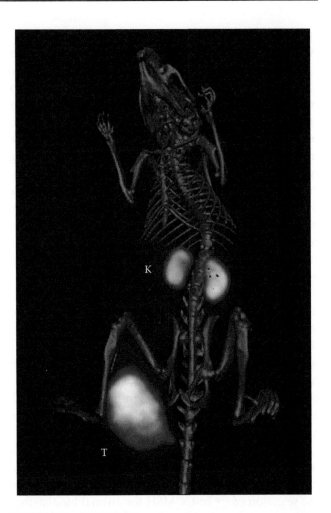

FIGURE 19.3 SPECT/CT imaging of a melanoma tumor-bearing mouse with ^{111}In-DOTA-Re(Arg[11])CCMSH. SPECT/CT images were collected 2 h postinjection of ^{111}In-DOTA-Re(Arg[11])CCMSH. High tumor (T) uptake (>17%ID/g) coupled with rapid whole-body disappearance of radioactivity from the normal organs and tissues, except for the kidneys (K), yielded well-defined melanoma tumor imaging.

CCMSH is a true matched pair imaging agent for the α-emitting [212]Pb-DOTA-Re(Arg[11])CCMSH (Miao et al. 2005). Biodistribution studies confirmed that [203]Pb-DOTA-Re(Arg[11])CCMSH accurately reflected the *in vivo* pharmacokinetics of [212]Pb-DOTA-Re(Arg[11]) CCMSH (Miao et al. 2005, 2008). Tumor uptake values for [203]Pb-DOTA-Re(Arg[11])CCMSH (SA 10 Ci/μM) were 11.87 ± 3.24, 9.86 ± 1.86, and 4.35 ± 0.24%ID/g at 2, 4, and 24 h postinjection (Miao et al. 2008). SPECT imaging studies demonstrated that melanoma tumors could be visualized using [203]Pb-DOTA-Re(Arg[11]) CCMSH 2 h postinjection (Figure 19.4).

The prospect of more sensitive and quantitative PET imaging drove the development of DOTA-Re(Arg[11]) CCMSH analogs labeled with positron-emitting radionuclides [64]Cu (McQuade et al. 2005; Wei et al. 2007a),

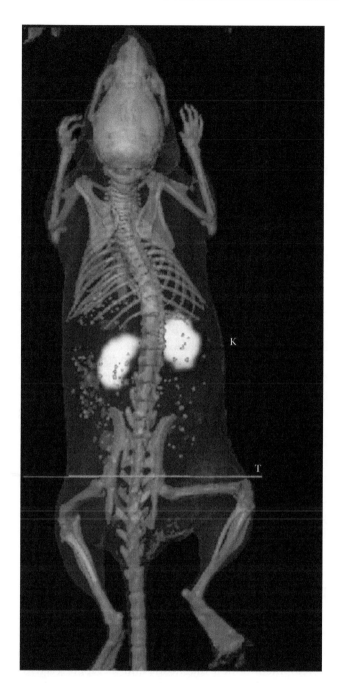

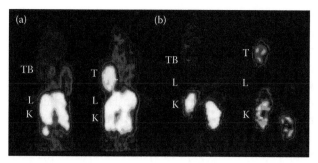

FIGURE 19.5 PET imaging of B16 melanoma-bearing mice with ^{64}Cu-labeled Re(Arg[11])CCMSH. (a) Melanoma tumor imaging with ^{64}Cu-DOTA-Re(Arg[11])CCMSH 2 h postinjection with (TB) and without (T) unlabeled peptide block. (b) Melanoma tumor imaging with ^{64}Cu-CBTE2A-Re(Arg[11])CCMSH 2 h postinjection with (TB) and without (T) unlabeled peptide block. Liver (L) and kidneys (K) are labeled.

FIGURE 19.4 SPECT/CT imaging of a B16 melanoma tumor-bearing mouse with [203]Pb-DOTA-Re(Arg[11])CCMSH. The melanoma flank tumor was clearly visualized 2 h postinjection of [203]Pb-DOTA-Re(Arg[11])CCMSH. Substantial tumor uptake (>11%ID/g) coupled with rapid whole-body clearance, apart from the kidneys, resulted in a readily identifiable flank tumor deposit.

[62]Cu (Yue et al. 2010), [86]Y (McQuade et al. 2005), and [68]Ga (Wei et al. 2007b; Cantorias et al. 2009). The most promising PET radioisotopes for peptide-targeted MC1-R-mediated melanoma imaging are ^{64}Cu and [62]Cu. Solid tumor uptake of [64]Cu[DOTA]-Re(Arg[11])CCMSH was 7.35 ± 1.47%ID/g at 4 h; however, liver uptake was nearly identical at 7.34 ± 1.79%ID/g

(McQuade et al. 2005). Imaging studies revealed excellent tumor visualization, and they reflected high liver accumulation of [64]Cu (Figure 19.5a) as well. The tumor images were very encouraging but the loss of ^{64}Cu from DOTA was concerning. A new chelator, CBTE2A, that had superior Cu-chelating abilities was conjugated to Re(Arg[11])CCMSH (Wei et al. 2007a). The tumor uptake of ^{64}Cu-CBTE2A-Re(Arg[11])CCMSH was identical to its DOTA homolog but the amount of radioactivity in the liver was significantly less (1.74 ± 0.52%ID/g) at 4 h, resulting in superior imaging (Figure 19.5b). Improved tumor uptake of ^{64}Cu-CBTE2A-Re(Arg[11])CCMSH was obtained by increased specific activity. For example, B16/F1 tumor uptake of ^{64}Cu-CBTE2A-Re(Arg[11]) CCMSH 2 h postinjection was 7.09 ± 3.20%ID/g at 600 mCi/mmol and 22.59 ± 2.99%ID/g at 5000 mCi/mmol underscoring the importance of high-specific activity for optimal imaging of low-density target receptors. HPLC purification of radiolabeled ^{64}Cu-labeled Re(Arg[11])CCMSH peptides was essential to obtain high-specific activity for optimal tumor uptake. Ideally, a one-step kit formulation for radiolabeled DOTA-Re(Arg[11])CCMSH would be preferable. Recently, the availability of very high-specific activity [62]Cu (Yue et al. 2010) allowed the one-step production of [62]Cu-DOTA-Re(Arg[11])CCMSH at a high-enough specific activity that did not require HPLC purification prior to imaging. Studies are underway to determine the viability of [62]Cu-labeled peptide imaging. The copper positron emitters [62]Cu and ^{64}Cu are currently the lead candidates of PET imaging.

Efforts to translate the ^{111}In and [203]Pb-labeled DOTA-Re(Arg[11])CCMSH are currently underway. The ^{111}In-DOTA-Re(Arg[11])CCMSH exhibited the best tumor-to-normal organ uptake ratios and could be

Chapter 19

formulated using readily available clinical grade [111]In. Lead-203-labeled DOTA-Re(Arg[11])CCMSH is the matched pair imaging agent for the α-emitting melanoma therapeutic [212]Pb-DOTA-Re(Arg[11])CCMSH and will be advanced to clinical trials on a parallel track. A strategy built on clinical translation of the melanoma imaging agent prior to the therapeutic agent was adopted to facilitate initial patient trials. The first task was to transfer the peptide synthesis technology, recyclization and purification, to a company that could perform pharmaceutical-grade synthesis according to current good manufacturing practices (cGMP). This entailed working with Bachem USA to synthesize and characterize DOTA-Re(Arg[11])CCMSH according to good laboratory practices (GLPs), which could be later scaled to cGMP. While cGMP quality peptide might not be necessary for early clinical trials, it will be necessary for final FDA approval.

Toxicity testing was performed on both In- and Pb-labeled DOTA-Re(Arg[11])CCMSH. Per FDA guidance, toxicity testing was performed on In-DOTA-Re(Arg[11])CCMSH and DOTA-Re(Arg[11])CCMSH at 100× the estimated dose based on an m^2 body surface area. Acute toxicity studies, with a 14 day rebound arm, were performed by Charles River Laboratories Preclinical Services. No adverse effects were observed. A 14 day toxicity study was performed with Pb-DOTA-Re(Arg[11])CCMSH and DOTA-Re(Arg[11])CCMSH in juvenile swine at the University of Missouri College of Veterinarian Medicine. Juvenile swine (20 kg) were administered Pb-DOTA-Re(Arg[11])CCMSH and DOTA-Re(Arg[11])CCMSH at 50× and 100× the projected human dose on a mg/kg basis. Physical parameters, blood work, and postmortem histopathology analyses demonstrated no adverse effects from compound administration. Results from toxicity testing in two animal species demonstrated that the metallated peptide or peptide alone were not toxic. Current efforts are focused on improving the tumor uptake to kidney retention ratio of radioactivity. We have demonstrated that coadministration of lysine will reduce kidney retention of radiolabeled ^{99m}Tc-CCMSH by ~50% (Chen 1999). In addition, reduction in kidney retention was reported for ^{111}In-DOTA octreotide by coinjection of Gelofusine (Vegt et al. 2006) and albumin fragments (Vegt et al. 2008), or by combining lysine infusion administration of Gelofusine (Melis et al. 2009). Optimizing the tumor-to-kidney ratio is critical to improving imaging sensitivity in the midgut region.

The impact of melanoma-selective imaging agents on patient care imaging lies in determining whole-body tumor localization and burden, monitoring melanoma treatment response, and in determining patient-specific dosimetry to support peptide-targeted therapy. Academic and industrial partnerships are being explored to support the first clinical trials with ^{111}In-DOTA-Re(Arg[11])CCMSH, while the translation of [203]Pb-DOTA-Re(Arg[11])CCMSH is linked to the progress of the melanoma α-therapeutic [212]Pb-DOTA-Re(Arg[11])CCMSH.

19.3 Gastrin-Releasing Peptide Receptor-Targeting Peptides

Gastrin-releasing peptide (GRP) receptors are expressed in very high numbers in human tumors, including breast, pancreatic, and prostate cancer (Cescato et al. 2008). Bombesin (BBN) receptors are G protein-coupled, 7-transmembrane proteins having the capacity to be endocytosed upon binding by an effective agonist ligand (Sun et al. 2000). Of the four known mammalian BBN receptor subtypes, BB1 (neuromedin B receptors, NMBR), BB2 (gastrin-releasing peptide receptors, GRPRs), BB3 (BRS-3), and BB4 (bombesin receptor subtype) (Jensen et al. 2008), design and development of radiolabeled agents for the GRP receptors have focused primarily on the highly prolific BB2 receptor subtype targeted by BBN peptide or BBN-like analogs. BBN is a 14-amino acid amphibian peptide analog of the 27-amino acid mammalian GRP. BBN and GRP share an amidated C-terminal sequence homology of seven amino acids, [–Trp–Ala–Val–Gly–His–Leu–Met–NH$_2$], which is essential for high-affinity receptor binding to GRPRs. The driving force for design and development of new diagnostic and therapeutic agents targeting GRP receptor-positive tumors is based primarily on the ability of GRPR agonists to specifically target GRPR-expressing neoplasias with minimal accumulation on collateral nontarget tissues. The ability of these targeting vectors to be rapidly internalized upon binding followed by prolonged retention in tumors has produced high-quality, high-contrast PET or SPECT images and continues to provide impetus for the development of TRT radiopharmaceuticals (Thomas et al. 2009).

In recent years, [BBN(7–14)NH$_2$] agonist has been the primary focus of many research groups for design and development of new diagnostic and therapeutic agents. For example, the clinical manifestations of BBN agonist ligands have been demonstrated for each diagnosis and therapy of human disease. First-generation diagnostic molecular imaging agents focused on targeting vectors of the general type: [^{99m}Tc-N$_3$S-X-BBN(7–14) NH$_2$] (Figure 19.6a). In these studies, Van de Weile et al. (2000) demonstrated the selectivity of these agents for GRPR in human patients presenting with either prostate or breast cancer. Their studies showed that [^{99m}Tc-N$_3$S-5-Ava-BBN(7–14)NH$_2$] localized in tumors with high specificity producing good tumor-to-normal tissue uptake ratios and high-quality SPECT images (Van de Wiele et al. 2000). Another successful [BBN(7–14)NH$_2$] GRPR-targeting agent is [^{177}Lu-AMBA] (AMBA = 1,4- paraaminobenzoic acid, Figure 19.6b), developed at Bracco for diagnostic imaging and treatment of metastatic disease. Biodistribution studies in human, prostate, PC-3 tumor-bearing mice showed [^{177}Lu-AMBA] having very low kidney accumulation and retention (2.95 and 0.91%ID/g, 1 and 24 h p.i., respectively) and high tumor accumulation and retention (6.35 and 3.39%ID/g, 1 and 24 h p.i., respectively) (Lantry et al. 2006). Currently, [^{177}Lu-AMBA] is undergoing Phase I clinical trials as a systemic radiotherapeutic agent for hormone refractory prostate cancer (Baum et al. 2007; Bodei et al. 2007; Thomas et al. 2009).

Interests in new ^{64}Cu-labeled [BBN(7–14)NH$_2$] radiopharmaceuticals for targeting the GRPR has recently reemerged due to improvements in the pharmacokinetics and clearance properties of the new radioligands. Conjugates of this type have been of interest for many years due to the ideal nuclear characteristics of ^{64}Cu ($t\frac{1}{2}$ = 12.7 h, β^+ = 0.651 MeV), making it suitable for *in vivo* molecular imaging via PET technology. Historically, ^{64}Cu-labeled peptides have suffered from demetallation *in vivo* and subsequent uptake and accumulation in nontarget tissue. However, recent reports by Hoffman, Prasanphanich, and Lane have shown GRP-targeting vectors of the general type [NO2A-X-BBN(7–14)NH$_2$], when radiolabeled with ^{64}Cu (Figure 19.6c), to produce high-quality, high-contrast micro-PET images with superior resolution of xenografted human tumors in prostate- and breast-cancer rodent models (Hoffman and Smith 2009; Prasanphanich et al. 2009). Results have shown tumor accumulation values of 6.05 ± 1.15%ID/g (X = AMBA) at 1 h p.i. in PC-3 tumor-bearing severe combined immunodeficient (SCID) mice (Lane et al. 2010) and 2.27 ± 0.08%ID/g at 1 h p.i. in T-47D tumor-bearing SCID mice (Hoffman and Smith 2009).

The clinical effectiveness of monomeric radiolabeled peptides can be potentially limited by receptor density, binding affinity, and pharmacokinetics of the targeting vector. High-quality, high-contrast PET or SPECT images with superior resolution, for example, require very high numbers of receptors to be present on tumor cells as compared to normal, collateral tissue. During tumor staging and development, the number of effective receptors available for radioligand binding may differ significantly, resulting in significantly reduced accumulation and retention of targeting vector, and hence, a compromised quality of the resulting PET or SPECT image. Furthermore, the binding affinity of monomeric peptides can be considered to be relatively low as compared to multimeric-targeting vectors (Liu et al. 2009c). Finally, radiolabeled monomeric peptides can be limited by their inherent pharmacokinetic profile. For example, clearance properties from normal tissue and the mode and rate of excretion of targeting vector may limit the diagnostic imaging utility of a specific monomeric probe (Liu et al. 2009c). For these reasons, multimeric or multivalent regulatory peptide probes have recently become a new and exciting approach for the development of diagnostic molecular imaging and TRT of tumors expressing either single or multitargetable receptors (Liu et al. 2009a–d; Shi et al. 2009; Yang et al. 2009).

Liu and coworkers have focused their recent research efforts on RGD/BBN multimeric peptide-targeting vectors for integrin/GRPR dual receptor imaging, as many GRPR-positive tumors are also positive for the $\alpha_v\beta_3$-integrin receptor (RGD = Arg–Gly–Asp $\alpha_v\beta_3$-targeting vector) (Liu et al. 2009a,c,d). Integrins are cell surface transmembrane glycoproteins existing as $\alpha\beta$ heterodimers. $\alpha_v\beta_3$ and $\alpha_v\beta_5$ integrin subtypes are expressed on the endothelial cells of tumor neovasculature during angiogenesis and form the basis of investigations for molecular imaging and TRT of angiogenesis and tumor formation *in vivo*. Liu and coworkers have focused on the ^{68}Ga- or ^{64}Cu-radiolabeled [NOTA-RGD-BBN(7–14) NH$_2$]-targeting vector (Figure 19.6d), where the RGD- and BBN-targeting motifs are linked by a glutamic acid (Liu et al. 2009a,c). Their studies showed [^{68}Ga-NOTA-RGD-BBN(7–14)NH$_2$] multimer and [^{68}Ga-NOTA-BBN(7–14)NH$_2$] monomer to have comparable accumulation of the tracer in PC-3-xenografted human tumors at all time points (Liu et al. 2009c). The [^{68}Ga-NOTA-RGD] monomer, on the other hand, showed much lower accumulation in the PC-3 tumor model. For example, at 2 h p.i., PC-3 tumor uptake was shown to be approximately 4.0, 3.2, and 0.80%ID/g for [^{68}Ga-NOTA-RGD-BBN(7–14)NH$_2$], [^{68}Ga-NOTA-BBN

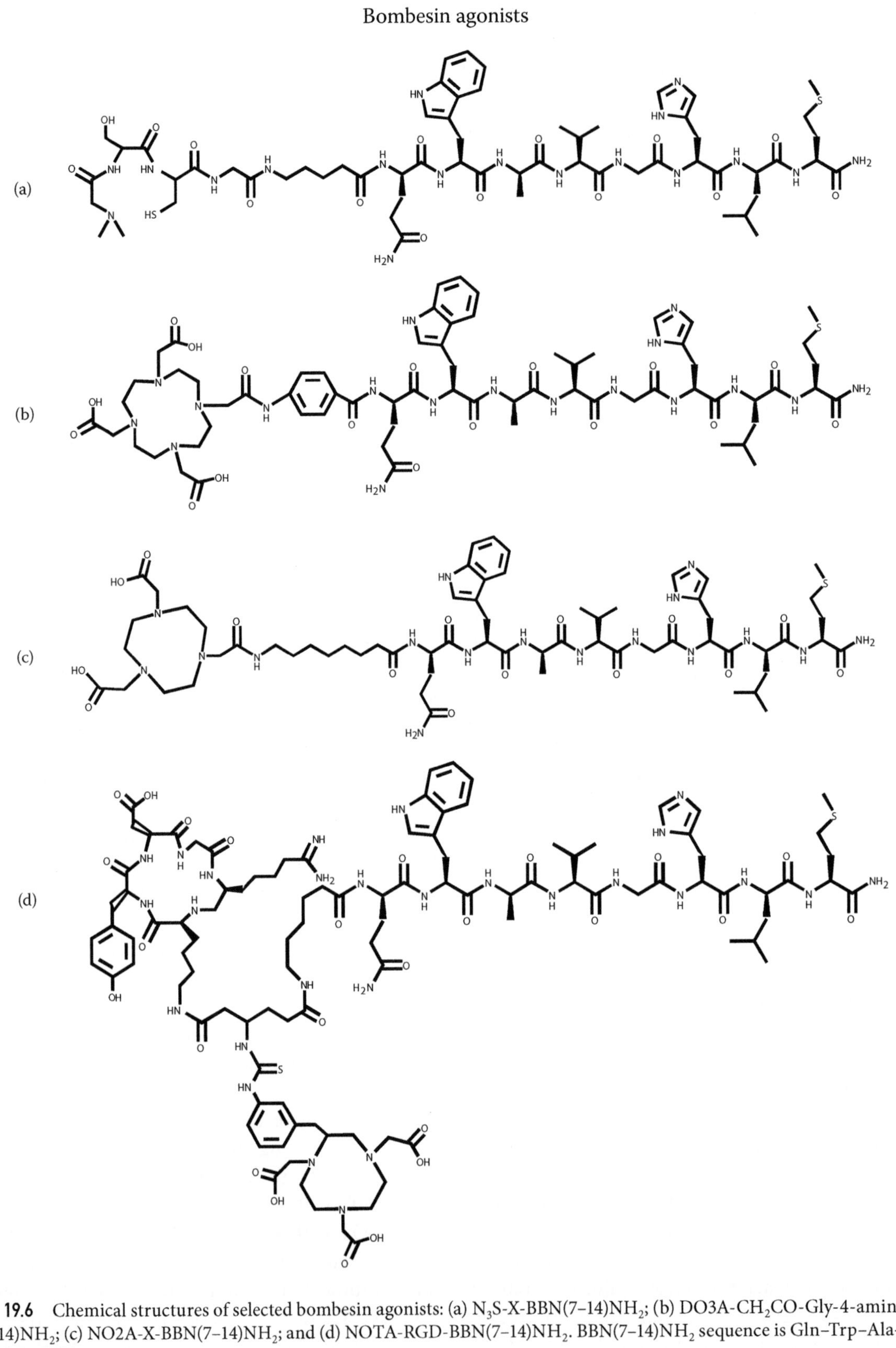

Bombesin agonists

FIGURE 19.6 Chemical structures of selected bombesin agonists: (a) N₃S-X-BBN(7–14)NH₂; (b) DO3A-CH₂CO-Gly-4-aminobenzoyl-BBN(7–14)NH₂; (c) NO2A-X-BBN(7–14)NH₂; and (d) NOTA-RGD-BBN(7–14)NH₂. BBN(7–14)NH₂ sequence is Gln–Trp–Ala–Val–Gly–His–Leu–Met–NH₂.

(7–14)NH$_2$], and [^{68}Ga-NOTA-RGD], respectively. ^{64}Cu-radiolabeled [NOTA-RGD-BBN(7–14)NH$_2$] demonstrated some degree of improvement in retention of the tracer at later time points. For example, at 20 h p.i., PC-3 tumor retention was 2.04 ± 0.35, 0.44 ± 0.39, and 0.55 ± 0.32%ID/g for [^{64}Cu-NOTA-RGD-BBN(7–14)NH$_2$], [^{64}Cu-NOTA-BBN(7–14)NH$_2$], and [^{64}Cu-NOTA-RGD], respectively. Each of the new ^{68}Ga- and ^{64}Cu-radiolabeled-targeting vectors exhibited clearance of the peptide heterodimer primarily via the renal/urinary excretion pathway. Furthermore, uptake and retention of conjugate in hepatic tissue for the radiolabeled heterodimers were significantly improved as compared to monomeric BBN- or RGD-targeting vectors (Liu et al. 2009a,c).

Until recently, preclinical evaluation of radiolabeled BBN peptides has been limited to BBN-targeting agonist ligands. BBN agonist ligands elicit a biological response from the cell and are effectively endocytosed and internalized upon binding to the receptor. BBN antagonist ligands, on the other hand, are not internalized and have previously been deemed inappropriate for targeted molecular imaging and peptide receptor TRT. Further impetus in support of the use of BBN antagonists is the lack of mitogenic properties which is directly associated with the use of BBN agonists. Recent studies with GRPR-targeting antagonist ligands, however, have indicated that the preferable use of agonist ligands for *in vivo* molecular imaging and TRT clearly bears reconsideration (Okarvi 2008; Reubi and Maecke 2008; Schottelius and Wester 2009; Schroeder et al. 2009, 2010; Beer et al. 2011).

In 2003, Nock and coworkers reported the synthesis of the first ^{99m}Tc-labeled BBN antagonist suitable for SPECT applications, ^{99m}Tc-Demobesin 1 [^{99m}Tc-N$_4$-Bzdig0,(D)Phe–Gln–Trp–Ala–Val–Gly–His–Leu–NHEt] (Nock et al. 2003). The complex (Figure 19.7a) was readily prepared in high yield through direct SnCl$_2$ reduction of ^{99m}TcO$_4^-$. The binding affinity for ^{99m}Tc-Demobesin 1 to the BBN receptor was determined to be in the subnanomolar range (0.7 ± 0.08 nM) in PC-3 cells. The report indicated that a small percentage of ^{99m}Tc-Demobesin 1 was internalized (~25%) which was noted as intriguing since the parent peptide [(D)Phe^6, Leu-NHET^{13}, des-Met^{14}]BN(6–14) was determined to be a potent antagonist in previous studies. Pharmacokinetic studies conducted in PC-3 cell-xenografted Swiss nu/nu mice demonstrated significant and sustained tumor uptake at 1, 4, and 24 h postinjection (16.2 ± 3.1, 15.61 ± 1.19, and 5.24 ± 0.67%ID/g, respectively). Further work by Cescato and coworkers

published in 2008 expanded on these studies through the direct comparison of the ^{99m}Tc-labeled antagonist Demobesin 1 with potent BBN agonists (Cescato et al. 2008). These later results showed *in vitro*-binding comparable to that of the agonists, [BBN(1–14)NH$_2$] and [N_4-[Pro^1,Tyr^4,Nle^{14}]bombesin] (Demobesin 4), in GRP receptor-transfected HEK293 cells, PC-3 cells, and human prostate cancer specimens. In PC-3 cells at 4 and 24 h, [^{99m}Tc-Demobesin 1] uptake was fourfold and twofold better than that of [^{99m}Tc-Demobesin 4], respectively. Similarly, this group demonstrated even higher tumor uptake and retention in PC-3-xenografted SCID mice than previously reported with PC-3 tumor uptake at 1, 4, and 24 h postinjection reported to be 24.61 + 1.98, 22.66 + 2.20, and 5.38 + 0.72%ID/g, respectively. In fact, when compared with PC-3 tumors, [^{99m}Tc-Demobesin 1] was shown to label GRP receptors more intensely and longer than any GRPR agonist (Reubi and Maecke 2008). As a result of these findings, Reubi and others have suggested that GRPR antagonists might be the future direction of focus in BBN radiopharmaceutical development (Reubi and Maecke 2008; Cescato et al. 2008).

Further BBN antagonist SPECT agent development involving the synthesis and characterization of an ^{111}In-DOTA-conjugated BBN antagonist was reported by Abd-Elgaliel et al. (2008). In these studies, ^{111}In-DOTA-aminohexanoyl-[D-Phe^6, Leu-NHCH$_2$CH$_2$CH$_3^{13}$, desMet^{14}] BBN[6–13] was evaluated (Figure 19.7b) *in vitro* using PC-3 cells and *in vivo* in PC-3-xenografted SCID mice. Although this compound demonstrated a high affinity for the GRP receptor *in vitro* (IC$_{50}$ of 1.36 ± 0.09 nM), the *in vivo* tumor targeting in PC-3 tumor-bearing-xenografted SCID mice was modest and comparable to agonist targeting data achieving only 3.72 ± 0.60 and 2.15 ± 0.36%ID/g at 1 and 4 h postinjection, respectively.

Additional SPECT and PET imaging agent development has focused on DOTA-conjugated BBN antagonist analogs that incorporated the potent BBN receptor antagonist (H-D-Phe–Gln–Trp–Ala–Val–Gly–His–Sta–Leu–NH$_2$) (Garrison et al. 2008). In 2009, Mansi and coworkers reported on the ^{111}In and ^{68}Ga complexes of DOTA conjugated to H-D-Phe–Gln–Trp–Ala–Val–Gly–His–Sta–Leu–NH$_2$ (Figure 19.7c) employing a glycine-4-aminobenzoyl-linking group (Mansi et al. 2009). The ^{111}In-conjugate demonstrated an IC$_{50}$ of 14 ± 3.4 nM/L for the BBN receptor. Tumor-targeting values obtained in PC-3 tumor-bearing nude mice revealed uptake and retention values of 14.24 ± 1.75, 13.46 ± 0.80, and 6.58 ± 1.14%ID/g at 1, 4, and 24 h postinjection,

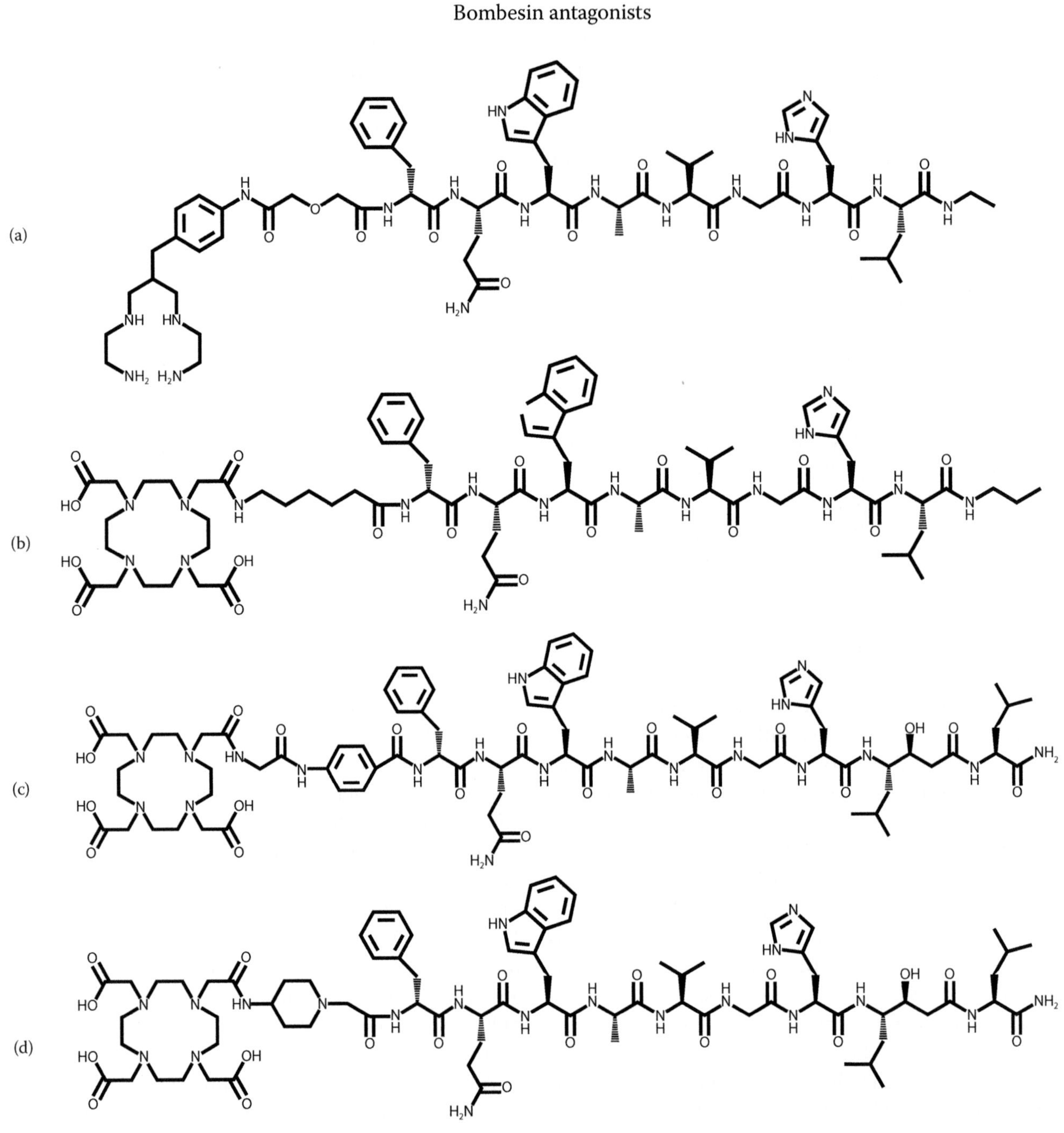

Bombesin antagonists

FIGURE 19.7 Chemical structures of selected bombesin antagonists: (a) -N₄-Bzdig0,(D)Phe–Gln–Trp–Ala–Val–Gly–His–Leu–NHEt; (b) DOTA-aminohexanoyl-[D-Phe⁶, Leu-NHCH₂CH₂CH₃¹³, desMet¹⁴] BBN[6–13]; (c) DOTA-Gly-4-aminobenzoyl-D-Phe–Gln–Trp–Ala–Val–Gly–His–Sta–Leu–NH₂; (d) DOTA-4-amino-1-carboxymethyl-piperidyl-D-Phe–Gln–Trp–Ala–Val–Gly–His–Sta–Leu–NH₂.

respectively. Tumor targeting of this ¹¹¹In-antagonist is nearly five times higher than the data reported for the ¹¹¹In-antagonist developed by Abd-Elgaliel and coworkers. Further work in 2011 from this same group explored the use of a positively charged linker, 4-amino-1-carboxymethyl-piperidine (Figure 19.7d), for the preparation of both an ¹¹¹In SPECT agent and a ⁶⁸Ga PET agent

(Mansi et al. 2011). Both the ¹¹¹In and ⁶⁸Ga conjugates were reported to be prepared in high yield and high-specific activity. Similarly, tumor-targeting values for the corresponding ¹¹¹In conjugate were again obtained in PC-3 tumor-bearing nude mice with uptake and retention values of 15.23 ± 4.78, 11.75 ± 2.43, and 6.84 ± 1.02%ID/g at 1, 4, and 24 h postinjection, respectively. Comparable high tumor-targeting values employing the same model were obtained for the ^{68}Ga conjugate with 14.11 ± 1.88 and $13.61 \pm 0.64\%$ID/g obtained at 1 and 2 h postinjection. The authors indicated that employing the 4-amino-1-carboxymethyl-piperidine spacer group was chosen with the intent to improve upon the pharmacokinetics of their previously published work employing the glycine-4-aminobenzoyl linking group. Although the intent was to employ a positively charged linking moiety to enhance overall pharmacokinetics, the only enhancements of significance noted were those of higher K_d and IC_{50} values.

Each of the radiolabeled BBN antagonist compounds studied to date has demonstrated relatively high affinity for the BBN receptor (nanomolar and subnanomolar range) and all have shown limited capacity to undergo receptor-mediated internalization. In spite of these observations, all of the BBN antagonist studies presented to date have reported data indicating that at least 25% of the radiolabeled antagonist that targets *in vivo* BBN receptor expressing prostate tumor xenografts

remains at the tumor site for at least 24 h postinjection when compared with the reported biodistribution values obtained at 1 h postinjection. The rationale for prolonged retention of these radiolabeled antagonists has been attributed to several possible mechanisms including *in vivo* internalization occurring at a slow rate, a multistep antagonist-binding process whereby the antagonist binds to a peripheral site on the receptor and then slowly migrates to a more central-binding site, strong receptor–antagonist interactions producing a more stable complex, and finally, the structure of the antagonist itself may simply be more stable to enzymatic protein degradation than the corresponding agonist counterpart (Mansi et al. 2009, 2011; Gu et al. 2011). Currently, the superior tumor targeting and retention properties, as well as the generation of high-quality preclinical SPECT and PET images of radiolabeled BBN antagonist tumor localization, are clearly an impetus to warrant evaluation of select candidates in Phase 1 human clinical trials for GRPR tumor targeting (Cescato et al. 2008; Mansi et al. 2009, 2011).

19.4 Radiolabeled ST$_h$(1–19), GC-C Receptor-Specific Radiotracers

The human guanylin/guanylate cyclase-C (GC-C) receptor provides a promising, well-defined receptor target capable of enabling selective *in vivo* targeting of colorectal cancers with diagnostic or therapeutic radiopharmaceuticals. The GC-C receptor is a member of the guanylate cyclase family of proteins that produce signaling via the intracellular GMP second messenger molecule (Forte 2004). The structure of the GC-C protein was initially defined by molecular cloning of cDNAs from rat intestine, which encodes a cell surface receptor that, upon expression in cells, is strongly activated by *Escherichia coli* heat-stable human enterotoxin (ST$_h$(1–19)), guanylin, and uroguanylin (Schulz et al. 1990; de Sauvage et al. 1991). The GC-C is abundantly expressed on the apical membranes of epithelial cells which comprise the intestinal mucosa. Virtually all the cell types found within the intestinal mucosal express high levels of apical GC-C receptors that bind ST$_h$(1–19) guanylin and uroguanylin with high affinity and selectivity (Krause et al. 1994; Scheving et al. 1996). A basic physiological function common to all epithelial cells in the intestinal mucosa may involve an influence of cGMP on the turnover of cells in the intestinal epithelium. Activation of cGMP production by uroguanylin, guanylin, or ST$_h$(1–19) elevates cGMP levels and induces apoptosis of human cancer cells (Shailubhai et al. 2000; Liu et al. 2001; Pitari et al. 2001).

Another biological action of uroguanylin, guanylin, and ST$_h$(1–19) is related to cGMP-mediated regulation of fluid secretion into the intestinal lumen (Field et al. 1978; Currie et al. 1992; Hamra et al. 1993; Forte 2004).

The rationale underlying research to develop radiolabeled ST$_h$(1–19) analog for molecular imaging and TRT applications stems from the observations by Waldman and colleagues which showed that cognate GC-C receptors for ST$_h$(1–19) continue to be expressed in high levels on primary and metastatic human colorectal carcinoma (Carrithers et al. 1996; Cagir et al. 1999). The ST$_h$(1–19) binds with high selectivity and affinity to GC-C receptors and exhibits exceptionally high stability in blood and other physiological fluids (Field et al. 1978; Carrithers et al. 1996). Radiolabeled ST$_h$(1–19) analogs have agonist properties when binding to cognate GC-C receptors and undergo receptor-mediated endocytosis producing a long-term deposition of the radionuclide reporter in the tumor cells. It is important to recognize that GC-C receptors are expressed only in the luminal side of normal intestinal mucosal cells making them essentially inaccessible to radiolabeled ST$_h$(1–19) molecules circulating in the bloodstream. On the contrary, invasive and metastatic adenocarcinoma cells have random orientations in tumors making GC-C receptors on their surface available to radiolabeled ST$_h$(1–19) molecules in the bloodstream and extracellular fluid.

Chapter 19

Considering all these factors in aggregate, it was envisioned that intravenous administration of radiolabeled $ST_h(1-19)$ analogs would result in a high level of specific accumulation of the radiotracer in human colorectal tumors with low levels of accumulation of radioactivity in normal tissues and organs to facilitate development of effective diagnostic and therapeutic radiopharmaceuticals that selectively target primary and metastatic colorectal adenocarcinomas in humans (Wolfe et al. 2002; Giblin et al. 2005a).

19.5   Design and Synthesis of Radiolabeled $ST_h(1-19)$ Conjugates

The primary structures of human guanylin, uroguanylin, and $ST_h(1-19)$ are shown in Figure 19.8. These three peptides bind with high specificity to GC-C receptors expressed on human colorectal cancer cells and are candidates for use as a basis to develop radiolabeled conjugate development of effective *in vivo* molecular imaging and or TRT applications. The $ST_h(1-19)$ moiety is the most promising targeting vector since it has the highest-binding affinity for the GC-C receptor (Forte 2004). The three disulfide bonds in the $ST_h(4-18)$ moiety provide for exceptional *in vitro* and *in vivo* stability. The GC-C-binding domain of $ST_h(1-19)$ consists of a sequence of 13 amino acid residues with six cysteines which form three disulfide bonds that are critical for the construction of the tertiary structure. The six cysteines can adopt 15 possible disulfide arrangements with only one (see Figure 19.8) exhibiting high GC-C receptor-binding affinity and biological activity (Gariepy et al. 1987; Carpick and Gariepy 1993; Sato et al. 1994). Thus, in the formulation of the ST analogs as potential imaging or therapy application, isolation of the $ST_h(4-18)$ moieties in the correct disulfide bond arrangement (i.e., $Cys_{6,11}$, $Cys_{7,15}$, and $Cys_{19,18}$) is critical. Production of high-purity $ST_h(1-19)$ conjugates with the proper disulfide bond arrangement and 3D conformation in reasonable yields is a tedious and technically challenging

endeavor. The most widely used methods for the synthesis of ST peptides and their conjugates involve simultaneous folding and disulfide bond formation of the fully deprotected linear peptides by air oxidation in aqueous solutions (Sato et al. 1994; Houghten et al. 1984; Aimoto et al. 1983). High-purity $ST_h(1-19)$ moiety preparations that exhibit full biological activity and binding to GC-C receptors with high affinity and specificity were produced in yields below 5% using this approach.

To achieve improved synthetic reproducibility, a new strategy was used that utilized three well-known orthogonal thiol-protecting groups (i.e., Trt, AcM, and t-Bu) to enable formation of the three disulfide bonds by successive reactions using 2,2′-dithiodipyridine (2-PDS), iodine, and silyl chloride–sulfoxide deprotection systems (Kamber et al. 1980; Koide et al. 1991; Akaji 1993). DOTA conjugated to the N-terminal end of the linear $ST_h(1-19)F^{19}$ peptides with thiol groups appropriately protected were synthesized on an automated peptide synthesizer and subsequently oxidized by this three-step deprotection method and purified by gradient HPLC separation to produce the final DOTA–$ST_h(1-19)F^{19}$ product in very high purity (Gali et al. 2002; Giblin et al. 2006a). Substitution of Phe^{19} for Tyr^{19} did not alter GC-C receptor-binding affinity or specificity (Gali et al. 2002). A series of radiolabeled $ST_h(1-19)$ and $ST_h(1-19)F^{19}$ conjugates were synthesized and subjected to gradient HPLC separations to produce highly pure analogs exhibiting high GC-C receptor-binding affinities and biological activities (Giblin et al. 2004, 2005a; Gali et al. 2006). The conjugates were used to prepare the respective radiolabeled $ST_h(1-19)$-based radiotracers with high-specific activities (Gali et al. 2002; Giblin et al. 2006a).

The general design of radiolabeled $ST_h(1-19)F^{19}$ conjugates synthesized and assessed is found in Figure 19.9. These conjugates have three components; the GC-C receptor $ST_h(1-19)$-binding region, a linking moiety, and a radiometal complex. In the example provided in Figure 19.9, a DOTA chelated to a radioactive metal is shown; however, a variety of other chelators and radiometals (Giblin et al. 2005b; Liu et al. 2009) can be

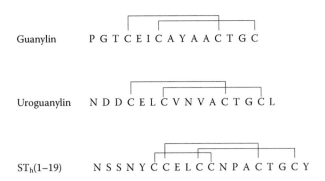

Guanylin   P G T C E I C A Y A A C T G C

Uroguanylin   N D D C E L C V N V A C T G C L

$ST_h(1-19)$ N S S N Y C C E L C C N P A C T G C Y

FIGURE 19.8 Primary structures of human guanylin peptides with amino acids abbreviated using single letter code. The $ST_h(1-19)$ peptide shows the unique paring of the six cysteine SH groups to form the three specific S–S bonds to produce the biologically active conformation with high-binding affinity to GC-C receptors.

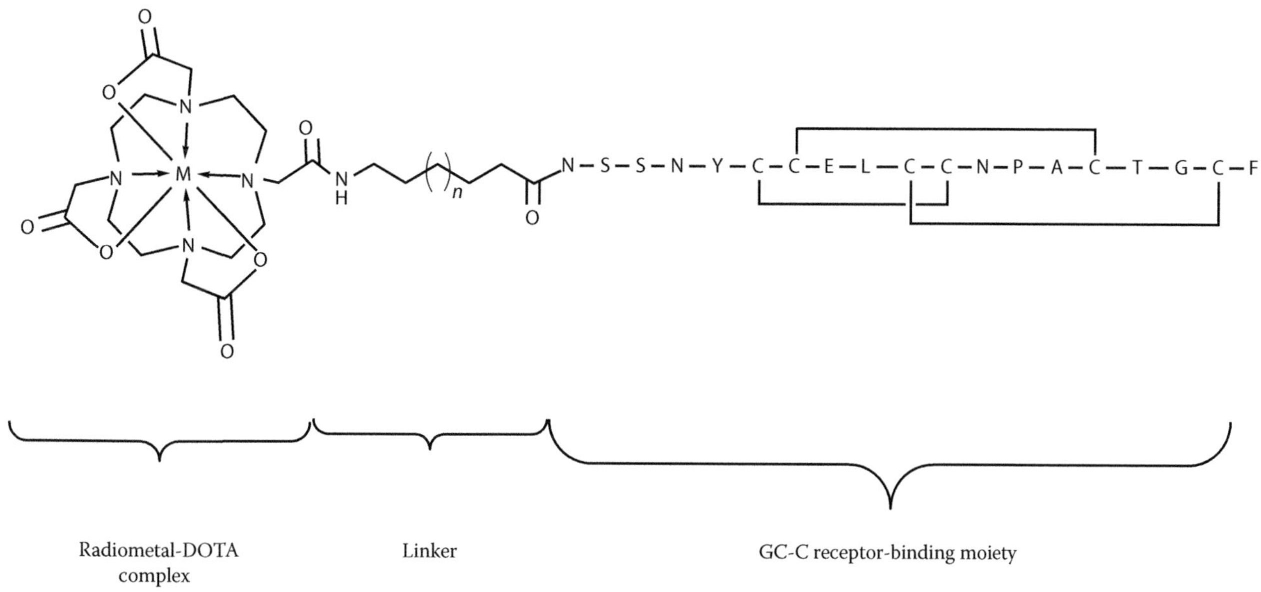

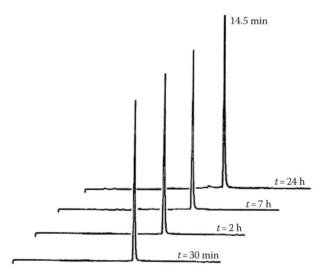

FIGURE 19.9 General structure of radiolabeled conjugates of $ST_h(1–19)$ radiotracers for targeting human colorectal adenocarcinomas. A metallated DOTA complex is appended to the N-terminal end of the $ST_h(1–19)$ linker-conjugated peptide in this example.

conjugated to the linker. The linkers used in our studies are either *n*-amino-carboxylic acids or a sequence of amino acids (Gali et al. 2001; Giblin et al. 2004).

The majority of the research in the past decade has focused on the utilization of DOTA–$ST_h(1–19)$ conjugate analogs with the goal of developing an effective radiopharmaceutical for *in vivo* imaging of metastatic colonic or rectal adenocarcinomas in humans. Selected DOTA–$ST_h(1–19)$ analogs have also been used to evaluate their potential as TRT agents when labeled with ^{177}Lu and [90]Y (Giblin et al. 2006b). The approach used to synthesize ^{111}In/In-DOTA–ST_h conjugates involved incubation of the ^{111}In/In-chloride at pH 5.5–6.0 in a 0.2 M tetramethylaminnim acetate solution concerning excess of the DOTA–ST_h conjugate at 80°C for 1 h (Gali et al. 2002; Giblin et al. 2004). ^{111}In-DOTA–ST_h analogs are routinely produced in high yields (>95%) using this approach and exhibit exceptional *in vitro* and *in vivo* stability (Gali et al. 2002; Giblin et al. 2004, 2006a). The *in vitro* stability of the ^{111}In-DOTA–$ST_h(1–19)$ conjugates is excellent as studies performed with these radiotracers show no evidence of degradation during 24 h incubation in human serum at 37°C (Figure 19.10). This high level of stability is not unexpected since the $ST_h(1–19)$ peptide is stable for long periods in stomach and intestinal fluids (Sato et al. 1994; Forte 2004).

The rigid 3D structure of the $ST_h(4–18)$ moiety in the $ST_h(1–19)$ conjugates is predominantly responsible for specific and high-affinity binding to GC-C

receptors (Gariepy et al. 1987; Carpick and Gariepy 1993; Carrithers et al. 1996). The five amino acid tail at the N-terminal of this moiety (i.e., N-N-S-S-Y) in the $ST_h(1–19)$ peptide (Figure 19.8) can serve as a linking group to which DOTA can be conjugated and not significantly reduce binding specificity and affinity (Gali et al. 2002; Giblin et al. 2006a). This provides important flexibility in designing radiolabeled STh(1–19) analogs that retain the desired-binding properties to cognate GC-C receptors in colon and rectal

FIGURE 19.10 HPLC chromatograms of ^{111}In-DOTA–$ST_h(1–19)$ F[19] incubated in human serum at 37°C, pH 7.4–7.8 under 5% CO_2 atmosphere for 0.5, 2, 7, and 24 h. The retention time of 14.5 min is unchanged for 24 h.

adenocarcinoma cells. As a result, several DOTA–ST_h(1–19) and In-DOTA–ST_h(1–19) conjugates have been synthesized that exhibit IC_{50} values equal to that of GC-C receptors expressed on cells in the single-digit nM range. For example, IC_{50} values for binding to GC-C receptors on human T-84 colon cancer cells versus ^{125}I-Tyr^5-ST_h(1–19)F^{19} ($K_d = 1.8$ nM) for In-DOTA–ST_h(1–19)F^{19}, DOTA–ST_h(1–19)F^{19}, DOTA–6Ahx–ST_h(1–19)F^{19}, and ST_h(1–19)F^{19} were found to be 3.3 ± 2.0, 0.6 ± 0.1, 1.6 ± 1.2, and 0.5 ± 0.1 nM, respectively (Gali et al. 2002). Note that all of these and other studies to determine IC_{50} values for ^{111}In-DOTA–ST_h(1–19) analogs using GC-C receptor expressing cells used "cold" In as a surrogate for ^{111}In (Gali et al. 2002; Giblin et al. 2004, 2006a). These studies as well as other chemical and *in vitro* experiments were performed with various ^{111}In/In-DOTA–ST_h(1–19) analogs to carefully document their binding characteristics with GC-C receptor expressing cells to identify the most promising candidates to initiate *in vivo* studies in normal and tumor-bearing animal models.

The ^{111}In-radiotracers administered intravenously were high-specific-activity formulations. The ^{111}In-

DOTA–ST_h(1–19) radiotracer eluted well ahead of the unlabeled DOTA-ST(1–19) reagent during gradient HPLC purification (see Figure 19.11). The tumor uptake and pharmacokinetics of a variety of ^{111}In-DOTA–ST_h(1–19) analogs that exhibited high-binding affinities with GC-C receptors were studied in SCID mice with human colon cancer xenografts with the goal of optimizing tumor uptake and retention and minimizing the uptake in normal organs (particularly the kidneys) (Gali et al. 2002, 2006; Giblin et al. 2004). Results from these studies indicate that ^{111}In-DOTA linked directly to the NNSSY peptide sequence normally occurring at the N-terminal end on ST_h(1–19) exhibited the most favorable properties for targeted imaging of human colorectal cancers. Table 19.1 shows the uptake of ^{111}In-DOTA–ST_h(1–19)F^{19} in tumors and normal tissues 1 h postintravenous injection (p.i.). The blocking study demonstrates that tumor uptake is related to the high-specific *in vivo* binding to GC-C receptors (Table 19.1). The %ID in the urine at 1 h p.i. is 95.8% illustrating the rapid clearance of activity from the body via the renal pathway. HPLC analysis of the urine shows that the only ^{111}In-species in the urine has an identical retention time as the ^{111}In-DOTA–

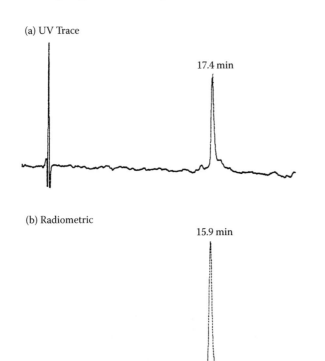

(a) UV Trace

17.4 min

(b) Radiometric

15.9 min

FIGURE 19.11 HPLC traces of (a) unmetallated DOTA–ST_h(1–19)F^{19} (UV: $\lambda = 220$ nM trace) and (b) ^{111}In-DOTA–ST_h(1–19)F^{19} (radiometric trace). The identical gradient conditions were used to generate HPLC chromatograms.

Table 19.1 Biodistribution (%ID/g) of ^{111}In-DOTA–ST_h(1–19)F^{19} in Human T-84 Tumor-Bearing SCID Mice at 1 h Postinjection

Tissue	%ID/g[a] ^{111}In-DOTA–ST_h(1–19)F^{19}	Blocking[b]
Blood	0.23 ± 0.14	0.69 ± 0.23
Lung	0.20 ± 0.07	0.61 ± 0.17
Liver	$0.08 \pm .009$	0.31 ± 0.10
Large intestine	0.27 ± 0.06	0.25 ± 0.08
Small intestine	0.76 ± 0.14	0.23 ± 0.06
Kidney	2.16 ± 0.31	16.4 ± 11.7
Muscle	0.03 ± 0.02	0.14 ± 0.07
Tumor	2.04 ± 0.30	0.81 ± 0.25
Urine (%ID)	95.8 ± 0.2	88.8 ± 6.1
Tumor/blood	8.9 ± 5.6	1.2 ± 0.5
Tumor/muscle	68.0 ± 46.4	5.8 ± 3.4
Tumor/liver	25.5 ± 4.9	2.6 ± 1.2

Source: Adapted from Gali et al. 2002. *Bioconj. Chem* 13:224–231.

[a] Values are mean ±SD.

[b] Coinjection of 100 µg of 60 Ahx-ST_h(1–19) with ^{111}In-DOTA–ST_h(1–19)F^{19}.

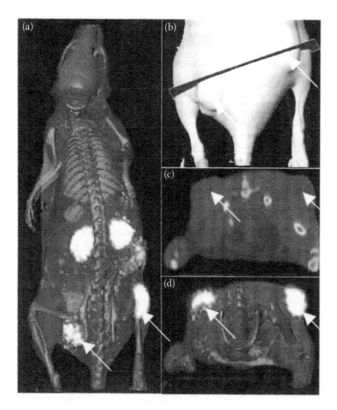

FIGURE 19.12 Micro-SPECT/CT images of ^{111}In-DOTA–ST$_h$(1–19)F^{19} in SCID mouse with bilateral T84 human colon cancer tumors at 1 h p.i. Arrows point to tumor locations. (a) Whole-body SPECT/CT. (b) Skin surface reconstruction. (c) Transaxial CT image. (d) Transaxial SPECT/CT image.

ST$_h$(1–19)F^{19} providing strong evidence that there is no significant *in vivo* degradation of this radiotracer (Gali et al. 2002). The *in vivo* SPECT/CT images shown in Figure 19.12 are consistent with the biodistribution results in Table 19.1. The images shown in Figure 19.12 demonstrate selective uptake of ^{111}In-DOTA–ST$_h$(1–19) at 1 h p.i. in the human T-84 colon cancer tumors in SCID mice. The images also show that the only other organs with significant uptake in the body are the kidneys. It is important to note that uptake in liver, lung, and other tissues is not visualized. These data provide strong evidence that metastatic colorectal adenocarcinoma will be specifically targeted in human patients following intravenous administration of ^{111}In-DOTA–ST$_h$(1–19)F^{19} and should be clearly imagable in organs (especially liver) or tissues where metastasis occurs. Despite the promising results obtained in these preclinical *in vitro* and *in vivo* studies in mouse models, securing funding to initiate Phase I clinical studies with ^{111}In-DOTA–ST$_h$(19)F^{19} has been a challenge even though current methods to diagnosis the presence of metastatic colorectal adenocarcinomas in humans utilize nonspecific radiopharmaceuticals (e.g., ^{18}F-FDG) or anatomical diagnostic approaches (e.g., CT or MRI).

Acknowledgments

We acknowledge the support by DOE 93ER61661, NIH R42-CA85106, NIH 1R43 CA11492, the NIH Imaging Center grant P50-CA-103130, and the MU Institute for Clinical and Translational Science.

References

Abd-Elgaliel, W.R., Gallazzi, F., Garrison, J.C. et al. 2008. Design, synthesis, and biological evaluation of an antagonist-bombesin analogue as targeting vector. *Bioconjug Chem* 19:2040–8.

Abdel-Malek, Z.A. 2001. Melanocortin receptors: Their functions and regulation by physiological agonists and antagonists. *Cell Mol. Life Sci* 58: 434–41.

AI-Obeidi, F., Castrucci, A.M., Hadley, M.E. et al. 1989. Potent and prolonged acting lactam analogues of alpha-melanotropin: Design based on molecular dynamics. *J. Med.Chem.* 32: 2555–61.

Aimoto, S., Watanabe, H., Ikemura, H. et al. 1983. Chemical synthesis of a highly potent and heat-stable analog of an enterotoxin produced by a human strain of enterotoxigenic *Escherichia coli. Biochem. Biophys. Res. Commun.* 112:320–26.

Akaji, K., Fugino, K., Tatsumi, T. et al. 1993. Synthesis of cystine-peptide by a new disulfide bond-forming reaction using the silyl chloride-sulfoxide system. *J. Chem. Soc. Chem. Commun.* 167–68.

Akazawa, H., Arano, Y., Mifune, M. et al. 2001. Effect of molecular charges on renal uptake of ^{111}In-DTPA conjugated peptides. *Nucl. Med. Biol.* 28:761–68.

Anderson, C.J., Welch, M.J. 1999. Radiometal-labeled agents (Non-Technetium) for diagnostic imaging. *Chem Rev* 99:2235–68.

Baum, R.P., Prasad, V., Mutloka, N. et al. 2007. Molecular imaging of bombesin receptors in various tumors by Ga-68 AMBA PET/CT. *J Nucl Med* 48:79P.

Beer, A.J., Eiber, M., Souvatzoglou, M. et al. 2011. Radionuclide and hybrid imaging of recurrent prostate cancer. *Lancet Oncol* 12:181–91.

Beneditt, E., Morelli, E., Accardo, G. et al. 2004. Criteria for the design and biological characterization of radiolabeled peptide-based pharmaceuticals. *BioDrugs* 18:279–95.

Bodei, L., Ferrari, M., Nunn, A.D. et al. 2007. ^{177}Lu-AMBA bombesin analogue in hormone refractory prostate cancer patients: A phase I escalation study with single-cycle administrations. *Eur J Nucl Med Mol Imaging* 34:S221.

Chapter 19

Boswell, C.A., Sun, X, Nin, W. et al. 2004. Comparative *in vivo* stability of copper-64-labeled cross-bridged and conventional tetraazamacrocylic complexes. *J. Med. Chem.* 47:1465–74.

Buchegger, F., Bonvin, B., Kosinski, M. et al. 2003. Stabilization of neurotensin analogues: Effect on peptide catabolism, biodistribution and tumor binding. *J. Nucl. Med.* 44:1649–54.

Buchsbaum, D.J., Chaundhuri, T.R., Yamamoto, M. et al. 2004. Gene expression imaging with radiolabeled peptides. *Ann. Nulc. Med.* 18:275–83.

Cagir, B., Gelmann, A., Park, J. et al. 1999. Guanylyl cyclase C is a biomarker for recurrent stage II colorectal cancer. *Ann. Intern. Med.* 131:805–12.

Cantorias, M.V., Figueroa, S.D., Quinn, T.P. et al. 2009. Development of high-specific-activity ^{68}Ga-labeled DOTA-rhenium-cyclized α-MSH peptide analog to target MC1 receptors overexpressed by melanoma tumors. *Nucl. Med. Biol.* 36:505–09.

Carpick, B.,W., Gariepy, J. 1993. The *Escherichia coli* heat-stable enterotoxin is a long-lived superagonist of guanylin. *Infect. Immun.* 61:4710–15.

Carrithers, A.L., Barber, M.T., Biswas, S. et al. 1996. Guanylyl cyclase C is a selective marker for metastatic colorectal tumors in human extraintestinal tissues. *Proc. Natl. Acad. Sci. USA*, 93:14827–32.

Cescato, R., Maina, T., Nock, B. et al. 2008. Bombesin receptor antagonists may be preferable to agonists for tumor targeting. *J Nucl Med* 49:318–26.

Chen, J., Cheng, Z, Owen, N.K. et al. 2001. Evaluation of an ^{111}In-DOTA-rhenium cyclized alpha-MSH analog: A novel cyclic peptide analog with improved tumor targeting properties. *J. Nucl. Med.* 42:1847–55.

Chen, J., Giblin, M.F., Wang, N. et al. 1999. *In vivo* evaluation of Tc-99m/Re-188-labeled linear alpha-melanocyte stimulating hormone analogs for specific melanoma targeting. *Nucl. Med. Biol.* 77:5788.

Cheng, Z., Chen, J., Miao, Y. et al. 2002. Modificatin of the structure of a metallopeptide: Synthesis and biological evaluation of ^{111}In-labeled DOTA-conjugated rhenium-cyclized α-MSH analogs. *J. Med. Chem.* 45:3048–56.

Cody, W.L. Mahoney, M., Knittle, J.J. et al. 1985. Cyclic melanotropins. D-phenylalanine analogues of the active site sequence. *J. Med. Chem.* 28:583–88.

Cone, R.D., Lu, D., Koppula, S. et al. 1996. The melanocortin receptors: Agonists, antagonists, and the hormonal control of pigmentation. *Recent Prog. Horm.* Res. 51:287–317.

Currie, M.G., Fok, K.F., Kato, J. et al. 1992. Guanylin: An endogenous activator of intestinal guanylate cyclase. *Proc. Natl.. Acad. Sci. USA* 89:947–51.

deJong, M., Bernard, B.F., DeBruin, E. et al. 1998. Internalization of radiolabeled [DTPA] octreotide and [DOTA₈Tyr₃] octreotide: Peptides for somatostatin receptor-targeted scintigraphy and radionuclide therapy. *Nucl. Med. Commun.* 19:283–88.

deSauvage, F.J., Camerato, T.R., Goeddel, D.V. 1991. Primary structure and functional expression of the human receptor for Escherichia coli heat-stable enterotoxin, *J. Biol. Chem.* 266:17912–18.

Drewett, J.G., Garbers, D.L. 1994. The family of guanylyl cyclase receptors and their ligands. *Endocr. Rev.* 15:135–162.

Duvernay, M.T., Filipeanu, C.M., Wu, G.Y. 2005. The regulatory mechanisms of export trafficking of G-protein-coupled receptors. *Cell Signal* 17:1457–65.

Eberle, A.N., Mild, G., Froidevaux, S. 2004. Receptor-mediated tumor targeting with radiopeptides. Part 1: General concepts and methods: Applications to somatostatin receptor-expressing tumors. *J. Recept. Signal Transduct Res.* 24:319–455.

Eckelman, W.C., Volkert, W.A., Bonardi, M. 2008. True radiotracers: Are we approaching theoretical specific activity with Tc-99m and I-123? *Nuci. Med. Biol.* 35:523–27.

Field, M., Graf, Jr. L.H., Laird, W.J. et al. 1978. Heat-stable enterotoxin of *Escherichia coli*: In vitro effects on guanylate cyclase activity, cyclic GMP concentration, and ion transport in small intestine. *Proc. Natl. Acad. Sci. USA* 75:2800–04.

Fleischmann, A., Waser, B., Reubi, J.C. 2009. High expression of gastrin-releasing peptide receptors in the vascular bed of urinary cancers: Promising candidates for vascular targeting applications. *Endo-Rel. Cancers* 16:623–33.

Forte, L.R. 2004. Uroguanylin and guanylin peptides: Pharmacology and experimental therapeutics. *Pharma. & Therapeutics* 104:137–62.

Fritzberg, A.R., Gustavson, L.M. et al. 1994. In: *Chemical and Structural Approaches to Rational Drug Design*, Weiner, D.B., Williams, M.V., eds. CRS Press, Boca Raton, Fl. p125–38.

Gali, H., Hoffman, T.J., Sieckman, G.L. et al. 2002. Chemical synthesis of *escherichia coli* ST_h analog for specific targeting of human colon cancers. *Bioconj. Chem* 13:224–231.

Gali, H., Sieckman, G.L., Hoffman, T.J. et al. 2001. *In vivo* evaluation of an ^{111}In-labeled ST-peptide analog for specific-targeting of human colon cancers. *Nuci. Med. Biol.* 28:903–09.

Gali, H., Sieckman, G.L., Hoffman, T.J. et al. 2006. Synthesis and *in vitro* evaluation of ^{111}In-labeled heat-stable enterotoxin (ST) analogue for specific targeting of gunaylin recpeptors on human colonic cancers. *Anticancer Res.* 21:2785–92.

Garcia-Garayo, E., Maes, V., Blauenstein, P. et al. 2006. Double-stabilized neurotensin analogues as potential radiopharmaceuticals for NTR-positive tumors. *Nucl Med Biol* 33:495–503.

Garcia-Garayoa, E., Blauenstein, P., Blanc, A. et al. 2009. A stable neurotension-based radiopharmaceutical for targeted imaging and therapy of neurotension receptor-positive tumours. *Eur J Nucl Med Imaging* 36:37–47.

Garg, P.K., Alston, K.L., Welsh, P. C. et al. 1996. Enhanced binding and inertness to de halogenation of α-melanotropic peptides labeled using N-succinimidyl 3-iodobenzoate. *Bioconjug Chem* 7:233–9.

Gariepy, J., Judd, A.K., Schoolnik, G.K. 1987. Importance of disulfide bridges in the structure and activity of *Escherichia coli* enterotoxin STIb. *Proc. Natl. Acad. Sci, USA*, 84:8907–11.

Garrison, J.C., Rold, T.L., Sieckman, G.L. et al. 2008. Evaluation of pharmacokinetic effects of various linking groups using ^{111}In-DOTA-X-BBN(7–14)NH₂ structural paradigm in a prostate cancer model. *Bioconjug Chem* 19:1802–12.

Giblin, M.F., Sieckman, G., Gali, H. et al. 2004. *In vitro* and *in vivo* comparison of human E. coli heat-stable peptide analogues using the ^{111}In-DOTA group and distinct linker moieties. *Bioconjug Chem* 15:1872–80.

Giblin, M.F., Sieckman, G.L., Owen, N.K. et al. 2005a. Radiolabeled *Escherichia coli* heat-stable enterotoxin analogs for *in vivo* imaging of colorectal cancer. *Nucl. Instrum. Methods Phys. Res. B.* 241:689–92.

Giblin, M.F., Sieckman, G.L., Watkinson, L.D. et al. 2006a. Selective targeting of E. coli heat-stable enterotoxin analogs to human colon cancer cells. *Anticancer Res.* 26:3243–52.

Giblin, M.F., Sieckman, G.L., Watkinson, L.D. et al. 2006b. *In vitro* and *in vivo* evaluation of ^{177}Lu-and ^{90}Y-labeled E. coli heat-stable enterotoxin for specific targeting of uroguanylin receptors on human colon cancers. *Nucl Med Biol* 33:481–8.

Giblin, M.F., Veerendra, B., Smith, C.J. 2005b. Radiometallation of receptor specific peptides for diagnosis and treatment of human cancer. *In Vivo* 19:9–29.

Giblin, M.F. Wang, N. Hoffman, T.J. et al. 1998. Design and characterization of alpha melanotropin peptide analogs cyclized through rehenium and technetium metal coordination. *Proc Natl Acad Sci USA* 95:12814–18.

Ginij, M., Zhang, H., Waser, B. et al. 2006. Radiolabeled somatostatin receptor antagonists are preferable to αgonist for *in vivo* peptide receptor targeting of tumors. *Proc Natl Acad Sci USA* 103:16436–41.

Ginj, M., Maecke, H.R. 2004. Radiometallo-labeled peptides in tumor diagnosis and therapy. *Metal Ions Biol Syst* 42:109–42.

Giron, M.C. 2009. Radiopharmaceutical pharmacokinetics in animals: Critical consideration. *Q J Nucl Med Mol Imaging* 53:359–64.

Gu, D., Ma, Y., Niu, G. et al. 2011. LC/MS evaluation of metabolism and membrane transport of bombesin peptides. *Amino Acids* 40:669–75.

Hamra, F.K., Forte, L.R., Eber, S.L. et al. 1993. Uroguanylin: Structure and activity of a second endogenous peptide that stimulates intestinal guanylate cyclase. *Proc. Natl. Acad. Sci., USA* 90:10464–68.

Harden, T.K., Boyer, J.L., Dougherty, R.W. 1998. Drug analysis based on signaling response to G-protein-coupled receptors. *Receptor-Based Drug Design*, ed., Leff, P., Marcel Dekker, New York, NY, pp. 79–105.

Heppeler, A., Froidevaux, S., Maecke, H.R. et al. 1999. Radiometal-labelled macrocyclic chelator-derivatised somatostatin analogue with superb tumour-targeting properties and potential for receptor-mediated internal radiotherapy. *Chem Eur J* 7:1974–81.

Hoffman, T.J., Smith, C.J. 2009. True radiotracers: Cu-64 targeting vectors based upon bombesin peptide. *Nucl Med Biol* 36:579–85.

Houghten, R.A., Ostresh, J.M., Klipstein, F.A. 1984. Chemical synthesis of an octadecapeptide with the biological and immunological properties of human heat-stable *Escherichia coli* enterotoxin. *Eur. J. Biochem.* 145:157–62.

Hruby, V.J., Sharma, S.D., Toth, K. et al. 1993. Design, synthesis, and conformation of superpotent and prolonged acting melanotropins. *Ann N Y Acad Sci* 680:51–63.

Jensen, R.T., Battey, J.F., Spindel, E.R. et al. 2008. Mammalian bombesin receptors: Nomenclature, distribution, pharmacology, signaling, and functions in normal and disease states. *Pharmacol Rev* 60:1–42.

Johnson, C.V., Shelton, T., Smith, C.V. et al. 2006. Evaluation of combined 177Lu-DOTA-8-AOC-BBN(7–14)NH2 GRP receptor-targeted radiotherapy and chemotherapy in PC-3 human prostate tumor cell xenografted SCID mice. *Cancer Biother Radiopharm* 21:155–6.

Kamber, B., Hartmann, A., Eisler, K. et al. 1980. The synthesis of cystine peptides by iodine oxidation of S-trityl-cysteine and S-acetamidomethyl-cysteine peptides. *Helv. Chim. Acta.* 63:899–915.

Knight, L.C. 2003. Radiolabeled peptides for tumor imaging. In *Handbook of Radiopharmaceuticals*, eds., M.J. Welch and C.S. Redvanly, John Wiley & Sons, Ltd., pp. 643–84.

Koide, T., Otaka, A., Suzuki, H. et al. 1991. Selective conversion of S-protected cysteine derivatives to cystine by various sulphoxide compounds/trifluoroacetic acid system. *Synlett.* 345–46.

Krause, W.J., Cullingford, G.L., Freeman, R.H. et al. 1994. Distribution of heat-stable enterotoxinguanylin receptors in the intestinal tract of man and other mammals. *J. Anat.* 184:407–17.

Krenning, E.P., Kwekkeboom, D.J., Bakker, W.H. et al. 1993. Somatostatin receptor scintigraphy with [111In-DTPA-D-Phe1]- and [123I-Tyr3]-octreotide: The Rotterdam experience with more than 1000 patients. *Eur J Nucl Med* 20:716–31.

Krenning, E.P., Kwekkeboom, D.J., Pauwels, S. et al. 1995. Somatostatin receptor scintigraphy. *Nuclear Medicine Annual*, ed., Freeman, L.M., Raven Press: New York, NY, pp. 1–50.

Krenning, E.P., Kwekkeboom, D.J., Volkema, R. et al. 2004. Peptide receptor radionuclide therapy. *Ann NY Acad Sci* 1014:234–45.

Kwekkeboom, D.J., Mueller-Brand, J., Paganelli, G. et al. 2005. Overview of results of peptide receptor radionuclide therapy with three radiolabeled somatostatin analogs. *J Nucl Med* 46S:62S–66S.

Landon, L.A., Zou, J., Deutscher, S.L. 2004. An effective combinatorial strategy to increase affinity of carbohydrate binding by peptides. Mol Divers 1:113–32.

Lane, S.R., Nanda, P.K., Rold, T.L. et al. 2010. Optimization, biological evaluation, and microPET imaging of copper-64-labeled agonists, [64Cu-NO2A-X-BBN(7–14)NH2], in a prostate tumor xenograft mouse model. *Nucl Med Biol* 37:751–62.

Lantry, L.E., Cappelletti, E., Maddalena, M.E. et al. 2006. 177Lu-AMBA: Synthesis and characterization of a selective 177Lu-labeled GRP-R agonist for systemic radiotherapy of prostate cancer. *J Nucl Med* 47:1144–52.

Li, Z.-B., Chen, K., Chen, X. 2008. 68Ga-labeled multimeric RGD peptides for microPET imaging of integrin $\alpha_v\beta_3$ expression. *Eur J Nucl Med* 35:1106–08.

Lin, J.Y., Fisher, D.E. 2007. Melanocyte biology and skin pigmentation. *Nature* 445:843–50.

Lister-James, J., Moyer, B.R., Dean, R.T. 1997. Pharmacokinetic considerations in the development of peptide-based imaging agents. *Q J Nucl Med.* 41:111–8.

Liu, D-J, Overby, D., Watkinson, L. et al. 2009. *In vivo* imaging of human colorectal cancer using radiolabeled analogs of the uroguanylin peptide hormone. *Anticancer Res.* 29:377–383.

Liu, I, Li, H., Underwood, T. et al. 2001. Cyclic GMP-dependent protein kinase activation and induction by exisulind and cp461 in colon tumor cells. *J. Pharmacol. Exp. Ther.* 299:583–92.

Liu, S. 2004. The role of coordination chemistry in development of target-specific radiopharmaceuticals. *Chem Soc Rev* 33:1–18.

Liu, Z., Li, Z.B., Cao, Q. et al. 2009a. Small-animal PET of tumors with 64Cu-labeled RGD-bombesin heterodimer. *J Nucl Med* 50:1168–77.

Liu, Z., Niu, G., Shi, J. et al. 2009b. 68Ga-labeled cyclic RGD dimers with Gly3 and PEG4 linkers: Promising agents for tumor integrin $\alpha_v\beta_3$ PET imaging. *Eur J Nucl Med Mol Imaging* 36:947–57.

Liu, Z., Niu, G., Wang, F. et al. 2009c. 68Ga-labeled NOTA-RGD-BBN peptide for dual integrin and GRPR-targeted tumor imaging. *Eur J Nucl Med Mol Imaging* 36:1483–94.

Liu, Z., Yan, Y., Chin, F.T. et al. 2009d. Dual integrin and gastrin-releasing peptide receptor targeted tumor imaging using 18F-labeled PEGylated RGD-bombesin heterodimer 18F-FB-PEG3-Glu-RGD-BBN. *J Med Chem* 52:425–32.

Maina, T., Nock, B., Nikolopoulou, A. et al. 2002. [99mTc]Demotate, a new 99mTc-based [Tyr3]octreotate analogue for the detection of somatostatin receptor-positive tumours: Synthesis and preclinical results. *Eur J Nucl Med Mol Imaging* 29:742–53.

Mankoff, D.A., Link, J.M, Linden, H.M. et al. 2008. Tumor receptor imaging. *J Nucl Med* 49:149S–63S.

Mansi, R., Wang, X., Forrer, F. et al. 2009. Evaluation of a 1,4,7,10-tetraazacyclododecane-1,4,7,10-tetraacetic acid-conjugated bombesin-based radioantagonist for the labeling with

Chapter 19

single-photon emission computed tomography, positron emission tomography, and therapeutic radionuclides. *Clin Cancer Res* 15:5240–9.

Mansi, R., Wang, X., Forrer, F. et al. 2011. Development of a potent DOTA-conjugated bombesin antagonist for targeting GRPR-positive tumours. *Eur J Nucl Med Mol Imaging* 38:97–107.

McQuade, P., Miao, Y., Yoo, J. et al. 2005. Imaging of melanoma using [64]Cu- and [86]Y-DOTA-ReCCMSH(Arg[11]), a cyclized peptide analogue of α-MSH. *J Med Chem* 48:2985–92.

Meares, C.F., McCall, M.J., Deshpande, S.V. et al. 1988. Chelate radiochemistry: Cleavable linkers lead to altered levels of radioactivity in the liver. *Int J Cancer* 2:99–102.

Melis, M., Bijster, M., de Visser M. et al. 2009 Dose–response effect of Gelofusine on renal uptake and retention of radiolabelled octreotate in rats with CA20948 tumours. *Eur J Nucl Med Mol Imaging* 36:1968–76.

Miao, Y., Benwell, K., Quinn, T.P. 2007. [99m]Tc and ^{111}In-labeled α-melanocyte stimulation hormone peptides as imaging probes for primary and pulmonary metastatic melanoma detection. *J Nucl Med* 48:73–80.

Miao, Y., Figueroa, S.D., Fisher, D.R. et al. 2008. [203]Pb-labeled a-melanocyte stimulating hormone peptide as an imaging probe for melanoma detection. *J Nucl Med* 49:823–9.

Miao, Y., Fisher, D.R., Quinn, T.P. 2006. Reducing the renal uptake of ^{90}Y and ^{177}Lu-labeled alpha-melanocyte stimulating hormone peptide analogues. *Nucl Med Biol* 33:723–33.

Miao, Y., Hylarides, M., Fisher. D.R. et al. 2005. Melanoma therapy via peptide-targeted alpha-radiation. *Clin Cancer Res* 11:5616–21.

Miao, Y., Shelton, T., Quinn, T.P. 2007. Therapeutic efficacy of a ^{177}Lu labeled DOTA conjugated α-melanocyte stimulating hormone peptide in a murine melanoma-bearing mouse model. *Cancer Biother Radiopharm* 22:333–41.

Miao, Y., Whitener, D., Feng, W. et al. 2003. Evaluation of the human melanoma targeting properties of radiolabeled α-melanocyte stimulating hormone peptide analogues. *Bioconjug Chem* 14:1177–84.

Miao, Y.-B., Figueroa, S.D., Fasher, D.R. et al. 2008. 203Pb labeled alpha-melanocyte stimulating hormone peptide as an imaging probe for melanoma detection. *J Nucl Med* 49:823–9.

Nock, B., Nikolopoulou, A., Chiotellis, E. et al. 2003. [^{99m}Tc] Demobesin 1, a novel potent bombesin analogue for GRP receptor-targeted tumour imaging. *Eur J Nucl Med Mol Imaging* 30:247–58.

Okarvi, S.M. 2008. Peptide-based radiopharmaceuticals and cytotoxic conjugates: Potential tools against cancer. Cancer Treat Rev 34:13–26.

Pitari, G.M., DiGuglielmo, M.D., Park, J. et al. 2001. Guanylyl cyclase C agonists regulate progression through the cell cycle of human colon carcinoma cells. *Proc. Natl. Acad. Sci. USA* 98:7846–51.

Prasanphanich, A., Narda, P.K., Rold, T.L. et al. 2007. ^{64}Cu-NOTA-8-Aoc-BBN(7-14)NH[2] conjugate: A novel targeting vector for positron emission tomographic imaging of gastrin releasing peptide receptor-expressing tissues. *Proc Natl Acad Sci USA* 104:12462–7.

Prasanphanich, A.F., Retzloff, L., Lane, S.R. et al. 2009. *In vitro* and *in vivo* analysis of [^{64}Cu-NO2A-8-Aoc-BBN(7–14)NH[2]]: A site-directed radiopharmaceutical for positron-emission tomography imaging of T-47D human breast cancer tumors. *Nucl Med Biol* 36:171–81.

Reubi, J.C., Maecke, H.R., Krenning, E.P. 2005. Candidates for peptide receptor radiotherapy today and in the future. *J Nucl Med* 46S:67S–75S.

Reubi, J.C., Maecke, H.R. 2008. Peptide-based probes for cancer imaging. J *Nucl Med* 49:1735–8.

Reubi, J.C., Schaer, J.C., Laissue, J.A. et al. 1996. Somatostin receptors and their subtypes in human tumors and peritumoral vessels. *Metabolism* 45:39–41.

Reubi, J.C., Wenger, S., Schminckli-Mauer, J. et al. 2002. Bombesin receptor subtypes in human cancers: Detection with the universal ligand [125]I-[D-Tyr[6], beta-Ala[11], Phe[13], Nle[14]] bombesin(6–14). *Clin Cancer Res* 8:1139–46.

Reubi, J. C. 2003. Peptide receptors in molecular targets for cancer diagnosis and therapy. *Endocrinol Rev* 24:389–427.

Sato, T., Ozaki, H., Hata, Y. et al. 1994. Structural characteristics for biological activity of heat-stable enterotoxin produced by enterotoxigenic *Escherichia coli*: X-ray crystallography of weakly toxic and nontoxic analogs. *Biochemistry* 33:8641–50.

Sawyer, T.K., Sanfilippo, P.J., Hruby, V.J. 1980. A [Nle[4]-D-Phe[7]] α-melanocyte stimulating hormone: A highly potent α-melanotropin with ultralong biological activity. *Proc. Natl Acad Sci USA* 77:5754–8.

Scheving, L.A., Russell, W.E., Chong, K. 1996. Structure, glycosylation, and localization of rat intestinal guanylyl cyclase C: Modulation by fasting. *Am. J. Physiol.* 271:G959–68.

Schottelius, M., Wester, H.J. 2009. Molecular imaging targeting peptide receptors. *Methods* 48:161–77.

Schroeder, R.P.J., Müller, C., Reneman, S. et al. 2010. A standardised study to compare prostate cancer targeting efficacy of five radiolabelled bombesin analogues. *Eur J Nucl Med Mol Imaging* 37:1386–96.

Schroeder, R.P.J., Weerden, W.M.V., Bangma, C. et al. 2009. Peptide receptor imaging of prostate cancer with radiolabelled bombesin analogues. *Methods* 48:200–4.

Schulz, S., Green, C.K., Yuen, P.S.T. et al. 1990. Guanylyl cyclase is a heat-stable enterotoxin receptor. *Cell* 63:941–48.

Shailubhai, K., Yu, H.H., Karunanandaa, K. et al. 2000. Uroguanylin suppresses polyp formation in the APC[Min+] mouse and induces apoptosis in human colon adenocarcinoma cells via cyclic GMP. *Cancer Res.* 60:5149–55.

Shi, J., Kim, Y.S., Zhai, S. et al. 2009. Improving tumor uptake and pharmacokinetics of ^{64}Cu-labeled cyclic RGD peptide dimers with Gly[3] and PEG[4] linkers. *Bioconjug Chem* 20:750–9.

Siegrist, W., Solca, F., Stutz, S. et al. 1989. Characterization of receptors for alpha-melanocyte-stimulating hormone on human melanoma cells. *Cancer Res* 49:6352–8.

Smith, C.J., Gali, H., Sieckman, G.L. et al. 2003. Radiochemical investigations of ^{177}Lu-DOTA-8-AOC-BBN(7–14)NH[2]: An *in vitro/in vivo* assessment of the targeting ability of this new radiopharmaceutical for PC-3 human prostate cancer cells. *Nucl Med Biol* 30:101–9.

Smith, C.J., Volkert, W.A., Hoffman, T.J. 2003. Gastrin releasing peptide (GRP) receptor targeted radiopharmaceuticals: A concise update. *Nucl Med Biol* 30:861–68.

Smith, C.J., Volkert, W.A., Hoffman, T.J. 2005. Radiolabeled peptide conjugates for targeting of the bombesin receptor superfamily subtypes. *Nucl Med Biol* 32:733–40.

Sun, B., Halmos, G., Schally, A.V. et al. 2000. Presence of receptors for bombesin/gastrin-releasing peptide and mRNA for three receptor subtypes in human prostate cancers. *Prostate* 42:295–303.

Thomas, R., Chen, J., Roudier, M.M. et al. 2009. *In vitro* binding evaluation of ^{177}Lu AMBA, a novel ^{177}Lu-labeled GRP-R agonist for systemic radiotherapy *in* human tissues. *Clin Exp Metastasis* 26:105–19.

Vaidyanathan, G., Zalutsky, M.R. 1997. Fluorine-18-labeled [Nle[4],D-Phe[7]]-alpha-MSH, an alpha-melanocyte stimulating hormone analogue. *Nucl Med Biol* 24:171–8.

Vanderheyden, J.-L. 2009. The use of imaging in preclinical drug development. *Q J Nucl Med Mol Imaging* 53:374–81.

Van de Wiele, C., Dumont, F., Broecke, R.V. et al. 2000. Technetium-99m RP525, a GRP analogue for visualization of GRP receptor-expressing malignancies: A feasible study. *Eur J Nucl Med* 27:1694–9.

Vegt, E., van Eerd, J.E., Eek, A. et al. 2008 Reducing renal uptake of radiolabeled peptides using albumin fragments. *J Nucl Med* 49:1506–11.

Vegt, E., Wetzels, J.F., Russel, F.G. et al. 2006. Renal uptake of radiolabeled octreotide in human subjects is efficiently inhibited by succinylated gelatin. *J Nucl Med* 47:432–6.

Virgolini, I., Traub, T., Leimer, M. et al. 2000. New radiopharmaceuticals for receptor scintigraphy and radionuclide therapy. *Q J Nucl Med* 44:50–58.

Virgolini, I., Traub, T., Novotry, C. et al. 2002. Experience with indium-111 and yttrium-90-labeled somatostatin analogs. *Curr Pharm Des* 8:1781–7.

Wei, L., Butcher, C., Miao, Y. et al. 2007a. Synthesis and biological evaluation of ^{64}Cu-labled rhenium-cyclized α-MSH peptide analogs using a cross-bridged cyclam chelator. *J Nucl Med* 48:64–72.

Wei, L., Miao, Y., Gallazzi, F. et al. 2007b. Gallium-68-labeled DOTA-rhenium-cyclized α-melanocyte-stimulating hormone analog for imaging of malignant melanoma. *Nucl Med Biol* 34:945–53.

Weiner, R.E., Thakur, M.L. 2005. Radiolabeled peptides: Role in diagnosis and treatment. *Biodrugs G* 19:145–63.

Wester, H.J., Kessler, H. 2005. Molecular targeting with peptides or peptide–polymer conjugates: Just a question of size? *J Nucl Med* 46:1940–5.

Wester, N.-J. 2007. Nuclear imaging probes: From bench to bedside. *Cli Cancer Res* 13:3470–81.

Whetstone, P.A., H., Meares, C.F. 2004. Evaluation of clearable (TYR3) octreotate derivatives for longer intracellular probe residence. *Bioconjug Chem* 15:647–57.

Wild, D., Schmitt, J.S., Ginj, M. et al. 2003. DOTA-NOC, a high affinity ligand of somatostatin receptor subtypes 2, 3 and 5 for labeling with various radiometals. *Eur J Nucl Med Mol Imaging* 30:1338–47.

Wolfe, H.R., Mendizabal, M., Lleong, E. et al. 2002. *In vivo* imaging of human colon cancer xenografts in immunodeficient mice using a guanylyl cyclase C-specific ligand. *J. Nucl. Med.* 43:392–99.

Wraight, E.P., Bard, D.R., Maughan, T.S. et al. 1992. The use of a chelating derivative of alpha melanocytes stimulating hormone for the clinical diagnosis of malignant melanoma. *Br J Radiol* 65:112–8.

Yang, J., Guo, H., Gallazzi, F. et al. 2009. Evaluation of a novel Arg-Gly-Asp-conjugated α-melanocyte stimulating hormone hybrid peptide for potential melanoma therapy. *Bioconjug Chem* 20:1634–42.

Yue, Z., Lu, B.-Y., Vazquez-Flores, G. et al. 2010. A new 62Cu generator for high specific activity peptide PET imaging. *J Nucl Med* 51 (Suppl 2):254.

Chapter 19

Index

Printed and bound by CPI Group (UK) Ltd, Croydon, CR0 4YY

24/10/2024

01778288-0020